2640
LM

AGRICULTURE IN THE TROPICS

SOME OTHER E.L.B.S. LOW-PRICED TEXTBOOKS

NOTE.—(1) The prices listed below are subject to alteration without notice
(2) The abbreviation "p" used below stands for new pence

ABERCROMBIE, M., HICKMAN, C. J.
and JOHNSON, M. L.
A Dictionary of Biology	*Penguin*	10p	2s. 0d.

ARTHUR, G. M.
Wright's Veterinary Obstetrics	*Baillière, Tindall and Cassell*	£1·40	28s. 0d.

COOKE, G. W.
The Control of Soil Fertility	*Crosby Lockwood*	£1·25	25s. 0d.

COBLEY, L. S.
An Introduction to the Botany of Tropical Crops	*Longman*	£1·00	20s. 0d.

ELTON, C. S.
The Ecology of Invasion by Animals and Plants	*Methuen*	60p	12s. 0d.

HEYWOOD, V. S.
Plant Taxonomy	*Arnold*	20p	4s. 0d.

IMMS, A. D.
A General Textbook of Entomology	*Methuen*	£1·40	28s. 0d.

PHILLIPSON, J.
Ecological Energetics	*Arnold*	20p	4s. 0d.

RICHARDSON, M.
Translocation in Plants	*Arnold*	20p	4s. 0d.

RUSSELL, E. W. and
RUSSELL, SIR E. J.
Soil Conditions and Plant Growth	*Longman*	£1·05	21s. 0d.

SOULSBY, E. J. C.
Mönnig's Helminths, Arthropods and Protozoa of Domesticated Animals	*Baillière, Tindall and Cassell*	£1·50	30s. 0d.

STREET, H. E.
Plant Metabolism	*Pergamon*	45p	9s. 0d.

UVAROV, E. B., CHAPMAN, D. R.
and ISAACS, A.
A Dictionary of Science	*Penguin*	10p	2s. 0d.

WIGGLESWORTH
Principles of Insect Physiology	*Methuen*	£1·50	30s. 0d.

AGRICULTURE IN THE TROPICS

by

C. C. WEBSTER

C.M.G., PH.D., B.SC.(AGRIC.), A.I.C.T.A.

Formerly Director, Rubber Research Institute of Malaya

and

P. N. WILSON

PH.D., M.SC.(AGRIC.), DIP. ANIMAL GENETICS, F.I.BIOL.

*Formerly Professor of Agriculture,
Imperial College of Tropical Agriculture,
University of the West Indies*

THE ENGLISH LANGUAGE BOOK SOCIETY

AND

LONGMAN

LONGMAN GROUP LIMITED
London
*Associated companies, branches and representatives
throughout the world*

© Longman Group Ltd (formerly Longmans, Green & Co Ltd) 1966

All rights reserved. No part of this publication may be reproduced, stored in a retrieval system, or transmitted in any form or by any means, electronic, mechanical, photocopying, recording, or otherwise, without the prior permission of the Copyright owner.

*First published 1966
Fourth impression 1971
First published for E.L.B.S. 1966, reprinted 1971*

ISBN 0 582 46641 5

*Printed in Great Britain by
Lowe & Brydone (Printers) Ltd., London*

CONTENTS

Chapter		Page
1	The influence of climate on agriculture in the tropics	1
2	Tropical soils (by H. Vine, B.Sc., Ph.D., A.I.C.T.A., lately Associate Professor, Department of Agricultural Chemistry and Soil Science, University of Ibadan)	28
3	Tropical vegetation	68
4	Some social factors affecting agriculture in the tropics	87
5	Soil and water conservation	98
6	Land clearing, tillage and weed control	126
7	The maintenance of fertility under annual cropping I. Shifting cultivation	151
8	The maintenance of fertility under annual cropping II. Modifications and alternatives to shifting cultivation	178
9	The maintenance of fertility under annual cropping III. Permanent farming systems associated with the production of swamp rice	202
10	The culture of tree and bush crops	217
11	Natural grasslands and their management	248
12	The use of natural grasslands by native pastoralists	272
13	Cultivated fodder crops and pastures	287
14	Classes of tropical livestock	320
15	Adaptation of livestock to tropical environment	347
16	Cattle management in the tropics	369
17	Livestock improvement	394
	References	449
	Index	479

List of illustrations

PHOTOGRAPHS IN COLOUR

Facing page 226
I   Lowland evergreen rain forest, Sabah
II   'Miombo' woodland, Tanzania

Facing page 227
III   *Acacia-Commiphora* thicket, Tanzania
IV   Grassland following clearing of *Acacia-Commiphora* thicket, Tanzania

Facing page 242
V   Terraced rice fields, Ceylon
VI   Rice field with mixed tree culture in background, Malaya

Facing page 243
VII   East African goats at Entebbe Livestock Station, Uganda. (*P. N. Wilson*)
VIII   Jamaica Hope cow from the Government Herd maintained at the Bodles Old Harbour Livestock Station, Jamaica. (*P.N. Wilson*)

PHOTOGRAPHS IN BLACK AND WHITE

Facing page 34
1. Highland grassland and forest, Kenya. (*A.V. Bogdan*)
2. *Acacia-Themeda* grassland, Kenya. (*A.V. Bogdan*)

Facing page 35
3. Clearing lowland rain forest, Sabah
4. Shifting cultivation, Sabah. (*E. J. Berwick*)

Facing page 50
5. Soil erosion in the Andes. (*P. N. Wilson*)
6. Gully erosion. (*A.V. Bogdan*)

Facing page 51
7. Vegetable growing on bench terraces, Malaya
8. Young rubber on modified bench terraces, with cover plants sown in the inter-rows, Malaya

List of illustrations

Facing page 98
9 Manually felled jungle, Malaya
10 Clearing for shifting cultivation, Sabah. (*E. J. Berwick*)

Facing page 99
11 *Left*, fore-mounted rake; *centre*, tree-dozer and stumper; *right*, tractor equipped with terracing blade in front and ripper at rear. (*Malayan American Plantations Ltd*)
12 Tree-dozer. (*Malayan American Plantations Ltd*)

Facing page 114
13 Stumper uprooting rubber tree. (*Malayan American Plantations Ltd*)
14 Fore-mounted root rake stacking timber in windrows. (*Malayan American Plantations Ltd*)

Facing page 115
15 Ripping equipment. (*Malayan American Plantations Ltd*)
16 Terracing blade. (*Malayan American Plantations Ltd*)

Facing page 146
17 Manual preparation of land for rice planting, Malaya
18 First ploughing of rice field with buffalo, Malaya

Facing page 147
19 *Foreground*, uprooting rice seedlings in nursery; *background*, puddling rice field, Malaya
20 Transplanting rice, Malaya

Facing page 162
21 Threshing rice by hand, Malaya
22 Rice fields lying fallow in the off-season, Malaya

Facing page 163
23 Vegetable growing in rice fields during the off-season, Malaya
24 Mixed tree culture, Malaya

Facing page 290
25 Mist spray propagation of rubber cuttings
26 Terminal stem cutting of rubber beginning to root

Facing page 291
27 Rooted rubber cutting in veneer tube ready for planting
28 Budding rubber; removing the bud patch from the budwood

List of illustrations ix

Facing page 306

29 Budding rubber; removing wood from the bud patch
30 Budding rubber; inserting the bud patch on the prepared stock

Facing page 307

31 Budding rubber; the bud patch tied and shaded after insertion on the stock
32 Budding rubber; scion growing from bud patch after cutting back stock

Facing page 354

33 Ravi type Water Buffalo cow, at the Okara Military Farm, West Punjab, Pakistan. (*C.B.A.B.G.*)
34 Crossbred Murrah type Water Buffalo steers emerging from a wallow at the University Field Station, St. Augustine, Trinidad. (*Trinidad Guardian*)

Facing page 355

35 Baggage Camel, or Dromedary, used for farm transport in the Canary Islands. (*P. N. Wilson*)
36 Mule, imported from the U.S.A. into the West Indies, being used for light farm transport at the University Field Station, St Augustine, Trinidad. (*Trinidad Guardian*)

Facing page 370

37 Ox cart in Northern India. (*Associated Press*)
38 East African Goat, at Serere Experiment Station, Uganda. (*P. N. Wilson*)

Facing page 371

39 West African Dwarf Goats, at Kumasi, Ghana. (*J. L. Jollans*)
40 Intensive grassland farming at the University Field Station, St Augustine, Trinidad. (*Texaco Trinidad Inc.*)

Facing page 402

41 Charolais cattle at the Star Farm, Point-a-Pierre, Trinidad. (*Texaco Trinidad, Inc.*)
42 White Fulani cattle at the Animal Health Livestock Farm, Accra, Ghana. (*West African Photographic Service*)

List of illustrations

Facing page 403

43 Improved N'Dama cattle at the Main Farm, Pong-Tamale Veterinary Station, Ghana. (*C.B.A.B.G.*)
44 Criollo cow on a milk ranch near Maracaibo, Venezuela. (*P.N. Wilson*)

Facing page 418

45 Kenana bull from the Sudan, at the Veterinary Research Station, Entebbe, Uganda. (*P.N. Wilson*)
46 Sahiwal Cow at the Indian Agricultural Research Institute, New Delhi, India. (*C.B.A.B.G.*)

Facing page 419

47 Angoni Zebu Bull at the Livestock Improvement Station, Malawi. (*C.B.A.B.G.*)
48 Red Sindhi cows in the Malir herd, Karachi, Pakistan. (*J. P. Maule*)

PREFACE

This book gives a general account of the basic factors affecting agriculture in the tropics and of the applications of existing knowledge of the principles of agriculture to the improvement of tropical arable farming, planting and animal husbandry. It deals with tropical climates, soils and natural vegetation types; the principles of soil and water conservation; methods of arable cultivation; the production of annual and perennial crops; the management and utilization of natural and cultivated grasslands; the classes of livestock specially important in the tropics and the adaptation of animals to tropical environments; animal husbandry and the improvement of tropical livestock by better nutrition, hygiene and breeding.

Efforts have been made to incorporate an up-to-date review of the findings of research relevant to the tropics. It does not deal in detail with those aspects of agriculture, such as, for example, pig and poultry husbandry, where temperate practice can be modified to suit the particular requirements of the tropics, but the book refers the reader to other texts in which such matters are adequately covered.

The primary aim has been to provide an outline of the subject for students reading for degrees or diplomas in tropical agriculture and for those who proceed to work, or study, in tropical agriculture after graduating in agriculture or similar related subjects in the temperate zone. It should be stressed that many students from tropical countries take their University degrees in temperate countries, and therefore require to adapt their temperate theory to tropical practice. It is hoped that this book will be specially valuable to such students in undertaking this re-orientation. It is, however, also hoped that this book will prove useful to many tropical planters and farmers and to others indirectly concerned with agriculture, such as administrators, development planners, educationalists, geographers, economists and veterinarians.

In some sections of the book it has been assumed that the reader has an elementary knowledge of soil science, plant and animal breeding, or some other applied science subject, such as the undergraduate student obtains in courses in pure or applied science having some bearing on agriculture. However, such knowledge is not essential to a proper understanding of the greater part of the book and, for those who do not possess it, an indication has been given of where it may be obtained by further reading.

The book does not set out to give a full and comprehensive treatment of any given tropical crop or tropical farm animal. It deals with principles rather than with specific detail. The reader is referred in the text to the standard works dealing with particular facets of tropical agriculture, and it is recommended that the study of this text should be followed up with more detailed reading in special fields of interest.

The list of references will be found to be unusually long for a general work of this type. The reason for this additional detail is that the authors are aware, through their own University teaching experience, that a reference list of this nature will assist teachers of tropical agriculture in the preparation of their courses, and will guide postgraduate students in their private reading. The general reader will probably not have the time or facilities to make use of the references, but even a glance through will serve to indicate where research work is currently being conducted in each field.

Thanks are due to numerous persons who have contributed, directly or indirectly, to the production of this book. It is difficult to single out persons for individual mention, but especial thanks are due to the following who kindly assisted by reading parts of the manuscript and suggesting various improvements to the material presented: Mr T. Chapman, Dr W. E. Coey, Dr K. Hill, Dr D. Horrocks, Mr T. R. Houghton, and Dr D. W. Stringer.

The authors also wish to record their appreciation of the help given by Miss Linda Chiew, Mrs Jean Herbert and Mrs Bunny Wilson in the typing, preparation and checking of the manuscript.

ACKNOWLEDGEMENTS

We are indebted to the following for permission to reproduce copyright material:

Cambridge University Press for extracts from an article by H. C. Pereira published in *Journal of Agricultural Science*, Volume 49, 1957; the Commonwealth Agricultural Bureaux for extracts from 'The Soil under Shifting Cultivation', *Com. Bureau of Soils: Technical Communication* 51; the Executive Committee of Junta de Investigações do Ultramar, Portugal, for extracts from *Carta geral dos solos de Angola*, 1, *Distrito da Huila, Lisbon,* by Botelho da Costa and Azevedo, 1959, and John Wiley & Sons Ltd for extracts from *Physical Geography* by A. N. Strahler.

CHAPTER 1

The Influence of Climate on Agriculture in the Tropics

The Main Climates of the Tropics

Climate has an important influence on the nature of the natural vegetation, the characteristics of the soils, the crops that can be grown and the type of farming that can be practised in any region. There are a number of diverse types of tropical climate and there is correspondingly great variation in the agricultural potential of different parts of the tropics. In the following brief description of the influence of the main types of tropical climate on agriculture a genetic classification based on the main causes of climate, as used by Strahler (1960), is followed. The primary factors giving rise to different climates are the planetary pressure belts and the resulting system of winds and their associated air masses; no attempt is made to describe these here, but they are discussed by Strahler (loc cit.). Within each of the main climates defined by this method of classification the several features of the climate vary within rather wide limits. Each zone dominated by a given type of climate also shows local variations from this type because of modifications due to such circumstances as altitude or the presence of great lakes. A brief general account cannot detail the variations within a climatic type, nor describe the local differences in different parts of large geographical regions, such as East Africa, but can only indicate broadly the influence of the main climates on agricultural potential and practice. More detailed information on the nature and classification of tropical climates will be found in publications by Blumenstock (1958), Beckinsale (1957), Garbell (1947), Riehl (1954), Strahler (1960), and Watts (1955). Kendrew (1953) describes the climates of most parts of the tropics and accounts for more limited areas have been given by Braak (1929), Netherlands Indies; Carter (1954), India and Africa; Chamney (1928), Ghana; Griffiths (1959), East Africa; Hamilton and Archibold (1945), Nigeria; Howe (1953), the Rhodesias and Nyasaland; and Miller (1952), Nigeria.

WET EQUATORIAL CLIMATES

Wet equatorial climates occur mainly within 5°N. and S. of the equator, where climate is dominated during most of the year by deep, moist, equatorial air masses, and heavy convectional rainfall is frequent. Total precipitation usually lies between 80 and 120 inches per annum (although in many places it exceeds the latter figure) and some rain occurs in all months of the year, but in some areas there may be periods of up to two or three months that are distinctly less wet than others. Rainfall in almost any period of two or three weeks is considerably in excess of evapo-transpiration. Temperatures show little seasonal variation, monthly means usually being around 82° F, and mean monthly maxima and minima about 87° and 77°. Day length varies little from twelve hours all the year round. This climate is therefore characterized by constant heat, rainfall and humidity; there is an absence of any seasonal rhythm, and the modest reduction of temperatures at night affords little relief to the daytime heat. The diagram for Singapore in Fig. 1 is an example of mean monthly rainfall and temperatures for this type of climate.

Climates of this kind occur over a large area in South America, in parts of Brazil, Surinam, French and British Guiana and Venezuela; in West Africa in the great basin of the River Congo, but only in a narrow interrupted strip along the northern coast of the Gulf of Guinea; and in the East over parts of Malaya, Sumatra, Java, Borneo, the Celebes, New Guinea and many of the Pacific Islands. They do not occur in equatorial latitudes in East Africa or on the west of South America.

Climax vegetation is normally luxuriant evergreen rain forest, although in some places, where two or three months are drier than the remainder, a proportion of deciduous trees may occur. There are no marked seasons for agricultural operations, and neither lack of rain nor low temperatures are significantly limiting to crop production at any time. The most suitable crops are those that flourish under continuous, hot, wet conditions and do not demand a pronounced dry season for harvesting. Commercial agriculture is largely based on perennials, such as rubber, oil palm, banana, *liberica* coffee and, to a lesser extent, coconut and cocoa. Some other perennials, such as *arabica* and *robusta* coffee, mango and citrus, which require a relatively high rainfall but also need a short drier (and preferably cooler) spell for satisfactory flowering and fruiting, are not well-suited to a wet equatorial climate. Subsistence native farming also often includes a plot of fruit trees and other perennials of economic value, but roots are commonly the most important food crops, especially yams (*Dioscorea* spp), cassava (*Manihot utilissima*), cocoyams, dasheen or eddo (*Colocasia* spp) and tannia (*Xanthosoma sagittifolium*). Rice is extensively grown in parts of the East where the equatorial climate is slightly modified by the occurrence of a brief dry season that permits

The Main Climates of the Tropics

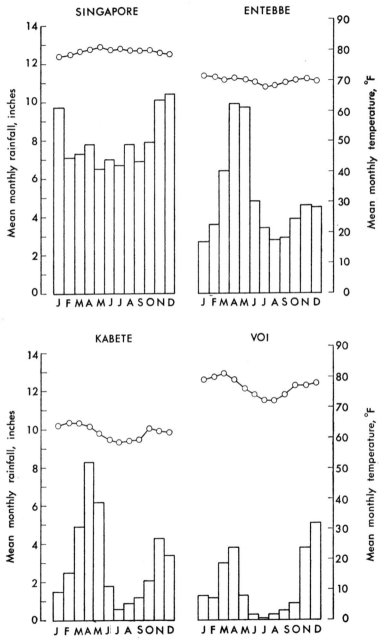

FIG. 1a. Mean monthly rainfall and temperature at various stations

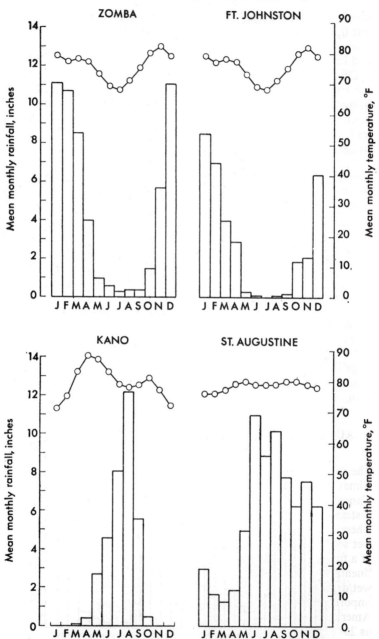

Fig. 1b. Mean monthly rainfall and temperature at various stations

satisfactory harvesting of the small acreage grown on a peasant farm, but the main rice producing zones have climates with a more pronounced dry season. Maize is the only other cereal grown appreciably, and its successful cultivation demands adapted varieties.

Efficient cultivation and weed control, especially in annual crops, are rendered difficult by the continuous wet weather, weed growth being extremely rapid and luxuriant. Soil fertility is commonly limited by nutrient deficiencies, owing to intense leaching under the high rainfall. In perennial crops trace element deficiencies occur quite frequently, as well as those of major elements.

Apart from pigs and poultry, livestock are unimportant under these climatic conditions, mainly because there are no extensive natural grasslands but also, in some places, on account of endemic diseases such as trypanosomiasis. The climate is suitable for the development of sown, or planted, pastures and fodder crops, but so far little progress has been made in the selection of suitable grasses and legumes, or in working out economic methods of management and manuring.

DRY TROPICAL CLIMATES OF THE CONTINENTAL TROPICAL AIR MASS SOURCE REGIONS

In many regions between latitudes 15° and 30° N. and S., climate is dominated by adiabatically heated, dry air masses descending and moving outwards from the sub-tropical high pressure cells over the large land masses. This results in hot, arid climates and desert areas, of no agricultural significance, such as those of North Africa, Arabia, Iran, North-west India, Australia and South America.

ALTERNATELY WET AND DRY TROPICAL CLIMATES (MONSOON CLIMATES)

The regions lying between the low latitude zones of wet equatorial climate and the belts of dry tropical climates centred on the tropics of Cancer and Capricorn are those over which the pressure and wind systems migrate with the apparent movement of the overhead sun. These intermediate regions therefore tend to have an alternation of a wet season controlled by intertropical convergence zone conditions at a time of high sun (summer) and a dry season controlled by continental tropical air masses at a time of low sun (winter). Such alternate wet/dry climates are experienced by vast areas of great agricultural importance lying roughly between 5° and 15° N. and S. in South America, West and Central Africa and Australia, but extending as far as 25° N. in parts of South-east Asia, where the heating of the large

Asiatic land mass during the summer causes the development of a powerful low pressure centre which pulls the tropical pressure and wind systems further to the north. Partly due to the influence of this Asiatic low pressure centre, which deflects the south-east trade winds, a similar type of climate is also found over a large part of East Africa in equatorial, as well as in higher, latitudes.

Although all these areas are similar in that they experience alternate wet and dry seasons of varying intensity at times of high and low sun respectively, it is not very satisfactory from an agricultural point of view to lump them all together. Apart from local differences due to altitude, presence of great lakes, etc., there are important variations in the number of wet and dry seasons per annum, in the duration of these seasons and in the amount of annual and seasonal rainfall. In places near the equator the two annual passages of the overhead sun are of relatively short duration and relatively widely spaced in time. Consequently there tends to be two rainy seasons and two intervening dry seasons each year, or at any rate, two periods of peak rainfall and two intervening spells of lower rainfall. Nearer the tropics of Cancer and Capricorn the two passages of the overhead sun are slower and closer together in time, so that the two periods of peak rainfall corresponding with these passages tend to coalesce, and there is only one rainy season and one dry season annually. There are also big differences from place to place in the total annual rainfall and in the duration of the wet and dry seasons. Over considerable areas, as for example in West Africa north of the equator, the rainfall tends to decrease, and the length of the dry season to increase, as one moves away from the equator, but there are too many exceptions to this for it to be regarded as a general rule.

Within the areas classified as having alternate wet/dry climates, there is in fact a continuously variable succession of climates blending into each other. At one extreme is a climate with high total precipitation falling in a two-peak rainy season occupying most of the year, with only short intervening drier spells of reduced rainfall. This obviously approximates closely to the wet equatorial climate commonly experienced in nearby regions. At the other extreme is a climate with low total rainfall in one short wet season, and a long dry season. This transition from wetter to drier conditions is reflected in a gradual transition in the natural vegetation from semi-evergreen forest through various types of deciduous seasonal forest to open woodlands of broad-leaved trees and, finally, to thorn woodlands and thickets.

Throughout this range of climates the earliest part of the dry (or less wet) season, tends to be the coolest time of the year, and then as the sun gets higher and the rains approach, temperatures rise appreciably, so that in most areas the period of from one to three months before the break of the rains is the hottest time of the year as well as the driest. This makes the work of preparing the dry, hard ground for planting particularly arduous for peasant farmers, most of whom possess only hand tools or primitive ploughs. Although the sun is higher during the rainy season, temperatures are lower because of the cloud cover and the cooling effects of rainfall. Where there are two dry and two wet seasons in the year, temperatures rise at the end of each dry season but do so to a greater extent at the end of the winter dry season. In general, the seasonal variations in temperature, as well as the length and severity of the dry season, tend to be greater as one moves away from the equator. Places farther from the equator have markedly cool periods during the early and mid dry season, with temperatures rising to high peaks just before the rains, for example Kano and Fort Johnston (Fig. 1). On the other hand, nearer the equator temperatures do not rise so high just before the rains, nor fall so low during the earlier part of the dry season.

Despite the fact that within the zones experiencing alternate wet/dry climates there is a gradual transition and blending of climates which preclude any effective geographical definition of areas possessing distinct types of sub-climate, it is worthwhile from an agricultural point of view briefly to consider four variants of this main climatic type.

1. *Areas with good rainfall, usually between 40 and 80 inches per annum, falling in two rainy seasons with only short dry seasons or troughs of reduced rainfall between them*

Such areas are mostly relatively near to the equator, so that there is no great seasonal fluctuation in temperature and no pronounced cool season. The rainfall regime renders conditions suitable for both perennial and annual crops and allows of two cropping seasons a year for the latter. Such conditions exist, for example, in parts of Uganda, with a rainfall regime such as that shown for Entebbe (Fig. 1) where perennial crops such as coffee, tea and bananas can be satisfactorily grown and where there are two cropping seasons a year for annuals, the first commonly devoted to food crops and the second to cotton. Where a similar rainfall regime is found in southerly parts of Ghana and Nigeria the chief perennial crops are oil palms and cocoa, while annuals are mainly maize, pulses, yams and cassava. Rice is the chief annual crop under similar climatic conditions in South-east Asia. Owing to the absence of good natural grasslands and, in some places, the presence of endemic diseases, livestock have hitherto been relatively unimportant in these areas, but the climate is suitable for productive grassland if suitable herbage species could be selected and methods of management developed.

2. *Areas with two shortish rainy seasons and pronounced intervening dry seasons*

Under this regime there is a considerable variation in the average total annual precipitation, but this commonly falls within the range 25 to 50 inches. Kabete and Voi in Fig. 1 are examples of stations with this type of rainfall regime. Such climatic conditions are far less suitable than the preceding type for perennial crops, which are only satisfactory where rainfall is relatively high and reliable, or where the perennial crops are markedly drought-resistant, e.g. sisal, cashew. Nevertheless, in some areas, and particularly at higher altitudes, certain perennial crops are capable of giving yields high enough to be commercially profitable, provided that precautions are taken to conserve moisture. An example is the large coffee industry, mostly at altitudes between 5,000 and 6,000 feet, in the East Rift area of Kenya, where below optimum moisture conditions are to a considerable extent counteracted by growing coffee without shade, by regular mulching and by pruning to regulate crop production. In some places there are two cropping seasons a year for annuals, but in others the low amount and reliability of the rainfall in one of the wet seasons restricts crop production to short-term drought-resistant species, or makes it altogether uncertain. The amount and reliability of the rainfall naturally determines the annual crops which can be satisfactorily grown, but generally a fairly wide range is possible. The commonest crops are maize, sorghum, finger millet, sweet potatoes, cassava, groundnuts, beans and other pulse crops.

The importance of cattle varies considerably, depending upon the presence or absence of tsetse fly and, in part, on the traditional agricultural systems of the people. Fair quality open grasslands with scattered trees are extensive and, as a rule, provided tsetse is absent or can be controlled, conditions are suitable for cattle keeping. In many places where the average annual rainfall exceeds 30 inches the establishment of temporary grass leys is practicable as they are able to survive the dry season and carry some stock through it, while they can be highly productive during the rains.

3. *Areas with one fairly long rainy season, with rainfall usually in the range 30 to 50 inches per annum, and one long dry season*

Fort Johnston and Zomba, in Malawi, are examples of places with this type of climate (Fig. 1). These areas are not generally suited to perennial crops unless such crops are markedly drought-resistant (for example, sisal) or deciduous and with a marked dormant period (for example, tung-oil trees). Nevertheless, where total annual rainfall is good, say of the order of 50 inches, relatively slight modifications of climate due to altitude and the incidence of a very little rain in the dry season may render commercial production of perennial crops possible, as is the case with tea and coffee in parts of Malawi, although the long dry season certainly reduces yields well below those obtainable under more favourable rainfall conditions. The fact that the rain falls in one long rainy season of five or six months' duration generally makes these areas more suitable for annual crop production than those of the preceding category, where the annual rainfall is split between two shorter wet seasons. The range of annual crops grown is wide, including all those mentioned in the previous section and, in addition, certain crops which require a longer or more reliable rainy season, such as yams or cotton and, in the wetter places, rice.

Natural grasslands grow vigorously in the wet season but there is an abrupt cessation of growth when the rains cease, and carrying capacity is usually limited to the low stock numbers that can be sustained on poor quality standing hay during the dry season. The long, severe dry season is not conducive to the continuous productivity, or even to the survival, of grass leys, and their possibilities are confined to limited areas of better rainfall, mostly at higher altitudes.

4. *Areas of one short rainy season and a long dry season*

Such areas, which occur, for example, in Northern Nigeria (see rainfall for Kano, Fig. 1), are unsuitable for perennial crops but satisfactory for relatively short-term and/or drought-resistant annuals such as sorghum, bulrush millet, sweet potato, groundnuts and sesame. Cattle, where kept, require relatively large areas per beast because of the lack of keep in the dry season and stock are mostly maintained by nomadic pastoralists.

WET CLIMATES OF TROPICAL WINDWARD COASTS

Climates of this type occur mainly along the east coasts of Madagascar, Central America (including the West Indian Islands) and the northern part of South America. These coasts are exposed for much of the year to the moist north-east and south-east trade winds, but owing to the presence of a strong inversion layer these moist air masses are shallow. Rainfall is partly convectional (promoted by wavelike disturbances in the moist air stream), and partly results from orographic disturbances. Although some rain occurs in all months of the year, there is usually a markedly drier season during that part of the year when the trade wind air streams are at their shallowest. For example, in Trinidad rather more than 80 per cent of the total annual rainfall falls in the seven months June to December, and during the remainder of the year three months can be expected in which the rainfall will not exceed 3 inches per month more than once in three years. Owing to the importance of orographic rainfall, and the effect on its geographical distribution of the position of higher land in relation to the direction of prevailing winds, there is marked local variation in the total annual rainfall and in the severity of the dry season. Thus in Trinidad the eastern, or windward, extremity of the northern range of hills receives 150 inches a year and experiences only a slight dry season, whereas on the leeward side of the island the rainfall may be only about 50 inches. The temperature range during the year is small; for example, mean monthly temperatures near sea level in Trinidad range only from 76° to 80°F, but the diurnal range is greater as the nights are always markedly cooler than the days.

As a result of the big local variation in total rainfall, a great variety of crops is grown in different parts of this climatic zone. In some parts the high rainfall and absence of a severe dry season permit the cultivation of a wide range of perennial and annual crops, such as rice, maize, yams, tannia, dasheen, sweet potato, arrowroot, sugar-cane, cocoa, coconut and nutmeg. The existence of a dry season renders conditions below optimum for crops ideally demanding high rainfall throughout the year, such as rubber and oil palms, but the climate alone would not preclude the cultivation of these crops. Perennials such as coffee and citrus, which do best under more moderate rainfall, and with a short dry season accompanied by lower temperatures, are commercially profitable despite the fact that conditions are not ideal. Crops requiring a lower rainfall and a more pronounced dry season for harvesting, such as cotton, tobacco or groundnuts, are difficult in the wetter areas, but are grown quite extensively in the drier parts of the zone.

Cattle have hitherto not been of great importance in these areas because natural grasslands are either limited in extent or of low productivity, but with recent progress in the establishment of improved permanent pastures, particularly of Pangola grass (*Digitaria decumbens*), and in the breeding of locally adapted cattle, their importance is likely to increase. Despite the fact that good rainfall and the absence of a really severe dry season permits a good growth of grass throughout most of the year, temporary leys in rotation with arable have not so far been used, as no really suitable ley grass species for these climatic conditions have yet been found. Goats, pigs and poultry are extensively kept, and although it is too wet for sheep in parts of the zone, they do well in some of the drier areas.

DRY CLIMATES OF TROPICAL AND SUB-TROPICAL WEST COASTS

The climates of all west coasts roughly in latitudes 15° to 30° N. and S., as for example in the north-west and south-west African coasts and the west coast of South America, are arid and relatively cool. These are therefore desert areas of no agricultural importance.

Rainfall

Rainfall is the most important climatic factor influencing agriculture in the tropics, as it generally has the biggest effect in determining the potential of any area, the crops which it is practicable to grow, the farming systems which can be followed and the nature, timing and sequence of farming operations. The agriculturist is primarily interested in rainfall as the supplier of soil moisture for crops and grassland.

The soil moisture supply, however, does not depend on rainfall alone, but also on various other factors concerned in the hydrological equation, which may be expressed as:

$$P = E + R + D + S.$$

In this equation, P = precipitation. E = evapo-transpiration, or moisture returned to the atmosphere by evaporation from the soil surface, or of rainwater intercepted by the leaves and branches of the vegetation, or by transpiration. R = surface run-off of rainwater which is unable to percolate into the soil, either because the soil is already saturated with water, or because the permeability of its surface layer is low owing to inherently poor structure, or to loss of structure resulting in compaction, or packing. In the tropics it can often happen that the intensity of the rainfall is so high that the amount falling exceeds the ability of the soil to accept it for infiltration. D = the rainwater which drains beyond the root range of plants down into the deep subsoil on to the parent rock, and S = the amount of water stored in the soil. The last will vary greatly, probably being at field capacity fairly frequently during the rains but possibly being much reduced to some depth during the dry season. The extent to which the rainfall in any area can provide soil moisture available to crops therefore depends not only on the total amount of rainfall per annum, but also upon its seasonal distribution, its reliability within and between seasons, its intensity and rate of infiltration into the soil, and on the balance between rainfall and evapo-transpiration from the crop and soil.

AVERAGE ANNUAL RAINFALL

Figures for average annual rainfall are of limited value to the agriculturist in most parts of the tropics for several reasons. First, they give no indication of the seasonal occurrence of rainfall, although this is of minor importance since it can usually be at least roughly ascertained without difficulty. Secondly, whereas in temperate regions annual rainfall may be a fair indication of soil moisture, since evaporation is only moderate and rainfall is usually of low intensity, in the tropics this is far from the case, as evaporation is much higher and a large proportion of the rain falls in storms of high intensity, with the result that, unless counter measures are taken, much of it runs off the surface of the soil. Thirdly, average annual rainfall figures are not necessarily a good indication of the likely expectation of rainfall during any one year, since in many parts of the tropics the rainfall varies greatly from year to year, and to a much greater extent than is commonly experienced in temperate zones. Manning (1958) has presented figures for a number of tropical and temperate stations which illustrate this fact.

Differences in rainfall distribution under the various types of tropical climates have already been broadly indicated. In regions experiencing a wet equatorial climate, or the wettest types of monsoon climate, insufficient rain is at no time a serious limiting factor to crop production, although it may be some disadvantage over short periods for certain crops on certain soils. Under all other types of climate the occurrence of dry seasons limits the production of many crops to part of the year, and the duration of the wet season, or the amount, distribution and reliability of the rainfall therein, may restrict the choice of crops that can be grown or limit the yields obtainable. This is of great importance in the vast areas of alternate wet-dry, or monsoon, climate, especially where the total rainfall is moderate or low. There are large areas with this type of climate, particularly in Africa, where low, ill-distributed or unreliable rainfall is severely limiting to agriculture. In such places a study of the distribution and reliability of the rainfall is likely to be helpful to the agriculturist.

RAINFALL RELIABILITY

The value of having some more-reliable indication of the likely expectation of annual rainfall may be illustrated by data for Kibaya in Tanzania, quoted by Glover and Robinson (1953). Kibaya has one rainy season per annum, all the rain falling in the seven-month period November to May, and the average annual rainfall for eleven years was 24·89 inches, but during this period it varied in individual years from 18·70 to 45·11 inches. Supposing that it were desired to make some assessment of the adequacy of the rainfall for annual crop production then it would not be unreasonable to suggest that crops of flint maize or sorghum might be grown at Kibaya on a rainfall of 20 inches, and the question would immediately arise: how often would the rainfall for the year be less than 20 inches? Since the frequency distribution of annual rainfall totals is normal, or nearly so, this can be answered by calculating

$$t = \frac{\text{Difference of 20 inches from the mean}}{\text{Standard deviation}} = \frac{24·89 - 20·00}{7·82} = 0·6253$$

and referring to Fisher's table of t (Fisher and Yates 1953), which shows that the value of 0·625 corresponds to a probability of 54·1 per cent, so that in 54 per cent of the years a deviation from the mean as great or greater than 4·89 may be expected, but in only 27 per cent of the years, or approximately one year in four, would a deviation of 4·89 below the mean, or a rainfall of less than 20 inches, be expected. Judged solely on the expectation of annual rainfall, therefore, one might expect to grow such crops successfully in three years out of four.

The calculation of the reliability of annual rainfall for a number of stations enables the preparation of rainfall reliability maps demarcating areas according to the frequency with which annual rainfall may be expected to fall below, say, 20, 25, or 30 inches. Such maps have been prepared for East Africa by Glover, Robinson and Henderson (1954), showing areas in which the chances of failure to receive either 20 or 30 inches are below 5, 5–15, 15–30, or over 30 years in 100. Such maps give a broad indication of agricultural potential. For instance, areas shown on the East African map in which less than 20 inches of rain may be expected more frequently than once in three years, may be regarded as unfit for crop production and suitable, at best, for extensive ranching. Areas where less than 30 inches is not expected more frequently than once in five years can be regarded as suitable for arable cropping. Intermediate areas are commonly marginal for crops.

The reliability of annual rainfall can also be expressed by calculating the fiducial, or confidence, limits for any level of probability, that is by showing the limits within which the rainfall for the year may be expected to lie for any given level of probability. Thus, if a probability level of $P = 0·20$ is selected, the 4:1 confidence limits (upper and lower) can be calculated, and rainfall outside these limits would be expected only once in five years, while a rainfall below the lower limit would be expected once in ten years. Similarly, with 1:1 confidence limits, a rainfall outside either limit would be expected once in two years, but a rainfall below the lower limit only once in four years. (See, for example, Fig. 2 which shows 4:1 confidence limits for *monthly* rainfall at Kibaya, according to Manning, 1956a.)

While an estimate of the likely expectation of the total season's rainfall will be of some value in indicating the possibility of growing certain crops in an area, obviously the distribution and reliability of the rainfall within the season is important, and knowledge of this may

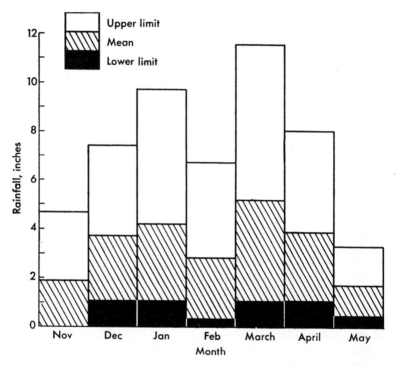

FIG. 2. 4:1 confidence limits of monthly rainfall at Kibaya (Manning, 1956)

be essential before deciding whether a crop can be grown. Total rainfall may be adequate, but poor distribution or low reliability in any one month may increase the chances of failure, or reduce yields. The probability of getting a given amount of rainfall in any one month can be calculated in the same way as we have calculated the probability of an annual rainfall, and the probability of getting a pattern of rainfall over, say, a five-month growing season can be calculated by multiplying the monthly probabilities together. It should be mentioned, however, that the reliability of monthly rainfall, or of rainfall over other short periods, can only be satisfactorily calculated provided that the monthly rainfall figures are independent. Furthermore, while the frequency distributions of annual or seasonal rainfall are normal, those of rainfall over shorter periods are usually skew, hence there is a need to transform data before performing the calculations. For details of the statistical methods involved reference may be made to Manning (1950, 1956a).

Referring to Kibaya again, we have seen that the possibility of growing crops of flint maize or sorghum is not ruled out on the expectation of total seasonal rainfall, but the question arises, what are the chances of getting sufficient rainfall in each month over a five month growing season? A glance at Fig. 2 is sufficient to show that the low average rainfall for November and May is likely to limit the growing season to the period December to April, and that the chances of getting a suitable pattern of rainfall in this period are small because of the low average for February. By making the necessary calculations Glover and Robinson (1953) were able to show that a total of even 18 inches of rainfall suitably distributed over a five-month growing season could only be expected in six years out of 100. Hence there is clearly a very poor chance of successful flint maize or sorghum production at Kibaya.

A knowledge of the within season distribution of rainfall and of rainfall reliability over monthly, or similar short periods, is not only of value in indicating what crops may be successfully grown in an area, but also as a guide to other matters of great importance to the agriculturist, such as the optimum time of planting, the best timing and sequence of farming operations, and the level of crop yields which the rainfall will usually permit.

The value of such studies may be illustrated by reference to Fig. 3, which shows the 1:1 confidence limits for three-week moving totals of rainfall at Namulonge in Uganda. Manning (1956b) has produced evidence that the calculation and plotting of such three-week moving totals provides the most useful description of the distribution and reliability of rainfall throughout a season. In this part of Uganda the rainfall exhibits a well-marked bimodal pattern with two rainy seasons, during each of which sowings of crops may be effected, but with a drier time between them in the period June to August. Food crops are normally planted in the first wet season, cotton being grown in the second rains and harvested in the dry weather from December to February. Native practice is to plant the main food crop, maize, on average, at the end of March or early in April, as it is believed that before this the early rainfall is too uncertain for successful establishment.

Maize planted at this time is not normally harvested in time to permit of preparation and replanting of the land with cotton before the end of July or August. Reference to the confidence limit curves, however, shows that July/August is a poor time for satisfactory crop establishment and that the expectation of rainfall thereafter is rather inadequate for the production of a good crop of cotton. Investigations have confirmed that it is desirable to plant cotton on the decline of the main rains in June, the optimum planting date being about the middle of that month. The longer the planting of cotton is delayed after that time, owing to the land being occupied by maize, the lower the yields

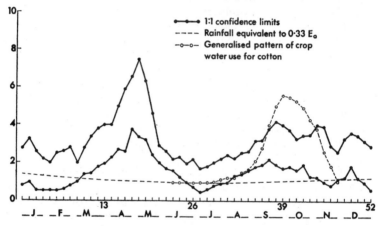

FIG. 3. 1:1 confidence limits of three-weekly moving totals of rainfall at Namulonge (Manning, 1958)

of cotton will be. It is therefore important that maize should be planted as early as possible.

Now the curves for the confidence limits do not suggest any marked period of unreliable rainfall or droughty conditions is to be expected in the early part of the rainy season after the end of February. A suitable time to plant maize would be when there is a reasonable expectation that the rainfall will exceed, and will subsequently continue to exceed, the evaporation losses from the bare soil, which in this investigation were assessed as being equivalent to 0.33 times the evaporation from a free water surface. Reference to the curve for the lower limit shows that as from the latter part of February the rainfall is not likely to fall below this required amount more than once in four years. Hence maize could satisfactorily be planted in the last week in February, with only the comparatively small risk of loss of seed not more than once in four years. A study of the rainfall distribution and reliability has thus indicated fairly clearly that more productive use of the land than that obtained by customary usage might be made if a suitable short-term variety of maize, maturing in about 3½ to 4 months, were introduced and planted in the last week in February, at least four weeks earlier than local practice. Maize could then be harvested in time to allow for preparation and planting of the land with cotton by the end of June, which, in the majority of seasons, would lead to appreciably better growth and yields of cotton than could be obtained by the hitherto accepted practice of planting in late July or August.

In the past a great deal of effort has been expended in endeavouring to determine the optimum planting time for various crops in the tropics by carrying out field experiments comparing different planting dates. Since there was obviously considerable year to year variation in the date of the commencement of the rains, and in the total amount and distribution of rainfall in the season, such trials were repeated over a number of years, in the hope that results would apply at least to average conditions. As the incidence of such average conditions was somewhat unpredictable, the results of such experiments were usually not of high accuracy, even if based on a large number of trials. Obviously data on the expectation or probability of rainfall throughout the season provide a much simpler, less laborious and more accurate means of assessing optimum planting dates. Furthermore, since nutrient uptake and spacing are interlinked with soil moisture availability, the results of fertilizer and spacing trials are unlikely to be capable of simple interpretation, as they will again vary from season to season depending on the incidence and distribution of rainfall. The planning and interpretation of such experiments can therefore also be greatly facilitated by a knowledge of rainfall expectation.

It must be mentioned that the extent to which use can be made of rainfall reliability studies is limited by two factors. The first is the lack of adequate rainfall records in many areas. The second is our very limited knowledge of the water or rainfall requirements of tropical crops, to which reference is made later.

RAINFALL INTENSITY

In most parts of the tropics a large proportion of the rainfall falls in heavy showers and storms of high intensity, much higher intensity than that normally experienced in temperate zones. For example, Evans (1955) found that at Masasi in southern Tanzania (Tanganyika) 45 per cent of the annual rainfall fell in heavy showers and storms of more than one inch per day, although he does not record the actual intensities. At Namulonge in Uganda in 1950/51 25 per cent of the storms had peak rates greater than 3½ inches per hour, 9·4 per cent had peak rates greater than 5½ inches per hour, and a maximum peak intensity of 10 inches per hour was recorded (Farbrother and Manning, 1952). In Trinidad at ten stations where intensities are recorded, intensities exceeding 3 inches per hour for periods varying from five to forty minutes were recorded fairly frequently, and peak intensities of up to 4·7 inches per hour were experienced.

The battering of the soil surface by the raindrops of high-intensity storms breaks up the soil aggregates, or crumbs, and the resulting very fine particles block the soil pores, causing the soil surface to become sealed, or 'capped', thus greatly reducing the infiltration of rainwater. As a result, run-off and erosion are increased, and the proportion of the rainfall that percolates into the soil and becomes effective for crop production is reduced. The sealing effects often go deeper than the surface layer because turbid water entering the soil through cracks or holes made by worms, termites, etc., will have its silt load filtered out at lower levels, resulting in deeper sealing. The energy involved in the impact of high intensity rainfall on the soil is very great. Ellison (1952) has calculated that the energy in 3 inches of rain falling in one hour with a drop velocity of 30 feet per second is equivalent to that required to plough the land twenty-nine times. Even storms in which 3 inches of rain falls within an hour are far from unknown in many parts of the tropics, and it is quite common for such intensities to be reached, and exceeded, over shorter periods.

The adverse effects of rainfall, in destroying soil crumbs, reducing infiltration and enhancing run-off, increase with increasing intensity. This is illustrated by Fig. 4, based on observations made at Namulonge, Uganda (Farbrother and Manning, 1952), which show a steep rise in run-off from bare soil with increasing rainfall intensity. Run-off is naturally reduced by protecting the soil with a cover of crops or other vegetation, or of mulch, which breaks the impact of the raindrops and slows up the movement of water over the soil surface. Fig. 4 shows that both the amount of the run-off, and its rate of increase with increasing rainfall intensity, are much less from land protected by a grass mulch than from bare soil. The basic percolation rate varied between 0·4 and 0·6 inches per hour on bare soil, but was increased to about 4·0 inches per hour by the application of a grass or stone mulch.

These results at Namulonge were obtained from plots situated on land with only a 2 per cent slope, i.e. agriculturally flat land, yet during a period in which 22·68 inches of rain fell, 39 to 64 per cent of this was lost from bare soil as run-off, depending upon whether the

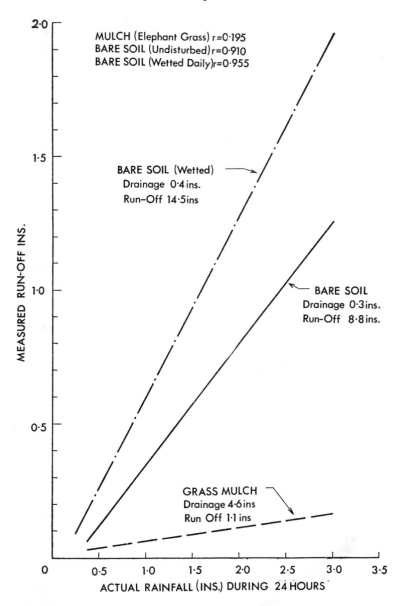

FIG. 4. Rainfall intensity and run-off at Namulonge (Farbrother and Manning, 1952)

surface of the soil was kept constantly wet or not. The major part of land cultivated will have slopes appreciably greater than 2 per cent, consequently the percentage of the rainfall which is lost by run-off may often be greater than is suggested by the Namulonge figures.

The occurrence of a considerable proportion of the rainfall in storms of high intensity is thus a matter of the first importance since it may greatly reduce the proportion of the rainfall which becomes available to the crop, and greatly increase run-off, with consequent erosion and loss of fertility. It is the main reason why it is so generally necessary to adopt agronomic and mechanical measures for soil and water conservation in the tropics.

THE BALANCE BETWEEN RAINFALL AND EVAPO-TRANSPIRATION

The extent to which the rainfall in any area can satisfactorily supply soil moisture for crop production depends, in part, on the balance between rainfall and evapo-transpiration. The rate of evapo-transpiration depends on the following factors:

1. The climatic factors of solar radiation, air temperature, wind speed and atmospheric humidity.
2. The soil moisture supply. The growth and transpiration rates of plants increase with increasing available water in the root zone until the soil moisture content reaches a level somewhat below field capacity, the precise moisture content at which maximum rates are attained varying with the plant species, its stage of growth and, probably, with soil texture. As the soil moisture content is further increased the growth and transpiration rates remain constant until they are adversely affected by lack of aeration in a very wet soil.
3. The plant cover. Under identical conditions plants of different species may transpire at different rates, due to differences in such characteristics as frequency of stomata, rate of conduction of water through the system, depth of root system.
4. The land management. This may directly affect the rate of evaporation from the soil surface (e.g. by mulching), or indirectly affect the transpiration losses by influencing the growth and size of plants (e.g. by cultivation and manuring).

Actual evapo-transpiration is therefore very variable, being dependent on the interrelation of a number of variable factors. Not only will it vary from place to place, and at different times of the year, because of the climate or the weather, but it can be readily altered by modifying the factors of soil moisture, plant cover and land management. The measurement of actual evapo-transpiration under field conditions is not only very difficult and laborious, but also can only give data applicable to the place and time at which it is investigated.

Penman (1948, 1949, 1950, 1956) has studied potential evapo-transpiration and its relationship to evaporation from a free water surface. Potential evapo-transpiration may be defined as the water lost when a green crop completely shades the ground and there is an optimum supply of soil moisture. It is the maximum rate of evapo-transpiration permitted by the amount of the solar radiation, which provides the energy needed to heat the overlying layer of the air and to vaporize the water. It can be determined by growing plants in sunken tanks filled with soil constantly supplied with moisture, and measuring the water added, the outflow of drainage water and the water retained in the soil; but its measurement is laborious and the results only apply to the site of the experiment. On the other hand, an open water surface is a reproducible surface of known properties; evaporation from it depends entirely on the weather conditions that influence evapo-transpiration losses from soil and crops, and can be measured at any place by means of an evaporimeter.

Penman made studies of the water use of a closely cut grass turf, which provided a complete leafy cover to the ground at all times, and which was always plentifully supplied with water. Under these conditions he found that the relationship between potential evapo-transpiration, E_t, and evaporation from a free water surface, E_o, was mainly governed by the length of day, and the evapo-transpiration rate varied between $0 \cdot 6 \, E_o$ in midwinter and $0 \cdot 8 \, E_o$ in midsummer. At other latitudes the seasonal values will also vary with the length of day, but towards the equator, where day length varies little, it should be constant all the year round at about $0 \cdot 75 \, E_o$. Penman also concluded that provided the water supply is always plentiful, provided that the crop forms a more or less complete cover over the ground and is growing in an extended area such as a field, and provided further that it does not change colour on ripening (which would influence the amount of solar radiation reflected and absorbed), then the water consumption is mainly dictated by the day length and the weather, and little affected by the particular crop, which 'behaves like a sheet of wet, green blotting paper which is covered up at night'. Estimates of potential evapo-transpiration in terms of the evaporation from a free water surface should therefore apply to all other crops and vegetation, provided the conditions postulated above obtain. Consequently, if measurements of E_o from an evaporimeter are available, it should be possible to calculate E_t, which is an indication of the water need of a crop for optimum growth, and, by balancing E_t against the rainfall, to obtain a rough assessment of the extent to which the rainfall will satisfactorily provide soil moisture for crop growth. In fact, evaporimeter records are not widely available in the tropics, but evaporation from a free water surface can be calculated, by a formula derived by Penman (1948), from weather data for mean duration of bright sunlight, mean air temperature, mean vapour pressure and mean wind speed. Estimates made by Penman's formula have been found to agree closely with recorded evaporation from a free water surface at a large number of stations, including some in the tropics.

THE WATER NEED OF CROPS

Calculations of potential evapo-transpiration are chiefly of value in estimating the water needs of certain crops when grown under irrigation, where the conditions aimed at, particularly that of optimum water supply, are not limiting to evapo-transpiration. Calculations based on Penman's methods have been found useful in planning some irrigation projects in the tropics, particularly with a perennial crop, such as sugar cane, which more or less completely covers the ground for the greater part of its growing period. The value of the method for estimating the water need of annual and rain-fed crops in the tropics is, however, limited for several reasons. First, it is only valid when water supply is not limiting; after the roots have used up immediately available soil moisture, soil factors, fertilizer treatment, crop management and crop type may greatly affect evapo-transpiration. Secondly, moisture reserves in the soil at the beginning of the growing season are not taken into account. Thirdly, annual crops differ markedly from a grass turf, or any other vegetation which provides at all times a continuous cover of uniform density and colour. In each season's growth of an annual crop there is a sequence of changing conditions from bare soil at planting time, through a time of increasing leaf area to a maximum of vegetative cover over the soil. This is followed by a period of decreasing leaf area as senescence sets in during which many crops also change colour on ripening, which alters transpiration rates by affecting the amount of radiation which is reflected, as well as by a change in the physiological activity of the leaf.

Consequently, estimates of potential evaporation are not very satisfactory indications of the water need of rain-fed crops, especially annuals, nor can they be very satisfactorily used in assessing the balance between rainfall and the actual evapo-transpiration losses of a crop, or in the application of studies of rainfall reliability. What is required is some more accurate estimate of the actual evapo-transpiration, or water need, of different rain-fed crops. For an annual crop this necessitates the measurement, throughout the growing season, of the variable amount of moisture lost by evaporation from the soil and by transpiration from the crop under controlled conditions providing for optimum, or near optimum, growth and yield. Under field conditions this necessitates, not only the measurement of water added as rainfall or irrigation and lost through drainage, but also frequent determinations of soil moisture content to the full depth explored by the roots of the crop. Since changes in the soil moisture content of cropped land cannot be followed with sufficient precision by the use of tensiometers or resistance blocks (Pereira, 1954a), they have hitherto had to be determined gravimetrically on soil samples taken frequently from different depths, a laborious and unsatisfactory procedure since it involves much soil and root disturbance. The actual evapo-transpiration determined will in any case only apply to the site and season of the experiment, but if it can be related to the measured evaporation from a free water surface, E_o, then a model curve for the water consumption of the crop studied throughout the growing season could be plotted in terms of E_o. This curve could then be applied to other areas where the crop is grown, provided that a reliable measurement or estimate for E_o is available for those areas. Superimposing such a model curve on a graph of, say the 1:1 confidence limits of three-weekly moving totals of rainfall, should show whether the expected amount and distribution of the rainfall would permit the satisfactory growth of the crop, and should indicate what planting date is likely to give the best fit of the pattern of expected rainfall and the curve of water need of the crop.

An attempt to carry out this kind of study was made with cotton by Hutchinson, Farbrother and Manning (1958) in Uganda. As no accurate measurements were made of soil moisture changes to the full depth of the cotton roots, their figures for evapo-transpiration losses may not be very precise, but their results do illustrate the pattern of water consumption through the growth period of an annual crop. The loss of water by evaporation from the soil fell from about 0·33 E_o at planting time, when the soil was bare, to zero when the leaf cover reached one acre per acre, while at the same time the use of water by the crop itself increased with increasing leaf area up to a maximum of one acre of leaf area per acre, after which it fell as vegetative and fruiting development ceased and senescence set in. The graph of the generalized pattern of crop water use for cotton has been superimposed on Fig. 3, which gives the 1:1 confidence limits of three-weekly moving totals of rainfall at Namulonge.

Humidity

Atmospheric humidity naturally tends to vary both with the absolute amount and the seasonal distribution of rainfall, being uniformly high throughout the year in wet equatorial and wet monsoon climates and seasonally falling to very low levels in places where there is a severe dry season. There is usually a marked diurnal fluctuation. Even in rainforest where the relative humidity throughout the year is always at or near saturation during the night, it may fall to as low as 70 per cent during daylight on dry days. The common occurrence of high humidities over long periods, combined with high temperatures, favours rapid development and spread of fungus diseases of crops and of moulds on stored produce. In some areas it is difficult without special drying equipment to dry out grain to a low enough moisture content to permit storage in good condition for any prolonged period.

Temperature

Temperatures are seasonally higher, and temperature ranges are greater, in the areas of drier monsoon and arid climates than they are in areas experiencing wet equatorial, wet windward coast, or wet monsoon climates. In the latter zones very high temperatures exceeding 100°F occur rarely, if at all; the diurnal temperature range seldom exceeds 20°F and is not often greater than 10°F, while the range of mean monthly temperatures is usually somewhat less. On the other hand, in the drier areas, especially those near the tropics of Cancer and Capricorn, maxima of over 100°F are common during the summer months but there is also a distinctly cool winter season, so that the range of mean monthly temperatures will commonly exceed 20°F, and a range of over 30°F is not unusual. Diurnal ranges are of the same order.

Low temperature is not normally a limiting factor to crop growth and production in the tropics except where it is markedly reduced by altitude. As a very general rule, temperature decreases by 3·3°F for every 1,000 foot rise, and this has a very marked effect on climate and agriculture in certain tropical countries where there is a considerable area of highlands. For example, in Kenya there is a sharp contrast between the tropical crop production of areas at, or near, sea level and the extensive semi-temperate farming of the highlands above 6,500 feet, which is based mainly on wheat, barley, oats and pyrethrum, and livestock production on natural grasslands or sown pastures. Above certain altitudes frost is regularly experienced at certain seasons of the year. Frost damage to certain crops occurs above about 7,500 feet in Kenya, above 5,500 feet in Malawi and above 5,000 feet in Rhodesia.

Some tropical crops can certainly stand higher temperatures than others, but this is largely a matter of drought resistance, as is the case with sorghum, bulrush millet or sisal. Adverse effects of high temperatures are commonly associated with water strain, for example, high temperatures and low humidity may kill maize tassels by dessication, cause boll shedding of cotton, or premature fruit drop with certain fruit trees. A more direct effect of high temperatures is local killing of areas of bark, commonly known as sun scorch. This usually occurs as a result of sudden exposure to full sunlight after the plant, or parts thereof, have been shaded. For example, if tea plants are taken from a shaded nursery and planted out into the field, they will suffer from scorching of the bark, unless they are shaded in the field. Similarly, the sudden removal, by pruning, of the leafy canopy of tung oil trees, or rubber, can result in scorching of the bark. High temperatures combined with long periods of bright sunshine can also cause scorching of certain fruits, notably of pineapples.

Certain crop plants require a period of low temperature, or winter chilling, for normal initiation and development of flower buds and fruit. In the absence of sufficient chilling, flowering may be irregular, flower buds may be shed, or embryo flower parts may fail to reach full development before the buds open. This feature is likely to be encountered in attempts to grow certain temperate and subtropical fruit trees even at fairly high altitudes in the tropics, where it is often responsible for poor or negligible crops. Fruit crops requiring chilling include apple, pear, plum, peach, apricot, almond, olive, pecan, grape and strawberry. There are marked differences in chilling requirements between species and between varieties within species. With pyrethrum, which is an important crop in Kenya, a period of low temperatures is required to initiate flower development and a flush of flowers appears about three months later. A number of varieties of other temperate crops require chilling, including cereals, grasses and clovers, but these are only encountered in very limited areas of the tropics at high altitudes.

The important effects of temperature and other climatic factors on livestock, especially on cattle of exotic breeds or cross breds with a proportion of temperate blood, are dealt with in Chapter 16.

Light

In the past it has commonly been considered in many, but not all, countries producing them, that certain tropical crops which are indigenous to forest zones, notably tea, coffee and cocoa, require some reduction in light intensity by the provision of shade if they are to do well. Reduction in light intensity is only one effect of shade trees, and the whole question of shade is discussed on pp. 239–244; but it may be mentioned here that, with the possible exception of tea, there is no evidence that reduction in light intensity *per se* is necessary for any of the major tropical crops.

At the equator the day length (from sunrise to sunset) is always twelve hours, and the maximum variation in day length at the tropics is just under three hours. Consequently the possible daily amount of sunshine is never less than nine hours, but the actual amount of bright sunshine is always much less, owing to the high degree of cloudiness experienced in most parts of the tropics. Cloudiness decreases north and south of the equator, reaching minima near the tropics.

Some crop plants are photosensitive, or photoperiodic, that is to say their flowering, and in some species other characteristics such as dormancy and the formation of tubers, is regulated by the length of day. Photoperiodism is a somewhat complex subject of which only the briefest mention can be made here. Photosensitive plants have a critical day length for flowering induction, i.e. a day length which is just capable of inducing flowering. Long day plants require a period of long days for flowering and will not do so if the day length is always shorter than their critical day length. Day length within the tropics is always below the critical day length of certain long day plants, such as hops, and chrysanthemum, and consequently they will not flower. Short day plants require a period of short days for flowering and will not do so if the day length is always longer than their critical day length, for example, some varieties of soya bean and tobacco. Other plants require alternate periods of short and long days in order to flower, and with many more, flowering is not affected by length of day.

In addition to these effects there are many varieties of plants which will flower in either long or short days but much earlier in one than the other, and their period from sowing to flowering is therefore affected by the time of the year at which they are planted, or the latitude where they are grown. This applies for example, to varieties of rice, maize, sorghum, soya bean and kenaf (*Hibiscus cannabinus*). Many varieties of rice are very sensitive to the shortening of day length and reach maturity only at certain seasons of the year irrespective of the time that they are sown. On the other hand, other varieties of rice are not photoperiodic, but take approximately the same time from sowing to flowering no matter at what time they are planted. Varieties of some crop plants may also be latitudinally adapted. For example, maize and sorghum varieties from the tropics will not flower when grown out of the tropics in America, but locally adapted varieties will flower and set seed even when the days are as long as sixteen hours. Selection and breeding can therefore produce day length neutral varieties, or latitudinally adapted varieties, of several species of photosensitive crops.

CHAPTER 2

Tropical Soils

H. VINE, B.SC., PH.D., A.I.C.T.A.

Soil Surveys and the Classification of Tropical Soils

Systematic soil surveys of many tropical areas have been carried out in the past twenty years, and in many cases these have been of direct value as a basis for selecting land for agricultural development or planning conservation work or other improvements. In the long run the best use can be made of the maps and soil descriptions if they are not used, uncritically, to pinpoint particular sites and read the generalized information on the soils indicated there on the map, but are regarded largely as guides to recognition and further evaluation and study of the soils in the field. It is not possible to map entirely homogeneous units, nor to plot boundaries between the soils with complete accuracy on a practical scale. There are a few areas where special features such as salinity or alkalinity afford a satisfactory basis for mapping; but generally the soil surveyor, in dividing the whole range of variations in the soil profiles which he finds into a manageable number of mapping units, is only able to do this on the basis of two things in combination – the morphology of the profiles and their relationships to the landscape. Thus the units customarily known as soil series, and labelled with a place name for convenience, are set up; and their distinguishing features are closely related to their geological origin and the processes of their formation.[1]

The morphological features and topographical relationships of soils also largely determine their ecological character, and the distinctions between soil series and their varying effects on vegetation or crops are often easily recognizable. Sometimes one series may differ most notice-

[1] It has not been possible to include further discussion of the formation of tropical soils and the relative significance of geomorphological history, parent rock and climate in the present account, but the reader is referred to reviews and papers by Brammer (1962), Carter and Pendleton (1956), Jessup (1961), Milne (1947), Mohr and van Baren (1954), Morison (1949), Nye (1954 and 1955), Paton (1961), Prescott and Pendleton (1952), Thorp *et al.* (1960), van Wambeke (1962) and R. Webster (1960).

ably from another in colour or some other feature that is of little agronomic significance, but it is nearly always possible to point to differences that can be expected to affect the growth of crops, the ease of working the land or its susceptibility to erosion. First, crop growth may be related to differences in 'root room', which is affected both by structural conditions in the upper layers and by the depth of the rooting zone (Hardy, 1951). The latter can be limited by dense or hard layers, lack of aeration, or salinity or other toxic conditions. Secondly, *groundwater* with a capillary fringe that remains within reach of the roots of a crop for some time after the end of a rainy season is sometimes an important feature. Thirdly, in soils without accessible groundwater (which are much the more extensive) differences in their *capacity to store available water* may be of great importance. Fourthly, there may be differences between soil series in *nutrient supplies, lime content*, and *acidity*, and in the amounts and nature of any *decomposing rock minerals* that are present within the reach of roots. The availability of particular nutrients, estimated by field or pot trials or laboratory tests, is too variable and dependent on treatment of the land in the immediate past to be used as a criterion in soil mapping.

The systematic mapping of large areas on small scales, such as 1/50,000 or 1/100,000, requires the use of defined 'associations' or 'complexes' of soils, which are geographical groupings of soil series that may be very different in their characteristics. The most extensive soil associations that are mapped are *repeating topographical sequences of soils in erosional landscapes*, i.e. catenas, using the term in the same way as Milne *et al.* (1936) and Milne (1947). The soils of a catena occur in a recognizable pattern within a more or less distinct region of hilly, rolling, or undulating ground and a particular bedrock or complex of bedrocks. They do not form perfectly regular zones, nor are all the series necessarily found along any one transect. Variability in the rock contributes to irregularity of the soils on the slopes, uneven deposition of local alluvium causes irregularity in the low ground, and the pattern across the valley of a small tributary is somewhat different from that across a major valley. Some workers divide the soil catena of a landscape into two or more sections and call each of these an association. There are also soil associations consisting of the soils formed from differing but related materials occurring in particular *constructional landscapes* of extensive alluvial deposits, dune sand left by the retreating fringe of a desert, lava flows, or volcanic ashes. Associations and complexes, after definition in the field, may largely be recognized on air photographs, and their boundaries mapped to a considerable extent by this means.

Standardized techniques and descriptive terms have been most fully worked out in the widely-used American Soil Survey Manual (U.S. Dept. Agric., 1951). It has always been the policy in the American Soil Survey reports to give, first, systematic descriptions of all the soils mapped, showing what their inherent features are, and then to discuss their agricultural relationships on the basis of 'interpretive groupings'. In modern surveys these groupings take account of subdivisions of soils into slope classes, as well as the internal properties of the soils. Generally a classification of the soils into standardized 'land-use capability classes' is given, and also a more elaborate set of 'management groups', with recommendations as to crops, methods of cultivation and fertilizer use. The recommendations made can be expected to require modification as economic conditions change and new farming methods or varieties of crops are developed, and the soils may then need to be regrouped. Provided that the basic units are clearly defined and mapped, revised interpretations can be made (for example, a new set of management groups could be drawn up) without the necessity to repeat all the field work, which would arise if this were originally done in an *ad hoc* manner only. Such considerations are important in view of the large amounts of time and money required for carrying out intensive surveys.

Table 1. *A classification of tropical soils*

Azonal Soils (little profile development shown).
1. Fresh alluvium, Lithosols, and Regosols.

Intrazonal Soils (effects of parent rock or drainage factor or shortness of time dominant over those of climate and vegetation).
2. Dark cracking clayey soils (grumusols and non-grumusols).
3. Allophane soils formed from volcanic ash and agglomerate.
4. Calcimorphic soils.
5. Hydromorphic soils.
6. Halomorphic soils.

Zonal Soils (effects of climate and vegetation acting over long periods dominant over those of parent-rock and drainage factors).
7. Pedocals.
8. Pedalfers: (a) latosols; (b) podsols (not including groundwater podsols); (c) fersiallitic soils (including some ferruginous siallitic soils).

In principle, soil series should be units of a general classification having 'families' or 'fascs' at the next level and 'great soil groups' or 'global soil types' at the level above that. A new American classification (U.S. Dept. Agric., 1960), which takes into account a great deal of recent detailed study and employs a new terminology, divides the soils of the world into ten *Orders*, of which eight are further divided into a total of twenty-nine *Sub-Orders*, made up of 106 *Great Soil*

Groups. The most extensive soils of the tropics, including latosols and kaolinitic hydromorphic soils, make up the ninth Order, to be known as 'oxisols', but this has not yet been definitely divided into Great Soil Groups. The tenth Order, 'histosols' (peat and bog soils), has also not been definitely subdivided. Other important schemes of classification are those of the Australian, French, Portuguese, and Belgian workers (Stephens, 1953; Aubert and Duchaufour, 1956; Botelho da Costa and Azevedo, 1960; Sys, 1960), that drawn up for a soil map of Africa (d'Hoore, 1960), and that of C. F. Charter (Brammer, 1962).

Table 1 gives a broad classification into Orders and Sub-Orders, based on the well-known older American scheme (Thorp and Smith, 1949), which can conveniently be used for a discussion of tropical soils at the present time. Some explanatory comments on the differences from Thorp and Smith's classification are required.

Sub-orders of the intrazonal soils added after 1949: the dark cracking clayey soils and the allophane soils formed from volcanic ash and agglomerate

'Margalitic soils' and 'grumusols' are alternative names for the majority of dark non-saline cracking clayey soils of humid to semi-arid and tropical to warm-temperate regions, which particularly include most or all of the '*regur*' or 'black cotton' soils of India. Modifying slightly the original statements of Oakes and Thorp (1950), grumusols are non-saline soils having the following features:

1. A very sticky consistence when wet and very pronounced shrinkage on drying and swelling on wetting, due to a high content of montmorillonoid clay.[1]
2. A medium to high content of exchangeable calcium and usually, also, $CaCO_3$ concretions at a depth of 2 or 3 feet or more.
3. Dark grey-brown to black colour of the surface soil, often extending into the subsoil – sometimes to a depth of several feet.
4. 'Self-mulching' of the surface, by its splitting into a loose layer of fragments when moistened after drying.

[1] Botelho da Costa and Azevedo (1959) describe the clay fraction of black earths of Angola as 'strongly siallitic' to 'very strongly siallitic', according to the following scale of silica/alumina and silica/sesquioxide ratios:

	SiO_2/Al_2O_3	$SiO_2/(Al_2O_3+Fe_2O_3)$
Very strongly siallitic	Not less than 3·5	Greater than 2·0
Strongly siallitic	3·4–2·4	,, ,, ,,
Weakly siallitic	2·4–2·0	,, ,, ,,
Fersiallitic	Greater than 2·0	Less than 2·0
Weakly ferrallitic	2·0–1·7	,, ,, ,,
Medium ferrallitic	1·6–1·4	,, ,, ,,
Strongly ferrallitic	Not greater than 1.33	,, ,, ,,

5. The formation of wide cracks (which implies that there must be a definite dry season during the year), so that surface soil falls and becomes incorporated into the subsoil sufficiently to prevent the formation of separate eluvial and illuvial horizons.

Oakes and Thorp added provisionally that grumusols were soils of areas of tall grass and savanna. Later work has shown that dark montmorillonitic clays found naturally under forest – which have a fairly high humus content – should be included. Grumusols were also thought generally to have a low humus content (1 to 3 per cent), but in virgin areas in the southern prairie region of the U.S.A. they have more recently been found to have about 5 per cent of humus to a depth of 6 inches or more. A 'hog-wallow' or 'gilgai' surface was recognized as a common feature, but there are good examples of grumusols that do not have this, and there are other kinds of poorly drained soils that have such surfaces.

Soils with a clay fraction consisting largely of allophane or allophane-like materials (rather indefinite hydrated amorphous aluminium silicates, with a SiO_2/Al_2O_3 ratio of about 1), formed from volcanic ash and agglomerate, have become recognized as another additional sub-order of Intrazonal soils. The term 'ando soils' was originally used for dark humic soils of this class, but is now often applied also to the related brown soils, which are more extensive.

Omission of any broad category of zonal soils under grassy vegetation

Thorp and Smith divided Zonal soils into several sub-orders, instead of only pedocals and pedalfers. They conceived of an intermediate world-wide grouping of soils of grassland and forest-grassland transition regions, of dark colour and strong development of structure due to accumulation of humus, such as the 'prairie soils' of the U.S.A. It is doubtful whether soils of this grouping can be recognized in the tropics. Some soils in the tropics have been referred to in recent years as 'reddish prairie soils', but they seem to be essentially latosols and fersiallitic soils of relatively high base status, associated with moderate rainfall and leaching, but of somewhat lower humus content than soils of similar origin under forest.

Restriction of the term 'laterite' to sesquioxidic material that becomes hard on exposure or is already hard

There has been a great deal of confusion in the use of the terms 'laterite', 'lateritic soils', and 'laterization' in the past. The word 'laterite' was first used in 1807 by F. Buchanan, to describe ferruginous mottled material of parts of south-west India which was of moderate hardness in place but became very hard after being cut out in building blocks with iron tools and exposed to the weather. Stephens (1961) has described the quarry sites now being worked at Buchanan's original type locality.

After some years the name 'laterite' was generally applied by geologists to ferruginous layers that were already hard, which have been found to be much more extensive than the original Buchanan laterite. These are often cellular or clinkerlike in structure, as a result of the partial or complete washing out of clayey material from tubular and other spaces, together with the deposition of hard concretionary coatings of iron oxide and filling of cracks with similar material.

Almost all soil scientists now agree that the word 'laterite' should only refer to the forms and varieties of sesquioxidic material in soils that either become hard on exposure or already have hardened, and may or may not be present in 'latosols' and can also occur in other soils (Kellogg, 1949). Recently the American Soil Survey decided to substitute a new term 'plinthite' for 'laterite' in this sense; but Sivarajasingham, Alexander, Cady, and Cline (1962) still refer to the material as 'laterite'.

Latosols (or ferrallitic soils)

In the 1930s most soil scientists had a quite different concept of 'laterite', based on chemical analysis, and all subsoils or weathering crusts of rocks that contained high proportions of free and combined sesquioxides (Al_2O_3 and Fe_2O_3) became known as 'lateritic' or, in extreme cases, 'laterites'. Very large areas of *deep friable* soils, possibly including certain amounts of ferruginous gravel but often none, came to be classified under these names. Robinson (1951) suggested 'ferrallitic soils' and 'ferrallites' instead, and these terms have now been widely adopted, whilst the name 'latosol' (Kellogg, 1949) is also commonly used.

Kellogg and Davol (1949) tentatively divided the latosols of the Congo into six great soil groups and gave field descriptions of about twenty profiles, with emphasis on the structure, consistence, and porosity, together with some analytical data. Generalizing, latosols were defined as a sub-order of Zonal soils having the following features (amongst which the colouration was too narrowly specified):

(*a*) low silica/sesquioxide ratio of the clay fraction;
(*b*) medium to low cation-exchange capacity of the clay;

(c) low 'activity' of the clay (presumably in the sense of its effect on stickiness, plasticity, and shrinkage);
(d) low content of primary minerals, except for the highly resistant ones (especially quartz);
(e) low content of soluble materials;
(f) a relatively high degree of aggregate stability;
(g) 'red colour or reddish shades of other colours'.

Of these criteria, (a), (b), and (c) mean that the upper layers contain no montmorillonoid clay; in *ferrallitic soils* the clay fraction usually consists of kaolinite (SiO_2/Al_2O_3 ratio 2·0) with about 8 to 18 per cent of ferric oxide, although there may sometimes be aluminosilicate minerals other than kaolinite present; on the other hand, in *ferrallites* the sesquioxides become dominant. With regard to (d), the low content of minerals such as biotite and lime-soda felspar is significant because this means a lack of reserves of nutrients; but the range in texture of latosols from clay to coarse or fine clayey sand, depending on their quartz content, affects their productivity considerably. Characteristic (e) emphasizes that latosols, as Zonal soils, only occupy slopes and summits, and are pedalfers. The next point, (f), is of the greatest significance, as it refers to the friability, permeability and generally high resistance to erosion of these soils; but, paradoxically, many soils that should certainly be regarded as latosols have since been described as having structureless subsoils. With regard to the last point, (g), it is recognized that the commonest iron oxide mineral, goethite, ranges from red to yellowish-brown in colour, and latosols vary accordingly.

Some writers have thought that the term 'latosols' should refer only to 'ferrallites' (or 'strongly ferrallitic soils'), with very low SiO_2/Al_2O_3 ratios. But latosols, according to the most widely held general concept, which is adhered to in this chapter, include the majority of the reddish and yellowish-brown soils of the humid and sub-humid tropics. Firstly they are pedalfers, formed on uplands and slopes by processes of decomposition, oxidation and leaching, such that there is 'a differentiation or tendency to differentiation of the clay complex in the different soil horizons, resulting in accumulation of sesquioxides' (Robinson, 1951, p. 448). Secondly, they are evidently the product of a process of combination of the colloidal silicates and oxides, which could more appropriately be said to give a stabilized microstructure than a structure of stable aggregates. A criterion pointed out by Bennema (1963) is that very little, if any, of the clay fraction of a 'latosolic B horizon' is brought into suspension by shaking with distilled water alone.

1 Highland grassland and forest, Kenya

2 *Acacia-Themeda* grassland, Kenya

3 Clearing lowland rain forest, Sabah

4 Shifting cultivation, Sabah

Controversial points in the classification and nomenclature of other pedalfers

Some podsols of temperate regions, occurring on slopes, are classified as pedalfers of the Zonal order; others are subject to waterlogging and are regarded as intrazonal hydromorphic soils. Both classes are formed of strongly weathered materials which are generally fairly thoroughly leached of bases. The special features of these soils are produced by the action of chelating and reducing substances dissolved out from the leaves of various conifers, heaths, etc., which appear to mobilize both alumina and clay particles, as well as iron, producing the well-known ashy-grey 'A2' horizon of eluviation and the coffee-coloured and rusty-brown horizons of deposition of organic material and iron oxide below. Groundwater podsols have been found in several parts of the tropics, and there may also be some podsols formed in conditions of free drainage; low temperatures are not necessary for their formation.

There are extensive soils in the warm humid region of the south-east of the U.S.A. which are known as 'red-yellow podsolic' soils. These have a mixed woody vegetation including conifers and it may be that they tend to be weak podsols, with bleached subsurface layers, on this account. They are moderately to strongly acid soils, commonly showing a clear division between a rather sandy 'A' horizon and a more clayey 'B' horizon, which consists of firm, coarse, blocky aggregates and is less friable and permeable than the subsoil of a latosol. Some reddish tropical soils (for example, some soils in Trinidad, formed from phyllite) that apparently have much hydrous mica in the clay fraction have these properties but do not have ashy-grey superficial layers. It is misleading to call such soils 'podsolic', and most of the soils that have been described as 'red-yellow podsolic' in the tropics can reasonably be included in a broad group of 'fersiallitic' soils; but some are sufficiently friable and permeable to be classified as latosols.

Botelho da Costa and Azevedo (1959, 1960) classified certain red to yellowish-brown soils of Angola having clay fractions of intermediate composition, formed from acid and basic igneous rocks, as 'fersiallitic' soils. They thought that the soils which Milne *et al.* (1936) had classified as 'non-laterized red earths' in East Africa, which generally seemed to have a SiO_2/Al_2O_3 ratio greater than 2·0 in the clay fraction and were described as 'clod-structured, with heaviness proportional to clay content', were similar to these. They stated that their concept of 'fersiallitic soils' approached more or less the 'rotlehm' of Vageler and 'red loam' of Marbut. Possibly all pedalfers of the tropics that are neither podsols nor latosols can conveniently be termed 'fersiallitic soils'; but some ferruginous 'siallitic' soils (on Botelho da Costa and Azevedo's scheme of classification of clay fractions according to SiO_2/Al_2O_3 and SiO_2/R_2O_3 ratios) would probably be included. 'Fersiallitic soils' in this sense are found in the same range of climatic conditions as the latosols, and the differences in composition and morphology can be attributed partly to differences in parent rock and partly to other factors. In the current classification of African soils 'sols ferrugineux tropicaux', corresponding roughly to 'fersiallitic soils', are held to occupy broad zones associated with less intense or prolonged action of soil-forming processes than in the zones of ferrallitic soils; this concept does not seem to be generally valid.

Soils of Uplands, Slopes and Low Tablelands

LITHOSOLIC AND REGOSOLIC SOILS

Lithosols of some agricultural significance in more-or-less humid regions include: (a) soils on the slopes of rocky hills of gneisses and igneous rocks, usually of high organic matter content, which may enrich considerably the soils of clefts in the hills and a zone around their base, by hill-creep; (b) recent volcanic ash soils. The latter, if deep, may have only a sparse cover of vegetation for many years; but if less than two or three feet of ashes are deposited forest may regenerate rapidly, and in a wet climate such as that of St. Vincent or Martinique it may be possible to replant tree crops or sugar-cane soon after an eruption (Hardy, Robinson, and Rodrigues, 1934).

Common types of lithosolic latosols are: (a) very deep, excessively gravelly and stony materials on hills of quartzites, quartz schists or rocks containing many quartz veins; (b) thin reddish and brownish soils over sheets of laterite or ferruginized sandstones or mudstones (sometimes with patches of grey seasonally-waterlogged soil). In parts of Nigeria areas of the second of these types are used for collecting firewood and for rough grazing, and the tracks which are formed concentrate run-off and cause severe erosion of the adjoining land.

Regosols occur as coastal dunes and some other sandy accumulations, but there are extensive soils that can be called 'regosolic' which show similarities to well-developed zonal or intrazonal soils but whose properties are mainly determined by their very high content of quartz sand.

DARK NON-SALINE CRACKING CLAYEY SOILS

The grumusols of parts of the southern U.S.A. are deep clays, mainly situated on undulating plains over *marls and calcareous clays*; they were mistakenly classified as rendzinas until the essential differences from the rendzinas of Europe were pointed out by Oakes and Thorp (1950). The calcareous 'Houston black' soil was regarded as the standard, but it is greatly exceeded in area by other grumusols that become non-calcareous and acid with increase of rainfall. The physical properties of these soils cause some difficulty in their use. Ploughing is possible with heavy equipment in dry conditions; subsequently planting, harrowing, etc. can only be done during short periods when the moisture content is suitable since the soil is exceedingly sticky when wet. Drainage does not seem to be defective under careful management, and the top 18 inches of the soil usually have a compound structure of 'coarse blocky peds' which crumble to 'granules' on moistening. It is significant that subsoil exposed to drying and moistening in newly dug drains and roadsides also soon breaks down to loose material (Templin et al., 1956). It seems possible that severe sheet erosion might occur in soils of this type, exposing the subsurface layers which might not be recognized as such, on account of rapidly breaking down to a structure and a consistence similar to those of the surface soil. This may have happened in large areas of grumusols in India.

Similar soils to the Houston Series were likewise formerly classified as rendzinas in the Caribbean region. Here they possibly formed the land of the highest natural productivity (Hardy, 1951). They developed under forest and uneroded profiles have been found to have about 8 to 10 per cent of humus in the top 3 inches and 4 per cent in the second 3 inches. In Puerto Rico, Roberts (1942) classified the shallower soils on hill-slopes, which had about 6 or 9 inches of very sticky dark greyish-brown granular clay and pale grey to yellow-brown subsoil over *soft limestone*, as rendzinas, and deeper colluvial soils, which were mostly placed in the Camaguey and Santa Clara Series originally mapped in Cuba, as chernozems. Neither sub-group differed much between the zones of relatively low rainfall (30 to 55 inches) and high rainfall (up to 100 inches). These soils were very largely used for sugar-cane, assisted by some irrigation in the drier areas. In Trinidad the older residual soil of calcareous belts in the dissected-peneplain region is the black clay of the Princes Town Series, which is similar to the Camaguey Series of Puerto Rico, but irregular. The younger parts of the landscape over the same or similar rocks have paler soils (Tarouba and Brasso Series).

Undulating ground in the northern Transvaal, under a mean annual rainfall of 22 to 32 inches, has typical grumusols over *basic igneous rocks*, but somewhat acid sandy loams with iron concretions and laterite over granite and other rocks. The grumusol profiles have about 3 feet of black, slightly calcareous clay, over rock and decomposing rock with tongues of brown clay containing $CaCO_3$ concretions and gypsum ($CaSO_4.2H_2O$) crystals (van der Merwe, 1941). In these profiles exchangeable sodium is much less than calcium and magnesium and evidently the soils have been sufficiently leached in present or past conditions for sodium salts to be removed. The topsoil contains about 3 per cent humus. Black clays over a belt of *basic gneisses* in the Accra Plains of Ghana have been shown to be very similar to those of the northern Transvaal (Brammer, 1962).

The most extensive *regur* soils of India are those of the sub-humid to semi-arid Deccan plateau, a region of gently undulating plains between 1,000 and 2,000 feet above sea-level. Simonson (1954a) has given precise profile descriptions of the main sub-types. All were somewhat alkaline and contained scattered $CaCO_3$ concretions. The most typical of the soils went down into decomposing *basic igneous rock* at a depth of about 4 feet or less, but there were deeper soils, more representative of grumusols as a whole, which had olive-grey clay with 'slick faces' (attributed to the movement of large blocks and wedges of soil against one another as they swelled and shrank with wetting and drying) at about 3 or 4 feet. The dark-coloured part of the profile usually contained 0·3 to 0·8 per cent organic carbon (i.e., 0·5 to 1·4 per cent humus), and 0·04 to 0·06 per cent nitrogen, decreasing only slightly with depth within the dark layers. Singh (1954) proved that the dark colour of the soil was due to organic matter. Simonson (1954b) discussed the much poorer yields of crops on these soils than those on the grumusols of the U.S.A., and suggested that the poor tilth that could be produced with small bullock-drawn implements was an important factor, together with nutrient deficiencies.

Kanitkar (1944) described this land as having been used in the past on a system of shifting cultivation, with resting periods for restoration of fertility. Resting land, as well as shallow soils that were seldom cropped, provided forage for the cattle and firewood. With great increases in numbers of people and their cattle, however, most of the land had come to be continuously cultivated, trees and bushes had been destroyed and the soil exposed to erosion. Kanitkar stated that the control of run-off and erosion was the most conspicuous problem of the region, and observations and measurements at several research stations showed that severe losses of water and soil occurred under the usual methods of cultivation. The main purpose of the improved 'Bombay Dry Farming Method' which was being tried out was to prevent these losses by bunding and by tillage to increase infiltration of rain.

Kanitkar showed that one of the chief difficulties with these soils was the very high moisture-retaining capacity of the top layer, which caused a very large proportion of the rainfall to be wasted by direct evaporation; light rains only moistened the surface without penetrating sufficiently for the water not to be drawn up again by capillarity unless more rain followed very soon. Field capacity was about 40 to 45 per cent (by weight), and the permanent wilting point about 16 to 20 per cent; during the dry season the top 6 inches dried out much below the wilting point, to about 6 per cent moisture. Less retentive red soils in similar conditions were, particularly, much more favourable for the growth of early millet, because of the smaller evaporation losses. Experimental results on the black soils do not seem to have shown that more thorough preparatory tillage or more surface cultivation during the early rains (before planting) or during the later rains (between the rows of the crop) increased the yields of sorghum much, provided that weed growth was controlled. The use of farmyard manure seems to have been mainly responsible for the increased yields obtained under the 'Bombay Farming System', but the prevention of erosion could be expected to be of the greatest importance in the long run.

Since the early 1940s the value of crops has increased greatly, whereas the price of fertilizers has not changed much, and the Deccan Plateau region is one where the use of fertilizers should now be feasible and would probably be of the greatest value in increasing yields. The increased amounts of plant residues resulting from use of fertilizers would also lead to substantial increases in the organic carbon and nitrogen contents of the soil, with some general improvement in structure and in nutrient reserves. Kanitkar's account and comparison with similar soils elsewhere (such as the Transvaal) suggest that the very low humus content of these soils is largely the result of continuous cultivation and erosion.

ALLOPHANE SOILS FORMED FROM VOLCANIC ASH AND AGGLOMERATE

Humic soils ('ando soils')

Ando soils, in the original sense of the term (Thorp, 1949; Mohr and van Baren, 1954; Kanno, 1956; Bramao and Dudal, 1958) are a distinct sub-class of volcanic ash soils, characterized by a very high humus content (8 to 30 per cent to a depth of 1 or 2 feet), looseness, very high porosity and nearly black colour. They occur under moderate to very high rainfall, are strongly leached and of low base status, and are found from near sea-level to heights of a few thousand feet. Their clay fraction probably always consists of allophane or allophane-like materials, which have been recognized as the usual first product of weathering of sub-basic volcanic ashes. The steeper and rougher areas of these soils should be left under forest; other areas seem to be well suited to various crops, though they are liable to be deficient in nutrients.

Brown earths with allophane

Hardy and co-workers (e.g. Hardy, Rodrigues, and Nanton, 1949; Hardy and Beard, 1949) recognized two subgroups of freely draining soils related to the brown forest soils of temperate regions in the volcanic islands of the West Indies, having dark upper layers (containing about 2 or 3 per cent organic matter) one or more feet thick, over 'yellow earth' and 'brown earth' respectively. The yellow earth is the product of fine-grained ashes, deposited in broad, moderately sloping beds, whilst the brown earth consists of clayey material and partly decomposed boulders and stones (the coarse ejecta which formed the steeper slopes of volcanic cones, or disintegrating agglomerate). These soils are remarkable for their unusual physical properties, which are evidently due to allophane-like material. This material occurs partly in the form of pseudomorphs of the felspar grains from which it is formed by hydrolysis, and it may be partly because of this that the soils have an exceptional capacity to hold water, whilst at the same time they are very permeable but also of somewhat firm consistency.

Because of these unusual physical properties it has been possible to stop erosion of yellow earth soils throughout most of the agricultural land in St Vincent by means of contour strips and banks, with occasional silt pits, whereas on most normal soils in similar conditions it would have been necessary to lead off excess water by means of graded furrows and banks.

Very similar soils to these yellow earth and brown earth soils are known to occupy large areas elsewhere. The genetic relationships between these soils and various geographically associated soils are complex, and changing conditions as well as time have been involved. In the West Indian volcanic islands the yellow earth and brown earth soils only occupy about 25 per cent of the area, on the whole; lithosols, zonal red earths, and soils containing varied dense siliceous layers occupy comparable areas and there are minor occurrences of the humic soils. According to brief reports, in some countries conditions have favoured the development of montmorillonoid clay and distinct cracking clay soils, in association with allophane soils. In general – as in Java (Dudal and Soepraptohardjo, 1960) – there appears to be a process of weathering, from volcanic-ash particles, through different forms of 'allophane' to the crystalline clay-mineral metahalloysite, and finally to kaolinite; and the older soils are latosols.

INTRAZONAL CALCIMORPHIC SOILS

Under this heading may be placed soils similar to the Brown Forest Soils other than those containing allophane. There may be considerable fertile areas in the tropics with soils closely related to those formed from diorite in Tobago, described by Hardy, Akhurst, and Griffith (1931), which probably contain a mixture of clay minerals. These are weakly acid to neutral loamy and clayey hill-creep soils, with decomposing rock at about 3 or 4 feet, and their occurrence is largely due to the steepness of the slopes and the shortness of time since the present cycle of weathering and erosion began. Another soil, which is formed from an unusual soft, impure limestone containing chlorite (from which iron oxide is formed), is the very productive Montserrat Series ('chocolate soil') formed in small areas in Trinidad. In spite of its inherent good qualities, remarkably heavy applications of mixed fertilizers are considered necessary in re-establishing high-yielding cocoa in old plantations on this soil.

HYDROMORPHIC ('GLEY') SOILS

Soils with prominent grey or mottled colouration of the sub-surface layers showing lack of aeration in rainy periods occur at upland and intermediate sites at various places in the tropics. Although these soils have mostly not been accurately described it seems that there probably are two fairly distinct main groups: (1) hydromorphic highly leached sedentary soils of uneroded parts of old surfaces of planation that have been subject to decomposition and the eluviation of bases, iron oxide and some clay from the surface layers for a very long time; (2) hydromorphic 'hill-creep' soils with formation of stiff clayey material (probably containing montmorillonite) over decomposing rock, marl or calcareous clay, or alluvium mixed with volcanic ash, at a depth of a few feet.

Group 1. Two soils of the Pleistocene peneplain of Trinidad (Piarco and Long Stretch Series) are examples of this group. Very similar soils have been classified as Low-Humic Gley soils in some places. They have a few inches of dark topsoil over grey fine sandy or loamy subsoil with brown staining and mottling, over a thick layer of red-and-white mottled sandy clay, which appears to have developed *in situ* in past conditions from unconsolidated Pliocene and Pleistocene sediments. They are very poor soils for agriculture, not only because of their high acidity and very low nutrient content, but also because of the restricted depth of rooting caused by lack of aeration in the rainy season and consequent drought conditions in severe dry seasons. However, some areas have been planted with sugar-cane, using fertilizers (especially to remedy severe phosphate deficiency) with a fair amount of success, and recently Pangola grass has been grown satisfactorily on these soils with liberal applications of ammonium sulphate.

In Trinidad these soils appear to have been occupied almost entirely by rainforest vegetation before there was human interference. But within the forest there were small treeless areas, of which only a small group (the Aripo savannas) have survived until now, which were edaphically determined and not the product of burning (Richards, 1952; Chap. 15). The Aripo savannas are situated on a variant of the Piarco–Long Stretch soils which had no drainage and was waterlogged almost to the surface for considerable periods.

Forest species have tended to be more rapidly destroyed by fires on the Piarco and Long Stretch series than on other soils. Richards thought that the Rupununi savannas of British Guiana and the Llanos plains of Venezuela probably were likewise associated with soils of impeded internal drainage, and an alternation of waterlogging and drought. Jones *et al.* (1958, 1959) have shown that this is not the case in the Rupununi region; poor savanna growth is evidently not necessarily an indicator of poor drainage in north-eastern South America, but appears to be associated with various highly leached, infertile soils of free to imperfect drainage.

Fine sandstones and shales in parts of Thailand (Pendleton, 1943) and Malaya (Arnott, 1957) have given rise to poor, highly leached soils of low tablelands, similar to the Piarco–Long Stretch soils. In Ceylon such soils occur chiefly over low-lying alluvium, but there are smaller areas on plateau sites also (Moormann and Panabokke, 1961). In these countries the mottled substratum has largely been hardened to a form of laterite; in much of their area the soils have poor drainage and can be described as 'groundwater-laterite soils'. But in Africa and south-west India the hardening of the mottled layer appears mostly to have been associated with much dissection and the lowering of the wet season water-table, and inspection would probably seldom show definite signs of waterlogging in the top 2 or 3 feet or more of the profile at the present time. Thus there are widespread soils that can be referred to as 'relict groundwater-laterite soils', but it is misleading if they are referred to as hydromorphic. One exception is a grey soil with great quantities of concretions and broken-up ironpan over stiff grey-and-yellow mottled clay over shales and mudstones of part of the Voltaian Formation in Ghana (Brammer 1962).

There are also groundwater podsols that have developed from highly leached materials on old surfaces of planation. Typical soil of the Valencia Series of Trinidad, which is similar in geology and geomorphology to the Piarco–Long Stretch soils, is an example. Richards (1952; Chapters 9, 10, 11) remarked that in the Guianas and Indonesia these soils were an outstanding instance of vegetation being closely determined by soil and topography, but more probably the formation of these podsols is attributable to particular species which produce reducing and chelating substances.

Group 2. This group includes some soils classified by Thorp *et al.* (1961) as 'planosols', in a region of about 40 inches of rainfall in south-western Kenya. They occur extensively on the gentle slopes of wide valleys, either formed from alluvial deposits containing volcanic ash or overlying decomposing granite, syenite, or basalt at a depth of a few feet, and consist of grey, sandy loam or loam with stiff clay in the subsoil. On the higher parts of the broad undulations there are reddish-brown sandy and loamy latosols containing large amounts of concretions and broken-up ironpan ('relict groundwater-laterite soils') over the granite and syenite. In the basalt areas the higher ground is occupied by 'grumusols intergrading to planosols'.

Another recorded example of soils that probably belong to this group is that of the 'anomalous grey soils' (nearly black, sandy clays to yellowish-grey, very coherent sands) on upper slopes and summits, over gneisses, in certain parts of the eastern foothills in Tanzania (Milne, 1947; pp. 196–200).

PEDOCALS

The term 'pedocals' refers to soils of dry regions where the leaching action of the rain is restricted to the top layers so that $CaCO_3$, and possibly also $CaSO_4$, tend to accumulate in the upper layers of the subsoil. The term has been applied more specifically to the zonal soils, and not to soils showing much hydromorphic or halomorphic influence, which are common in many dry areas. In this sense the pedocals of northern continents appear to show a simple climatic zoning of well-defined Great Soil Groups: grey and red desert soils and sierozems; brown and reddish-brown soils of the steppes; chestnut (dark brown) and reddish-chestnut soils; chernozems. Although these are formed from calcareous materials, which predominate in the regions concerned, it has often been supposed that zones of very similar soils were to be found in the drier parts of the tropics, but little information has been available until recently, except in Australia. They could perhaps have been expected more often to resemble soils formed from non-calcareous materials in dry temperate regions described by Kubiena (1953), but these have themselves received little attention. It is evident now that the soils are a complex assortment and that many of them consist of the altered remains of much decomposed and leached soils formed in pluvial periods in the past, or of the remains of soils with $CaCO_3$ and $CaSO_4$ deposits which accumulated on old surfaces in arid periods that have now been more or less dissected. It also appears very improbable that there are any soils having any close resemblance to chernozems – of high humus content – in the drier parts of the tropics or subtropics, although early information on some of the grumusols suggested this.

In Australia (Stephens, 1953) the most extensive upland and slope soils of regions with less than about 20 inches of rain have been described as pedocals having '*deflated* neutral to alkaline eluvial horizons and calcareous and/or gypseous illuvial horizons', much eroded by wind and water or otherwise much altered over long periods. They include 'desert loams', with surface layers of sand and stones left by the wind, over calcareous and gypseous subsoils, 'desert sandplain soils' and 'stony desert tableland soils' formed from profiles with lateritic or siliceous concretion layers, and 'desert sandhills' of red sands, which have become partly stabilized. There are also skeletal soils of tablelands and ranges. The soils of uplands and slopes that may be regarded as of normal development as pedocals, 'red-brown earths' and 'brown soils of light texture', are less extensive.

The CCTA soil map of Africa (d'Hoore, 1960) shows 'desert soils' in regions having less than 4 inches of rain and almost devoid of vegetation, consisting of sands, clay-plain deposits, pebbles and calcareous materials, as well as some areas of 'rocks and rock debris'. The map shows 'brown tropical soils of arid and sub-arid regions' in areas with up to about 20 inches of rain, in an east–west strip south of the Sahara from Senegal to beyond the Nile and in extensive areas in East Africa, the Kalahari and the Lowveld of southern Africa, as complexes with other soils – desert soils, lithosols, and 'ferruginous tropical soils' (pedalfers).

The post-desert landscape south of the Sahara is a belt about 400 miles wide, consisting mainly of gently undulating, sand-covered plains but also including great areas of clays that appear to have been deposited as outwash from higher ground to the south. In the northern part of this belt the vegetation is *wooded steppe with abundant Acacia and Commiphora*,[1] with a marginal strip of *tropical sub-desert steppe*,[1] and the sands typically form a somewhat coherent reddish soil with a pH of about 7 and containing up to 10 per cent of clay

[1] Terms used by Keay (1959).

(Worrall, 1961). This soil has been referred to as 'dior' or 'goz', and may be regarded as a regosolic chestnut soil. However, in the southern part of the belt the rainfall is about 30 inches, and the sandy soils are regosolic latosols and the vegetation which covers this area consists of savanna woodland of a relatively dry type.

In the central part of the Kalahari region there are deep reddish sands similar to the 'dior' or 'goz' and the vegetation they support is again the *Acacia-Commiphora* wooded steppe. Sandy deposits continue far to the north, over parts of Angola, Rhodesia and the Congo. In south-western Angola the soils have been classified at the level of 'groupings' (very similar to fascs) and vary with rainfall from acid under relatively moist savanna woodland ('miombo'), to neutral under a complex of formations dominated by *Acacias* or, more often, by *Colophospermum mopane* (Botelho da Costa and Azevedo, 1959). Typical profiles have about 0·8 to 1·5 per cent of humus in the top 6 inches, with C/N ratios between 10 and 13, which are probably indicative of formation of humus and production of nitrate in balanced amounts. This appears also to be the case in the Sudan Zone savannas of West Africa, in contrast to the tall grass savannas, where C/N ratios are high and nitrate deficiency is severe.

Pichi-Sermolli (1955) summarized the available information on the soils of the dry parts of East Africa. In the most arid regions, with less than about 15 inches of rainfall (e.g., much of Somalia), layers of $CaCO_3$ and $CaSO_4$ seem to be common and pH values always to be above 7, and soils of the Australian 'deflated' types predominate. Hemming and Trapnell (1957) have given fuller information on such an area in north-west Kenya. Pichi-Sermolli mentioned that 10 to 15 inches of rainfall were sufficient to support a thick growth of xerophilous open woodland (*Acacia-Commiphora* wooded steppe), made up largely of bushes and perennial grasses, but that in some places overgrazing together with fires had caused much exposure of the soil, wastage of a large proportion of the rain by run-off, accelerated soil erosion and deterioration of the vegetation to poor scrub growth.

The regions with about 15 to 30 inches of rainfall in East Africa, extending from Somalia across Kenya into Tanzania, were provisionally recorded by Milne *et al.* (1936) as having a predominance of calcareous or non-calcareous 'plains soils', described as light-coloured pedocals, usually forming catenary sequences with black calcareous clays in the broad depressions. In some areas they were interrupted by 'red earths' and 'plateau soils' (soils with iron concretions) on and around inselbergs. The calcareous soils were found chiefly in Kenya, where they were of a pale reddish colour and had a layer of $CaCO_3$ accumulation similar to that of chestnut soils. In Kenya the non-calcareous soils were pale grey to pale brown and formed of transported materials mixed with small amounts of volcanic ash. In Tanzania the group particularly included red-brown sandy loams that were recognized as having been developed by prolonged decomposition and leaching under higher rainfall in the past and contained insufficient calcium for $CaCO_3$ layers to appear, but could be expected to show some affinity to pedocals; these occurred in the area around and to the north of Kongwa, for example. The term 'plains soils' has sometimes been used subsequently in East Africa, but without being clearly defined. Mohr and van Baren (1954) observed that it had been particularly applied to soils with compact subsoils that appeared to have rather high amounts of exchangeable sodium.

LATOSOLS AND FERSIALLITIC SOILS

a. Mineral composition and physical properties

It will be convenient to consider first four fairly distinct groups of latosols and then a broad group which includes the most widespread latosols, provisionally referred to as 'red and red-yellow latosols'; the distinctions are based on an interim classification used by Bennema (1963), and take account of differences in the composition of the clay fraction, consistence and structure.

The fersiallitic soils will then be considered in two broad groups according to base saturation. The first of these is intended to include Bennema's 'lateritic podsolic soils with lower base saturation in the B horizon', and also soils with low base saturation containing larger proportions of hydrous mica, mixed-layer or montmorillonitic clay, which Bennema does not mention. The second group is intended to include the soils which Bennema describes as 'lateritic podsolic soils with medium to high base saturation in the B horizon', occurring in moist-deciduous and evergreen forest climates, 'red and yellow Mediterranean soils', occurring in semi-arid to sub-humid climates with 'deciduous forest or shrub' vegetation, and other soils of similar composition that he refers to, briefly, as 'reddish prairie soils'.

There is still very incomplete knowledge of the composition of the clay fractions of these soils and the effects of differences in the clay fraction on coherence, plasticity, structure and porosity. Silica/sesquioxide and silica/alumina ratios give some indication of the mineral composition, but have not been determined in a great many of the soils; and (according to Russell, 1961, Chapter 6) the results of direct identifications by means of X-ray diffraction, differential thermal analysis and the electron microscope, seem often not to be quite as complete and certain as has sometimes been thought.

The characteristic stable structure or microstructure of latosols has generally been thought to be very largely due to a cementing effect of ferric oxide, and Fripiat and Gastuche (1952) and d'Hoore, Fripiat and Gastuche (1954) had evidence that this effect was related to the ferric oxide becoming finely disseminated and overlaid upon particles of kaolinitic and other minerals. But results of Despande, Greenland and Quirk (1964) appear to show that ferric oxide has very little, if any, effect on soil structure. Aluminium oxide was shown to have a pronounced binding effcet.

1. *Pale yellow latosols with low amounts of sesquioxides.* According to Bramao (1962) these *weakly ferrallitic soils* are predominant in the forests of the Amazon, where they are very sandy. They are associated with various hydromorphic soils and there are intergrades with these. Soils that may be placed in this group are probably fairly common elsewhere and they may include more clayey types; but the group has not been clearly distinguished from the broad group of 'red and red-yellow latosols', of medium iron-oxide content, in which many (probably most) dull yellow to slightly reddish yellow soils of tropical Africa are better placed. Bennema specified an iron oxide content less than about 10 per cent of the clay fraction.

2. *Latosols with very low amounts of silicate clay minerals.* These *strongly ferrallitic soils* attracted much attention in the past and came to be regarded, mistakenly, as the end-product of decomposition and leaching, to which other soils tended to develop in hot, humid conditions. The Nipe Series of Cuba and Puerto Rico belongs to this group and has often been referred to, but seems to be a very unusual soil; consisting mainly of iron oxide and only formed in a small area over a rather uncommon type of rock. The chief soils of the group are those which have a predominance of aluminium oxide (mainly as the mineral gibbsite), but also more iron oxide in the clay fraction than most latosols (about 20 to 25 per cent). Most soils of this group and the next are formed from basic rocks or limestones and consequently contain little or no quartz sand. Both have a very well developed finely granular structure in topsoil and subsoil, but the topsoil is liable to become very powdery and easily eroded after cultivation for a long time (Colwell, 1958).

3. *Other latosols of high iron-oxide content.* In these soils silicate clays such as kaolinite are predominant over free aluminium oxide. They include the soils known as *terra roxa legitima* in Brazil, formed from basic igneous rocks on gentle slopes under evergreen and semi-deciduous forest, which have been found suitable for intensive

use. Some of the soils of Hawaii have been correlated with them, as well as some 'krasnozems' of Australia. They seem clearly distinguishable from the *terra roxa estruturada* of Brazil and from soils that have been described as 'terra rossa' elsewhere.

4. *Brown latosols.* This group includes some latosol-ando soil intergrades, containing certain amounts of allophane, in parts of Indonesia and Hawaii, and some soils that occur at somewhat higher altitudes than other latosols in Brazil, over various materials, that seem similar in their physical properties.

5. *Red and red-yellow latosols.* This name refers to the most typical soils of the tropics, in which the clay fraction consists mainly of a mineral of the kaolinite type with a medium amount of iron oxide (about 10 to 15 per cent); there may be little or no free aluminium oxide (in *weakly ferrallitic soils*), or moderate amounts (in *medium ferrallitic soils*). Bennema remarks that these soils occur 'under a wide range of climatic conditions and under a great variety of natural vegetation'. It should be accepted that soils of the specified mineral composition should be included without restrictions as to the base saturation, which some definitions specify as being below, or 'normally below', 40 or 50 per cent in the subsoil. The organic matter content varies with rainfall and soil texture.

The texture varies from sand to clay, depending on the nature of the parent rock and whether the soil particles have been sorted during erosion and transportation in the processes of formation of the landscape and the soils. Many of the sedentary soils have mottled layers of a firmer and denser consistence than the overlying material, ferruginous concretions formed from mottles or from highly decomposed residues of parent rock, or large masses of hard or rubbly laterite. Quartz gravel and stones also may or may not occur near the surface or at certain depths where they appear as 'stone lines' in pits or road cuttings. Most 'relict groundwater laterite' is sufficiently broken up not to prevent the penetration of water and roots, and where the soil material with it is clayey it is common to find satisfactory cultivation of annual crops by hand methods, although areas where there are many large blocks of laterite near the surface have to be avoided in tractor cultivation. Mottled layers are seldom so near the surface that there is any likelihood of hardening being brought about by using the land for agriculture; however, small eroded patches are occasionally found (such as old village sites) where the material has developed into a hard layer.

Clays of the 'red and red-yellow latosols' group formed from basic rocks and limestones are probably more extensive than the clayey latosols of high sesquioxide content, to which such rocks also give rise,

Soils of Uplands, Slopes and Low Tablelands

which have already been mentioned. With thorough decomposition of felspars and other minerals, or solution of the carbonates of the limestones, very little sand-size material is left and the soils have an almost uniform texture down the profile, unless modified by extraneous material in the surface layers; structure and consistence may also be very uniform or may vary. The Catalina and Cialitos Series, occurring under a rainfall of about 80 inches in the uplands of Puerto Rico (Roberts, 1942), consist of hill-creep material formed by weathering of volcanic rocks and have a brownish, finely granular topsoil, about 2 feet of redder, firmer subsoil and below that many feet of bright reddish, friable, granular clay. In both soils the whole profile is readily permeable to water and roots, and there is little run-off or erosion except on unprotected slopes exceeding 10 or 15 per cent. Smaller areas of latosols (e.g. the Matanzas and Coto Series), which are the direct product of weathering of hard limestones, occur in the coastal plains of Puerto Rico.

In East Africa there are very extensive porous red clays in the highlands between Nairobi and Mount Kenya, derived mainly from sub-acid lavas and originally developed under forest. The typical soil has been known as 'Kikuyu red loam', which is unfortunate, because such soils have very commonly been referred to as 'red earths' whilst less well aggregated and more plastic soils have been grouped as 'rotlehm' or 'red loam'. Milne *et al.* (1936) stated that the Kikuyu soil was typically a 'laterized red earth' in its morphology and field properties but had a silica/alumina ratio of 2·0 in the clay fraction; it seems to be very similar to the Catalina Series of Puerto Rico. Pereira (1957) describes it as having been formed by 'pluvial era' leaching, and as of such depth that profiles were studied to 15 feet in investigating the moisture relationships of the arabica coffee of the region. Most of the roots were in the top 10 feet but they extended beyond this after prolonged droughts, during which 10 feet of soil were dried out to the wilting point. Table 2 summarizes the data for mechanical composition, porosity and moisture retention for a typical group of pits, sampled in one-foot layers. An important feature is the rather low available-water capacity, which Pereira calculates to be equivalent only to 12 inches of rain water in the whole of the top 10 feet, remarking that only the great depth of soil and the extensive root range of the coffee crop make this a usable store of water.

Bennema (1963) remarks that latosols may dry out rapidly, even if not very sandy, and 'it may be that the available water is not so high as the field capacity suggests; much water may still be present at the wilting point'. Texture is clearly important in this connection, but the matter requires investigation. For one thing, little attention has been given to the probable importance of silt-size particles in forming capillary spaces that hold water in the pF range in which it is available (between 4·2 and approximately 2·5) in tropical soils.

Table 2. *Physical properties of Kikuyu soil* from H.C. Pereira's data for a typical group of seven pits

Depth (feet)	0–1	1–4	4–7	7–10
Sand (%) (2·0–0·02 mm)	12·1	12·9	14·4	8·9
Silt (%) (0·02–0·002 mm)	13·3	8·0	8·4	7·5
Clay (%) (less than 0·002 mm)	70·2	74·1	73·0	81·2
Percentage by volume				
Total pore space	60·9	57·1	55·5	54·1
Field capacity	37·2	40·0	42·3	44·0
Wilting point	24·2	26·7	29·7	34·5
Available water capacity	13·0	13·3	12·6	9·5
Equivalent amount of available water storage, in inches per foot depth	1·56	1·60	1·51	1·14

Sandy clays, clayey sands or sandy loams, and sands of the 'red and red-yellow latosols' group cover great areas of sedimentary, metamorphic, and granitic rocks in Africa and South America. The patterns of soils in the Congo outlined by Livens (1949) illustrate their general occurrence. Firstly, the Kwango-Kasai region, near the boundary with Angola, includes a dissected upland surface formed of sedimentary sands and grits, covered by brownish-red sandy soils with about 15 to 20 per cent clay, similar to very large areas of Tertiary and Cretaceous sediments in West Africa and in parts of East Africa. In Nigeria such soils always increase in clay content with depth, typically having a loose slightly clayey sand grading into firm clayey sand to sandy clay below and of negligible silt content throughout, and though very permeable they have no noticeable structure. They occur as an old colluvial mantle and vary in texture according to the proportions of clay beds intercalated with the sands (Vine, 1954, 1956). Further south-west, Livens describes pale yellow Kalahari sands with less than 5 per cent clay, deposited over the older sands and grits, bearing only *grass steppe* (Keay, 1959) on the high ground. These regosols contrast with the red-brown and orange-brown very sandy latosols formed from wind-sorted sands in the Sudan Zone of Northern Nigeria, which are generally fairly productive, but especially so in the Kano region where they are thin and overlie the remains of old, more clayey soils, and where decomposing rock is often within reach of tree roots. Secondly, the great region which Livens describes as the Cuvette Centrale consists of deep clayey sands to sandy clays (with 20 to 40 per

5 Soil erosion in the Andes

6 Gully erosion

7 Vegetable growing on bench terraces, Malaya

8 Rubber on modified bench terraces, with cover plants sown in the inter-rows, Malaya

Soils of Uplands, Slopes and Low Tablelands 51

cent clay), over sediments, very similar to some of the more clayey Nigerian sedimentary soils mentioned above. Typical profiles of these soils at the research headquarters at Yangambi were described by Kellogg and Davol (1949) as 'red-yellow latosols'.

Thirdly, the Uele-Ituri region, formed of the crystalline basement-complex rocks, includes several patterns of soils that are well known in other parts of Africa. The Uele district, along the boundary with the Sudan and Oubangi-Chari, consists of an old peneplain surface that has been dissected in several stages and includes part of the Nile–Congo watershed. The soils throughout this district form catenary sequences (although Livens questioned the use of the term 'catena'), and include latosols with varying quantities of concretions or remains of laterite sheets and small areas of grey soils in low ground. Livens describes the most typical soils, without laterite sheets, as forming the following sequence, from upper to lower slope: '(i) red-brown, very dull at the surface, with 50–65 per cent of clay and a high proportion of concretions often at no great depth, (ii) orange-brown, with concretions, and a clay content of 40–45 per cent, and (iii) orange-yellow, seldom possessing concretions and with only 20–35 per cent of clay.' The Ituri district consists of a mountainous area near Lake Albert and a series of plateaux, about 3,000 to 4,000 feet high, joining this to the Uele peneplain and Livens describes the soils as forming a catenary sequence from shallow soils on rocky hills to expanses of red and brown soils with 50 to 70 per cent clay and rarely any concretions (occurring as colluvial drifts or mantles), and poorly drained soils in low ground. The same sequence of soils is predominant in Uganda between Lake Albert and Lake Victoria, as shown by Radwanski and Ollier (1959) and Radwanski (1959). The red soils of Tanzania are also formed from mantles of drift, as in the catenas at Tabora and Ukuriguru and also at Kongwa and Nachingwea (Muir, Anderson, and Stephen, 1957), and this appears also to be the case in Rhodesia. Catenas with sedentary soils on the upper slopes, similar to those of the Uele district, are predominant in southern parts of Nigeria and Ghana, over the crystalline rocks; but drift soils have been described in some parts of Ghana over these rocks as well as over sedimentary rocks (Brammer, 1962).

In the classification of soils developed in the course of comprehensive surveys in the Congo in the 1950s (Sys, 1960), latosols were divided into:

(a) *Ferralsols* – with granular and feebly developed subangular blocky structure, low amounts of silt, and very little sign of 'clay-skins' deposited on the surfaces of aggregates.

(b) *Ferrisols* – with one or more of the following features: (i) fairly well developed blocky structure in the subsoil, with 'clay-skins'; (ii) a silt/clay ratio greater than 0·15 or 0·20; (iii) over 10 per cent of weatherable minerals in the fine sand fraction.

In general the ferrisols were considered more productive soils than the ferralsols. Presumably they would tend to have a greater available water capacity or to have significant reserves of K, Mg and possibly other nutrients.

Soils containing substantial amounts of weatherable minerals, but otherwise having the properties of latosols, may occur as hill-creep soils with decomposing rock residues; Botelho da Costa and Azevedo (1960) included these in their 'paraferrallitic soils'. The fertility of old latosols which have received additions of fine volcanic ash has also been mentioned by several writers.

In Ceylon, Moormann and Panabokke (1961) only considered a limited area of soils, on an old coastal terrace, to be 'red and red-yellow latosols'. They correlated the dominant soils of the 'wet zone' (mostly having over 80 inches of rainfall) with the red-yellow podsolic soils of the U.S.A.; but it seems that a large proportion of these soils, which consist of residual and colluvial material from quartzose crystalline rocks and some old river terrace deposits, may be similar to the latosols over similar rocks in the forest zone of West Africa. For some of the soils it is recorded that the SiO_2/Al_2O_3 ratio of the clay may be 1·7 or less and the cation-exchange capacity of the clay less than 10 mg-equivalents per 100 g. The soils are relatively sandy in the upper layers and more clayey and red to yellowish-brown below. They mostly lack a sedentary mottled layer, but there is a sub-group which has either a mass of mottled material that would harden if exposed ('cabook'), or red mottles that would harden, at a depth of 4 feet or more; the overlying material contains hard concretions or, sometimes, sheets or blocks of laterite.

The group of 'reddish-brown lateritic soils' of Ceylon seems also to consist largely, but not entirely, of 'red and red-yellow latosols'. These soils are formed from some of the less siliceous rocks in the wet zone and are described by Moormann and Panabokke as consisting mainly of old slope colluvium in rolling to hilly terrain, having excellent drainage and a high degree of aggregate stability, rarely having any *in situ* laterite, and having no mottled layer; these features are in accordance with recorded values of less than 2·0 for the SiO_2/Al_2O_3 ratio. But cation-exchange capacity figures of 25 to 40 mg.-equivalents per 100 g of clay fraction, and observations of strongly developed clay coatings, are also mentioned and it must be assumed that these relate to fersiallitic soil profiles.

6. *Fersiallitic and ferruginous siallitic soils with low base-saturation in the subsoil.* According to general descriptions, soils that have been classified as 'red-yellow podsolic' in South America and South-east Asia, as well as in Ceylon, not only show deposition of clay in a B horizon, heavier in texture than the layers above, but also have markedly less stable aggregation than latosols of a similar textural profile, lower permeability and a well-developed blocky structure, except in the more sandy types, and a higher cation-exchange capacity of the clay fraction (Bennema, 1963). Presumably a considerable proportion of the 'red-yellow podsolic soils' of Ceylon are of this nature, as well as those of the 'reddish-brown lateritic soils' that have clay fractions of high cation-exchange capacity and strongly developed clay coatings. A practical feature of such soils is that they are more susceptible to erosion than latosols in general.

A number of soil series of Puerto Rico were classified by Roberts (1942) as red-yellow podsolic. Those of undulating to gently rolling inland areas were typified by the Fajardo Series. This soil was formed from shales of low base content and had a brownish topsoil which formed a good tilth in cultivated fields, over compact grey, yellow and red-mottled clay at a depth of a foot or less, which interfered with root development and slightly hindered the percolation of water. Iron oxide concretions occurred in the surface and weathered shale fragments throughout the profile. The clay fraction had a silica/alumina ratio of 2·9 to 3·4 and a silica/sesquioxide ratio of 2·2 to 2·8 in a typical profile. The closely related Moca Series had a stickier, more plastic and more heavily mottled subsoil, whilst the Lares Series, formed from transported clayey material from the areas of Catalina and Cialitos Series, was much less plastic and fairly permeable and well drained, though prominently mottled brown, yellowish-brown and red. In contrast, the extensive Los Guineos Series of steep rocky slopes in the uplands had a rather simple profile with about one foot of greyish and yellowish clay over red clay described as plastic but permeable.

In Trinidad the dissected peneplain underlain by marine clays and fine sandstones has large areas of mottled plastic clay and sandy-clay soils, similar to the Fajardo and Moca Series of Puerto Rico, and also extensive reddish and yellowish loamy sands and sandy loams with certain amounts of mottling in the subsoil, in all of which the cation-exchange capacity of the clay fraction is about 50 mg.-equivalents per 100 g. The Northern Range of Trinidad consists of phyllites (with varying proportions of fine quartz and of micaceous minerals), and the soils are mostly rather featureless hill-creep sandy loams to sandy clay loams which show little structure or marked downward leaching of clay, but are very leached and acid. These seem to contain a largely micaceous clay and therefore to be 'fersiallitic' or 'ferruginous siallitic'.

It seems that, generally, physico-chemical properties associated with the type of clay should be regarded as of more significance than structure and morphology, in dividing these non-latosol pedalfers from the latosols.

7. *Fersiallitic and ferruginous siallitic soils with medium to high base-saturation in the subsoil.* Soils meeting this description include the group which Moormann and Panabokke (1961) have termed 'reddish-brown earths' in Ceylon. They are stated to be the most extensive soils of Ceylon, occupying most of the lowland dry zone, with a rainfall of 50 to 70 inches, and generally corresponding to the groups of 'non-lateritic red and reddish brown loams' previously described by A.W. Joachim. They are formed from Archean rocks containing considerable amounts of ferromagnesian minerals, and to a smaller extent from limestone. They are described as having a moderately to strongly developed fine subangular blocky structure in the B horizon, and as having pH values between 6 and 7 (tending to be lower near the surface than below, and sometimes being as low as 5·5 in the topsoil in the higher rainfall areas). The cation-exchange capacity (measured with reference to saturation at pH 8·2, giving somewhat higher results than the most common procedures) is reported as usually 45 to 55 mg-equivalents per 100 g of clay, and the silica–alumina ratio of the clay fraction as greater than 2·2. The red soil over gneiss at Nizamabad, India, (Nagelschmidt, Desai and Muir, 1940) with a rainfall of 37 inches, seems comparable with these soils, as do the yellowish-brown, orange and red fersiallitic soils found in small areas in the Huila district in Angola, described by Botelho da Costa and Azevedo (1959).

From their physical properties the *terra roxa estruturada* soils, which produce most of the coffee crop of Brazil, seem to belong to this group rather than to the latosols, but Bennema indicates that they have mainly 1:1-lattice clay minerals. Like the other type of *terra roxa*, these soils are derived from basic igneous rocks. The subsoil has blocky structure, with clay-film deposits, and the top-soil consists of extraordinarily coarse dark red aggregates. Other soils over basic rocks have been correlated with these, in Hawaii and Abyssinia.

Various plastic and sticky soils that apparently contain certain amounts of montmorillonitic clays, of reddish and yellowish colours and weakly acid to alkaline in reaction, occur in Cuba and Puerto Rico. Some of the Puerto Rico soils were described by Roberts as 'reddish prairie soils'. A similar soil occurs over much of the higher ground in Barbados.

b. *Humus content and its effect on soil structure*

A great many analyses have shown that normal tropical soils do not contain much less humus (i.e., organic matter present after picking out visible pieces of roots, leaves, etc.) than typical freely-drained agricultural soils of temperate regions. Importance is rightly attached to the chemical and physical effects of humus. It has a cation-exchange capacity of over 200 mg-equivalents per 100 g, and the retention of calcium, magnesium and potassium against leaching in the top layers of these soils is largely due to this. Humus is itself the main source of mineral nitrogen supplies for plant growth, as microbial breakdown continues; it also provides a proportion of the phosphate and sulphate supply, as well as certain amounts of micronutrients. The state of aggregation of the topsoil is to a great extent dependent on humus and microbial activity associated with it, and this especially affects the infiltration of rain water and the lessening of surface run-off and soil erosion.

The residues left in, or deposited on, the soil are largely consumed by insects (especially termites) and other fauna, but there is also direct attack by micro-organisms. The products become thoroughly mixed with mineral material in the soil surface. Leaf litter does not remain on the surface for long, and layers of humus or partly humified organic material, such as are to be seen in many temperate forests, do not occur in these soils. Estimates of the additions of plant debris under tropical forest more than ten times too great have been current, and this appears to be one reason why the rate of decomposition of litter, and (by inference) the rate of decomposition of humus, have been greatly exaggerated, as has been pointed out by Nye and Greenland (1960); for it is apparent that a steady state becomes established under forest or savanna where addition and loss of organic material during the year are equal. Nye and Greenland quote actual measurements of litter fall of about 4 tons of dry matter per acre per annum under well-grown forest and estimate that the root system adds about half this quantity of residues. In savanna areas they estimate the total amount to be not much more than one ton, because the majority of the leaf and stem growth is destroyed by fire.

If there is sufficient moisture, the cellulose, proteins and most other components of the plant residues become digested by the various heterotrophic organisms of the soil. The sugars, amino-acids, etc. which are produced are rapidly absorbed and used in respiration and in building up protoplasm and cell walls, while fungal hyphae and bacteria thoroughly penetrate the spaces between mineral particles, leaving, in turn, films of colloidal organic material as their remains are attacked by other organisms. On the other hand, it is generally agreed that the lignins, which form a large proportion of the plant material, are much more resistant to hydrolysis into simple molecules which can be assimilated and that they are chiefly affected by oxidation and combination in some way with ammonia and with proteins. The result is that a very large proportion of the carbon is given off as CO_2 and the remainder, together with most of the nitrogen, phosphorus and sulphur, and certain amounts of other elements, is left in the form of a mixture of colloidal substances which continue to be acted upon slowly by the soil organisms. This colloidal material is not a stable end-product; the fact that large amounts of such organic material (commonly 2 to 4 per cent in the surface layer) can remain in the soil and only gradually tend to disappear is not fully explained, but some kind of 'protective' combination with clay minerals and iron, and possibly other cations, seems to be responsible, and the humus content of soils is directly related to their clay content. Studies by F.E. Broadbent in California have shown that when readily decomposable matter is added to soils, producing a burst of microbial activity, the humus present in the soil is decomposed at an increased rate and, although the residues of the organisms make some addition, the final result may be a decrease in humus content (Russell, 1961; p. 269). This demonstrates the unstable nature of the humus and also – from the practical point of view – fits in with the finding that incorporating fresh plant material, such as green manure, may be ineffective in increasing the humus content of soils.

Smith, Samuels and Cernuda (1951) showed that humus and nitrogen contents were relatively high in the tropical climate of Puerto Rico and that the depth of penetration of humus in the profiles of permeable soils was an important factor in their productivity. They gave some evidence, also, that the humus and nitrogen contents of exposed subsoil could be built up remarkably rapidly in some conditions with dense covers or mulches. Birch and Friend (1956a) found that in East Africa there was in general an apparent inverse relationship between humus content and temperature, but that this was due only to the increase of rainfall with altitude; increasing temperature no doubt favoured the rate of decomposition of organic material, but it also favoured its rate of production (under a natural cover of vegetation or a good planted cover). Rainfall was the main factor governing the humus and nitrogen content. For example, thirty observations on the top foot of soil, at sites with a mean temperature of 66 to 69°F (19 to 21°C) and mean annual rainfall 23 to 38 inches, gave roughly a straight-line graph for humus content: 2·0 per cent with 24 inches, 2·5 per cent with 30 inches, 3·0 per cent with 36 inches. On the other hand, no correlation with temperature was found when the data for sites with approximately the same rainfall were picked out.

Birch and Friend (1956b) made the further observation that if incubation of soil samples is interrupted by periods of air-drying and remoistening there is, each time, a flush of microbial activity and CO_2 production, so that on the whole considerably more humus may be decomposed than in samples kept steadily in a moist (but well aerated) condition. It has since been shown that drying the soil produces a quantity of very easily decomposed material, probably both by the death of a large proportion of micro-organisms and by physical changes in the films of organic colloids. It seems possible that in practices such as mulching, shading and the use of planted or natural fallows the prevention of very frequent drying and rewetting of the surface may be more important than keeping down the temperature in conserving or increasing the humus content of the soil. Wetting and drying also influence the nitrate content of cultivated soils; this fluctuates greatly and is affected by leaching and by uptake by microbes and by crops, but in general there tends to be a maximum for some weeks when the soil is remoistened by rain after a pronounced dry season.

The greater part of the humus has relatively little direct effect in stabilizing the structural aggregates of the soil. The natural soil conditioners that have this effect are gums produced by various bacteria, but they are substances that are fairly readily decomposed by the general microbial population and the stabilizing of structure requires continued activity of the bacteria. Additions of fresh organic material, such as litter and root residues under cover crops or planted fallows, as substrate for these organisms, are therefore the most effective means of promoting soil structure (Bradfield and Miller, 1954). But the majority of the upland soils of continental areas of the tropics are too sandy to develop lasting structure. In shifting cultivation, root-holes, insect burrows and the looseness maintained for a while by coarse plant remains are probably much more significant in aiding infiltration and preventing erosion than soil structure in the usual sense.

In East Africa grass fallows have, until recently, been thought to have a beneficial effect on the soil through the improvement of structure, but measurements reported by Pereira, Chenery and Mills (1954) and Pereira (1955a) have shown that the physical effects of grasses do not last long, even on the more clayey soils, but largely disappear during the first year of cultivation. But mulching has been found markedly beneficial in improving surface soil structure. In an experiment on the rehabilitation of compacted and eroded 'Kikuyu red' soil of an old coffee plantation, Pereira and Jones (1954) found that plots which had either been mulched with grass for two years, or kept under a growing elephant grass cover, gave rainfall acceptance values (in a laboratory test) of nearly 100 per cent, but those where the grass had been uprooted after two years and coffee replanted gave values no better than untreated plots – about 70 per cent. A stable surface structure had been built up under the mulch by microbial activity aided by the mechanical protection of the soil from the beating rain.

c. Leaching and soil acidity

In most of these soils there is a close relationship between the percentage saturation with 'exchangeable bases' (Ca^{++}, Mg^{++}, K^+, and some Na^+) and the pH value, as ordinarily determined. This is not so in strongly ferrallitic soils and there may be other exceptions.

In well-grown forest and deeply weathered soils there is an almost closed cycle of absorption, movement and return of these cations. This has recently been studied much more fully than before, on a quantitative basis, especially by Nye and Greenland (1960), together with the losses by leaching and removal in harvested products when the land is cultivated. The quantities involved are exemplified by the total additions to the surface of the soil in a plot of high forest in Ghana – 262 lb of K^+, 283 lb of Ca^{++}, and 63 lb of Mg^{++} per acre per annum (of which, it is surprising to notice, 75 per cent of the K^+, 9 per cent of the Ca^{++}, and 25 per cent of the Mg^{++} were added as rain wash from the leaves).

Generally, the higher the rainfall the greater is the tendency for cations to be leached out of the profile. A rough division can be made between latosols of excessive leaching, with pH values less than 5·0 throughout the profile, and those of more moderate leaching. In Ghana these have been called 'oxysols' and 'ochrosols' respectively (Brammer, 1962), but a zone of oxysol-ochrosol intergrade soils is recognized. In Nigeria there is a fairly clear division between a zone of yellowish-brown latosols with rainfall over 80 or 90 inches, and a much more extensive zone in which reddish colours are prominent in the upper members of catenary associations; the difference seems to be mainly due to geomorphic history, but it happens also to divide the soils very roughly into those of excessive leaching and those of moderate to strong leaching.

The nature of the anions that must balance the cations leached out of the soil has generally not been carefully considered. Nye and Greenland have pointed out that in acid soils only nitrates are likely to be leached in substantial amounts; it is only above pH 6 that significant amounts of bicarbonates can be formed, even when the concentration of CO_2 in the gas phase is high. Very high rainfall does not necessarily, therefore, cause much leaching in soils that are already acid; this is dependent on nitrification. Nye and Greenland mention that actively growing plants reduce leaching losses both by transpiring and thus reducing percolation and by absorbing anions, especially nitrate, from the soil solution, but may also have a great effect through repressing the formation of nitrate. This latter process operates especially in some types of grassland; the Southern Guinea savanna of West Africa, where the dominant grasses are *Andropogon* and *Hyparrhenia* species, seems to be a good example. There is usually severe nitrate deficiency when this land is cleared, and the amount of nitrate present when it is under grass must be extremely small. This would explain why the soils are weakly acid to neutral, although they are porous and mostly sandy and the rainfall averages 45–50 inches, with several very wet months in succession.

The base saturation can be expected to change rapidly if conditions are altered, and particularly if they are made favourable for rapid nitrate formation, in sandy kaolinitic soils of rather low cation-exchange capacity. For this reason, it seems evident that relatively high base saturation cannot be a satisfactory criterion for classifying any soils as non-latosols ('reddish prairie soils', 'ferruginous tropical soils', or others).

Experiments on the effects of liming acid soils in the tropics have been in disfavour for some years, but this has not been well justified. In various trials in Southern Nigeria there have been considerable benefits from small amounts of lime applied to sandy latosols with a pH below 5·0, and the effect has appeared to be mainly due to increased production of nitrate. The relationships of nitrogen transformations to pH are complex and require much investigation, especially if the long-term objective of increasing the humus content of the soil is considered and not only that of immediate increases in crop yields. Several recent investigations have shown that tropical legumes are capable of fixing large amounts of nitrogen from the air; they do not require a high base status, and Norris (1956) has made it clear that it is only a few specialized legumes of temperate regions, such as clovers, that are not tolerant of acidity. The genus *Beijerinckia*, of bacteria similar to but slightly larger than *Azotobacter*, has been found widespread in more or less acid soils, especially latosols. These bacteria are active non-symbiotic fixers of nitrogen. The possibility exists that nitrogen fixation and other nitrogen transformations may be adversely affected by low availability of molybdenum at pH values below 5·0.

In fersiallitic and ferruginous siallitic soils of low pH the part of the cation-exchange capacity that is due to 'permanent charge' on the clay particles becomes occupied by aluminium ions, not by hydrogen as in humic colloids or typical kaolinite; and large amounts of exchangeable aluminium prevent proper root growth in many crops (e.g. sugar-cane). As Ignatieff and Lemos (1963) mention, it should then be necessary only to add sufficient lime to replace the exchangeable aluminium. Raising the pH to 5·5 or 6·0 should be satisfactory for this purpose. In latosols, which have very little permanent charge cation-exchange capacity, aluminium toxicity cannot be expected, even at very low pHs of 4·0 to 4·5, and any beneficial effects of liming must be related to some other factor.

d. *Phosphate in latosols and fersiallitic soils*

Some years ago a much-emphasized theory grew up that phosphate became 'fixed' in the form of very insoluble iron and aluminium compounds in these soils, especially in strongly acid conditions. This implied that small dressings of superphosphate would be ineffective, that the fixation capacity of the soil required to be satisfied (at least locally, around the bands or spots where the fertilizer was applied) before crops would show a response, and that residual effects would be very small. Field evidence on these points shows the concept to be wrong, as mentioned by Kurtz (1953), Tempany and Grist (1958), and Ignatieff and Lemos (1963). Kurtz was concerned with bringing together information from many sources clarifying the questions of the reactions and the absorption of phosphate in soils. The mistaken ideas seem to have arisen largely because of the low values obtained for 'available phosphate' in tests using weak acids as extractants, and because there were inadequate attempts to relate these figures to field responses. Probably, also, there were instances where the lack of response to phosphate was attributed to fixation of the fertilizer in unavailable form, whereas it was actually due to there being an adequate amount of phosphate already present, as in 'latosols and krasnozems' in Australia (Stephens and Donald, 1958).

The newer approach to the study of phosphate supply in the soil distinguishes between the *quantity* of 'labile' $H_2PO_4^-$, which is usually measured as that which exchanges reversibly with added phosphate labelled with the radioisotope P^{32}, and the *intensity* of supply; the latter is closely related to the concentration of $H_2PO_4^-$ or the thermodynamic potential of calcium phosphate in the soil solution (which are approximately the same as may be measured in an N/50 calcium chloride extract). Some measure of the proportions of more readily and less readily available phosphate may also be given by using different periods for exchange between a soil sample and labelled phosphate added in solution. The results obtained with various extracting agents that have been proposed from time to time are now more clearly understood. In the case of a quick-growing annual crop requiring to obtain phosphate at an optimum rate for a short period, tests which mainly reflect the intensity factor are most applicable (within a limited range of variations in the soil type and management practices); but with pastures and long-term crops tests which are more related to the quantity of labile phosphate are likely to give better indications. The amounts extracted in these tests, and the amounts obtained by plants growing in the soils, depend very much on solution from very thin layers on Al_2O_3 and Fe_2O_3 films and on the action of complexing agents (colloidal or water-soluble organic substances in the soil in the field, fluoride ions, etc. in laboratory tests) in blocking Fe^{+++} and Al^{+++} (Russell, 1961, Chapter 26).

These newer concepts have not yet been applied on a large scale in the tropics, nor have extensive data on the amounts of labile phosphorus in typical soils, differing in parent rock, texture, clay minerals, and vegetation, been obtained. Little is yet known of the rate at which organically-combined phosphorus is mineralized (Acquaye, 1963), or of the possibility that this is much affected by drying and wetting. Pronounced phosphate deficiency seems much more common in savanna areas than in forest areas in Africa, and such differences need to be explained. Stephens and Donald remarked that the very low amounts of phosphate in old highly weathered subsoils of Australia could not be ascribed to leaching, but possibly to long-continued extraction by plants, together with loss of the top layers by erosion in changing climatic conditions. Effects of different species in extracting phosphorus (Nye and Foster, 1958; 1960) require further study in connection with such problems.

Some Soils of Alluvial Flats and Low Terraces

AZONAL SOILS (FRESH ALLUVIUM, LITHOSOLS, REGOSOLS)

These soils include the following:

i. Alluvium accumulating in inland areas in papyrus swamps.
ii. Alluvium accumulating in saline conditions in mangrove swamps.
iii. Sand and stones, with a little humus, on coral atolls (where much copra is produced).

iv. Raised beaches and banks containing varying proportions of quartz and calcareous shell or coral fragments, especially between coastal creeks and lagoons and the sea.

Clearing mangrove swamps for cultivation has given different results in different areas. When clay is predominant, as in the 'Frontland Clay' of British Guiana, the nutrient content is likely to be high, and another advantage is that not much seepage of salt water under the banks occurs in empoldered land; coarse sands with fibrous *Rhizophora* root residues, which probably form a large proportion of the coastal swamps of Nigeria, are the opposite in both respects.

In Sierra Leone certain large mangrove areas were successfully brought into use many years ago; these are areas that are naturally flooded with brackish water during the dry season but with fresh water when the rivers come down in spate, and the mangrove vegetation has been cleared and rice is grown during the flood period without any control of the movement or level of the water (H.D. Jordan, 1963). When Government action was taken to empolder some of the more extensive swamps along the coast – as had been done successfully on a large scale in Portuguese Guinea – good results were obtained only in limited areas, chiefly those that had been occupied by *Avicennia*. Toxic conditions were found to develop in soil which contained a great amount of fibrous material, characteristically formed under *Rhizophora*, when it became exposed to oxidation; formation of sulphuric acid from sulphides brought the pH down from about 6 or 7 to 2 or 3 or even less. Coastal deposits which develop such toxicity are termed 'acid sulphate soils' by Moormann (1963), who records that they are common in South-East Asia. Edelman and van Staveren (1958) remarked that coastal muds that contained considerable amount of pyrites (FeS_2) and polysulphides were those that developed toxicity, and that they tended to form in weakly saline conditions, where plant growth was vigorous and there was much vegetable matter for sulphate-reducing bacteria to act upon. In contrast, where alluvium collected in strongly saline conditions plant growth was more restricted, less vegetable matter was available to support bacterial activity, and only a small amount of ferrous sulphide (FeS) was formed.

In Sierra Leone liming alone to remedy the toxic conditions would be prohibitively expensive, but leaching out of toxic substances by rain water may take ten or fifteen years or more even though the rainfall is very heavy. A solution to the problem has now been found in the use of sea water for leaching, which is far more effective, mainly because of cation exchange (Hart, Carpenter, and Jeffery, 1963). The procedure is to empolder and drain the land, leave it for some months for the smallest (and most active) particles of sulphides to oxidize, then allow it to be washed again with salt water moving with the tide for about two months, and then drain off the salt water and allow the remaining salt to be washed out by rain-water before planting rice. Indications are that if this cycle of processes is carried out for two or three years the sulphides then remaining will only oxidize very slowly and sea water can then be permanently excluded. The operation of the system must depend on the control of the water by sluice gates, and this is presumably much facilitated by there being about 15 feet between high and low tide in Sierra Leone (which is much greater than the range along many coasts).

SOILS OF CLAY PLAINS AND LOW GROUND IN THE DRIER REGIONS

There are large areas of fine-grained alluvium and colluvium, within the semi-arid to sub-humid regions of the tropics, which seem to have accumulated during Late Quaternary times on account of changes in rainfall and erosion and continued movements of depression or uplift. The soils formed from these parent materials probably could be classified as hydromorphic and halomorphic soils of various types, non-saline cracking clays (or, more specifically, grumusols), zonal soils (pedocals), and intergrades between these genetic sub-orders. However, they are usually only considered in one general grouping.

In Australia the most extensive of these soils are the 'grey and brown soils of heavy texture', of the zone of 10 to 25 inches of rainfall. Descriptions given by Stephens (1953) show that these differ in structure and consistence, as well as in colour, from the 'black earths', which are typical grumusols. Difficulty in describing or recognizing such differences causes uncertainty in making comparisons between the properties, potentialities and present use of dark clay soils in various places. However, in many places the constitution of the soil may not be as great a limitation as the difficulty of controlling floods or of draining low ground.

Milne (1947) mentioned some of the problems of bringing areas of these soils into use in East Africa, but stressed that they were potentially of relatively high fertility, and for some areas suggested the concept of integrated development of a whole catena of soils in such ways as to 'bring fertility uphill' – for example, by a mixed farming system in which fodder and silage were produced on low-lying clays ('mbugas') and adjoining weakly halomorphic clay-pan ('ibambasi') soils, and animal manure was used on the soils on the slopes, which were less fertile but better suited to most annual crops.

Further distinctions made by W. E. Calton between the larger areas of these soils in Tanzania for a generalized soil map of the country (Department of Lands and Surveys, Tanganyika, 1956), could well be adopted or modified for use elsewhere. Areas were mapped as follows:

a. Areas with soluble salts or alkali.
b. Areas with $CaCO_3$ but little or no soluble salt
 : dry arid phase
 : wet sump phase.
c. Areas not much affected by lateral illuviation.

Soils formed from lake-bed sediments (e.g., forming the low-ground at Kongwa) would form a large part of the last of these units, but elsewhere in Africa some other soils would be added. In Tanzania the most saline and alkaline soils appear to have formed from alluvium which included additions of easily weathered volcanic ash – for example, in the Pangani Valley (Dames, 1959).

The greatest region of clay plains is that of the Sudan, extending from the footslopes of the Abyssinian highlands and the Nile–Congo watershed to the fringe of the desert beyond Khartoum. Soil conditions vary considerably, with rainfall and other factors (Worrall, 1961; K. M. Barbour, 1961), but the soil consists, almost everywhere, of montmorillonitic clay with small amounts of silt and fine sand, and cracks widely on drying. Irrigated farming has been developed with great success, especially in the Gezira, which is the northern part of the land between the Blue Nile (from which the irrigation water is drawn, at the Sennar Dam) and the White Nile. The smoothness of the ground, with a very gentle slope from east to west, permitting perfect control of water through canals and furrows, has been one of the main factors in the success of the Gezira Scheme. The profile is very similar throughout; it contains about 50 per cent clay and is very alkaline (pH 9·0 to 9·5), with varying amounts of $CaCO_3$ at all depths. It is dark brown and low in salt content to about 2 or 3 feet, where it becomes grey, and crystalline deposits of gypsum appear; the content of sodium salts increases abruptly at about 3 or 4 feet. Long-staple cotton is grown in rotation with sorghum, *Dolichos* (a leguminous fodder crop), and grazed fallows.

Jewitt and Middleton (1955) remarked that the possibility of an increase or rise of salt in the Gezira had been feared many years previously, but their results proved that irrigating annually or biennially over a period of twenty-two years had caused *downward* movement

Some Soils of Alluvial Flats and Low Terraces

both of sodium salts and of calcium sulphate. Calculations also confirmed that the amounts of salts added in the irrigation water during this time were much less than those originally present in the soil.

The success of the Gezira Scheme is in obvious contrast to the ruination brought about by irrigation in some countries, by salt accumulation associated with high water-tables and inadequate provision for drainage, and by irrigation water of too high a sodium content. The washing down of salts reported by Jewitt and Middleton is also significant in that it shows that the subsoil is not completely impermeable. Greene (1928) and Greene and Snow (1939) had previously found water to penetrate considerably, evidently through cracks between 'slickensided' clay surfaces. They found that applications of calcium sulphate to the surface had very remarkable effects in increasing the penetration of water, which clearly supported the view that the deflocculation of the clay in the top layers, due to a large proportion of exchangeable sodium, was of dominating importance in these soils. The ratio of sodium ion to clay content has for many years been taken as the chief criterion in assessing the potentialities of other clay areas in the Sudan, and since 1957 the Managil Extension, almost equal in area to the original Gezira Scheme, has been developed after selecting land on this basis; productivity has in fact proved to be closely related to the sodium/clay ratio within the developed area (Randell, 1964). Further west and north in the Gezira, the soils, on account of some difference in their origin, were too high in sodium for development without expensive treatment. Finck and Ochtmann (1961), whilst suggesting that the sodium/clay ratio was of more limited application than was usually thought, also specified a limit of 0·15 per cent salt in the top foot of soil. However there appears to be agreement that the physical condition is the important factor in developing such soils as these, not the toxic effect of salt concentration. Waterlogging is observed to have only very localized and temporary effects on the cotton in the Gezira Scheme. The percolation of water, to become available to roots in the subsurface layers, appears to be of the greatest significance, but the formation and uptake of nitrate are evidently also much affected by the physical condition and variations in water content of the soil, especially those resulting from different treatments of fallow growth in rotations (Crowther, 1943).

South and east from the Gezira Scheme the clays become low in exchangeable sodium. The better-drained land seems to consist mainly of typical grumusols, and there is some settled peasant cultivation, especially near to hills which emerge above the plain in some places. An attempt to introduce mechanical cultivation in one of these localities (Gedaref) has made some progress, but the mean rainfall is only 25 inches there and evaporation losses are great, as in the *regur* soils of the Deccan (see p. 39), and as reflected in the drier type of vegetation found on clays than on sandy soils under similar rainfall in the Sudan itself. It is remarkable that the greatest density of unirrigated cropping is in a zone of still lower rainfall, and that this involves a traditional practice of conserving water by means of earth banks ('terus'). Great areas of the plains, including soils subject to seasonal flooding, are used by nomadic herdsmen.

Investigations of the soils and the possibilities of agricultural development have been in progress for some years in areas of clay plains to the west, south, and east of Lake Chad. Further west, between Ségou and Timbuktu, there is a great area known as the 'inland delta' of the Niger, formed largely of finer grained alluvium, where a scheme for bringing nearly 600 square miles into use for production of cotton and rice has been developing during the past thirty-five years. Accounts of the soils have been given by Dabin (1951, 1954). They are far more variable than the Gezira soils, but exchangeable sodium is low throughout. The occurrence of waterlogged patches after irrigation, due to unevenness of the surface, has been one of the difficulties with the cotton crop, and for this and other reasons rice has been the more successful crop to date. Dabin stated that nitrogen was the most deficient nutrient, and that rice yields were related to the total nitrogen percentage in the soil. Much work had been done with green manures in rotation with cotton, but their value remained uncertain.

INTRAZONAL AND ZONAL SOILS OF FLATS AND TERRACES IN THE WETTER REGIONS

Minor amounts of hydromorphic soils occur in valley bottoms in rolling and hilly landscapes in regions of moderate to high rainfall. Alluvial terraces of various ages are often present along larger streams; these may have fertile intrazonal calcimorphic soils formed from relatively young alluvium, such as those which have been sought after for banana production in Central America in the past, but old and leached soils, classifiable as non-hydromorphic pedalfers and hydromorphic soils, are more common and extensive. Typically the older types of soil have strongly mottled layers in the subsoil.

The larger flood plains and areas of low ground in the wetter parts of the tropics, inland from the mangrove swamps, probably also have, on the whole, a great predominance of soils of low nutrient content, partly because of the fact that the parent materials were already thoroughly decomposed before deposition, and partly because of subsequent prolonged leaching, especially in areas that have been left above flood level during changing conditions.

The hydromorphic soils with smaller amounts of organic matter (gley soils) vary from coarse sands to heavy clays. In some larger areas, such as the flood plains of the Niger and some tributaries within Nigeria, an important limitation to agricultural development is the irregularity of the surface, which includes parts of several terraces that have been dissected by minor streams and meanders. These soils are mainly of interest in connection with swamp rice. Their properties are affected by the flooding and puddling which are usually done before the crop is transplanted. Numerous investigations have been made of the availability of nutrients to swamp rice, in soils which would be very poor under normal upland crops. These form a special topic that cannot be dealt with more fully here.

The hydromorphic soils include those in which organic matter has accumulated, in permanently swampy conditions, which Coulter (1950) termed as a group 'bog soils' in Malaya, dividing them into: 'peat', with less than 35 per cent mineral matter; 'muck', with 35 to 65 per cent mineral matter; and 'organic clay', with 65 to 80 per cent mineral matter. Peat soils are reported to occupy over 1,000 square miles in Malaya, and to be very extensive also in Sumatra, Sabah and Sarawak. Smaller areas occur in Surinam and British Guiana, over Pleistocene clays which are flooded by fresh water from the interior, lying inland from the fertile 'frontland clay' (which is geologically recent). Some areas of the shallower peat soils have been used for various crops in Malaya. Certain areas, characterized by a particular species of tree ('gelam') become highly toxic when exposed to oxidation (Coulter, 1957); this problem also arises in British Guiana and appears to be due to the presence of highly active sulphides.

CHAPTER 3

Tropical Vegetation

This chapter does not attempt any detailed account of the very varied vegetation of the tropics, but endeavours to give a brief general description of some of the more widespread formations and to note their features that are of agricultural importance. The terms describing various types of vegetation are here used in the sense employed by Burtt-Davy (1938). Any collection of plants growing together which has, as a whole, a certain individuality is a plant community. Several communities may be classified together as an association, which is a floristic unit of vegetation recognizable by the species, or at least by the dominant or codominant species, of which it is composed. A group of associations which resemble each other in general physiognomy, and in climatic or edaphic habitat, more closely than they resemble any other group of associations, is termed a formation. Unlike an association, the same formation may be composed of different species in different parts of the world, although it will have a similar structure and physiognomy wherever it occurs. For example, tropical rainforest in Asia is composed of different species to those comprising it in Africa, but in both continents it is a luxuriant forest consisting wholly, or predominantly, of evergreens with several storeys of trees.

A study of vegetation is of value to the agriculturalist in two main ways. First, since the nature of the vegetation is, to a considerable extent, controlled by climate, soil and topography, which are the principal physical factors determining agricultural potential, it is often a useful indication of such potential. Secondly, some types of vegetation possess features of direct agricultural significance. For example, some kinds of forest or woodland are much more difficult to clear than others and, after clearing, it is relatively easy to maintain pastures in some vegetation zones, whereas in others this is difficult because of persistent regeneration of woody species.

The extent to which vegetation is a useful indicator of agricultural potential varies in different places. Although the nature of the vegetation is largely determined by soil, climate and topography, there is an interrelation between all four factors, which form one integrated ecosystem. For example, the vegetation itself influences the formation and nature of soil, both by the organic matter it deposits on death or defoliation and also, as pointed out by Jacks (1934), by the degree to which the living cover affords protection to the soil from the climate in reducing insolation, moderating soil temperatures and breaking the battering effect of rainfall. Vegetation, especially extensive forests, may also influence climate, though probably only to a limited extent. Furthermore, in many places the situation is complicated by the introduction of a fifth interacting factor into the ecosystem, namely the influence of man and his domestic animals, which has certainly modified vegetation and soil, and even climate in some degree, over vast areas.

The degree to which vegetation reflects soil type is especially variable. In many places the same type of vegetation may occur on several soils, or different types of vegetation may occur on the same soil, although the latter is commonly due to modification of the vegetation by man. On the other hand, there are large areas where there is a marked correlation between soil and vegetation, and where the soils can be readily recognized by the nature of the bush upon them, as was observed for example, by Trapnell and Clothier (1937) and Trapnell (1943) in his extensive soil and ecological surveys in Northern Rhodesia. Despite these limitations, a vegetation survey is often a useful aid in assessing agricultural potential and possibilities, especially in undeveloped areas where soil surveys are frequently not available and meteorological records may be meagre or absent.

The most important way in which climate, soil and topography affect the vegetation is by their influence on the moisture available to plants, taking into consideration, on the one hand, the rainfall and the moisture stored in the soil and, on the other, the losses from surface run-off, subsoil drainage and evaporation. Rainfall is obviously of paramount importance, not only in its total amount but also in its seasonal distribution, since the length and severity of the dry season, if such occurs, or of periods during which potential evapo-transpiration exceeds precipitation, are of special significance. The other climatic factors of temperature, atmospheric humidity, duration of bright sunlight and exposure to strong winds, all affect the moisture relations of plants by their effect on evapo-transpiration. Soil type appears to be most important in respect of those physical features which decide its moisture relations and it may thus modify the effect of rainfall. Topography exerts its influence mainly by affecting soil drainage. On steep slopes a high proportion of the rainfall may be lost by surface and subsurface run off; on low lying sites water may accumulate to produce water-logged or swampy conditions.

Beard (1944 and 1955) has given a good account of the interplay of these factors as affecting vegetation, and of their relation to the classification of natural climax vegetation formations. Optimum conditions for the development of vegetation occur (i) in areas at low to medium altitudes possessing a wet equatorial climate with good rainfall, high humidity and relatively high temperatures throughout the year; (ii) on relatively level, well-drained land with deep, permeable soil, neither subject to inundation nor to seasonal drought; (iii) in sites sheltered from violent winds. Moisture is thus available in sufficient amount throughout the year. Under such optimum conditions, or in places where they approach these optima, the natural climax vegetation is the most luxuriant type of all – tropical lowland evergreen rainforest.

Conditions less favourable than this optimum, excluding saline and alkaline soils, may be classified into five categories. Thus, including the optimum, or rainforest, conditions, there are six categories mainly based on the ability of the habitat to supply water to plants and, correspondingly, vegetation may be classified into six formation series, as follows:

Habitat	*Formation series*
1. Optimum, as described above.	1. Lowland evergreen rainforest.
2. Well-drained lands with seasonal lack of available moisture due to ill-distributed rainfall.	2. Seasonal formations.
3. Well-drained lands with a more or less constant lack of available moisture, evaporation exceeding moisture supply for much of the year.	3. Dry evergreen formations.
4. Elevated lands, exposed and cold.	4. Montane formations.
5. Ill-drained lands subject to flooding.	5. Swamp formations.
6. Ill-drained lands subject alternately to waterlogging and dessication.	6. Seasonal swamp formations.

Within each of the above categories except the first there is a range of conditions from nearer the optimum to markedly adverse, and, correspondingly, the formations within each series range from more luxuriant types of vegetation to those that are sparse and/or specialized. For example, the nature of the seasonal formations within type 2 above, is mainly determined by the length and severity of the dry season. As the effect of a dry season increases, the climax vegetation changes successively from evergreen seasonal forest (having only a small proportion of deciduous species), to semi-evergreen seasonal forest (with a larger proportion of deciduous species), and moist deciduous forest, all these being closed forest types. Thereafter there is a series of open woodland types, those at the wetter end of the scale being predominantly of broad-leaved species, and those at the drier end having narrow-leaved thorn trees dominant. Finally, the vegetation is reduced to various types of subdesert scrub. Similarly, in other categories, conditions vary from good to adverse according to the effect of increasing altitude or degree of waterlogging. Thus, within each formation series, each formation may often shade by imperceptible stages into the next formation, so that there are considerable areas of vegetation that is in the nature of a transition between two formations.

Within each of the above six climatic-edaphic categories, there are areas where man has changed the natural climax vegetation by clearing and cultivating, by grazing livestock and, above all, by the use of fire. These derived formations, which may be maintained more or less permanently, thus owe their nature only partly to the effect of soil and climate, and partly to the influence of man. Because of the great importance of fire, man's influence on the vegetation has been greatest in the zones of the drier seasonal forest and woodland formations, where his uncontrolled fires have annually swept fiercely over vast areas during the dry season. In these zones the destruction by annual burning of all but a limited number of fire resistant species of trees and shrubs, sometimes combined with soil impoverishment due to shifting cultivation or overgrazing, has converted huge tracts of forest and woodland into grassland with scattered trees, known as savanna, or into grassland without any trees, sometimes referred to as steppe.

The extent of induced vegetation types and their relative permanence must be largely due to the prolonged period over which man's influence has been exerted. This is especially true of his use of fire which, since it was uncontrolled, must have spread to affect much larger areas than his other activities. There are historical records of bush fires in the tropics dating back over 2,000 years, but it seems evident that vegetation was regularly burned by man long before historical times. Stewart (1956) reports archaeological evidence of the widespread use of fire for agricultural and pastoral purposes ever since the domestication of plants and animals about 8000 B.C. He considers that the uncontrolled destruction of vegetation probably occurred long before this, as a result of camp fires being abandoned without being extinguished, and of fires started to facilitate hunting.

The Main Climax and Derived Formations and their Features of Agricultural Significance

In the following account of the main vegetation formations reference is made to the man-induced or derived formations as well as to the natural climaxes: grasslands are described in more detail in Chapter 11. The basis of the classification is similar to that of Beard (1944, 1955).

Numerous papers have been published which give more detailed classification and description of the vegetation of various parts of the tropics. Burtt-Davy (1938) and Cain and Castro (1959) have dealt with the classification of vegetation in the tropics as a whole. For the continent of Africa general accounts have been given by Burtt-Davy (1938), Keay (1959), Phillips (1959), Shantz and Marbut (1923); and for various regions thereof by Andrews (1948), Sudan; Aubreville (1938, 1949), West Africa; Burtt (1942), East Africa; Edwards (1956), Kenya; Foggie (1947), Ghana; Gillman (1949), Tanzania; Henkel (1931), Southern Rhodesia; Harrison and Jackson (1955), Sudan; Keay (1949 and 1951), Nigeria; Langdale-Brown (1959a and b), Uganda; Phillips (1930), Tanzania; Trapnell (1943), Trapnell and Clothier (1937), Northern Rhodesia (Zambia); Trapnell and Griffiths (1960), Kenya.

For Asia there are accounts by Champion (1936), India and Burma; Garfitt (1941), Malaya; Holmes (1958), Ceylon; Kostermans (1958), Borneo; Rosayro (1950), Ceylon; Stamp (1925), Burma; and van Steenis (1958b), Malaysia. For tropical America accounts have been given by Asprey and Robbins (1943), Jamaica; Asprey (1959), the Caribbean area; Barbour (1942), tropical America; Beard (1944, 1955), tropical America; Beard (1946), Trinidad; Beard (1949), Windward and Leeward Islands; Fanshawe (1952), British Guiana; Gleason and Cook (1926), Puerto Rico; Stèhlé (1935, 1939, 1941), the French Antilles; and Stevens (1942), British Honduras.

LOWLAND EVERGREEN RAINFOREST

True rainforest, in the sense of a maximum development of vegetation occurring in areas possessing the optimum conditions described by Beard (1944), is probably not of very widespread occurrence.

Normally the term is used in a wider sense, but authors differ in what they include in this formation, those in Africa tending to include less luxuriant forest, growing under somewhat drier conditions and perhaps extending to rather higher altitudes, than do those reporting from Asia (van Steenis, 1957). Here the term lowland evergreen rainforest is used in the sense employed by Burtt-Davy (1938), to cover the luxuriant climax forest growing on normal soils of tropical lowlands up to an altitude of 2,000 to 3,000 feet, depending on latitude, in areas where the rainfall is heavy and there is, at most, only a very slight drier season or period of reduced rainfall. Total annual rainfall will rarely be below 80 inches, is commonly between 80 and 120 inches but may be as much as 250 inches, and normally no month will have less than 4 inches of rain.

Excellent comprehensive accounts of tropical rainforest have been published by Richards (1952) and Cain and Castro (1959). This is the most elaborate of all plant communities in structure and the richest in species. There is always an overwhelming majority of woody plants all, or almost all, of which are evergreen. Trees are the dominant elements but numerous climbing woody lianas are a conspicuous feature, as also are epiphytes.

There are usually three, sometimes four, tree strata but these are by no means always well defined or easy to recognize by casual observation. The topmost storey commonly consists of scattered, huge trees over 100 feet high, not themselves forming a continuous cover but emerging here and there from the canopy formed by the next storey below. The lower tree storeys occur at about 75 and 45 feet. Below the tree storeys is a layer of undergrowth, comprising young seedling trees and shrubs, and below this a layer of herbs and undershrubs. Owing to the reduction in light intensity, herbaceous vegetation is scarce, the undergrowth is not dense, and in the undisturbed forest it is possible to walk about without following paths. (Plate I.)

SEASONAL FORMATIONS AT LOW TO MEDIUM ALTITUDES

1. *Evergreen seasonal forest*

Evergreen seasonal forest, which occurs under very similar climatic conditions to lowland evergreen rainforest, except that there is a short drier season of reduced rainfall, differs little from the latter formation and by most writers is not separately classified. It usually has three storeys of trees, the top one comprising discontinuous, emergent trees over 100 feet tall. A small proportion of the trees of the top storey are facultatively deciduous, having a short leafless period lasting from a few days to a few weeks, depending on the weather during the short drier part of the year, but the lower storeys are virtually entirely evergreen. Lianas and epiphytes are very conspicuous and the ground vegetation is often more abundant than in true rainforest.

2. Semi-evergreen seasonal forest

Semi-evergreen seasonal forest again occurs under similar climatic conditions to the two preceding types, although normally with a more distinct, but still in no way really severe, drier season. Annual rainfall is usually from 80 to 100 inches. According to Beard (1944) in tropical America this forest has two tree storeys with the upper one forming a closed canopy, at least in the rainy season, at 60 to 80 feet and a lower one at 20 to 45 feet. The upper one contains a fair proportion of both species and individuals which are deciduous but most of them are facultatively deciduous, the amount of their leaf fall varying from year to year depending on the intensity of the drier period. Lianas are very abundant and strongly developed but epiphytes relatively scarce and, although there is a marked shrub layer of woody species, ground vegetation is rather scanty.

Induced vegetation types derived from lowland evergreen rainforest, and evergreen or semi-evergreen seasonal forest. If land in these forests is cleared and then abandoned after only a short period of cultivation, fire is not usually an important factor influencing the vegetation. The land will quickly revert to secondary forest, an initial growth of herbs and shrubs being followed by short-lived trees of medium height which grow very quickly. These trees have efficient means of seed dispersal and quickly colonize any clearing, but are light-demanding and are subsequently replaced by other longer-lived species as the vegetation again develops towards the climax. Abundance of small climbers and young saplings usually makes the secondary forest very dense. On the other hand, if the land is cleared and then cultivated for a prolonged period, or if the secondary forest succession is frequently interrupted by further periods of cultivation, then the impoverished land may become dominated by coarse grasses, notably *Imperata cylindrica*. As these grasses will burn readily during the drier part of the year, the area becomes subject to annual fires and may also be grazed. This prevents the regeneration of trees, consequently the vegetation becomes more or less permanently degraded to poor, coarse grassland with scattered shrubs, although if fire is excluded for a number of years it may be reinvaded by the moist forest species.

The types of forest so far described, together with their derived formations, occur in tropical America in a huge area in the Amazon basin, in the Guianas, and in the lowland parts of the West Indies and Central America. In Africa they occupy a great block in the Congo Basin, extending into (formerly) French Equatorial Africa and the Cameroons and with a narrow western prolongation along the Guinea

Coast. In Asia they occur mainly in parts of Malaya, Indonesia, Borneo, New Guinea, Thailand and the Philippines.

Agricultural features of areas of lowland evergreen rainforest and evergreen or semi-evergreen seasonal forest. In these luxuriant forest areas, with their dense stands of trees, (many of which are of great size and possess extensive, ramifying root systems and huge buttresses) and with the massive development of climbers and lianas, the felling and clearing of land for cultivation is very heavy and difficult work. Native methods normally comprise slashing the undergrowth, leaving the largest trees untouched, cutting down the remainder at any convenient height, making no attempt to remove stumps or roots and effecting removal of as much of the debris as possible by burning. This is adequate to permit shallow hand cultivation and cropping for a few years. The natural vegetation, however, has a tremendous capacity for rapid regeneration, and usually the cultivator has difficulty in coping with the resulting regeneration and weed growth if he attempts to crop the land for longer than two or three years.

Preparation of the land for mechanical cultivation necessarily involves practically full clearing, stumping and removal of the larger roots. Methods of land clearing are discussed in a later chapter but by whatever means it is done, whether manually or mechanically, it is inevitably a major and costly operation in these heavily forested lands. It has been found profitable to clear large areas for plantations of tree crops, such as rubber and oil palms, but this has not commonly involved the removal of roots and the mechanical ploughing and subsequent cultivation of the land. Mechanical cultivation of annual and short-term perennial crops in these areas is more difficult to achieve economically, not only because of the initial cost of root removal, but also because the high and frequent rainfall makes the greater part of the land continuously wet and difficult to work through most of the year.

The soils found in regions occupied by these forest types in different parts of the tropics are naturally very variable. There are large areas of high fertility, particularly where the soils are of alluvial or volcanic origin. In general, however, the predominant soils of the moist evergreen forest regions are deeply weathered and highly leached, red and yellow latosols, of poor to moderate fertility (see Chapter 2, pp. 48–53).

The growth of luxuriant forests on these relatively poor, leached soils is largely maintained because in the forest there is a nicely balanced, closed cycle of nutrients between vegetation and soil. When the forest is cleared and the land put under cultivation with annual crops this cycle is broken and, although satisfactory crops can be grown for two or three years by the native methods of shifting cultivation, more prolonged annual cropping quickly leads to a marked decline in yields and fertility (see pp. 169–74). The permanent and productive use of the land for perennial tree and bush crops presents less difficulty since, once established, they themselves protect the soil from the battering effects of rainfall and effect circulation of nutrients in the same way as forest does. The commercial agriculture of these regions is largely based on such perennials, and even the native shifting cultivators commonly possess small plantations of a mixture of fruit trees and other trees of economic value. The crops grown are mainly those mentioned under 'wet equatorial climates' in Chapter 1. Livestock are unimportant because of endemic diseases and the absence of good natural grasslands.

3. *Deciduous seasonal forest*

Moist deciduous forest occurs in areas of monsoon, or alternate wet/dry, climate experiencing an annual rainfall mostly within the range of from 40 to 80 inches, with one or two dry seasons per annum. Typically there are two tree storeys. The upper one, which may be as high as 80 feet, is usually discontinuous, and two-thirds of its species are obligately deciduous, but they shed their leaves for short periods at different times and the storey as a whole is never completely leafless. The lower storey forms a canopy at 10 to 40 feet and is mainly evergreen. Lianas and arboreal epiphytes are rare, palms unusual and the ground vegetation is often scanty.

The deciduous seasonal forest represents a gradual transition from the semi-evergreen seasonal forest to the much drier woodland and savanna regions. Corresponding with decrease in rainfall from about 80 to about 40 inches, and increasing severity of the dry season, there is a gradual change from relatively luxuriant closed forest, which is not readily penetrated by fire, to less luxuriant forest with an incomplete canopy and a greater development of herbaceous undergrowth, which is more readily swept by fire. Patches of savanna fire sub-climax therefore tend to develop towards the drier forest margins.

The soils of the moist deciduous forest zones are predominantly latosols (see Chapter 2, pp. 46–54) but they tend to be more fertile than those of the wetter forest formations, mainly because the lower rainfall results in less leaching. The upper horizons are neutral to slightly acid and tend to be higher in total nitrogen, calcium and magnesium than the latosols of wetter regions. Clearing of land and its maintenance under cultivation (particularly the mechanical cultivation of annual crops) is less difficult than in the wetter forest zones. Except where the rainfall falls below 50 inches, these regions are well

Main Vegetation Formations

suited to a variety of perennial crops. Conditions are below optimum for rubber and oil palm (although both these crops are profitably produced), but well suited to cocoa, coconut, banana and coffee. There is a great variety of annual food and cash crops. Rice is extensively grown in the wetter parts of these zones in Asia and elsewhere maize is an important cereal. As in the wetter forest regions, livestock are, so far, relatively unimportant.

4. *Broadleaved woodlands and derived savannas*

Under this heading are included a number of rather diverse vegetation types occurring in regions having an average annual rainfall within an approximate range of 25 to 55 inches and usually experiencing four to seven dry, often virtually rainless, months in the year, which may form one long dry season, or be divided between two shorter dry seasons. These somewhat diverse types have common features. The climax vegetation is woodland with broad-leaved, deciduous trees dominant and, as the trees are rather widely spaced, the canopy is open or discontinuous with a continuous ground cover, often mainly of grass but sometimes partly of woody shrubs. Fire readily passes through these more open, drier types of woodland, burning the above-ground parts of herbs and most of the shrubs, destroying tree seedlings and damaging trees. Consequently vast areas which originally had climaxes of these woodland types, but which have been regularly subjected to burning for long periods and usually also to varying degrees of shifting cultivation and grazing, have been converted to savanna, which consists of grassland with trees scattered at very variable density.

Corresponding with the rather wide range of rainfall conditions, there is natually a considerable range and diversity of agricultural conditions within the areas falling in this category. At one extreme a good total rainfall and moderate dry season allow the cultivation of certain perennial crops, such as coffee, together with a variety of annuals. At the other extreme, with low rainfall and a long, severe dry season, the moisture supply may be marginal even for the production of rather drought-resistant annuals. Between these extremes there is a wide range of intermediate conditions. In so far as Africa is concerned, it seems possible to give some indication of the range of agricultural potential by reference to two types of vegetation which are extremely widespread and are associated with distinct rainfall conditions.

In the better rainfall areas, which usually have between 35 and 50 inches per annum falling in either one long rainy season or in two shorter periods with three to five dry months, there commonly occurs vegetation of the type described by Keay (1959) as 'Undifferentiated – Relatively moist types', and broadly defined as woodlands in which the three genera *Brachystegia*, *Julbernardia* and *Isoberlinia* are absent or rare. In East Africa the type described by Edwards (1956) as 'scattered tree grassland: low tree-high grass, or *Combretum-Hyparrhenia*', which occurs over large parts of Uganda, Kenya and Tanzania, appears to be similar. Both these types have a relatively dense ground cover of tall grass.

This type of vegetation usually indicates areas of considerable agricultural importance. Clearing involves no great difficulties, bush regeneration in pastures is not a major problem, and the climate and soils do not place any great difficulty in the way of mechanization. In areas of better rainfall perennial crops such as *Arabica* and *Robusta* coffee, bananas and sugar-cane grow very satisfactorily, and there may be two seasons a year for the production of annual crops, which include all the common food crops and also groundnuts, sesame, cotton and tobacco. Cattle are extensively kept wherever tsetse fly has been eliminated. Further mention is made of these formations, and also of the 'miombo' woodlands (see below), in Chapter 11, with reference to the utilization of the grasslands.

The second, very widespread, type of vegetation falling in this category in Africa is that commonly known as 'miombo'. This consists of woodland characterized by a large proportion of trees belonging to one or other of the genera *Brachystegia*, *Julbernardia* and *Isoberlinia*, with which species of *Uapaca* are often associated. The ground cover is again mainly of tall grasses, but the grass is not usually as tall or as dense as in the preceding type. Annual fires commonly destroy the above-ground portions of the grass and defoliate most of the trees, leaving the ground with virtually no cover until towards the end of the dry season when the trees characteristically produce new foliage some time before the rains break. (Plate II.)

Miombo woodland and savanna appears to be confined to regions experiencing only one rainy season a year and a severe dry season of usually not less than six months' duration. The total annual rainfall may be as high as 50 inches in limited areas but for the most part it lies in the range 35 to 25 inches. The rainfall is also commonly somewhat unreliable and a high proportion of the total precipitation falls in storms of high intensity. Natural water supplies are often scanty and ill-distributed. It is therefore obvious that the agricultural potential of miombo areas is distinctly less than that of the moister woodlands and savannas previously described. Perennial crops, unless they are markedly drought-resistant, such as sisal or cashew, are only practicable in limited favoured areas. Over much of this type of country the most suitable food crops are those with some drought resistance, such as sorghum, bulrush millet, sweet potatoes, cassava and various pulses. Maize, groundnuts, cotton and tobacco are widely grown but often inadequate moisture reduces yields of the first three. Considerable areas of drier miombo country are marginal for crop production and only suitable for ranching.

A list of references concerning savanna and woodland soils in various parts of Africa is given by Nye and Greenland (1960), who also present typical analytical data. The soils are generally shallower than the forest soils. Rotting rock often occurs within 10 feet of the surface and, furthermore, over large tracts of savanna an indurated iron-oxide-cemented pan, usually called laterite, is present in the soil above the rotting rock. The pH of the surface soil is slightly acid to acid and falls slowly, if at all, with depth, The organic matter and total nitrogen contents of the top soil are considerably lower than in forest soils. The content of total and available phosphorus and of exchangeable potassium, calcium and magnesium is lower than in soils of the semi-deciduous and deciduous forest, but often higher than in the oxysols of the moist evergreen forest regions.

5. *Thorn woodlands, thickets and savannas*

This category includes a number of rather diverse vegetation types occurring in areas having a rainfall of 15 to 30 inches per annum, which may fall in one or two wet seasons, usually the former, with seven or eight dry, virtually rainless, months in the year. Trees are almost all deciduous, fine-leaved and thorny. They may be relatively tall trees, – up to 50 feet, – which may form an open woodland with a ground cover of relatively short grass or, in the derived savannas, be widely scattered in open grassland. On the other hand there are also associations with predominantly small trees, only 10 to 20 feet high, usually forming dense thickets (in which grass is practically absent), but sometimes rather more widely scattered in derived savannas. Both these types of woodland and their derived savannas are very widespread in Africa. In East Africa the savanna type with taller trees is typified by Edwards's (1956) 'scattered tree grassland and open grassland: *Acacia-Themeda*', which consists typically of flat-topped *Acacia* trees widely scattered in an even cover of grass 3 to 4 feet high. (Plate 2.) The second type is characterized by the prevalence of species of *Commiphora*, by the presence of more shrubs, and by a poorer, sparser grass growth. (Plate III.)

Even the most favourable parts of these areas must be regarded as marginal for crops, owing to the low and very unreliable rainfall which occurs in only a short wet season. It is pre-eminently lack of moisture that limits crop production rather than infertile soils. Reference to the soils of these zones is made in Chapter 2, pp. 44-6. Native shifting cultivators grow limited areas of the more drought-resistant food crops such as sorghum, bulrush millet, sweet potatoes, beans, cowpeas and even short-term maize varieties, but often obtain extremely low yields owing to the frequent occurrence of seasons of very poor rainfall. Large-scale commercial production of annual crops is very difficult, especially as the cost of clearing the trees, which have very strong ramifying root systems, is high. These regions are predominantly used for the non-intensive ranching of cattle or sheep and goats, and further reference to the nature and utilisation of the grasslands is made in Chapter 11.

6. *Sub-desert scrub and grass*

This type, which usually does not occur under a rainfall much exceeding 10 inches per annum, consists of scattered bushes, cactus and stunted trees that are mostly drought dormant and leafless except during brief periods of rain. Much of the ground between the bushes and trees is bare, ephemeral grasses and herbs springing up only after rain, and perennial grasses are insignificant. Such country is capable only of supporting a sparse population of nomadic pastoralists.

DRY EVERGREEN FORMATIONS

The dry evergreen formations occur in places where there is no effective seasonal variation in moisture supply but where it is fairly constantly insufficient for luxuriant growth. Their occurrence is rather limited and is generally due to high evaporation rates caused by high temperatures and/or strong winds, or to excessively freely drained soil. Vegetation types referred to by Champion (1936) as dry evergreen forest in India and Burma, and stated to be similar to the dry zone forest of Ceylon, do not properly belong to this category as they occur in areas which have a marked dry season and are probably secondary seral types produced by human interference some centuries ago (Holmes, 1958). Dry evergreen formations in the sense used by Beard include such diverse types as littoral woodlands and thickets subject to high winds containing salt spray, and dry variants of rain-forest on excessively well drained soils. They are limited in extent and usually of little agricultural importance.

MONTANE FORMATIONS

The montane formations reflect the well-known fact that different types of vegetation are encountered on ascending mountains, due principally to progressive reduction in temperature and increasing exposure to strong winds, though the changes may also be influenced by steep topography and shallow soils. In many parts of the tropics the lower slopes of mountains receive higher rainfall than the lowlands, coupled with only a slight reduction in temperature, resulting in the occurrence of a lower montane evergreen forest not very dissimilar from lowland rainforest. Above this the lower temperatures produce a change to a very different upper montane evergreen forest, referred to by some authors as subtropical or temperate montane forest. Higher still the lower temperatures and the effect of increased exposure in distorting and dwarfing trees produce, first, higher montane woodlands and thickets of low stature and then, above the limit of tree growth, tropical Alpine meadows or high altitude grasslands. On the highest peaks this is succeeded by zones of tundra and of perpetual snow.

1. *Lower montane evergreen forest*

This formation is very similar to lowland evergreen rainforest and by most authorities is not separately classified. Agricultural conditions are so similar in the two zones that separate discussion is not merited. The same major crops are grown in both, but in the lower montane forest their growth is usually rather slower and their yields may be slightly lower. These differences are mainly due to the factors of topography, shallower soils and greater exposure, and only to a limited extent because of the lower temperatures.

2. *Higher montane evergreen forest*

There is a gradual transition from lower to higher forest types, and the altitude at which typical higher montane evergreen forest will occur varies with latitude and also with local climatic conditions, there being a tendency for it to develop at lower elevations on small islands than on larger land masses. Beard (1944) states that it occurs at as low as 2,600 feet in Trinidad. According to Burtt-Davy (1938) it is found between 5,000 and 7,000 feet in Malawi, Zambia and Southern Rhodesia, between 6,000 and 8,000 feet in northern Tanzania and between 7,000 and 10,000 feet in Kenya, while Champion (1936) refers to its occurrence in the hills of south India above 5,000 feet and in Assam and Burma between 6,000 and 9,500 feet. Holmes (1958) places it between 4,500 and 8,000 feet in Ceylon. Total annual rainfall is very variable, usually between 50 and 100 inches but in parts of India it may be over 200 inches. The number of 'dry' months, with less than 2 inches of rain in the month, does not exceed four.

The forest consists of relatively tall, smooth-barked trees, often with pronounced plank buttresses, branching much lower than those of lowland rainforest and with dense rounded crowns forming a closed canopy. It has a more strongly developed shrub layer and more herbaceous vegetation on the forest floor than lowland rainforest. Ferns, mosses and epiphytes are conspicuous. Over large areas where clearing has been followed by cultivation, grazing and regular burning, the forest has been replaced by grassland composed of short grass species. These grasslands are dealt with in Chapter 11. (Plate I.)

These highland areas are usually of very good agricultural potential. In general, the annual rainfall is good and reliable; dry seasons are relatively short and their effects are moderated by cloudiness, mists and lower temperatures. In some places, however, excessive rainfall and the absence of any reliable dry season may render the cultivation and harvesting of some annual crops difficult or impossible. The majority of the soils of these highland forest zones are relatively fertile, although available phosphate levels may be low in some areas. Humus content may be higher than that of neighbouring lowland soils, but this is mainly due to an increase of rainfall with altitude (see Chapter 2, p. 56). Birch and Friend (1956a), who examined a large number of soil samples from sites in Kenya ranging from sea level to 10,000 feet, found that most samples from the top 6 inches of highland soils contained 5 to 8 per cent of organic matter and 0·25 to 0·40 per cent total nitrogen.

Conditions are, naturally, subtropical or temperate, and consequently many tropical crops cannot be grown. The only major tropical perennials which are extensively grown commercially are tea and coffee, and these are not found above about 7,000 feet. Local food crops include maize, finger millet, sweet potatoes, cassava and a variety of pulses. Bananas will only grow at the lower fringes of these regions and maize is unsatisfactory at the higher elevations. Introduced crops which, for example, form the basis of much European-style farming in the highlands of East Africa, include wheat, barley, oats, potatoes and pyrethrum. The natural, induced, grasslands are amongst the most productive found in the tropics, and as leys can be established with temperate, or selected local, species of grasses and clovers, intensive mixed farming is practicable.

3. *High altitude woodlands, thickets and grassland*

None of the areas occupied by these vegetation types is of much agricultural importance, being relatively limited in extent, inaccessible, exposed and used only for extensive grazing of cattle, sheep and goats.

SWAMP FORMATIONS

Where the soil is waterlogged or inundated for the whole, or the greater part, of the year various types of swamp forest, swamp woodland or thicket, or herbaceous swamp vegetation occur. The occurrence of these formations is not directly affected by climate, since they not only occur in climatic regions which have rain throughout the year, but can also occur in seasonally dry regions if local conditions of water supply and topography permit the development of swamps.

1. *Swamp forest*

Where saline or brackish conditions exist, as they do where silt or sand accumulates in the tidal estuaries of rivers or along sea shores, the typical swamp forest is that known as mangrove. Mangroves occur on coasts where rainfall is either year-round or seasonal but not on arid coasts, such as are found in south-west Africa or the west coast of South America. Mangrove forests, which are evergreen, possess many characteristic features, including stilt roots, knee roots and pneumatophores. They are regularly inundated by tides, sometimes daily, sometimes only at normal high tides and sometimes only at spring tides. They may form tall forests or low thickets, the height of the vegetation and the species present depending on a number of factors, including the velocity of currents, the depth of water, the depth and nature of the soil or mud and the degree of salinity.

The main agricultural use to which mangrove areas may be put is for the cultivation of swamp rice, and reference to their reclamation for this purpose in Sierra Leone has already been made in Chapter 2, p. 62.

Freshwater swamp forests are commonly characterized by a single storey of trees, which are predominantly evergreen, and many of which possess prominent plank buttresses or stilt roots. Root systems are shallow, widespread and often with numerous pneumatophores. The ground vegetation is sparse and although palms are nearly always present they do not form a large proportion of the vegetation. On normal mineral soils there is a moderate accumulation of organic matter and the water is more or less neutral. On such soils agricultural utilization depends mainly upon the extent to which drainage is practicable. Where good drainage can be provided a variety of crops can be grown. Since these forests occur most extensively in the region of lowland rainforest, evergreen seasonal forest and semi-evergreen seasonal forest, the crops will be those already mentioned as suitable for those zones.

In some parts of the tropics there is an extensive occurrence of peat forests, where the soil consists largely of organic matter and is highly acid. According to Kostermans (1958) peat swamp forests are characteristic of large areas of Borneo and Sumatra and occur to a lesser extent in the Celebes. They are also found extensively in Malaya and in parts of lowland tropical America as, for example, in British Guiana, but do not appear to exist in Africa. The use of such areas for agriculture depends, first, on obtaining adequate drainage and, secondly, on the depth of the peat. In some places the peat may be quite shallow, or there may be a thin organic layer that is not well defined after clearing and cultivation, as on the 'pegassy clays', or muck soils, of British Guiana. Given adequate drainage, such soils will grow good crops of rice, sugar-cane, maize and even tree crops such as oil palms, rubber, cocoa and coconuts, provided that the acidity is not too great. It is common, however, for difficulties to be experienced due to high acidity and the occurrence of toxic amounts of iron and aluminium. On the other hand, if the peat is deep – and in Malaya it may be many feet thick – the choice of crops is much more limited as many annuals will not grow and few perennials can obtain sufficient anchorage. Oil palms and *liberica* coffee will grow satisfactorily on peat of moderate depth, but on deep peat pineapple appears to be the only crop extensively grown. Furthermore, heavy machinery sinks into deep peat and therefore cannot be used for clearing and cultivation, nor can heavy lorries be used to evacuate produce unless heavy expenditure is incurred on road-making. A final difficulty is that after drainage, clearing and cultivation the peat shrinks, thus tending to expose crop roots.

2. *Swamp woodland and thicket*

This formation series occurs under wetter conditions than that just described and covers a variety of vegetation types but is generally characterized by the presence of fewer or more widely spaced trees, and in many areas by a much higher proportion of palms. It includes "tropical palm swamps", described by Burtt-Davy (1938) as consisting of palms with a grassy ground cover, and also the dense stands of Nipa palms mentioned by Kostermans (1958) and van Steenis (1958b) as covering very large lowland areas in Malaysia. In this category also are certain types of riparian woodlands and thickets, dependent upon constant high soil moisture, and occurring along water courses, around lakes and on some alluvial flood plains. Owing to the varied conditions no generalizations can be made about the agricultural use of such areas except that it is dependent on major drainage operations.

3. *Herbaceous swamp*

Herbaceous swamps consist of grass and herbs growing on land which is permanently shallowly inundated, or of herbaceous vegetation floating on top of water. The use of such areas for agriculture demands a complete transformation of conditions by major flood control and drainage works.

SEASONAL SWAMP FORMATIONS

These formations occur where, owing to impeded drainage caused by rock, concretionary ironstone or heavy clay just below the surface, the soil is waterlogged or inundated for part of the year, but seasonally becomes markedly drier. It follows that they can only develop in areas of seasonal rainfall, but they can do so in places where the dry season is so short that the climax vegetation on well-drained soils differs but little from evergreen rainforest.

1. *Seasonal swamp forest*

These usually consist of one or two storeys of predominantly evergreen trees of moderate height with a proportion of palms and a fairly well developed ground cover.

2. *Seasonal swamp woodlands and thickets*

These are commonly low, rather dense woodlands with fairly numerous palms. As a rule neither of these types is of much agricultural value owing to the fact that the soil is either a thin layer over rock or ironstone, or consists of a shallow, sandy layer (which becomes desiccated in the dry season) over a heavy clay that is very difficult to drain economically. (See under 'Hydromorphic, "gley" soils,' Chapter 2, pp. 41–2.)

3. *Savannas*

Savannas, consisting of pure grassland or grassland with scattered trees, may occur under conditions of seasonal rainfall and impeded drainage where alternation of excessively wet and rather dry soil conditions prevents the growth of trees, or restricts them to better drained patches. These savannas, however, exist under a rather wide range of conditions. At one extreme shallow soil, or flooding for a part of the year, may render them unsuitable for any use except grazing during the dry season. This is true of large areas of the Venezuelan llanos which are flooded annually for some months, in parts to a depth of 7 feet. At the other extreme are areas where flooding or waterlogging are so limited that simple drainage improvements may permit cropping even in the wet season, or where drainage is adequate and sufficient soil moisture is retained to allow of cropping in the dry season. Into this category fall the numerous low-lying areas of heavy, dark coloured, relatively fertile soils, carrying grass with or without scattered trees, which are known in various parts of Africa as 'toich' (Sudan), 'deseck' (Somalia), 'mbuga' (Kenya and Tanzania), 'dambo' (Malawi and Zambia), or 'vlei' (South Africa and Southern Rhodesia). Although some of these depressions are of considerable size, the majority are individually of limited extent, but in total they form a large acreage of useful land. They are customarily used by Africans for dry season grazing and, to a more limited extent, for dry season cultivation, crops commonly being grown on mounds or ridges to enhance drainage. One of the chief difficulties in their more extensive use for crops is that they can only be given preparatory cultivation by hand tools or ox-drawn implements over a limited period when the moisture content of the soil is suitable. For most of the time the soil is either too wet and sticky, or too dry and hard. Since they are usually more fertile than higher surrounding lands it seems likely that tractors, which could extend the period of cultivation, coupled with some drainage works, could be employed to increase the utilization of these areas.

CHAPTER 4

Some Social Factors Affecting Agriculture in the Tropics

Although it is the physical features of their environment that primarily determine a peoples' mode of life and the general nature of their agriculture, racial or tribal differences have an important secondary influence. Agriculture is generally affected by the differing characteristics of various peoples, and in particular by their stage of advancement in technology and their social structure, beliefs, traditions and customs.

Level of Technology

The level of agricultural technology attained by indigenous peoples in various parts of the tropics varies enormously. There are still a limited number of people, such as the Australian aborigines and certain pygmies and bushmen in Africa, who are primarily primitive food collectors and hunters. The majority of tropical people, however, base their use of land on either pastoralism, shifting cultivation, or some more permanent system of farming such as that involving the continuous cultivation of swamp rice. These three main systems are dealt with in Chapters 7, 9 and 12. They illustrate the very marked influence of the stage of technological advance attained by different peoples on their land use and farming systems. The comparatively poor natural grasslands of the tropics are readily ruined by the lack of skill of the majority of the pastoralists. Shifting cultivators may exhibit greater skill, but generally achieve low productivity at the expense of some damage to the land and other natural resources. Systems involving permanent rice cultivation, as practised by more advanced peoples, although demanding much labour, have proved capable of supporting dense populations over many centuries while at the same time the fertility of the cultivated land has been fairly well maintained.

Social Structure, Traditions and Beliefs

Among most tropical peoples agriculture was until recently practised primarily for subsistence, and farming systems and practices formed an intrinsic part of the culture of the people, being intimately bound up with other elements of that culture. This influence of the general cultural pattern on agriculture has to a considerable extent persisted in modern times and, despite the widespread production of crops for sale, farming in many countries is still in the nature of a traditional occupation rather than a business. Farming systems may be directly influenced by this interrelationship between agriculture and other cultural elements. Religious beliefs have an obvious influence in many parts of the world. For example, cattle are sacred to the Hindu and cannot be killed, a fact which has a profound influence on agriculture in India, resulting in much land being burdened with large numbers of unproductive cattle. Again, pig keeping is a most profitable and useful enterprise in combination with market gardening in the system extensively practised by Chinese in Malaya, but the native Malays, being Muslims, will not keep pigs. A further example may be taken from the Celebes where the cultivation of other crops on the same land after the rice crop was desirable and advantageous but was not practised, and was difficult to introduce, because of a religious taboo on planting maize and certain other crops after rice.

Other customs and features of their mode of life may have a marked effect on peoples' mode of land use. For example, cattle keeping and land use by the majority of pastoral tribes in Africa is profoundly affected by the fact that it is customary for each individual, or family, to aim at keeping as many animals as possible, irrespective of the carrying capacity of the land or of the quality of the stock. This they do partly as a matter of prestige, partly for the discharge of certain customary tribal obligations (such as the provision of the 'bride price'), and partly under the mistaken impression that possession of a large number of animals forms an insurance against times of famine, drought and disease. It is a custom which results in gross overstocking and consequent soil erosion, in poor quality stock and in very low productivity.

Development of sounder and more productive land use does not depend merely on technical improvements, such as better pasture management or better animal husbandry. It demands as a first step that the people shall abandon their basic customs in regard to cattle. Instead they must develop a commercial attitude and raise cattle for sale and profit. This is but one example of a commonly encountered fact that improvements in primitive systems of agriculture cannot be brought about merely by attempting to change agricultural techniques, but necessitate a process of general education and guided social change. People accustomed to regarding agriculture as a traditional occupation, providing only subsistence and influenced by custom and superstition, often cannot make progress unless they are persuaded to change their whole attitude and to regard farming as a business to be undertaken for profit, requiring advancing knowledge, skill and management. (See also pp. 369–72.)

The existence of a rather rigid form of group or tribal life has affected the development of native agriculture in some parts of the tropics, particularly in Africa, but also in parts of Asia and Oceania. Until recent times the majority of Africans lived for centuries, not as individuals in the European sense, but as members of a group, clan or tribe, living strictly in accordance with the customs of that group. The individual was expected to act according to tribal custom; there was little scope for initiative or deviation from an accepted pattern, because individual initiative was frowned upon and commonly incurred reprisals.

As a result there were virtually no agricultural innovators or experimentalists, and practically no exceptionally good farmers from whom others might learn. Consequently agricultural practices tended to change very slowly. There is no doubt that they were periodically adapted or modified, albeit slowly, as a result of external influence, perhaps by the gradual adoption of a method employed by a nearby tribe, or by the introduction of a new crop. The fact that many food crops widely grown in Africa, such as maize, sweet potatoes, cassava, groundnuts and some kinds of yam, are not indigenous but were introduced to the continent by the Portuguese, is evidence of this ability to absorb something new.

In recent times tribal customs and authority have been declining, group life has been disintegrating, and people are living more and more as individuals. Nevertheless the customs and habits of centuries cannot be changed in a few years, and in rural areas, particularly in remoter parts, group custom and influence are still strong. Consequently land use is still mainly undertaken by methods which differ little from the indigenous systems practised for centuries, and attempts to introduce improvements meet with strong conservatism and resistance to change.

Population Density and Distribution

Despite uncertainty as to the precise rate of growth, there is no doubt that there has been a big increase in the population in most parts of the tropics during this century, and that this increase is continuing at an accelerated rate. Rates of natural increase mostly vary between one and three per cent per annum, reasonably reliable estimates being, for example, 1·0 per cent for Uganda, 1·5 per cent for Tanzania, 1·6 per cent for Kenya (East Africa Royal Commission, 1955), 2·7 per cent for Trinidad and 3·2 per cent in Malaya. The growth and density of the population naturally varies from one district to another within a territory; fertile, healthy areas are now mostly fairly heavily populated while the more sparsely populated areas are usually, but not invariably, those that are infertile or suffer from tsetse infestation, lack of water or seasonal flooding.

Population density and increase may markedly affect farming systems, or their effects on the natural resources. In Africa the formerly relatively sparse population, practising shifting cultivation solely for subsistence, was in balance with its environment. Latterly the increase in population, coupled with greatly increased production of cash crops and larger numbers of livestock, has caused severe pressure on the land in certain areas. As a result restorative fallows become impracticable, or so much reduced as to be ineffective, and the soil rapidly deteriorates, yields decline with the result that the land is no longer capable of adequately supporting the people. Similarly, in many pastoral areas increasing human and stock populations have accelerated the ill effects of poor farming. Unfortunately in most parts of Africa population increase has not been accompanied by the evolution of improved farming systems, and the common picture is one of deterioration of the land and other natural resources. On the other hand, among the more advanced peoples of Asia an increase in population has commonly led to the development of more intensive systems; consequently, although the standard of living sustained in densely populated areas may be low, the land itself has not suffered anything like the degradation to be seen in many parts of Africa.

Bearing in mind the low yields per man and per acre at present obtained by primitive methods of farming, it is evident that very great problems are posed by the need for a higher standard of living and nutrition for the rapidly increasing population of the tropics. These problems are highly complex, but, from an agricultural point of view, the remedies must lie in the evolution of better farming systems and husbandry techniques. These include better soil conservation practice, and a greater use of fertilizers, fungicides and insecticides, coupled in some places with a degree of mechanization.

Human Health and Nutrition

Over a great part of the tropics agricultural improvement and development are retarded by the existence of numerous endemic diseases, of low standards of hygiene (often partly due to scarcity of water supplies), and of widespread malnutrition. The list of tropical diseases of man is a long one and though medical science has found means of

Human Health and Nutrition

preventing or curing many of them such known control measures have not yet been fully implemented, while for many other tropical diseases medicine has yet to find a satisfactory answer.

The widespread occurrence of many debilitating diseases naturally reduces the energy, initiative and mental capabilities of the people. In many areas, and particularly at certain seasons of the year, this limits the amount and intensity of the labour that can be applied to agriculture. Furthermore, disease may prevent the agricultural utilization of certain areas, or limit their use to certain seasons of the year. In Africa large areas were formerly rendered unsuitable for habitation by the presence of tsetse flies, which are vectors of trypanosomes that cause sleeping sickness. Another example is that some malarial areas may be unusable, or can only be used at certain seasons of the year perhaps for the dry season grazing of cattle. According to Farmer (1957) endemic malaria had a tremendous effect in retarding the agricultural development of the 'dry zone' of Ceylon until it became possible to control mosquitoes by the introduction of regular DDT spraying in 1945. Disease and malnutrition are often interrelated, because malnutrition frequently reduces a person's resistance to disease, delays his recovery and renders him more liable to relapses. Many diseases, particularly those due to parasitic infestation, reduce a man's power to produce his food, or to benefit by what he consumes.

The nutrition of tropical peoples is an extremely important subject on which a considerable literature exists, but it is only possible to make brief reference to it here. An excellent survey of the position in the tropical territories of the British Commonwealth was published in 1939 and still holds good in regard to most major matters (Economic Advisory Council, 1939). More recently a number of reports on nutrition in the tropics have been published by several United Nations organizations (see for example, FAO 1948a and b, 1952, 1953 and FAO–WHO 1950–1953).

The diet of the majority of tropical people is predominantly vegetarian; only relatively small quantities of animal products are consumed, except among certain pastoral tribes, and in riverine, lakeshore and coastal areas where fish is available. By European standards an unusually high proportion of the energy value of the diet is derived from carbohydrate, with the result that diets are usually bulky relative to their nutritive value. Moreover, people are very commonly dependent upon a single staple crop as their main supply of food. In Asia this is usually rice; in Africa it may be maize, millets, bananas, yams or occasionally cassava; in the Pacific it is usually rice, yams or tannia. Commonly, something of the order of 60 to 70 per cent of the calorific supply is derived from one staple food and as much as 90 per cent may come from carbohydrates. Other elements in the diet are often in the nature of relishes, and as a rule the diet is markedly lacking in fresh fruits and green, leafy vegetables.

As a result diets are generally below satisfactory standards of quality in several respects and there is widespread evidence of malnutrition. Total protein intake is commonly inadequate and there is especially a lack of proteins of high biological value, such as those derived from animal sources, although this may be somewhat ameliorated where appreciable quantities of beans and peas are eaten. Except where coconut and oil palm products are largely consumed fats are also low, which is an important defect since they are known to facilitate the absorption of certain vitamins, especially vitamin A. Vitamin deficiencies and diseases resulting therefrom are of frequent occurrence. Lack of vitamin A (the main sources of which are animal fats, yellow and red fruits and roots, fresh vegetables and palm oil) is widespread, and results in night blindness and various skin troubles. Deficiency of vitamin B_1, producing beriberi disease, and of the other vitamins of the B-complex, producing pellagra, is common in rice-eating countries. True scurvy, caused by lack of vitamin C (ascorbic acid), is not often reported, but other skin diseases (and swelling and haemorrhage of the gums) due to vitamin C deficiency are frequent. Some diets are also lacking in minerals. Calcium and phosphate are quite commonly deficient and the intake of iron and sodium may also be below optimum. As a result of this and other dietary deficiencies, anaemia is very common and sometimes proper bone formation does not occur.

The nature and extent of the deficiencies naturally varies in different places according to the diet of the people. Thus rice diets suffer particularly from insufficiency of the B-group of vitamins, and frequently from lack of vitamin A and calcium. They are also low in fats and though there is no firm evidence that rice diets of adequate calorific value are deficient in protein it seems fairly certain that a greater intake of protein, especially protein derived from foods other than rice, would be beneficial (FAO, 1948b). On the other hand, people in parts of West Africa who consume considerable quantities of palm oil are not likely to lack fats, vitamin A or protein.

Women and children are more liable to suffer from the poor quality of the food than men. Women when pregnant require above normal supplies of vitamins and minerals, and when lactating need extra protein as well. Infants are often unthrifty because no adequate substitute for breast feeding is provided. For example, in East Africa when a supplement to breast milk is needed it is usually provided by a pap made from the usual protein-poor staple such as maize, millets, bananas, etc. Children frequently do not have a diet to meet the demands of growth and to aid in resisting attacks of infectious diseases. Thus it is reported for Asia and the Far East (FAO–UNICEF, 1958), that child mortality is high, that school-children suffer from numerous malnutrition diseases, and that their growth and development lag behind normal standards.

Dietary deficiencies are often accentuated by the existence of tribal food prejudices and taboos which vary greatly from place to place but frequently reduce the consumption of protein and protective foods by women and children. Thus amongst many tribes in East Africa cow's milk or eggs are prohibited to girls and women, and in some places pregnant women are not allowed to have milk, meat or salted fish. Such prejudices and taboos are tending to disappear as education advances.

Quite apart from the unsatisfactory quality of the diet, the quantity of food available is often inadequate. In densely populated areas, especially in the East, a large proportion of the population may fairly constantly have less food than they really need. Elsewhere, although actual famines due to failure of the rains or attacks of locusts usually no longer occur unrelieved, seasonal shortages are still common. Not infrequently, following upon a poor season, people experience an acute shortage for two or three months before harvest. In areas where there is a marked dry season it quite often happens that just before the rains begin, although supplies of the staple foodstuff may be holding out, there may be serious deficiencies due to the lack of fresh fruit or vegetables, either wild or cultivated. This reduces the energy and capability of the people just at the time when they are required to undertake their most arduous work: the preparation of their land for planting.

This widespread inadequacy of both quality and quantity of food from which the people of the tropics so commonly suffer, is a matter of particular importance to those concerned with planning agricultural improvement and development. In land use and farm planning, and in devising improved farming systems, the need for increased quantity and better quality of food must be constantly borne in mind. A greater diversity of cropping may often be desirable, both to increase quantity and to improve quality by providing a more varied diet. Generally it will be desirable to encourage a greater production of fruit and fresh vegetables. More beans and pulses, particularly soya beans and groundnuts, are desirable in order to provide more high quality vegetable protein. Above all there is a great need to develop increased production of the nutritionally highly valuable animal products, meat, milk and milk products, and eggs.

Land Tenure and Inheritance

Many forms of communal and individual land tenure are found in tropical countries. Communal tenure, which is characteristic of somewhat primitive peoples, is by far the commonest system. It is characterized by rights in land being divided between the community and the individuals belonging to it. Amongst pastoral communities the use as well as the ownership of the land is usually communal, and various factors, particularly the sparse distribution of water supplies, may make the introduction of individual rights very difficult. In cultivating communities there is communal ownership of all land by the tribe or clan but, though individuals do not own land, they have user rights in cultivated land, usually including that lying fallow, which are permanent so long as they continue to use the land. A user right in land does not necessarily confer any right to property in established trees thereon, or even to the produce of such trees. Customs in this respect vary; probably the commonest situation is that wild, or semi-wild trees are regarded as communal property, but if a man plants a tree it belongs to him, and he may even retain the ownership of trees which he has planted after forfeiting his user rights in the land. But custom may forbid a man who has user rights in land from planting trees thereon, as this is regarded as an attempt to acquire permanent ownership of the land. Where people have both crops and stock, it is customary for there to be individual user rights in arable and fallow land, and communal user rights in grazing land, but in some places communal grazing rights may extend to fallow land and stubbles. Inheritance most commonly involves the subdivision of the land over which a man holds user rights amongst his sons; but customs vary, and in some tribes surviving wives may have at least a life user right, while in matrilineal tribes the rights pass to the sons of the holder's sister.

While it has certain advantages, communal tenure among the cultivating communities generally places difficulties in the way of improved farming and land use. In association with group life, in which everyone tends to be regarded as equal, it militates against individual initiative. A farmer does not have the same interest in the care and development of his land as he would if it were his own property especially if he knows that, should the fertility of his plot decline, he can move to new land. The land cannot be used as security for development loans. Customs regarding the ownership of trees may discourage people from planting profitable perennial crops. Land use planning, farm planning and the introduction of better farming systems are rendered difficult by this form of tenure.

Forms of individual tenure include freehold and leasehold and

Land Tenure and Inheritance

various types of feudal, or modified feudal, tenure. The chief disadvantages of unrestricted individual ownership (apart from fragmentation, which is mentioned below) are the tendency to abuse the land, or to hold land without using or developing it, the liability of smallholders to become chronically indebted by mortgaging their land, and the development of unsatisfactory landlord–tenant relationships. In the tropics the majority of peasant tenants do not hold their land under any written contract but under the vague provisions of custom. Where this is associated with population pressure leading to scarcity of, and competition for, land, and with lack of capital on the part of the tenant, it leads to several abuses. Landlords charge exorbitant rents and, as tenants are thereby unable to rent adequate land, many rented holdings are of uneconomic size. In the expectation of being able frequently to raise rents landlords will grant tenancies only on a year to year basis, which results in insecurity of tenure and a lack of incentive for good land use and improvement.

Fragmentation, or the existence of holdings comprised of a number of scattered pieces of land, is common in many places. Conditions which encourage fragmentation are a growing population and a law of inheritance which provides for subdivision of holdings among heirs. It is widespread under most systems of tenure, because the Muslim, Buddhist, Hindu, and most African tribal laws all provide for subdivision amongst heirs. It is disadvantageous in that it makes it impracticable for a man to use and tend all his lands efficiently. Fragments distant from the homestead cannot be given dressings of farmyard manure by people who do not possess wheeled transport nor can they be used for the production of crops requiring frequent attention or liable to be stolen. In the absence of fencing, fallow fields that are surrounded by other people's standing crops cannot be grazed, and this discourages the planting of improved pastures. The small size and irregular shape of the plots often results in waste of land and hampers cultivation by any other means than hand tools. Fragmentation also contributes to the difficulties of getting effective soil conservation measures implemented and increases the labour and time required in agricultural advisory work.

It will thus be seen that customary forms of land tenure, especially communal tenure, may place difficulties in the way of agricultural development and improvement. Under the influence of increasing population density and the introduction of a cash economy involving the production of cash crops, many of which are perennials, the disadvantages of communal tenure are being increasingly appreciated by the more progressive men in communities where it has hitherto been customary, and there is in many places a strong tendency towards the development of individual land ownership. This often leads to a conflict of views within the community, and to a risk that the unfavourable features of certain forms of individual tenure, mentioned above, may develop. Consequently, in order to clear the way for agricultural improvement, there is in many territories a need for governments to control the process of change, or even to step in to provide a more satisfactory land tenure system by administrative and legislative action.

Lack of Capital and Equipment

The poverty of tropical peasant farmers is well known. The majority are almost entirely lacking in capital, not only for major requirements in the improvement of a holding, such as livestock, buildings, fencing, water supplies, etc., but even for small recurrent items, such as fertilizers, insecticides and fungicides. The credit facilities at present available in many countries are very inadequate, but the problem of making them satisfactory is an exceedingly difficult one in view of the vast scale on which credit is needed, the lack of security on small farms (especially if land tenure is communal), the hazards of agricultural investment, and the difficulty of effectively supervising the proper use of credit by very large numbers of smallholders.

Linked with the peasants' poverty is the limited number, and generally primitive nature, of their implements. This is so throughout the tropics, but it is especially true of Africa. In the East peasant farmers often have a variety of hand and ox-drawn implements, but in Africa they may be literally limited to an axe, a hoe and a knife. The wheel was unknown in most parts of tropical Africa before the advent of Europeans, and even today there are large areas where it is rare to see any form of wheeled transport, other than the bicycle, possessed by peasants. The use of animal power for tillage is widespread (although far from universal) in eastern countries, but in Africa it is restricted to limited areas. In many parts the use of cattle is entirely precluded by the prevalence of tsetse flies, but even where they can be kept cattle are commonly little used for work. Mechanization so far plays a minor part in peasant agriculture. In some places limited use is made of tractors by a few large farmers and contractors, but the effect on peasant farming as a whole is slight, and there are many difficulties in the way of extending their economic use.

As a result of this lack of capital, equipment and power, peasant farming is largely geared to hand labour. Cultural operations, preparation of food and, to a great extent, processing of produce, are carried out slowly and laboriously; productivity per man and per acre is low.

Injection of capital appears essential to effect any real improvement in this state of affairs, but in many predominantly agricultural territories the extent to which capital can be found within the country is limited, because the surplus produced by largely subsistence methods of agriculture is small and consequently both Government revenues and local savings are slender.

CHAPTER 5

Soil and Water Conservation

Measures for the conservation of soil and of water are largely interdependent. Soil erosion by water results from the surface run-off of rainfall in excess of the amount that the soil can absorb. Obviously where run-off occurs part of the rainfall fails to infiltrate into the soil, where it can be stored in the profile for use by plants, or may gradually drain away to maintain the perennial flow of streams or to accumulate in subsurface aquifers. Methods of land use, or agronomic and mechanical measures suitable for the control of water erosion, aim principally at reducing run-off by increasing the proportion of the rainfall percolating into the soil and are consequently beneficial in conserving water. Wind erosion only occurs when the soil is dry, hence one of the chief measures for its control is the conservation of water in order to raise the moisture content of the soil.

Erosion by water has become a tremendous problem during the recent rather rapid development of the tropics, because of the employment of methods of land use unsuitable for intensive application in the tropical environment. Two of the widespread indigenous farming systems – pastoralism and shifting cultivation – have proved highly conducive to erosion under modern conditions, when they have been operated intensively owing to the increased human and stock populations and the increased cultivation of cash crops. Formerly, extensive erosion was also caused by the unsatisfactory methods used by expatriates on their plantations, but most of these planters now fully appreciate the need for conservation farming. In contrast to erosion by water, wind erosion, which occurs mainly in arid regions unsuitable for agriculture, is a relatively restricted problem.

Erosion Caused by Water

While land remains under undisturbed forest or grass its surface is protected from the impact of rainfall and the good structure of the top soil, combined with the presence of leaf litter, allows a high proportion of the rainfall to infiltrate so that there is little run-off. Clearing and cultivating the land, or overgrazing of pastures, exposes the soil to the battering effect of raindrops which break down the soil aggregates and seal the surface so that percolation of rainwater is diminished and run-off correspondingly increased. Excessive

9 Manually felled jungle, Malaya

10 Clearing for shifting cultivation, Sabah

11 *Left:* fore-mounted rake; *centre:* tree-dozer and stumper; *right:* tractor equipped with terracing blade in front and ripper at rear

12 Tree-dozer

cultivation of arable land, or heavy trampling of pasture by stock, can also cause mechanical breakdown of soil aggregates, thus further reducing water-absorbing capacity and enhancing run-off.

When water runs off it becomes capable of work in proportion to the square of its velocity, and carries away soil when it reaches a certain speed. The erosive capacity of the run-off increases with the amount of silt in suspension. Consequently erosion generally starts slowly and then rapidly accelerates, especially as the initial removal of a surface layer of relatively good structure and organic matter content often exposes subsoil less capable of accepting and absorbing rainfall.

The damage occurring first is usually of the kind known as *sheet erosion* which, as the name implies, is the removal of soil particles by run-off flowing more or less uniformly over a wide area. Considerable losses of soil can occur by this, the most widespread, type of erosion without being noticed by the inexperienced eye, especially as cultivations tend to conceal what is going on. Sheet erosion can result from a very modest amount of run-off on quite gentle slopes, and even from run-off that does not attain erosive velocity but acquires soil-transporting energy as a result of turbulence produced by the impact of large raindrops driven into the ground at high speed.

As sheet erosion proceeds, small channels soon appear in which the run-off is concentrated, giving rise to *rill or shoestring erosion*. If the volume and velocity of the water becomes sufficient to deepen and widen these channels, *gully erosion* starts. Gullies may also develop from depressions, or channels, resulting from up- and down-hill ploughing, from wheel ruts, cattle tracks, or footpaths. When the water begins to deepen a gully a small waterfall is formed at its upper extremity and the increased water velocity leads to *head erosion*, by means of which the gully gradually eats its way up the slope to the top of the watershed. The deepening of the bed may also lead to undercutting and caving in of the banks and, as the material which falls is rapidly washed away, vertical banks are left to be undercut by the next floods. The flow of water over small depressions at the top of the banks may also lead to further head erosion and the development of branches of the gully. Once started, gullies may become greatly enlarged in a very short time. (Plate 6.)

FACTORS AFFECTING WATER EROSION

The disposal of the rainfall between infiltration and run-off, the velocity of the run-off and hence the degree of erosion, are affected by a number of factors, the most important being: (1) the amount, distribution and intensity of the rainfall; (2) the slope and nature of the land surface; (3) the vegetative cover; (4) the type and fertility of the soil; and (5) the land use and farming practice.

The amount, distribution and intensity of the rainfall

Obviously, the higher the rainfall intensity the greater will be the volume of run-off water per unit time, and the higher the velocity of the run-off. This would be so even if higher intensities were unaccompanied by any deterioration in the ability of the soil to accept rainfall, but it has already been mentioned (p. 18) that the effect of raindrop impact in destroying soil aggregates and sealing the surface increases with increasing rainfall intensity. The effect of raindrop impact in reducing the ability of the soil to absorb rainfall is of major importance. This is illustrated by the results of a trial in Southern Rhodesia (Hudson, 1957) in which the following treatments were compared on land with a 4½ per cent slope:

1. Permanent dense sward of grass, giving protection from raindrop impact as well as impeding run-off flow.
2. Two layers of mosquito gauze placed on a frame 6 inches above a bare soil surface, giving complete protection from raindrop impact but not impeding surface flow.
3. Bare soil, no protection.

Over a period of three years the bare soil plots lost 350 tons of soil per acre, the grass plots 3·3 tons and the gauzed plots only 3 tons, thus indicating the overwhelming importance of raindrop impact in causing erosion. Intensity is thus the most important rainfall characteristic affecting erosion. It is common experience in the tropics that a large proportion of the annual erosion loss occurs during relatively infrequent storms of high intensity.

Erosion is usually worse in regions with alternating wet and dry seasons than in those having more evenly distributed rainfall, where the soil is nearly always moist, and a cover of vegetation is more continuously maintained. The first rains after a dry season are liable to cause much erosion, both on pastures, where the herbage has often been greatly reduced by over-grazing and burning, and on arable land, which is exposed after preparatory cultivation until the crop forms an effective cover. In some monsoon regions it may also be that a high proportion of storms of high intensity occur in the early part of the rainy season but this is not always so; for example, Hudson (1957) has shown that in Southern Rhodesia high intensity storms are more frequent in the middle of the wet season. Because of the effects of rainfall distribution and intensity there is not necessarily any close relationship between the total annual rainfall and the annual amount of run-off and soil loss. The dominant factor is usually the frequency of high intensity storms.

Slope and nature of land surface
On a smooth surface of uniform slope the velocity of the run-off is theoretically doubled if the slope is increased four times, and the longer the slope the greater is the volume of water which accumulates and the higher the velocity it attains. In theory, then, water and soil losses might be expected to increase in proportion to the degree and length of slope. In practice the relationship is variable, being affected by such factors as the amount and intensity of the rainfall, the permeability and moisture content of the soil and the nature of the surface. Water losses usually increase with slope, but not necessarily in proportion. Soil loss also generally increases with length and degree of slope and may do so more markedly than run-off, possibly because of the increased erosive capacity of silt-charged water moving at relatively high velocity. Run-off is less severe over a rough surface than over a smooth one. Any obstacles across the slope, such as natural surface irregularities, large clods of soil, stones, plants or ridges, slow up the rate of flow and give more time for infiltration.

Vegetation
The most important single factor affecting run-off and erosion is the nature and amount of vegetative cover over the land. Closed forest provides the most effective protection. The canopy of the trees and shrubs, together with any herbaceous ground cover, break the impact of rainfall. The forest litter also reduces raindrop impact, improves soil structure by adding organic matter as it decomposes, impedes water movement and acts as a filter through which water is slowly transmitted to the soil, free of any silt load which might cause sealing of soil pores. The surface soil is thus maintained in a highly absorptive condition and percolation of rainwater to greater depth is aided by cracks and channels opened up by living and dead roots. For effective soil and water conservation steep slopes are usually best left under forest, where this is possible.

A dense grass sward in good condition is little inferior to forest in conservation properties. The herbage breaks the impact of rainfall, dead leaves and stems provide surface litter, the living fibrous roots help to improve soil structure and dead roots leave channels to aid percolation of water. These beneficial effects, however, are dependent upon keeping the sward in good condition. Overgrazing, or indiscriminate burning, exposes bare soil, and excessive trampling by stock destroys soil structure with consequent reduction in infiltration.

Perennial crops vary in their effect on run-off and erosion according to species and management. Cocoa normally affords good protection. In the establishment phase it is usually interplanted with other plants to provide shade and ground cover. A mature cocoa plantation, with its permanent shade trees, provides a complete canopy and abundant leaf litter and is thus little inferior to closed forest. Mature rubber affords good protection but the incomplete canopy of immature rubber does not do so. Unless good cover crops and adequate mechanical conservation works are provided during the establishment phase, erosion occurs and, owing to deterioration in the absorptive capacity of the soil in the early years, may persist under the mature stand. Citrus, coconut and other crops that do not form a complete canopy when mature never afford full protection to the soil and require the provision of additional conservation measures. With the exception of sugar, most of the shorter-term perennial crops, such as sisal and pineapples, do not provide a good soil cover. Unless they are planted on the contour, and other conservation measures adopted, severe erosion may occur.

The protective effect of arable crops varies greatly with the type of crop and the method of cultivation normally associated with it. Within any given species the differences arising from varying the spacing can be striking, being due much more to variation in the extent to which the crop density breaks rainfall impact than to any effect in impeding the flow of run-off. In the absence of conservation measures, soil and water losses under intercultivated row crops, such as maize or cotton, are usually considerable on sloping land. However, they can be reduced both by planting at a high density in the rows and by practising trash mulching, as can be seen by a comparison of the figures for maize at 10,000 plants per acre and maize, trashed, at 15,000 plants per acre in Table 3. Crops that are drilled in close rows, or broadcast, and receive little cultivation after planting, such as wheat, finger millet or creeping varieties of groundnut, form a reasonably good protective cover. Established stands of fodder grasses that are periodically cut afford excellent protection, as can be seen from the figures for Napier grass in Table 3. Fodder grasses of tufted habit, however, will only give good protection if grown in rows on the contour. Green manure crops of creeping legumes give good protection if planted early, but not if sowing is delayed so that the soil is exposed during the early rains, as is indicated, for example, by figures reported by Roche and Joliet (1953) for *Dolichos lablab* and *Phaseolus lunatus*. Erect legumes are less satisfactory, and quite ineffective if planted late, as indicated by the figures for sunnhemp in Table 3.

Table 3. Effect of cropping practice on soil loss and run-off on Tatagura clay loam, Rhodesia
Hudson (1957)

	Cropping System	6¼% Slope				4½% Slope				3% Slope		
		Cont. Maize	Maize/ Green manure	Maize trashed	Napier grass	Cont. Maize	Maize/ Green manure	Napier grass		Cont. Maize	Maize/ Green manure	Napier grass
1953/4	Crop (plants/acre)	Maize 10,000	Sunnhemp	Maize 15,000	Napier grass	Maize 10,000	Sunnhemp	Napier grass		Maize 10,000	Sunnhemp	Napier grass
	Soil loss, (tons/acre)	4·0	23·4	—	1·4	2·5	7·5	1·7		2·8	5·9	0·4
	Run-off (inches)	10·4	14·5	—	2·7	10·0	9·7	5·2		6·7	11·1	0·9
	Run-off (% total rain)	29	40	—	8	28	27	14		19	31	3
1954/5	Crop (plants/acre)	Maize 10,000	Maize 10,000	Maize 15,000	Napier grass	Maize 10,000	Maize 10,000	Napier grass		Maize 10,000	Maize 10,000	Napier grass
	Soil loss, (tons/acre)	1·7	2·3	0·3	0·2	0·6	1·0	0·2		1·1	1·1	Nil
	Run-off (inches)	7·0	10·0	0·7	1·1	4·3	4·8	0·8		8·0	7·2	0·3
	Run-off (% total rain)	16	22	2	2	10	11	2		18	16	1
1955/6	Crop (plants/acre)	Maize 10,000	Sunnhemp	Maize 15,000	Napier grass	Maize 10,000	Sunnhemp	Napier grass		Maize 10,000	Sunnhemp	Napier grass
	Soil loss, (tons/acre)	4·2	7·6	0·9	0·7	1·5	3·6	0·9		1·0	1·5	0·3
	Run-off (inches)	7·8	6·8	2·8	0·7	5·2	7·2	0·7		5·0	4·4	0·1
	Run-off (% total rain)	22	19	8	2	15	20	2		14	12	Nil

Soil type and fertility

The characteristics of the soil that are of primary importance in respect to resistance to erosion are its ability to accept the infiltration of rainfall and to allow of drainage. These in turn are dependent on texture, structure, porosity, organic matter and colloid content, and on the depth and nature of the profile. Of the first importance are texture and structure, the latter being enhanced by high content of organic matter or colloids, which tend to bind the soil particles together to form stable aggregates. Fine-textured sandy soils are readily eroded because of their limited and unstable aggregates. Clay loams and ferruginous clays, with aggregates that break down only slowly, are the most resistant types. The capacity of a soil to absorb water and to permit drainage clearly depends on the depth of the profile and the permeability of the subsoil. Shallow soils over impermeable clay, or hard pan, quickly become saturated and are unable to accept further infiltration of rainfall, with resulting run-off. As a rule erosion speeds up as the more absorptive, humic surface layer is washed away to expose less humic and more impermeable subsoil.

A decline in fertility is a common cause of erosion. Newly cleared soil is usually of relatively good structure and humus content but, in the absence of special measures to maintain its fertility under cultivation, both humus and structure, and hence ability to absorb rainfall, are gradually reduced. A decline in nutrient status of the soil, which also commonly accompanies cultivation, may indirectly increase erosion by reducing the growth and density of cover of the crops, and the amount of organic matter returned to the soil in crop residues. But although lowered fertility is a major cause of erosion, even newly cleared land with fertile soil of good structure can rapidly be eroded on steep slopes.

Land use and farming practice

From what has already been said about the variation in the protective value of different types of vegetation and the destructive effects of unsuitable farming systems, it is evident that faulty land use can be an important cause of erosion. For example, the use of very steep slopes for the production of annual crops by native methods has caused widespread erosion in Africa. Steep land should be left under forest, or, in suitable circumstances, used for certain tree crops, such as cocoa, or for permanent pasture. The location of crops where they can be safely cultivated without risk of soil loss, and the adoption of suitable systems of farming, are essential for conservation.

A number of faulty methods of cultivation, currently or formerly practised in the tropics, are also conducive to erosion, such as the cultivation of crops on mounds, or on ridges running up and down the slope, or the clean weeding of plantations of trees. The ill-advised use of the plough up and down hill instead of on the contour has often led to this implement causing erosion.

THE EFFECTS OF WATER EROSION

The loss of top-soil and decline in fertility accompanying erosion usually result in lower crop yields. Although the reduction in yield is usually easily observed, especially if erosion exposes unproductive clay, sandy subsoil, or hard pan, quantitative information of the effect of erosion on yields in the tropics is lacking. Numerous experiments in America comparing yields on top-soil and subsoil showed that the decline on the latter was rarely less than 40 per cent and often much more. Yield reduction might be appreciably less after removal of the top-soil from some tropical soil types with deep and rather uniform profiles, but usually subsoils can only be rendered productive with difficulty. Erosion not only reduces yields on denuded slopes but may also seriously lower the productivity of flatter lands below as a result of floods depositing large quantities of coarse sand, gravel or stones, which bury the surface soil of the lower land. Discharge from large gullies can be particularly bad in this respect as it often fans out to cover a considerable area.

Erosion is often accompanied by a marked deterioration in water supplies. Where a large proportion of the rainfall is dissipated as run-off, instead of percolating through the soil to feed springs and sub-surface aquifers, streams cease to flow for part of the year and wells dry up, so that lower areas at some distance from the site of the erosion may be deprived of their water supply. This may be serious in areas of relatively low rainfall because, as erosion and run-off increase, a progressive process of desiccation sets in. Streams and wells are dry for long periods, the water table is lowered, for want of water the vegetation becomes sparser or changes its character and semi-desert conditions are eventually reached.

The flood water of streams resulting from excessive run-off may cause damage to water supplies, communications and property over a considerable distance. Dams and reservoirs built to conserve water, or to operate hydroelectric schemes, may become silted up. The deposition of silt in rivers and harbours may impede navigation. Heavy floods may undercut river banks and damage bridges, or enlarge stream and river beds to a size much greater than that which would obtain under a more regular flow with the consequence that culverts and bridges have to be made correspondingly bigger. Scouring and undercutting of river banks may destroy good agricultural land or property, rivers may jump their banks and change their courses and floods may damage crops and buildings, or wash away sections of road and railway embankments.

THE CONTROL OF WATER EROSION

Effective control of erosion demands the proper use and treatment of various types of land over the whole of a catchment area, in accordance with the conservation requirements of each different portion of any size. In the first place each type of land should be used only for such agriculture or forestry to which it is suited without undue risk of erosion. Thus the steepest slopes that are unsuitable for cultivation may be left under forest or permanent pasture; less steep land may be used for highly protective tree crops that provide a full canopy of foliage, such as cocoa; easier slopes may be planted with less protective tree crops, such as citrus or coffee; and the gentler slopes used for arable crops and temporary grass. Secondly, on each type of land appropriate soil-conserving methods of husbandry must be employed, such as contour planting and cover plants in tree crop plantations, or tie-ridging and strip cropping for arable crops. Thirdly, it may also be necessary to apply appropriate mechanical conservation measures, such as silt pits or narrow-based ridge terraces.

It cannot be too strongly emphasized that these three things – suitable land use, good husbandry and appropriate mechanical measures – should be applied in combination to a whole catchment area. Mechanical measures alone have a purely local effect and may be quite ineffective if applied to only part of the land, especially if higher portions of the catchment are improperly used. Good husbandry is essential, both for the maintenance of fertility and to provide, as far as possible, self-protecting land use, thus minimizing the cost and waste of land involved in the use of earthworks for erosion control.

The work of soil conservation usually involves three phases which succeed each other in the following order. First, some kind of physical survey is needed to enable different types of land to be mapped and classified as suitable for different purposes and methods of cultivation. Secondly, mechanical measures, such as the construction of earthworks, may have to be put in, and thirdly, appropriate agronomic practices must be employed.

Physical survey and classification of the land according to use capability
The survey of land and the classification of various areas into different land use categories is usually undertaken with the object of preparing

an overall land use plan, in which soil conservation is only one consideration and many other factors have also to be borne in mind. For conservation purposes the mapping of at least three land characteristics will usually be needed: soil type, slope, and degree of erosion. The soil survey should preferably include, for each soil type, observations on the profile, texture and water relationships, and should show phases according to depth, stoniness, presence of rock outcrops etc. In the under-developed countries lack of trained staff and the urgency of erosion problems may make it necessary to plan conservation on the basis of less detailed surveys. The approximate average slope of topographical units of reasonable size can be determined by use of an Abney level and marked on a map with a suitable slope-class symbol. The degree of erosion is ascertained by surface observations and auger borings and is also marked on the map with a symbol.

In addition to the three land characteristics already mentioned it may be desirable to map the main types of existing vegetation or land use, such as forest, pasture, tree crops or arable crops. This is likely to be necessary where considerable erosion has already occurred and

Table 4. *The most intensive suitable uses for land in each land capability class*
Vernon (1958)

Land Capability Class	Most Intensive Suitable Use
I. A and B slopes of good soils	Suitable for cultivation (tillage) with almost no limitations.
II. Mainly C slopes of good soils	Suitable for cultivation (tillage) with moderate limitations.
III. Mainly D slopes. Some gentler slopes of less favourable soils	Suitable for cultivation (tillage) with strong limitations
IV. Mainly E. slopes, some D slopes	Suitable for tree crops, grasses and very limited cultivation.
V. Mainly E and F slopes	Not suitable for cultivation, but suitable for planted forest, tree crops or improved grass.
VI. Mainly steep rocky land or dry climate	Not suitable for cultivation. Suitable for poor forest.
VII. Rock outcrops, riverwash, etc.	Little or no productive use.

	A	B	C	D	E	F
Degrees of slope:	0–2°	2–5°	5–10°	10–20°	20–30°	over 30°

	+	0	1	2	3	4	5
Erosion categories:	Accretion	Nil	Slight	Moderate	Severe	Very Severe	Extremely Severe

where its control may require complete reorganization and replanning of land use over a whole catchment.

The final stage is to establish and map land capability classes, which will be done from information obtained in the survey combined with knowledge of the local climate and agriculture. The land capability class will indicate the suitability of an area for certain types of crop (or other use) and will also broadly indicate the kind of agronomic and mechanical conservation measures that will be needed. It will not necessarily, or usually, show what species of crop should be grown or the precise conservation measures to be applied. These are matters depending partly on local economic and other factors and would be determined in the more advanced stage of producing detailed farm plans. An example of the kind of survey required is that made in Jamaica by Vernon (1958). By way of illustration, the slope and erosion categories and capability classes used in this survey are shown in Table 4.

MECHANICAL MEASURES FOR SOIL CONSERVATION

General principles

Apart from specialized methods of tillage, which are dealt with in a later section, mechanical measures for the control of erosion consist mainly of earth banks, with associated channels immediately above them, constructed at suitable intervals across the slope. These earthworks break up the slope, intercept the run-off before its volume and velocity become sufficient to cause serious erosion and divert the water down safe grades to suitable discharge points. It is essential that the concentration of the run-off into defined channels brought about by such systems of earthworks should be minimized, and that the banks and channels should be designed and spaced so that they have sufficient strength and capacity to deal safely with the maximum expected run-off, conveying it slowly across the land for discharge through suitable outlets. The outlets must be chosen and designed so that the disposal of the accumulated water from the system does not cause erosion on other land. Unless these requirements are met, and unless the whole system is properly maintained, more damage may be caused than if no earthworks had been constructed at all. For example, gullies may be started by water breaking through weak or ill-maintained earth banks, or by the scouring of channels and outlets which have too steep a grade. The vertical or horizontal intervals between the contour earth works, and their type, size and length, are matters which must be carefully determined in relation to the maximum run-off expected to occur. This is done on the basis of local experience and with the aid of standard tables and formulae, for which other works should be consulted, such as Gorrie (1946), Cormack (1951), Bennett (1959).

The channels may be constructed strictly on the contour, or with a constant gradient, or with a gradient which increases in the direction of the outlet. When rainfall is low in total amount and intensity channels are usually level, or with a constant gradient of less than one per cent. Under high rainfall, or when storms of high intensity occur, channels may be given a variable gradient to cope with the increasing volume of water that has to be carried nearer the outlets. In choosing and building conservation works the main consideration is to be sure that they will be effective in controlling erosion on the land in question. In addition, the type of structure should as far as possible be selected, and the system so designed and constructed, to minimize the amount of land which must remain uncropped, to interfere as little as possible with the use of farm machinery, to permit easy access to the land and to provide for cheap and easy maintenance.

In the following paragraphs a brief description is given of the main features of the common types of conservation works, but no attempt is made to go into details of construction and layout, for which reference should be made to standard works such as Gorrie (1946), Ree (1954), U.S. Dept. Agric. Soil Cons. Service (1954), Stallings (1957), Bennett (1959), Haws (1959).

Types of soil conservation works

The *hillside ditch* (Fig. 5) is simply a small ditch, made with a gradient of half to one per cent, with the earth removed from the furrow placed on the lower side. These ditches, which should have a minimum cross-sectional area of $1\frac{1}{2}$ square feet and a maximum length of 600 feet, are extensively used on permeable soils in the West Indies in conjunction with the peasant cultivation of land of 20 to 40 per cent slope where the construction of larger devices is impracticable. They are best used with a plant barrier, such as a strip of grass, planted along the upper side of the drain. In parts of the West Indies an interrupted row of short lengths of these drains is made on the contour without any gradient or outlets, and these are known as 'blind drains', but in most other parts of the world these sections would be called *silt pits*. Silt pits are commonly used in plantations of tree crops and some semi-permanent crops, such as bananas. They may be protected by a vegetative barrier on the upper side to filter out some of the silt, but they will still require cleaning out periodically.

The narrow-based ridge terrace, or contour ridge (Fig. 5), is a ridge of earth, from 4 to 12 feet wide at the base and 1 or 2 feet high, built along a contour or on a slight grade, with a channel on the upper side

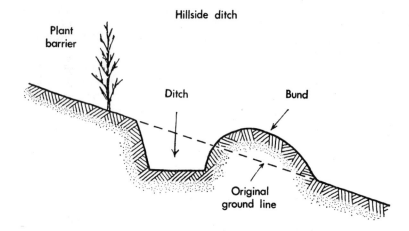

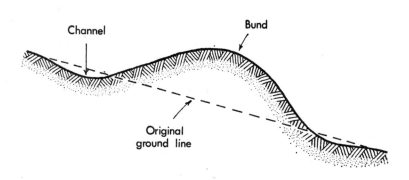

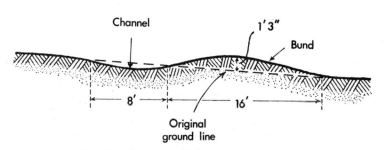

FIG. 5. Soil conservation works – diagrammatic sections

and often also on the lower side. Such ridges, which are suitable for fairly deep soils on moderate slopes of 10 to 20 per cent, can be constructed by hand, with horse or tractor drawn ploughs or with terracers or graders. The vertical interval between them is usually 4 or 5 feet, the gradient should not usually exceed three-quarter per cent and they should not usually be built to drain more than 1,000 feet in one direction. The ridges and channels are not usually cropped, the former normally being protected by a planted cover of grass.

The broad based ridge terrace (Mangum or Nichols terrace) (Fig. 5), is a much larger, but relatively lower, ridge, with a base width of 10 to 40 feet and a height of $1\frac{1}{2}$ to $2\frac{1}{2}$ feet and with a broad shallow channel above it, designed to permit the whole of the land to be tilled and cropped. These are mostly used on arable land with relatively deep soils and slopes of 2 to 12 per cent. Because of the large amount of earth moving required they are usually constructed with terracers or graders. Under low rainfall intensity and on absorptive soils they may be level but under other conditions they are graded, and the fall may be increased by 1 per cent for each successive 100 or 200 feet of channel up to a maximum of 5 per cent. Vertical intervals vary from as little as 2 feet on slopes of less than 3 per cent to 4 feet on slopes of 12 per cent, and horizontal intervals correspondingly from about 200 to 50 feet, standard recommendations differing in different countries. The maximum permissible length for such terraces is usually regarded as being about 350 yards but in Basutoland variable graded terraces of up to 1,000 yards are used (Tempany, 1949).

The oldest mechanical conservation measure is the use of *bench terraces* (Fig. 6) which consist of a series of steps cut into the slope on the contour, the forward edges, or banks, which are never cultivated, being constructed as nearly vertical as possible. If protected by stoloniferous grasses these banks can be given a slope of 2:1, but where stone is available the terraces may be faced with vertical masonry walls. Bench terraces are often given a slight slope from the front to a drain at the back, the latter having a gradient not exceeding 1 per cent. For the purposes of soil conservation these are used on steep, or very steep, slopes. The difficulty and cost of their construction nowadays usually limits their use to places where high-priced crops can be grown, or where they are the only practicable means of protecting steep land for use by a dense population. (Plate 7.) Bench terraces are also used, often on much less steep slopes, primarily for the purpose of water conservation and irrigation, for example, for the hillside cultivation of wet rice. (Plate V.)

Modified bench terraces (Fig. 6) are employed in the contour cultivation of tree crops on relatively steep slopes. Each terrace bench,

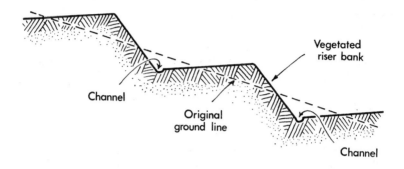
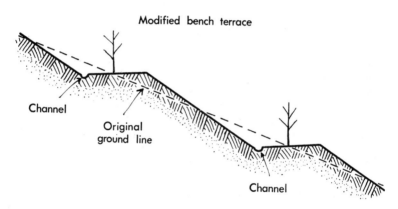
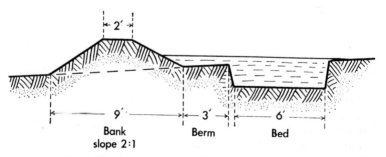

FIG. 6. Soil conservation works – diagrammatic sections

usually 4 to 6 feet wide, supports a single tree row, the interval from centre to centre of adjacent terraces being approximately equal to the normal row spacing of the crop. The sloping land between the benches is protected by planted cover crops or natural regeneration. The benches are cut with a slight inward slope towards a small channel at the back which may either be given a gradient of up to 1 per cent, or made level. (Plate 8.)

Individual bench terraces are discontinuous short lengths of modified bench terrace, each of which forms a small platform, 3 or 4 feet square, cut out of the slope and providing a planting place for one tree. They can be prepared along the contour lines before planting and may subsequently be joined up to form continuous terraces. Unless accompanied by other measures, such as hillside ditches and a good ground cover, individual bench terraces are not usually very effective.

Storm drains, or diversion ditches (Fig. 6). In addition to measures of the kind described above, which are designed to ensure the interception and safe disposal of run-off from areas actually under crops, it may be necessary to protect such areas from run-off from higher land which may be uncultivated. This might be necessary, for example, on an arable field lying below a steep slope consisting largely of rocky outcrops. Such protection may be provided by storm drains, designed to carry the maximum run-off to be expected from the catchment above them. These are usually relatively wide and shallow channels, with the excavated earth placed on the lower side to form a substantial embankment. The drain must have a non-scouring gradient, normally between a quarter and one per cent, and both the channel and the bank are best protected by being grassed over.

Outlets

It is imperative that outlets are provided in such a manner that the water discharged will not cause erosion elsewhere. Sometimes the water can be disposed of over rocks, or over gentle slopes under woodland, grass or other stable vegetation. In such circumstances the aim must be to discharge the water at as low a velocity, and over as wide an area, as possible, by reducing the gradient and increasing the width of the terrace channels at their lower ends. Stable natural waterways can also be used for the disposal of run-off from conservation systems, provided that the latter do not markedly increase the area of the catchment draining into the stream.

In the absence of natural outlets, special drainage ways will have to be provided. The simplest form is a broad, shallow, grassed waterway, roughly 1 foot wide for every acre of catchment, and best situated in natural depressions. The grass should preferably be mown for hay, but may be subjected to carefully controlled grazing; burning and overgrazing must be avoided. Where such waterways are impracticable, and particularly where land is valuable, it is necessary to construct narrower and deeper channels, which should be stabilized by planting them with grass and should have a non-scouring gradient. When the average slope of the land along the length of the channel exceeds a non-scouring gradient it becomes necessary either to make the whole channel of stone or concrete, or to construct a series of concrete or masonry steps, known as drop spillways, at intervals along the channel. (For details of these see U.S. Dept. Agric. Soil Cons. Service (1958).)

The dimensions, shape and size of a discharge channel or storm drain must be such that it has the capacity to carry the maximum run-off to be expected from the portion of the catchment lying above it. Their design requires calculations involving the degree and length of the slope, and estimates of the 'run-off coefficient' (i.e. the percentage of the rain that runs off), of the maximum rainfall intensity to be expected for periods of varying duration, and an assessment of the 'time of concentration of the watershed'. The latter is the time taken for water to travel from the top to the bottom of the section of the watershed under consideration and is thus the time for which rain must fall before the entire section of the watershed will be simultaneously contributing water to the waterway below it. Further information on these matters may be found in standard works on soil conservation such as Gorrie (1946), Ree (1954), and Bennett (1959).

Construction of the conservation works

In the actual construction of conservation works the first step is to provide outlets and main discharge channels. If the latter have to be planted with grass it must be remembered that in some places it may take a whole season for the grass to form a full cover. The second step is to dig any storm drains that may be required. Finally, the construction of the contour terraces etc. will be put in hand, starting from the top of the slope and working downwards.

Control of gully erosion

The existence of gullies usually indicates that sheet erosion has been going on for some time, consequently any work done on the gully itself should be accompanied by agronomic and mechanical measures designed to ensure the maximum infiltration of rainfall on the surrounding land. Apart from this the first step should be, as far as possible, to divert water flowing into the gully by the construction

13 Stumper uprooting rubber-tree

14 Fore-mounted root rake stacking timber in windrows

15 Ripping equipment

16 Terracing blade

of a storm drain or a ridge terrace across the slope above its upper end. At the same time the head of the gully should be protected and prevented from eating further up the slope by either converting it into a wide, grassed channel or by the construction of a masonry drop structure with an apron below it. With small gullies these measures, together with the exclusion of livestock and, possibly, the establishment of grass strips at the top of the banks on either side, may be sufficient to check further erosion and to permit the natural revegetation of the gully.

With larger gullies further work will be needed to ensure the safe passage of the remaining run-off through the channel, and to bring about the stabilization, vegetation and gradual in-filling of the gully. Vertical banks, subject to undercutting, may be cut to stable slopes and planted with grass. Temporary or permanent check dams or drop spillways may be constructed across the gully to reduce the velocity of flow, check channel erosion and hold enough soil and water for the growth of protective vegetation. Temporary check dams may be made of brushwood, planks or rocks held in position with wire netting. Permanent concrete or masonry spillways may be placed at points where there is an abrupt fall, so that the velocity upstream is checked and water passing over the weir falls on to a stone apron. Alternatively, concrete flumes may be built at these points of steep gradient. Details of the design and construction of both temporary and permanent structures for gullies are given by Bennett (1959) and Gorrie (1946).

AGRONOMIC MEASURES FOR SOIL CONSERVATION

Tillage

Ploughing, or any other form of cultivation which breaks a surface soil crust, will temporarily increase rainfall infiltration and reduce run-off. However, the beneficial effect is usually only transitory and in the long run the more cultivation that is done the more speedily will soil aggregates be broken down and the soil rendered less permeable. Consequently it is generally wise to keep cultivation to the minimum required to produce a satisfactory seed-bed and to control weeds. Subsoiling may aid water penetration and reduce run-off on land with an impermeable subsurface layer, especially if the latter is a hard pan, but where the less permeable layer is of clay the benefit of subsoiling is usually short-lived.

Ploughing and cultivating on the contour is to be generally recommended as one of the simplest and cheapest conservation measures. This practice not only reduces erosion but also has the merits of interfering very little with farming operations, and of saving power and wear on machinery. By itself contour cultivation will only check erosion on the gentler slopes and it is usually accompanied by other measures. Many crops grow as well on ridges as on the flat, and contour ridging is a further simple conservation practice for use either by itself or in conjunction with other measures.

In areas of moderate rainfall and for many, but not all, annual crops, tie-ridging is useful for soil conservation. This involves growing crops on ridges made roughly on the contour, adjacent ridges being joined at regular intervals (usually of 5 to 12 feet) by barriers or ties of the same height as the ridges, made with soil scraped from the furrow. The series of basins so formed holds the rainwater where it falls, allowing it to infiltrate into the soil and preventing all run-off, except in the most intense storms. Ties render ridges that are roughly on the contour effective without risk of breakage and thus obviate the need for making them strictly level or on a given grade. Except on steep land tie-ridging eliminates the need for other conservation measures, such as terraces or vegetative barriers, which take up space that might be cropped. Furthermore, there is no difficulty about the disposal of surplus water down drainage ways, which is often a problem. The effect of tie-ridging on water conservation and on the yield of various crops is discussed on pp. 123–5 and 142–5.

Contour strip cropping

Suitable rotations may be employed in the practice of strip cropping, in which relatively narrow strips of different crops are planted on the contour, strips of erosion-susceptible crops being separated by strips of close-growing, protective crops. Thus there may be successive strips of (1) a wide-spaced, row-tilled crop, such as maize or cotton; (2) a dense untilled crop, such as clover or grass; (3) a close-spaced crop receiving little or no cultivation after planting, such as finger millet or a creeping variety of groundnut. The densely planted strips slow up the velocity of the run-off and cause it to deposit much of its silt load, thus reducing erosion and affording protection to a more susceptible strip below. It is desirable that no two intercultivated strips, nor two strips having the same planting and harvesting times, should be adjacent. Strip cropping is often used in conjunction with contour terracing, the width of the strips being adjusted to the terrace interval.

Vegetative buffer strips and barriers

Permanent contour strips of grass or shrubs may be used either alone or with mechanical conservation measures. Provided that they give a dense and continuous cover, they slow up run-off and cause deposition of silt, so that in time the accumulation of silt behind the barriers gradually achieves a bench terracing effect. But they are often not fully effective because gaps develop in the vegetation and allow the passage of run-off, causing gullying. They also have the disadvantages that they tend to spread, thus necessitating labour to check them, and that they may depress the growth of crops adjacent to them over a distance of several feet. As the sole conservation measure on land used for arable cropping they are satisfactory only on gentle slopes. To be effective on steeper land the strips have to be so wide, or spaced so closely together, that a large proportion of the land is put out of cultivation. Vegetative barriers are, however, widely used in conjunction with hillside ditches and ridge terraces; strips of grass, or hedges of shrubs, being planted immediately above and parallel to the ditch or channel.

Plants most commonly used for these barriers are tall grasses, which are periodically cut back to 6 or 12 inches to promote tillering, the cut grass and trash being placed above the barrier. Since grazing speedily renders the strips ineffective, unpalatable grasses, such as Vetiver (*Vetiveria zizanioides*), *Cymbopogon citratus* or *C. nardus*, are preferable, in areas where stock are present, to palatable species. Short grasses that form a dense sward and bind the soil with vigorous roots and rhizomes, such as *Paspalum notatum* or *Cynodon dactylon*, are also used but are more liable to spread. Numerous other plants have been used as contour hedges – for example, *Tephrosia candida*, *T. vogellii*, *Leucaena glauca*, *Euphorbia tirucalla*, *Coleus* spp – but most of these are liable to develop gaps at ground level.

Cover plants

Control of erosion in plantations of tree crops on land of any considerable slope usually demands the maintenance of a good ground cover, whether or not mechanical conservation measures are employed, unless the latter are full bench terraces. During the establishment of the plantation, when the trees themselves afford little protection to the soil, it is usual to plant cover crops in the interrows. The creeping legumes commonly used as cover crops provide a dense protective cover, and also enhance rainfall infiltration by means of a layer of leaf litter and by improving the organic matter content of the soil. Planting leguminous shrubs, or allowing controlled growth of natural regeneration consisting mainly of shrubs, does not usually provide as good a ground cover, nor control erosion as effectively, as creepers do. Natural regeneration mainly comprising grasses is effective for soil conservation but liable to compete with the trees, although this may be countered by periodic mowing and fertilizing. Planted covers usually die out sooner or later on account of weed competition or because they are intolerant of the shade of the developing trees. The common procedure is then to encourage the growth of a ground cover of the more desirable indigenous species by selective weeding (see pp. 233–4). The amount of ground cover needed in a mature plantation depends, amongst other things, on the slope, the density of the canopy provided by the trees and the amount of leaf fall which the latter provides.

Mulching and stubble mulching

Mulching with cut grass or other vegetable trash (which is discussed on pp. 234–6) is effective in reducing run-off and erosion since it protects the ground from the impact of rain, slows up the movement of water over the surface, and improves the permeability of the soil. This is done by enhancing the soil organic matter content and by promoting increased termite and earthworm activity, which provides many channels for the percolation of water. It is not, however, much practised in the tropics for erosion control because of the labour and expense of cutting, carrying and applying the mulch and the rapidity with which it subsequently decomposes.

Stubble mulching, or trash farming, which is mainly applicable to arable lands and consists in leaving the whole or part of the crop residues and weeds as a protective covering on the surface of the soil, has similar effects to ordinary mulching. It is carried out by working the land after harvest to a depth of only 2 or 3 inches with a special plough or cultivator that severs the plant roots, thus leaving the stubble and weeds on the surface. Alternatively, the stubble may be only partially buried by shallow discing. So far, trash farming has been little used in the tropics but may deserve more attention, especially if accompanied by high density planting. In Rhodesia, Hudson (1957) found that dense planting and trashing of maize considerably reduced run-off and soil loss (see Table 3), while at the same time improving the moisture content and ease of working of the soil and aiding weed control.

Wind Erosion

Movement of soil by wind does not occur unless the surface soil is dry and unprotected by vegetation, and is favoured by a flat, or gently undulating, topography across which the wind can blow unimpeded. Wind erosion in the tropics is mainly confined to arid regions and to extensive cleared areas in places experiencing a long dry season. It is of minor agricultural importance compared to water erosion but it may be of local significance. It can do considerable damage in a short time, because the abrasive action of soil transported from a small area may rapidly destroy vegetation and tear more soil from the land surface over a much wider area.

Soil particles greater than about 1 mm in diameter are resistant to wind erosion, but smaller particles can be moved by wind of sufficient velocity in three ways. First, and most important, the most erodible particles – those in the middle of the diameter range of 0–1 mm – are moved by saltation, proceeding over the ground surface in a series of jumps and bounces. Secondly, the finest particles may be carried away in suspension in the air stream, but this fine dust is highly resistant to being picked up directly from the soil surface by wind and mostly becomes airborne as a result of an initial impact from particles in saltation. Thirdly, there is a more limited movement, mainly of the larger particles in the 0 to 1 mm diameter range, actually on the ground, known as surface creep, which is also mainly induced by the impact of particles in saltation. The movement of particles over and on the soil surface breaks down larger aggregates and renders them liable to erosion; the relative importance of this abrasive action increases with the size of the area over which wind erosion operates. The proportions of soil moved in the three ways will vary with the soil type; in the United States it is reckoned that between 50 and 75 per cent of the weight of soil moved is carried in saltation, 3 to 40 per cent in suspension and 5 to 25 per cent in surface creep (Chebnil, 1957).

The principal soil feature which affects susceptibility to erosion by wind is the dry aggregate structure and its stability which, in turn, is affected by texture, organic matter, calcium carbonate content and tillage practices. The coarser textured soils, lacking sufficient silt and clay to bind the sand grains into aggregates, are very susceptible to wind erosion. With medium to moderately heavy soils (loamy sands, loams and clay loams) resistance to erosion increases with the proportion of silt and clay, but the heaviest clay soils are less resistant because their aggregates are less stable. Raw vegetable residues have a protective covering effect, and decomposing organic matter decreases erodibility by aiding in soil particle aggregation. Free calcium carbonate in excess of 1 per cent reduces the stability of aggregates and therefore increases erodibility. Frequent tillage pulverizes the soil and renders it more liable to erosion; leaving the soil in cloddy condition makes it more resistant, numerous small clods being more effective than fewer larger ones.

A rough surface reduces wind velocity at ground level and tends to trap particles moving in saltation or by surface creep. Cloddiness is beneficial in this way and so are operations which produce a rough or irregular surface, such as ridging, or mulching with crop residues. A vegetative cover acts in the same way as surface roughness and, in addition, the binding effect of its roots on the soil is helpful; its beneficial effect increases with its density and, up to a point, with its height. Wide-spaced intercultivated crops are not well suited to the control of wind erosion but grass is one of the best covers, both on account of its density above ground and because its root system anchors it well and binds the soil.

Measures for the control of wind erosion aim principally at reducing wind velocity at ground level and at increasing the moisture content of the soil. The production of stable aggregates big enough to resist erosion by wind is an ancillary objective. Cultivation should preferably be done as soon as possible after rain and should be kept to the essential minimum in order to avoid pulverizing the soil. Ridging at right-angles to the direction of the prevailing wind is advantageous in reducing surface wind velocity. Those mechanical measures of water erosion control that also conserve moisture, such as broad-based ridge terraces made on the level, or contour tie-ridging, will aid in preventing wind erosion. The form of strip cropping known as 'wind stripping' is also helpful. By this procedure the land is divided into straight-sided and parallel strips placed with their long axes at right-angles to the wind, and alternate strips are used for crops that are resistant, or susceptible, to wind erosion. This may be used in conjunction with the fallowing of land in a wet season to store water for a crop in the subsequent season (see p. 125), alternate strips being cropped and fallowed.

Where it is practicable, one of the most effective means of countering wind erosion is the planting of windbreaks, or shelter belts, at intervals across the path of the wind. These obstacles will slow up wind velocity and cause the deposition of soil particles already in movement, thus preventing them from exercising their abrasive action further afield. As windbreaks are much more commonly used in the tropics for other purposes they are dealt with on pp. 244–5.

Water Conservation

Water conservation, in common with the control of erosion due to water, depends primarily on obtaining good infiltration of rainfall into the soil and minimizing run-off, but is also affected by the water consumption of vegetation. Methods of land use and management for water conservation may be considered under two headings, first, with respect to headwater catchments and, secondly, with regard to agricultural field practice.

LAND USE IN WATER CATCHMENTS

Sound land use is of vital importance in the catchment areas where perennial streams originate. This is especially so in many seasonal rainfall regions of the tropics where the major catchments are comparatively small highland areas receiving relatively high and reliable orographic rainfall, and are surrounded by much more extensive, lower and drier lands. The latter, which commonly experience rather unreliable rains of short duration, rely on rivers arising in the mountains for their water supply during a considerable part of the year. Agricultural and industrial development in such areas consequently depends very largely on the efficient management of the limited, well-watered, highland catchments. This situation obtains in East Africa where, as stated by Russell (1962), only about 4 per cent of the land area can expect to receive as much as 50 inches of rain in four years out of five. Since land covered with actively growing vegetation is likely to use at least that amount of water annually, it is only from this small proportion of the total area that perennial streams will normally originate. Moreover, 55 per cent of East Africa is unlikely to receive 30 inches of rain in four years out of five, so that in this much larger area 'perennial rivers are almost literally the life blood of the country'. To a greater or lesser extent, the same sort of situation exists in many other parts of the tropics – for example, in central and southern Africa, in parts of India and Burma, and in some of the West Indian islands.

While the catchments remain under their natural forest cover, infiltration is good and run-off slight. The result is that, despite the consumption of water by the vegetation, a proportion of the rainfall percolates into the deep soils and porous rock strata and will either feed springs and maintain a regular flow in rivers throughout the year, or accumulate in subsurface aquifers. Unfortunately, however, in many countries there has been wholesale and indiscriminate clearing of the forest over large tracts of these highlands, followed by use of the land for crops or pasture without any attempt to employ measures for the conservation of soil and water. This results in erosion and loss of much of the rainfall as run-off, with the consequence that streams cease to flow regularly, come down in periodic floods (often causing damage to lower lands) and dry up completely during the dry season. Since the occurrence of flash floods makes it difficult or impossible to store water by means of dams, and as the greatly reduced percolation precludes the accumulation of subsurface ground water which might be utilized by pumping, the lower lands have commonly been deprived of water supplies for much of the year.

Considerable portions of these highland catchments, especially where they are steep and broken, should be kept permanently in forest, and in most countries extensive, protective forest reserves are maintained by legislation. On the other hand, there are large forested areas of less steep slope which, by reason of their good rainfall and fertile soils, are of high agricultural potential. In view of increasing human population and the pressing need for economic development, there is a natural desire to use such areas for purposes more profitable than protective forestry if this can be done without undue harm to water resources. At present, as pointed out by Pereira (1954a), decisions on land use for such catchments have to be largely based on mere opinions, since little factual information exists on the effect on water resources of a change from natural forest to other forms of land use. Such information is now being obtained in East Africa from a series of experiments which seek to measure the effects on water consumption by vegetation, and on stream flow, of the application to head water catchments of four land use patterns, namely plantation forestry, plantation agriculture (tea), peasant cultivation of field crops and tribal herding, in comparison with the effect of maintaining the natural vegatation. So far, the picture is far from complete but a recent interim report (Pereira *et al.*, 1962) gives a valuable description of the methods adopted and contains some results of considerable interest.

The first of these studies (Pereira, 1952; Pereira *et al.*, 1962) is indicating that closed canopies of quick-growing cypress and pine, planted at 8,600 feet in the Aberdare Mountains, draw about the same amount of water from the soil as the indigenous bamboo thicket, and that total stream flow is increased during the first three years after clearing bamboo and establishing a clean, weeded pine plantation. In view of the fact that transpiration is reduced during the early growth of the softwoods (provided they are kept as cleanly weeded as possible) and again during periodic felling and replanting, it seems likely that there will be a distinct improvement in water supplies by replacing bamboo by the more profitable softwoods.

The catchment studies evaluating the effect of replacing tall forest by tea in the Kericho District of Kenya have so far demonstrated that the former is highly effective in minimizing run-off and regulating stream flow; total storm flow has amounted to only 1·3 per cent of the rainfall and even in the most severe storm recorded – 3 inches in fifty minutes – run-off was only 2·5 per cent. In the early stages of the development of tea gardens there has been an increase in run-off, but water use by transpiration has decreased, with a resultant increase of 15 per cent in stream flow. On steep slopes in the Mbeya District of Tanzania, where the forest is also highly effective in preventing run-off and erosion, the effect of two years' peasant cultivation has been a great increase in maximum run-off rate (76 cusecs per square mile compared with 12 cusecs per square mile from forest) and a doubling of the steady stream flow sediment load.

These studies should yield information of the highest value and enable sound decisions to be made in East Africa on appropriate land use and management for catchments. There is a great need for similar research in other countries to cover different climatic conditions, soils, vegetation types and land use patterns. Until more quantitative knowledge is available on the effects of other forms of land use on water resources it would clearly be wise to exercise the utmost caution in the clearing of forest or other protective natural vegetation from important catchment areas.

Despite efficient land use and management in catchment areas, there will naturally be big differences between rainy season and dry season stream flow in many places. This necessitates the provision of dams and reservoirs for storage in order to achieve a better distribution of available water throughout the year. Cormack (1958) gives information on the construction of the small storage structures that are of interest to agriculturists.

AGRONOMIC AND MECHANICAL MEASURES FOR WATER CONSERVATION IN THE FIELD

Measures to ensure the maximum conservation of rainfall as soil moisture available to crops and grass must, again, aim mainly at improving water infiltration and reducing run-off. Such measures are of especial importance in areas of low, or moderate, seasonal rainfall.

As indicated in Chapter 11, a large proportion of tropical grasslands occur in semi-arid regions, characterized by erratic rainfall distribution, high intensity storms, high temperatures and potential evapotranspiration totals in excess of the total annual rainfall. In such areas the misuse of pastures has led to catastrophic erosion and to an increasing scarcity of water supplies. The effect of overgrazing and the trampling of stock on moisture supply is illustrated by the results of studies on semi-arid pastures in the Karamoja District of Uganda. It has been shown that about 40 per cent of the rainfall is lost as run-off and that, under a rainfall of 26 inches in seven months (including several falls of more than one inch), the water never penetrated into the soil to a depth as great as 18 inches (Dagg, 1962).

The importance of agronomic measures to conserve rainfall for annual crops in seasonal rainfall areas is well illustrated by the results of experiments with cotton in Northern Nigeria (Lawes 1961). In this area the rainfall should usually be adequate for a good crop of cotton but in practice very low yields, of the order of 200 lb of seed cotton per acre, were usually obtained. The reason for this was that a large proportion of the rainfall was lost as run-off from a readily capping soil surface. Without conservation measures only 31 per cent of the rainfall percolated into the soil, and in a season when 42 inches of rainfall fell the soil at 30 inches depth was never wetted (Table 5). However, when run-off was prevented by tie-ridging the soil moisture content after the rains was increased to the maximum depth examined, namely, 48 inches. Once measures were taken to improve infiltration and prevent run-off, responses were readily obtained to fertilizers and manures and high yields of over 2,000 lb seed cotton per acre were achieved.

Table 5. *Soil moisture content at the beginning and end of the rains where run-off has, or has not, occurred*

Depth (inches)	Soil moisture, percentage oven dry weight		
	Before rains	After rains	
		Run-off allowed	Run-off prevented
6	2·0	18·1	18·7
12	5·0	19·5	19·9
18	11·9	18·4	21·1
24	13·8	16·2	21·1
30	13·5	13·4	21·3
36	13·9	13·9	20·9
42	13·6	13·6	20·9
48	13·6	13·5	20·1

The majority of the mechanical measures and tillage operations already mentioned for the control of erosion will also aid in water conservation since they reduce run-off and slow up water movement over the soil surface, giving more time for infiltration. This applies, for example, to narrow- and broad-based ridge terraces, bench terraces, ridging or tie-ridging on the contour, strip cropping, and mulching.

Ridging or tie-ridging on the contour has been shown to be an effective measure for water conservation at a number of places in the tropics. An example from Nigeria has already been quoted but mention may also be made of the results obtained by Walton (1962b), who studied soil moisture contents at various depths under ridge and flat cultivation over several seasons in Uganda. Over a period of several months during the rainy season he found little difference between ridged and flat plots in the moisture content to a depth of one foot. From the second to the sixth foot, however, there was consistently more moisture under the ridges on uncropped land, the difference between the two treatments increasing with depth. The steeper the slope the greater was the increase in moisture content obtained by ridging compared with flat cultivation, because of the increasing effect of the ridges in checking run-off.

The once-favoured practice of frequent surface cultivation after rain, which aimed at producing a mulch of stirred soil thought to check evaporation, has been shown to have little merit beyond its effect in controlling weeds. Weed control is, of course, important, especially as many species are deep rooted and capable of drawing water from the soil to considerable depth, but it should be effected with a minimum of cultivation, since pulverizing the soil will hasten surface sealing.

In some arid regions land is fallowed in alternate wet seasons in order to store rainfall in the soil for the growth of a crop in the following wet season, the fallows usually being clean cultivated in order to avoid depletion of soil moisture by weed growth. The proportion of the rainfall stored by such fallows is rather low; measurements made in several parts of America have indicated a moisture storage efficiency varying between 15 and 30 per cent (Evans and Lemon, 1957) but the high evaporation rates in dry tropical regions might well result in lower figures. Nevertheless, fallowing may be worth while if the rainfall stored permits crop production in an area otherwise too dry for this purpose, or even if it gives total yields somewhat below those obtainable by cropping in every season, since the cost of a fallow is appreciably less than that of growing a crop.

A clean-tilled fallow suffers from the disadvantage that the bare soil is exposed to sun, wind and rain for a considerable period, with consequent risk of erosion by wind and water. It may be preferable to provide some protective cover by stubble mulching, by killing weeds chemically and leaving them on the surface, or even by growing a vegetative cover which demands little water. In a trial at Kongwa, in Tanzania, Pereira *et al.* (1958) found that a sowing of 20 lb per acre of teff grass (*Eragrostis abysinica*) provided a shallow-rooted, protective cover, which suppressed weeds, allowed storage of subsoil water and enabled a yield of 400 lb per acre of groundnut kernels to be obtained in the following season in which there was only $8\frac{1}{2}$ inches of rain. Groundnut yields following a bare fallow, which did not achieve satisfactory soil conservation, were not significantly different. On the other hand, a volunteer weed fallow dried out the soil to a depth of 6 feet, using more water than a crop of maize, sorghum or groundnuts.

CHAPTER 6

Land Clearing, Tillage and Weed Control

Land Clearing

Methods of clearing land naturally vary greatly, depending on the nature of the natural vegetation to be removed, the topography, the equipment available to the farmer and the purpose for which the land is to be used.

NATIVE METHODS

Peasant farmers do not usually attempt full clearing and stumping of the land. As most farmers possess only simple tools such as cutlasses, axes and hoes, full clearing would be extremely laborious. Moreover, as the land is generally to be used for shifting cultivation with hand tools, the eradication of stumps and roots is neither necessary nor desirable. Those who practise shifting cultivation in forest or woodland country normally slash shrubby undergrowth just above ground level with cutlasses, and cut down trees with axes at any convenient height. Small and medium-sized trees are usually cut at two or three feet above the ground but larger trees, especially if buttressed, are cut appreciably higher in order to save labour, or are felled by maintaining a fire of brushwood at their base for some days. The largest trees are commonly left standing, as their removal is not thought worth the considerable effort involved, despite the fact that they reduce the yield of crops grown under or near them. As a rule no attempt is made to get out the stumps and roots of the trees, which regenerate when the land goes out of cultivation a few years later. (Plate 9.) In savanna country the roots of grasses are normally dug out with hoes.

Even when they intend using the land permanently for perennial tree crops, peasant farmers do not usually attempt full stumping, but merely remove stumps and roots adjacent to the planting holes, and then gradually effect more complete clearing over a period of several years by slashing sucker regeneration and by burning rubbish around tree stumps. When certain crops that require shade (e.g. cocoa) are to be planted, selected forest trees, suitably spaced, may be left standing to serve this purpose.

Slashed undergrowth and felled trees are almost invariably burned, this being the only practicable means of freeing the land of debris prior to cultivation for those who possess only hand tools. Provided that a thorough burn is obtained, burning also has the advantage that it enables the farmer to start with a weed-free seedbed. In savanna country the grass is also usually burnt at the time of clearing but in some places it is customary to bury it in mounds or ridges.

ESTATE METHODS

If land to be cleared from forest or woodland is subsequently to be cultivated mechanically, which would normally be the case if annual crops, or short-term perennials such as sugar or pineapples, are to be grown, then the clearing operations must include grubbing out the stumps and main roots of the trees. On the other hand, with some tree crops, such as cocoa and rubber, it is possible that little or no mechanical cultivation may be contemplated, in which case it may only be necessary to do limited stumping and root removal along the lines where the tree rows are to be planted.

Manual Clearing

In some parts of the tropics clearing by hand is still cheaper than mechanical clearing, especially if full stumping and root eradication are not needed. Manual clearing of jungle is still usually done for rubber planting in Malaya. The first step is usually to provide a fire break around the area to be cleared. The undergrowth is then slashed at ground level and trees felled with axes at any convenient height. Despite the fact that it may have certain disadvantages (which are discussed later, p. 133), every effort is usually made to obtain a thorough burn after clearing. This has practical advantages in freeing the land of debris and weeds, thus minimizing subsequent expenditure on further clearing and on the establishment and weeding of the leguminous cover crops that are normally sown between the planting rows immediately after burning. A good burn is not easy to get in the wet climate of Malaya, as the felled trees require six to eight weeks without heavy rain in order to dry out sufficiently to burn well. Care needs to be taken in piling lighter brushwood around heavy timbers, and in organizing the actual burning to best advantage. In countries with a marked dry season there is no difficulty in obtaining a complete burn of felled jungle. Stumping is frequently limited to removing the stumps of all trees up to 6 inches diameter in the planting rows and the removal of larger ones in the vicinity of the planting points. (Plates 3 and 10.)

Winching

One of the most economical methods of felling and stumping, in places where wage rates are still moderate, is by the use of a hand-operated winch and cable such as the Trewella winch. The winch is anchored to the base of a neighbouring tree and the cable attached to the tree to be felled as high above the ground as is convenient. As a preliminary it is usually desirable to clear the undergrowth and small saplings by hand. Winching is best done when the soil is thoroughly moist; under dry conditions trees often tend to break near the base, necessitating removal of the stumps by hand. For maximum removal of the roots of all but the smallest trees it is best to pull the trees partly over from one side and then to move the winch to pull from the opposite side, so that the roots on both sides are pulled out. A winch of this kind can deal satisfactorily with most kinds of trees up to about 8 inches diameter, and larger trees can be pulled over by using two winches. This method can give good results at relatively low cost.

Ring-barking and poisoning

Where certain root diseases, especially that caused by the fungus *Armillaria mellea* are prevalent, ring-barking, which consists in removing a ring of bark 8 to 18 inches wide from the trunks of the trees, usually at 3 or 4 feet above ground level, is sometimes done as a means of killing trees before clearing. The trees then die slowly, mostly taking one to two years to do so, exhausting the carbohydrates in their roots on which the fungus depends. In so far as *Armillaria* is concerned, it has been shown that the incidence of the disease in a following susceptible crop is reduced, although not entirely eliminated.

Poisoning with sodium arsenite has long been used, to a limited extent, in jungle clearing and for killing old stands of coconuts, oil palms or rubber before replanting. The commonest method is to make a frill girdle round the trunks of trees one or two inches deep with an axe, and pour a 50 per cent solution of the poison into the wound so made. Probably a better method is to ring-bark the trees 8 or 12 inches wide, and apply a paste, containing 5 lb sodium arsenite and 6 ozs tapioca starch in a gallon of water, to the wood so exposed. Trees poisoned with sodium arsenite tend to fragment and fall in pieces. The great disadvantage of sodium arsenite is its toxicity to humans and stock. This does not apply to the butyl ester of 2,4,5-T, which in recent years has been found effective for killing most trees when sprayed or painted on to the bark from ground level up to a height of about 15 inches. Trees treated with 2,4,5-T tend to fall almost entire, since the tops do not fragment, but the decay of the base and root

Land Clearing

system is more rapid when 2,4,5-T is used than when sodium arsenite is employed.

Mechanical clearing

In many places full clearing and stumping with hand labour is now prohibitively costly, and this is likely to become increasingly so as wage rates rise. All the operations involved in full clearing can now be done mechanically, albeit with varying efficiency and expense, and it is often most satisfactory to clear partly mechanically and partly manually.

Clearance of undergrowth. Where the trees are sufficiently widely spaced to allow the passage of the equipment, the undergrowth can be cleared with a bulldozer before felling the trees. There are also available specially designed brushcutters, consisting of a heavy V blade, shaped like a snow plough and fore-mounted on a tractor, which cut through the bush at ground level with the sharpened lower edge of the blade as the tractor moves forward.

Felling. Probably the commonest method of felling and stumping in one operation is by bulldozer. Its efficiency compared with other methods depends on the nature of the vegetation. It is ineffective in heavy forest, and in both 'miombo' woodland and *Commiphora* thicket in Tanzania the Overseas Food Corporation (1956) found bulldozing inferior to chain clearing on account of the time wasted in frequently reversing, and because the felled trees lay haphazardly on the ground, thus making the subsequent piling operation more difficult. On the other hand, bulldozing is effective in much secondary forest, and was found to be cheaper and better than the chain method for clearing old oil palms (Hartley, 1949) or old rubber that had virtually become secondary jungle (Murray, 1954). Treedozers are efficient for felling and stumping, since they have a boom that is raised to contact the tree 5 or 10 feet above the ground and push it away from the tractor, while immediately afterwards a lower V blade is driven into the base of the tree and subsequently raised to uproot it, the tractor moving forward all the time. (Plates 12 and 13.)

In chain clearing two heavy tractors, linked by a heavy anchor chain which they tow, move forward together on a parallel course about 50 feet apart. As they do so, the loop of the chain pulls trees out of the ground with part of their main roots, which snap below ground. Thus the tractors are able to fell and stump a swathe 50 feet or so wide as they move continuously forward. For satisfactory results the clearing must be done when the soil is moist, since under dry conditions the trees tend to break off at or above ground level. Occasional larger trees, which resist the chain and hold it up, can be dealt with by applying additional force with a tree-dozer tractor, following behind the chain. In extensive experiments in Tanzania, the Overseas Food Corporation (1956) found that when the soil was moist the chain method was the most economical and efficient means of clearing 'miombo' woodland and *Commiphora* thicket, although it was not completely satisfactory in dense *Commiphora* thicket, which tended to bend rather than pull out. Chain clearing is impracticable in forest containing numerous really large trees, but Francis (1957) reports the success of the method in the fairly heavy jungle of the Gal Oya area in Ceylon where four tractors (two hitched in tandem at each end) were used to pull the chain, and two tree-dozing tractors followed behind. Chain clearing normally facilitates the subsequent operation of piling felled timber into windrows since the chain brings all the trees together in rows facing in one direction.

Stumping. If it has not been accomplished in the process of felling, stumping may be done with a hand- or tractor-operated winch, a bulldozer, or a special stumping tool, such as the 'Fleko' stumper.

Piling before burning is most commonly done with a bulldozer, but this is not an entirely satisfactory implement for the purpose as it causes considerable disturbance to the surface soil and deposits part of it in the rows of piled debris. In wet climates this may make it difficult to get a good burn; in any case it is undesirable to have top-soil removed from the greater part of the area and concentrated in ridges or mounds that probably require subsequent levelling. A more satisfactory implement for moving and piling felled trees and debris is the heavy rake foremounted on a heavy tractor. This does a cleaner and faster job than a bulldozer, and moves far less soil. The soil passes between the tines and the operator is also able to shake the rake as he moves a load with it, which loosens soil clinging to the roots. (Plate 14.)

Root cutting, ripping and raking. If mechanical cultivation is intended, then felling, stumping and clearing debris must be followed by cutting up and removal of all roots of a size which will impede or damage agricultural implements. Root eradication can be done with heavy towed rakes, such as the Le Tourneau road-ripper, but the specially designed fore-mounted rakes mentioned above are more satisfactory. Cutting of the roots is often necessary before raking, for which purpose heavy root cutters are available, usually consisting of a V blade towed by the tractor. (Plate 15.)

Possible disadvantages of mechanical clearing. Clearing with heavy machinery may have disadvantages in that there is much disturbance of the surface soil and destruction of its structure, often accompanied by undue compaction, especially if the work is done in wet weather. The bulldozer, unless very skilfully used, may lead to concentration of top-soil in some places and exposure of subsoil in others. However, despite some possible disadvantages, an increasing use of machinery for land clearing is likely to result from rising wage rates, and adverse effects can be minimized by the proper selection and skilled use of equipment.

THE EFFECTS OF BURNING

Burning after felling destroys much of the organic matter in the vegetation and in leaf litter accumulated on the ground, but does not usually result in appreciable loss of the humified organic matter in the surface soil. Where the soil is largely left untilled after clearing (as it often is when the land is used for shifting cultivation in the forest zones or for the production of perennial tree crops) the destruction of the leaf litter is a physical disadvantage. If it remained it would continue for some time to protect the soil surface from the direct impact of raindrops, increasing water infiltration and reducing run-off. When the land is cultivated after clearing, the protective layer of litter is removed by incorporation in the soil. Burning also results in the loss of almost all the organic carbon, nitrogen and sulphur in the vegetation and litter, but the mineral nutrients are added to the surface soil in the ash. In the forest zones the amounts of these nutrients contributed by the ash depend on the soil type, the nature and development of the vegetation (especially on the amount of regrowth during the fallow period where shifting cultivation is practised) and on the thoroughness of the burning. The limited data available (Nye and Greenland, 1960) indicate that in forest areas burning results in a considerable increase in the phosphate and exchangeable cation content of the top-soil. In the savannas and woodlands the amounts of nutrients released by burning vary greatly, depending mainly on the density of trees and the proportion of trees that is felled and burned, but are very much less than in the forest, especially in respect of calcium.

Part of the ash is liable to be washed away by the first rainstorms that fall on the bare ground after clearing, and there will be some further loss of nutrients by leaching, at least until a crop has become established. The extent of these leaching losses has not been accurately determined, but Nye and Greenland (1960) consider that they are likely to be limited by the absence of significant quantities of anions that would hold the cations in the soil solution. However, the high level of nitrate in some forest soils could lead to considerable leaching of cations if the nitrate were not absorbed by crops quickly established after burning. Since less nitrate is produced in savanna soils leaching should be less, but late planting of crops, which is common on peasant holdings, may lead to leaching of cations along with nitrate at the beginning of the rains.

Corresponding with the increase in exchangeable cations following burning, there is a rise in the pH of the surface soil. This is important on acid forest oxysols, where the decrease in acidity resulting from the addition of exchangeable bases in the ash is probably mainly responsible for an increase in the rate of nitrification which occurs after clearing and burning. As the quantity of ash added to savanna soils by burning after clearing is much smaller than in the forest, the rise in pH is correspondingly less.

The direct effects of heat on the soil when vegetation is burned are probably mainly confined to limited areas where timber and debris are piled for burning; elsewhere the rise in temperature below the immediate surface of the soil is slight. Heating soil may affect the microbial population, the physical condition of the soil and the availability of nutrients, but when burning is done after clearing it is difficult to disentangle the effects of heating from those due to the removal of the vegetative cover, the addition of nutrients in the ash, and the rise in pH.

Temporary increases in the microbial population of the soil and in the rate of mineralization of nitrogen have been observed after burning in the forest zones by a number of workers, including Corbett (1934), Dommergues (1952), Focan *et al.* (1953), although Griffiths (1947) found in India that nitrifying bacteria were destroyed and that the available nitrogen supply was negligible, for a time, after burning. The effects observed by the majority of workers might be caused by a partial sterilization of the soil resulting from a moderate increase in temperature, but seem more likely to be due to the increase in pH and available nutrients, and the observations of both Corbett and Dommergues favour the latter explanation. In the savanna zones, where the quantities of nutrients added by burning, and the rise in pH, are much less, Dommergues (1954) in Madagascar and Berlier *et al.* (1956) in the Ivory Coast have observed an increase in the numbers of nitrifying bacteria after burning. On the other hand, in Kenya Meiklejohn (1955) found that in a completely dry month following burning the number of micro-organisms was much reduced, aerobic nitrogen-fixing and nitrifying organisms being killed, and though anaerobic nitrogen-fixers survived they were not very effective at any time. The differences between these observations may possibly arise because an increase in nitrifiers, and in mineralization of nitrogen, may only occur if the grass is completely dug out (which was not the case in Meiklejohn's trials) and when rain falls to wet the dry, burned soil. Birch (1958) has shown that a flush of humus decomposition and nitrate production occurs each time a soil is rewetted after drying, the magnitude of the flush depending in part on the effectiveness of the drying. Burning at the end of a dry season, by making the soil still drier, may lead to increased nitrate production, but this will only occur when the soil is wetted again.

Burning may improve the physical condition of some soils, especially clays, rendering them more friable, easier to cultivate and better able to accept infiltration of water. Such beneficial effects have been reported from the Philippines by Conklin (1957), from parts of India by Gourou (1953) and Griffiths (1947), from Madagascar by Riquier (1953), and from parts of the Sudan where the 'hariq' system (in which several years' growth of grass is allowed to accumulate before burning) is practised (Tothill, 1948). It is also the experience of Chinese cultivators in Malaysia that burning improves the physical condition of the heavier soils. However, the improvement is unlikely to last long. Even under a temperate climate, Scott (1956) observed that improved infiltration rates after burning lasted only a year, being gradually nullified by raindrop impact. Under the higher rainfall intensity of the tropics the improvement is likely to disappear even more rapidly, and has been reported to do so by Griffiths (1947) and Riquier (1953).

It will be seen that on the whole burning has an adverse effect on the fertility of the land. Any improvements in the microbial population or the physical condition of the soil are small and transitory, and against the latter must be set the loss of the protective layer of leaf litter. The rise in pH is only beneficial on certain acid soils. Most of the organic carbon, nitrogen and sulphur in the vegetation and litter are destroyed, and although the ash enriches the surface soil in phosphate and exchangeable cations a proportion of the latter is lost by leaching. The deposition of mineral nutrients in the ash, however, is an important factor in the practice of shifting cultivation, since it represents part of the transfer of nutrients from the subsoil, via the fallow vegetation, to the top-soil, thus replacing nutrients removed from the surface horizons by crops during the period of cultivation. Quite apart from this, burning is usually a necessity for shifting cultivators, since it is the only practicable means they have of freeing the land from debris before cropping. Provided that a good burn is obtained, it also destroys all grasses and herbs and a proportion of their fallen seeds, thus greatly assisting the farmer by enabling him to start with a comparatively weed-free seedbed.

When preparing forest land for planting tree crops that are not subsequently to be mechanically cultivated, it is possible to avoid the disadvantages of burning and to leave the felled vegetation to rot, piled between the planting rows. An alternative that is sometimes considered desirable is to have only a light burn of the felled jungle. This is done in order to minimize the destruction of leaf litter and other organic matter, and also to avoid killing out all the indigenous broad-leaved herbs. These plants, if allowed to survive and selectively weeded, will form a good ground cover, but, if destroyed, will be replaced by colonizing grasses that are more competitive and more difficult to eradicate. The great disadvantage of no burning, or of a light burn (which leaves much of the timber unburnt), is that the land is left encumbered with a mass of debris. Even if this is piled between the planting rows it makes the movement and supervision of labour difficult, may create a problem if certain pernicious weeds such as lalang (*Imperata cylindrica*) become established among it, and harbours vermin, particularly rats, which damage young trees. In addition there is a danger of subsequent accidental fires in any areas subject to a dry season. Consequently estate practice is usually to burn thoroughly in order to minimize subsequent expenditure on further clearing and on maintenance of the plantation in its early years.

MINIMIZING THE EXPOSURE OF BARE SOIL

If the land is subsequently to be mechanically cultivated, there will usually be no alternative to full clearing and stumping, either by hand or mechanically. This has the disadvantage that the surface soil is thoroughly disturbed, with consequent deterioration in its physical condition and its organic matter content. These disadvantages are enhanced if the soil is left bare for any length of time, so that soil temperatures are raised and rainfall impact further destroys structure, sealing the surface and increasing the risk of run-off and erosion. It is therefore important to provide a vegetative cover and to put in any necessary soil conservation works as soon as possible after clearing. The crop, whether annual or perennial, should be planted with a minimum of delay, and with tree crops it is desirable to establish a cover crop as quickly as possible. If shade plants are needed as they may be, for example, with cocoa, coffee or tea, they too should be planted as soon as possible after clearing.

If certain tree crops are to be grown without mechanical cultivation it may be possible to minimize soil disturbance during clearing by restricting stumping to the planting rows. Depending on the nature of the land and the natural vegetation, it may also be found that the most effective way of speedily establishing a good ground cover is to encourage the regeneration of the softer and least competitive indigenous herbs by a process of selective weeding. The aim in all cases should be to reduce the period in which soil is exposed to the elements to the absolute minimum.

Drainage

After clearing, the next step is to put in the soil conservation and drainage works which may be needed. The former have already been dealt with in Chapter 5, and the subject of drainage will not be dealt with in detail because the principles are the same as in the temperate zone and practice is largely dictated by local circumstances of soil, slope, crops grown and land values.

Over much of the drier tropics the chief object of the agriculturist is to conserve water, not to get rid of it. However, even where rainfall is moderate and a marked dry season occurs, it may be necessary to improve drainage in the wet season. In the wet tropics there are large areas, such as, for example, the coastal alluvial plains of Malaya, where the nature of the soil and topography render crop production impracticable without considerable drainage works.

Subsurface drains of tile, rubble, or brushwood are uncommon in the tropics. They are costly to install, are liable to be disrupted or blocked by the roots of many perennial crops and, in monsoon climates, they may easily dry out the land excessively at times and yet be inadequate at others. Open surface drains are cheaper and easier to construct, and can be made mechanically with scrapers, bulldozers, drainage ploughs or dragline excavators, the latter generally being considered more economical for making main drains while the tractor drawn implements are better for smaller, field drains. On the other hand, open drains have disadvantages in that they take up land, inconvenience mechanical tillage operations, tend to harbour obnoxious weeds and rodents, may be a danger to, or be broken down by, stock and require constant maintenance if they are to remain effective. However, cleaning and maintenance of open drains can to a considerable extent be mechanized: drag lines can be used for large drains and machines for the regular maintenance of smaller field drains have been developed (Boa, 1958).

The grade on drains must be sufficient to prevent rapid silting but not so great as to cause scouring, and the whole system of drains and outlets must be so constructed as to prevent gulleying and erosion (see Chapter 5). Field drains can often be constructed as part of the soil conservation system, running across the slope with only a slight grade.

Reference has been made on page 86 to the desirability of making more use of the lower lying, seasonally waterlogged areas of heavy, relatively fertile soils known in Africa as vleis, mbugas, etc. This involves drainage but, as pointed out by Cormack (1953), the necessary drainage works need very careful planning as otherwise considerable damage may be done. These vleis commonly receive and regulate the flow of run-off from surrounding higher land, and also form the sources of watercourses that are important water suppliers to lower areas. The improvement of drainage in the vleis must therefore be carried out in such a manner that (1) run-off from surrounding higher ground is removed without gulleying or erosion in the vlei; (2) the value of the vlei as a water source is maintained and due regard paid to the hydrological regime downstream; (3) the vlei is sufficiently well drained to permit of cropping in the wet season, but not so thoroughly drained that it prevents cropping by drying out completely in the dry season.

Land Preparation for Annual and Short Term Perennial Crops

OBJECTS AND METHODS

The main objects of preparatory cultivation are:

1. To loosen and break up the soil in order to increase aeration and the infiltration of water, the latter being of special importance where rainfall may be limiting to crop production.
2. To prepare a seedbed of suitable tilth for the crop to be grown.
3. To aid in weed control by keeping weeds in check while the young crop is germinating and beginning to grow, this being particularly important because of the rapid and luxuriant growth of weeds in the tropics.
4. To incorporate surface vegetation and crop residues into the soil, in which connection it should be noted that the residues of many tropical crops are considerably more bulky than those of temperate crops.

The preparatory cultivation done by native farmers is usually shallow and superficial, inevitably so in many places because of the primitive implements available and the presence of many tree roots in land only roughly cleared from forest. Sometimes the seeds are merely sown into loose forest soil and ash without any cultivation, or after a mere scratching of the surface soil with digging sticks or hoes. Most commonly, especially in savanna country where grass roots must be dealt with, hand hoeing is done to a depth seldom exceeding 4 inches. Alternatively, land may be prepared with ox-drawn wooden or light iron ploughs, possibly followed by a light harrow, usually locally made. Ploughing is usually shallow, often only about 4 inches deep and rarely deeper than 6 inches. In some places mounds or ridges may be prepared.

Mechanical cultivation of annual and short-term perennial crops is

Preparatory Cultivation

mainly, but not entirely, confined to regions experiencing a pronounced dry season, owing to the difficulty of mechanizing the production of such crops in the ever-wet areas. An exception to this is swamp rice, but consideration of the cultivation of this crop is deferred to Chapter 9. As in the temperate zones, there are broadly three methods of preparing land with tractor-drawn implements.

First, it may be done with a plough, followed by disc or tine harrows. The plough cuts and inverts, or partially inverts, a furrow, effecting some breaking up of the soil, although further comminution and possibly consolidation with harrows is usually needed. Mould-board ploughs, which give a more complete inversion of the furrow, may be desirable where the control of certain weeds is required, but usually those with digger bodies, or disc ploughs, are preferred. Disc ploughs are the most commonly used, principally because they cut through or ride over roots and stumps that would arrest or break mould-board ploughs. They will also cut through surface trash (which would choke mould-board ploughs), achieve greater penetration under hard soil conditions and are easier to operate and control.

A second method of land preparation is with heavy disc harrows or tined cultivators, possibly followed by lighter harrows. These break up the soil without inversion. Usually ploughing will be desirable when first breaking land, or when there are troublesome perennial weeds or bulky crop residues to be dealt with, but when land has been brought into regular cropping it may only be necessary to plough once every three or four years.

Thirdly, land may be prepared for crops with a roto-tiller, which produces a seedbed in one operation. The smaller types of roto-tillers cannot be used in first breaking land and are also ineffective when there is heavy weed growth or bulky crop residues, but heavier machines are available which can cope with such conditions.

Although preparatory cultivation will increase the infiltration and percolation of water, except where the surface horizons are loose and of excellent structure, the effect is usually transitory. In the long run all forms of cultivation tend to aid the rainfall in breaking down aggregates, sealing the surface and increasing run-off. In investigations with coffee in Kenya, Pereira and Jones (1954b) showed that the hand hoe, especially a fork hoe, caused less damage than the mould-board plough or disc harrow, while the ordinary roto-tiller, which rotates at a high speed, pulverized dry soil to a fine structureless dust.

TIMING OF PREPARATORY CULTIVATIONS

In regions of alternating dry and wet seasons, weed growth is naturally minimal or absent in the dry season, but once the rains break weeds grow rapidly and luxuriantly and are difficult to control. It is, therefore, desirable to prepare land before the rains break, in order to obtain a clean seedbed under dry weather conditions, and thus to reduce weed competition during the early stages of crop growth. Early land preparation is also desirable because numerous experiments, in many parts of the tropics with a variety of annual crops, have shown that early planting generally gives the highest yields. As a rule, yields are markedly and progressively reduced the longer planting is delayed after the onset of the rains. Apart from suffering less weed competition, early sown crops benefit by receiving the full season's rainfall, and because the soil nitrate content is high at the beginning of the rains, whereas later it is reduced by the slower rate of humus decomposition and by leaching.

For hoe farmers, and also to a great extent for those with ox-drawn ploughs, it is difficult to open up land from forest or bush fallow, or to do preparatory cultivations, during the dry season when many soils set very hard. Normal native practice is to wait until the earliest showers or 'grass rains', and then try to complete the work within a few weeks of the end of the dry season, which usually means that the land is not prepared early enough for the best results. In some places the pressure of work at the end of the dry season is reduced by the practice of clearing the land from forest or savanna towards the end of the previous rainy season, but as a rule there is a general tendency for land preparation and planting to be done later than the optimum time.

Tractor cultivation is a great advantage in this respect since it provides the power needed to plough hard soils in the dry season and to speed the preparation of land for sowing. Especially on heavier soils, good tillage is usually a matter of taking all possible advantage of the aid of the weather. The combination of wetting and drying plus cultivation at the right time will always produce the best seedbed. The weather produces weaknesses in the clods and the implement breaks them down because of these weaknesses. Ideally, land should therefore be broken with ploughs or heavy disc harrows before the end of the dry season, allowed to await the first showers of rain and then, when the right stage of dryness is reached, a lighter implement used to produce a seedbed. However, in many places, once the rains break conditions are liable to remain continuously wet for some time and in such circumstances it is better to complete seedbed preparation before the end of the dry season.

The above remarks apply particularly to monsoon areas where the duration of the rainy season is no longer than that needed for the production of one crop, and where the rainfall pattern is unimodal.

Preparatory Cultivation

In places with a longer, bimodal rainy season, planting early in the rains may not be so important, and for some crops there may be considerable latitude in the time of sowing. For example, Walton (1962a) found that the optimum sowing date for cotton at Serere in Uganda was 12 May, but, on average, sowing at any date between 22 April and 26 May made little difference, and at several places in eastern Uganda the yield loss incurred by delaying planting for fifteen or thirty days after the optimum date was small (Table 6). Where two successive crops can be grown in one rainy season it is usually desirable to prepare the land and plant the second crop as speedily as possible after the first has been harvested, in order to minimize the exposure of bare soil.

Table 6. *Percentage cotton yield lost through delay in sowing after optimum planting date*

Walton (1962a)

Days delay in sowing	Location in Uganda		
	Serere	Ngetta	Usuku
15	2·5	3	3
30	9	11	9
45	22	23	24
60	40	35	45

DEPTH OF CULTIVATION

Native farmers using hoes or ox-drawn implements usually only cultivate shallowly. Tractor-drawn implements give deeper and more thorough cultivation, and it is possible that this would have certain advantages.

Investigations in the temperate zone have produced little evidence in favour of really deep tillage. In reviewing the subject Russell (1956) pointed out that the results of experiments in Britain and other parts of Europe, while not altogether concordant, generally indicated that crop yields are only increased a little, if at all, by deep ploughing and/or subsoiling. Similarly, the general experience of American investigators, in both the wheat and the corn belts, has been that crops are generally insensitive to depth of tillage, that ploughing to about 8 inches is most convenient, and that deeper ploughing or subsoiling rarely gives any benefit.

The very few investigations carried out in the tropics do not indicate any general benefit from deep tillage although it may have advantages under special circumstances. For example, Ducker and Hoyle

(1947) compared no digging, digging with hoes to 4 inches and deeper digging to 8 inches, on both a red friable loam and a heavier black soil over several years, with a rotation of crops including cotton, groundnuts and maize. Only maize showed any response to deep digging, and the increase in yield was insufficient to pay for the extra cost. The Overseas Food Corporation (1956) found no definite advantage in deep ploughing apart from the fact that it was desirable to plough to 10 inches every two or three years in order to break the pan formed by disc implements cultivating to shallower depth. Lock (1957) found no benefit from deep ploughing for sisal and Hawkins (1960) considers that mechanization rarely leads directly to an increase in yields, although it may do so occasionally by subsoiling where circumstances make the operation desirable. Occasionally, yield increases have been reported from deep ploughing; for example, with tobacco in Rhodesia (Garmany 1954). The scanty data available provide little evidence of any marked benefit from deep ploughing or subsoiling unless one or more of the following circumstances obtain: (1) presence of a definite hard pan at shallow depth; (2) need to eradicate deep-rooted, perennial weeds; (3) heavy clays are being cultivated and steps have been taken to improve drainage, as in the cambered bed system mentioned below; (4) need to deal with bulky residues of a previous crop.

TILTH AND FREEDOM FROM WEEDS IN THE SEEDBED

Generally speaking, evidence from temperate countries is that preparatory cultivation should be no more than that required to produce a satisfactory tilth and freedom from weeds in the seedbed. Furthermore, experiments have indicated the paramount importance of good weed control in the seedbed, tilth being of lesser importance, and repeated cultivations to produce a fine seedbed are often unnecessary, (Russell and Mehta 1938), Russell and Keen (1941)).

The scanty evidence from the tropics, coupled with practical experience, suggest that in these regions the above conclusions apply with equal, or greater, force. A fine tilth is commonly less necessary than in the temperate zone because many tropical crops have large seeds which germinate and grow rapidly, or are transplanted, or are propagated vegetatively. Since high intensity rainfall usually causes breaking down of clods and loss of structure in cultivated soil, it is not necessary to produce a fine tilth by tillage, especially as the additional cultivation will accentuate the tendency to loss of crumb structure.

On the other hand, weeds are usually a greater menace in the tropics than in temperate regions, since they grow more rapidly and luxur-

Preparatory Cultivation

iantly. Consequently it is important that preparatory cultivation should achieve as much weed control as possible. Dry season ploughing is of particular value for the control of relatively deep rooted perennial weeds, such as some grasses. Sethi (1951) describes how 'Kans' (*Saccharum* sp) can be dealt with by ploughing two or more times during the dry season to a depth of 12 inches, preferably with a mould-board plough, the last ploughing being done at least a month before the rains break in order to ensure thorough desiccation of the roots. Alternate ploughing and harrowing in the dry season is of value in controlling many pernicious grasses, such as *Imperata cylindrica*, *Cynodon dactylon*, *Brachiaria mutica*, *Paspalum fasiculatum*, *Digitaria scalarum*, etc. It is, of course, a method impracticable for the hoe farmer, and one of the advantages of mechanical land preparation is that it often reduces the work of weeding during the growing season.

RIDGE AND FURROW CULTIVATION: CAMBERED BEDS

On heavy soils with impeded drainage the land may be formed into broad ridges or beds, with a suitable slight grade along the ditches, or furrows, between them, in order to render drainage and aeration satisfactory for crops. A well-known method of doing this is the 'ridge and furrow' system, in which the land is ploughed by turning all furrows towards a centre line midway between ditches spaced at intervals. Several ploughings in this manner form ridges with a considerable crown.

A modification of this is the cambered bed system which has been extensively employed on heavy, ill-drained soils in parts of the West Indies for the cultivation of sugar-cane, and to a lesser extent for coconuts, cocoa and citrus. It has also been used in cultivating 'vlei' soils in East Africa, as described by Robinson (1955, 1959a and b). These beds are constructed by first marking out the lines of the drainage ditches or furrows, which are commonly from 20 to 35 feet apart. Overlap ploughing is then done, starting from the centre of the bed and working towards the drain lines. This should be effected as deeply as possible (usually to 10 or 14 inches on West Indian sugar estates), and forms an appreciable crown. Subsoiling is then carried out to a depth of 18 to 24 inches. It is essential that this be done when the clay is relatively dry as the object is to open up and shatter the soil in order to provide a deep layer of loosened, better aerated and drained soil in which roots can develop vigorously. Subsequently the drains are taken out with an implement designed to throw the soil well up towards the centre of the bed. This is usually done with the Cuthbertson drain age plough, which is a double mould-board plough with long extension arms. Finally, the beds may be finished off by grading with a terracer blade. The result is a bed with a section as indicated in Fig. 7, in which some water is shed off the cambered surface of the bed (A), and some percolates through the cultivated, free draining layer (B), to move away laterally to the drains over the cambered surface of the uncultivated, impermeable clay subsoil (C). In Trinidad the loss of the artificial structure after planting with sugar-cane is rapid in the surface 2 or 3 inches, causing a fair proportion of the rainfall to run

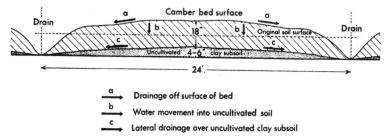

FIG. 7. Cambered bed – diagrammatic section

off the surface of the beds, but deterioration is more gradual in the remainder of the cultivated layer, in which the cane roots grow vigorously, whereas root development is much more restricted in the untilled clay immediately below.

GROWING CROPS ON RIDGES AND TIED RIDGES

In certain areas of most tropical countries native farmers customarily grow some of their crops on ridges. As a rule, ridges are chiefly used for root crops, but in some places they are traditionally employed for the majority of crops. Ridges may serve a variety of purposes. They provide a simple means of burying grass, weeds and crop residues, thus preventing the regrowth of weeds and making use of their manurial value. They are extensively used for this purpose in certain savanna areas in Malawi and Tanzania, where the farmers do not burn the grass when clearing land for cultivation, but bury it (see p. 167). By placing humic or well-manured surface soil near the roots, ridges may provide better physical and nutrient conditions in the rooting zone than flat cultivation. The harvesting of roots and groundnuts may be facilitated by ridging. In places with heavy rainfall and/or impeded drainage, ridges may be beneficial in locally improving drainage. The value of ridging and tie-ridging for soil and water conservation has already been mentioned in Chapter 5.

Preparatory Cultivation

Where they are used to improve drainage in areas subject to waterlogging, ridges naturally increase yields, but in other areas their effect on crop yields has been found to be variable. For example, the figures in Table 7 (Empire Cotton Growing Corporation, 1951–56) show that whereas ridging fairly consistently gave higher yields with several crops at Lubaga and Ukiriguru in the Lake Province of Tanzania, there was virtually no difference between ridge and flat cultivation for cotton at Namulouge, in Uganda. Walton (1962b), reviewing the results of ridge v flat trials carried out over many years in Eastern Uganda, reported that in some seasons ridges gave higher yields than flat, and in other seasons the reverse, mean yields over the years being little different. The results of a large number of trials with a variety of crops in Kenya were similarly variable, but, as illustrated by Table 8, the tendency was for ridges to be beneficial on light soils in the drier areas, although often of no advantage on loams under better rainfall.

Table 7
Effect of ridging on crop yields

Station	Soil	Mean rainfall (inches p.a.)	Period of expt. years	Crop	Mean yields (lb per acre)	
					Flat	Tie-ridges
Lubaga	Heavy, drainage impeded	31	12	Cotton	526	729
Lubaga	Light, free draining	31	9	Sorghum	734	1046
Ukiriguru	Light, free draining	33	6	Bulrush millet	618	665
Ukiriguru	Light, free draining	33	7	Cotton	616	759
Namulonge	Medium loam, drainage free	45	4	Cotton	793	782

The effect of ridging on yields may be influenced by the soil, the topography, the rainfall regime, the rainfall distribution in any given season, the crop grown and the date of planting. An obvious example of the sort of seasonal variation that may occur is that if sowing is followed by a spell of dry weather the surface of ridges will dry out quicker than on the flat, thus reducing germination, but if sowing is followed by very wet weather the improved drainage provided by the ridges may give a better germination than on the flat.

Walton (1962b and c) has explained the variable results obtained in trials comparing ridge and flat cultivation of cotton in Uganda. He showed that over a period of several months during the rainy season there was little difference in the moisture content of the top foot of soil under ridge and flat. However, after a wet spell there was a little more water under the ridge, and after a dry spell a little more under flat cultivation, the more rapid drying out of the ridged land being attributable to the greater surface area exposed. From the second to the

Table 8
Effect of ridging on crop yields

Station	Season	Crop	Yields per acre		Units
			Flat	3 ft. tie-ridges	
A. *Low rainfall, light free draining soil*					
Kampi ya Mawi	SR 1952	Beans	310	369	lb per acre, seed
Kampi ya Mawi	LR 1953	Bulrush millet	1,982	2,299	lb per acre, straw
Kampi ya Mawi	SR 1953	Beans	270	352	lb per acre, seed
Kampi ya Mawi	LR 1954	Bulrush millet	172	185	lb per acre, grain
Kampi ya Mawi	SR 1953	Sorghum	1,327	1,509*	lb per acre, grain
B. *Good rainfall, well drained loam*					
Embu	LR 1953	Maize	58·3*	33·8	lb per plot, grain
Embu	LR 1954	Maize	5378	5847*	lb per acre, grain
Kakamega	LR 1953	Maize	20·6	19·3	200 lb bags maize grain per acre

SR = short rains season.
LR = long rains season.
* Denotes significant increase.

sixth foot there was consistently more moisture under the ridges. Now it has been shown that if evaporation losses reduce the water available to the cotton plant below the minimum required to maintain normal exponential growth, this will reduce yields. If such a deficit occurs during the first eight weeks after planting, when the cotton roots are mainly in the top foot of soil, it will occur first in plants growing on ridges, and the earlier departure from exponential growth will result in lower yields than on the flat. If the deficit occurs after eight weeks, when the roots are at, or below, 2 ft it will be felt less on the ridged plots, which have more moisture from the second to the sixth foot, and will give higher yields than on the flat. Finally, if the deficit occurs at about the eighth week, when most of the roots are between 1 and 2 ft deep, then the lower moisture content in the top foot of soil under the ridge is about balanced by the greater moisture content of the second foot, hence both ridge and flat have about the same amount of water in the root zone and give similar yields.

It seems probable that similar considerations will apply to other crops, and that ridging will give better yields than flat cultivation wherever a crop is liable to experience a water deficit at a time when most of its roots have developed below the top foot of soil. This is likely to occur on freely drained soils in areas of low or moderate rainfall, where dry spells are experienced after the first few weeks of the growing season.

CULTIVATION AFTER PLANTING: WEED CONTROL

Tillage after planting does not materially conserve moisture by forming a surface soil mulch, as was formerly supposed, but is mainly beneficial by controlling weeds. Keen (1949) reported extensive investigations in Britain which showed that cultivating more frequently, or deeper, than is needed to control weeds does not increase yields but often reduces them. Many trials with maize and market-garden crops in America have also shown that maintenance of a soil mulch in the absence of weeds is of little or no importance.

Experimental evidence on the effects of frequency and depth of after-cultivation on tropical annual crops is scanty, but what there is suggests that the findings in the temperate zone probably apply, i.e. tillage after planting should be no deeper nor more frequent than is needed to control weeds. More than this only aids the rain in breaking down structure and sealing the soil surface. On certain soils that pack and seal readily under the influence of the weather, shallow cultivation may sometimes temporarily aid the infiltration of rain, but in the long run merely increases the tendency to loss of structure and sealing. If it is practicable, mulching (see p. 234–6) is a better method of dealing with the problem. The results obtained by Hofmeyer (1942) in an experiment with maize (Table 9) are typical. It will be seen that whether the rainy season was good or poor there was no advantage in frequent cultivation to maintain a soil mulch as compared with cultivating only as needed to control weeds, but that leaving the weeds to grow, or even cutting them, resulted in a marked drop in yields.

Table 9. *Maize yields as percentage of general mean yield*

Treatments	Poor rainy season 1939–40	Good rainy season 1940–41
1. Frequent cultivation after each rain to maintain 'soil mulch'	131·0	109·2
2. Clean cultivation whenever weeds appeared	139·4	103·7
3. Ridging	111·4	118·1
4. Weeds cut by hand	105·2	97·5
5. Nothing	13·0	71·5
Significant difference, $p = 0·05$	30·4	6·2

In all probability weeds are a more important limiting factor in tropical crop production than they are in temperate climates. They grow very rapidly and luxuriantly in the rainy season, competing strongly with the crop for water, nutrients and light. Poor weed control is one of the major causes of low yields, especially on small holdings. Ashby and Pfeiffier (1956) suggest that whereas yield increases from weed control in temperate climates are usually of the order of 25 per cent, in the tropics they are frequently 100 per cent.

It is especially necessary to weed early, while the crop is young. There is evidence from experiments with many crops, especially with maize, cotton and groundnuts, that crops are severely checked in their early growth stages by even a moderate cover of weeds and that this check reduces final yields. No amount of subsequent cultivation will offset this check. Conversely, if the crop is clean weeded in its early stages, quite heavy weed infestation in its later stages does not affect yields. For example, the results of Ducker and Hoyle (1947), given in Table 10, show that if weeding was delayed until thinning time, instead of being done two or three weeks earlier, there was a big drop in yield. Two subsequent weedings at a later stage in crop growth did little more than one late weeding to counteract the bad effect of the initial delay in weed control.

Native peasant farmers very commonly do not start weeding their crops early enough. Having waited for the first rains to soften the land

17 Manual preparation of land for rice-planting, Malaya

18 First ploughing of rice field with buffalo, Malaya

19 *Foreground:* uprooting rice seedlings in nursery; *background:* puddling rice field, Malaya

20 Transplanting rice, Malaya

Table 10. *Cotton yields expressed as percentage of control*

	One weeding after thinning	Two weedings after thinning	Mean
Weeded before thinning	84	100	92
Weeded at thinning	26	40	33
Mean	55	70	

before starting preparatory cultivation, they are naturally late with planting and by the time this has been completed the weeds are well away on the earliest planted part of the holding, so that the farmers are unable to catch up and do weeding at the optimum time. Mechanization has the great advantage that it makes it much easier to prepare the land early before the rains break, to plant early, and thus to weed at the appropriate early stage in crop growth. By enabling all these operations to be done more speedily at the right time, mechanization should result in increased yields.

CHEMICAL WEED CONTROL

Some of the older, non-selective weedkillers, such as sodium arsenite and sodium chlorate, have long been used to a limited extent in plantations of certain tree crops in the tropics but, while the development of the modern selective herbicides has led to their routine use on a vast scale in the temperate zone, the use of these chemicals in the tropics is still relatively small. They have mainly been employed for weed control in perennial crops in places where labour is costly; for example, in sugar in the West Indies and Hawaii, in pineapples in Hawaii and in coffee in Hawaii and Central America. Their use in annual crops is largely confined to a comparatively small number of large producers, such as the European farmers of East Africa and Rhodesia, and even there they are not applied on a scale comparable to that in the temperate zone. Peasant farmers have hitherto made negligible use of herbicides for either annual or perennial crops, partly owing to the cost of materials and spraying equipment, but mainly because they lack knowledge of the somewhat precise techniques required for their successful use.

Since labour costs are likely to rise, and herbicide costs to fall, it may be expected that in future the larger producers will make much greater use of herbicides. As research leads to safer and more efficient herbicides, they will probably also be more extensively employed on small holdings, possibly on a cooperative basis, or under schemes organized and supervised by governments.

It should be possible for chemicals to control weeds more effectively than hand or mechanical cultivation. Repeated cultivation for weed control has adverse effects on soil structure, leading to reduced water infiltration and increased risk of erosion. Moreover it has already been mentioned that it is difficult for hoe farmers to weed annual crops at the most vulnerable stage, soon after planting. Even with mechanization, the weather and other circumstances may often make it difficult or costly to control weeds really early by cultivation. The great advantage of herbicides is that they provide a means of doing this job speedily at the right time.

Herbicides are often best applied when the weeds are young, since weeds are most susceptible at this stage, particularly in the tropics where there is evidence that as growth advances they acquire tolerance to chemicals more quickly than corresponding weeds do in temperate regions (Russell, 1957). Furthermore, less chemical is required at this stage and less damage is caused to the crop by spraying equipment. Herbicides are therefore most commonly applied either in the early stages of crop growth (post-emergence spraying), or prior to the emergence of the crop above ground (pre-emergence spraying). The latter technique is of particular advantage for really early weed control and has been found to be very useful with a number of tropical crops. For example, pre-emergence application of 2,4-D or Simazine is the most effective method for weed control in maize. This obviates the difficulty often encountered when heavy rain immediately after planting makes tractor spraying impracticable.

However, a lot more research is needed to provide a sound basis for the expansion of chemical weed control in the tropics. In the first place there is a need for basic research on such matters as the ecology of the more important weeds, on the biochemistry of weeds, and on the mode of action of herbicides. Very little is known about the weeds themselves, and clearly the more that is known about the conditions that favour the growth of the various species in different types of agriculture, the greater will be the possibility of devising cheaper and more effective methods for their control.

Secondly, there is a need for much more investigation into the practical aspects of the use of herbicides. Relatively little is known about the tolerance of difference crop species to various herbicides. Some crops are so highly susceptible to damage from certain types of herbicide as to render the use of the latter impossible. For example, cotton is spectacularly damaged by even traces of 2,4-D and tobacco is severely damaged by the majority of weedkillers. Other crops, while showing susceptibility to certain chemicals, can be sprayed with those chemicals without loss provided that the herbicide is applied at a certain stage in the growth of the crop and at an appropriate concentration. For example, all cereals are to some extent susceptible to 2,4-D and MCPA, although less so than many non-gramineous weeds, but rice can be sprayed with these chemicals at concentrations adequate to kill weeds if this is done at the tillering stage, some weeks after planting. Earlier spraying causes damage and reduces yields. On the other hand, maize, which is more susceptible than the small cereals, can only be sprayed by the pre-emergence technique or when it is about 2–6 inches high, later spraying causing severe malformation of the plant. Sorghum is even more susceptible than maize to 2,4-D and MCPA applied at the wrong stage of growth (Duthie, 1957). It is therefore of the first importance to know for each crop, and probably for different varieties of a given crop, the chemicals that can be used and at what stage of growth and at what concentration they can be safely applied. The matter is further complicated by the fact that the susceptibility of both crops and weeds to many herbicides is affected by the weather at and shortly after spraying, and by differences in soil type. The effectiveness of most pre-emergent herbicides is reduced with increasing contents of clay and humus in the soil, and residual effects are least on soils that are high in organic matter, well aerated and moist. In sum, the action of herbicides is so much affected by plant species and environment that sound recommendations for their use in any locality can usually only be made on the basis of experimentation in that particular area.

The formulation of herbicides may also have an important effect on their performance, and those suitable to temperate conditions may not be the best for the tropics. This involves, not only the question of whether it is best to use certain chemicals, e.g. 2,4-D, in the form of the sodium salt, the ester or the amine, but also the use of certain additives which may help to get the active constituent into the plant tissue or to prevent it being washed off by rain.

One of the most important problems is to find chemicals to control certain species of perennial rhizomatous grasses and sedges that form some of the most pernicious weeds of the tropics, especially in areas where the absence of a marked dry season makes their control by ploughing and cultivation difficult or impossible. The selective weed-killers have little or no effect on most of these grasses: in fact, they may encourage grasses by removing the competition of non-gramineous weeds. TCA (trichloro-acetic acid) is only partially successful against these grasses, since it has to be applied at expensive heavy dosages, may have marked residual effects and does not achieve complete eradication of some species. Dalapon (dichloro-propionic acid) is more useful, but again often has to be applied at high rates, may have residual effects, and is not effective against all grasses. Aminotriazole controls some species that are unaffected by dalapon. Gramozone appears to hold promise as a general grass killer.

The use of herbicides does not reduce the need for good husbandry and proper cultivation. As Russell (1957) points out, if weedy land is the consequence of bad farming and the farmer uses herbicides without improving his standard of farming, he cannot possibly get the most out of these herbicides. In fact, he may do more harm than if he did not use any, for since each selective chemical kills its own group of weeds and does not affect others, he may merely spend money to change from one group of weeds to another. Moreover, any husbandry measures that aid in the production of a close, vigorous stand of the crop are likely to render the use of a herbicide more effective, since the herbicide only needs to check weed growth for a short time and the strongly growing crop will, itself, continue the work of suppressing weeds. On the other hand, if the crop stand is thin and patchy, then any weeds that are merely checked and not killed by the herbicide will have a good chance of recovering and making further rapid growth.

CHAPTER 7

The Maintenance of Fertility under Annual Cropping:
I – Shifting Cultivation

The General Problem of Maintaining Fertility

The chief factors contributing to soil fertility are nutrient supply, organic matter content, soil reaction, absence of injurious substances, and the physical characteristics of texture, structure, depth and nature of the soil profile. Soil depth is of particular importance in areas where the rainfall is seasonal and unreliable, since only a deep soil can hold an appreciable amount of water to supply crops in a dry spell, but it is a characteristic that can rarely be altered by the farmer. Soil acidity is not, as a rule, a major problem; most tropical soils are moderately acid, but with most crops responses to lime are only obtained on highly acid soils. Infertility associated with the presence of injurious substances is not widespread, occurring only in limited areas of saline soils with high concentrations of soluble salts, or on certain unusual soil types characterized by high concentrations of aluminium, manganese, or other metallic ions. In maintaining fertility the farmer is therefore mainly concerned with keeping the soil in a satisfactory physical condition, and with maintaining the nutrient supply. On most tropical soils it is difficult for the farmer to achieve either of these objectives if he attempts to use the land permanently for the continuous production of rain-fed annual crops.

A fertile soil must be in a physical condition that is favourable to root growth. This, as Russell (1958) has emphasized, depends upon the size and distribution of the pores, which determine the permeability of the soil to rainfall and to roots, and the ability of the soil to supply water and air to the roots. Roots can ramify freely through the soil only if there are coarse pores and cracks through which they can grow; their penetration may be limited if most of the large pores have been destroyed by incorrect cultivation or heavy consolidation. Roots absorb oxygen and produce carbon dioxide, and in order that carbon dioxide may be removed and oxygen replaced by a process of diffusion there must be a continuous system of pores containing air extending from the soil surface throughout the root zone. In a well-drained wet soil (which is a requisite for good growth) pores of a size greater than about 0·06 to 0·03 mm are full of air; smaller pores down to a size of about 0·003 mm hold water available to crops, but in still finer pores the water is held so tightly that roots are unable to absorb it. A fertile soil must have a fairly even distribution of pores of the sizes capable of holding air and water available to crops. In addition, there must be a proportion of interconnected larger pores, of a least 0·5 mm in size, if water from heavy rain storms is to percolate into the soil and not run off the surface. Except in very coarse sandy soils a system of pores larger than 0·03 mm can only exist if the soil particles are aggregated into crumbs. With most tropical soils the maintenance of this crumb structure is difficult under fairly continuous arable cropping. The impact of high intensity rainfall breaks down the crumbs on the soil surface, and the resulting fine particles block the coarse pores, forming a surface skin largely impermeable to water. This prevents gaseous diffusion between soil and air, until evaporation removes enough water to allow air to enter the pores again. Cultivation also breaks down aggregates, and exposure and disturbance of the soil accelerates the decomposition of organic matter, which, while present, aids in maintaining the porosity of the soil.

The maintenance of good crumb structure is assisted by providing a protective cover of crop vegetation, or mulch, and by minimal and timely cultivation, but under annual cropping it is generally impracticable to keep the soil fully covered throughout the rainy season, or to avoid some structure-destroying tillage. The consequent deterioration in the physical condition of the soil under cropping usually necessitates a periodic vegetative fallow to restore structure. It should be mentioned, however, that loss of structure under cultivation is comparatively slow on certain red soils in which the finer particles are bound together into rather stable aggregates by precipitated and hydrated iron oxides. On such soils there is less need of restorative fallows, and it is possible to practice fairly continuous cultivation for some years without complete loss of structure.

Although trace element deficiencies are encountered in some soils, the general problem of maintaining the nutrient supply of the soil is to keep up adequate levels of available nitrogen, phosphorus, potassium, calcium and magnesium. Unless there is heavy use of fertilizers – which the vast majority of peasant farmers in the tropics cannot afford – the maintenance of an adequate supply of these nutrients is primarily a matter of preserving a good level of soil humus, which will release mineral nutrients on decomposition. The soil organic matter is

The General Problem of Maintaining Fertility

particularly important because, in addition to containing the whole of the nitrogen reserves and part of the phosphorus reserves, it also has a high cation exchange capacity.

The organic nitrogen contained in the soil organic matter is gradually mineralized and made available to crops by micro-organisms which decompose the organic matter. In regions where the rainfall is well distributed throughout the year there is a steady slow release of nitrogen as nitrate, and the level of nitrate in the soil shows only comparatively small fluctuations, increasing somewhat during periods of moderate rainfall and decreasing, owing to greater leaching, during spells of very wet weather. On the other hand, in places experiencing a pronounced dry season there is a marked seasonal variation in the amount of nitrogen released as nitrate. During the dry season mineralization proceeds slowly and there is a gradual increase in the soil nitrate level until the moisture content falls to wilting point. The wetting of the dry soil at the beginning of the rains results in a flush of organic matter decomposition and nitrate release, possibly due to some partial sterilization effect of the previous drying of the soil (Birch, 1958). Consequently nitrate levels rise markedly at the beginning of the rains but soon fall again due to leaching. Crops planted early in the rains benefit from this flush of nitrate production, but in soils low in organic matter or mineralizable nitrogen they may later suffer from nitrogen deficiency.

Part of the soil phosphorus is that derived from the weathering of rocks and held in the soil in combination with calcium, iron and aluminium in very sparingly soluble inorganic forms, from which it is slowly released to maintain a low equilibrium concentration of available phosphorus in the soil solution. A further part is the organic phosphorus contained in the soil organic matter, which is gradually released in mineral form by microbial decomposition. The majority of tropical soils contain too little phosphorus for good yields of crops, and the organic phosphate not uncommonly forms an appreciable proportion of the total supply that may eventually become available to plants. For example, Nye and Stephens (1962) state that in most Ghana soils organic phosphorus comprises 20 to 30 per cent of the total phosphorus in the top six inches.

The soil reserves of potassium, calcium and magnesium are partly held in mineral combination, or as cations in the inner layers of the clay mineral complex, from which they are only very slowly released in a form available to plants. All the nutrient cations that are available to plants are held in the soil solution or in the cation exchange complex of the soil; that is, in the exchange positions on the soil colloids, where they compensate for negative charges. The weathering of rocks under humid tropical conditions results in the predominant clay minerals being kaolins, which have a low cation exchange capacity; consequently in many tropical soils most of the exchange positions are associated with the organic matter. The exchange capacity of the soil, or its ability to supply nutrient cations to plants, is therefore to a large extent dependent on the maintenance of soil organic matter.

Under annual cropping it is difficult to prevent a marked decline in the level of soil humus, and it is extremely difficult to achieve any considerable increase in that level by any practicable measures. In general, any addition of organic matter to the soil will increase the level of readily decomposable humus. However, raw organic matter added to the soil decomposes very rapidly and only a fraction of it is converted to humus, the remainder being lost, mainly by oxidation. Nye and Stephens (1962) state that the addition of 5 tons per acre of dry, organic mulch to each crop in fertilizer trials that ran for seven years only gave an increase of 0·08 per cent in the organic carbon content of the 0–12 inch layer of the soil. Very considerable amounts of organic matter must therefore be added to raise the humus level appreciably. It is possible to maintain a satisfactory humus level for continuous cropping over a long period by relatively heavy applications of farmyard manure. It may also be possible to do this by heavy use of fertilizers and the production of vigorous crops that leave large residues for incorporation in the soil, but this has not definitely been established. The best that can be done by most tropical farmers is to add comparatively small amounts of organic matter as farmyard manure (or mulch), to return all possible crop residues to the soil, and to limit the rate of decomposition of soil organic matter by minimizing soil disturbance and exposure. As this is generally insufficient to maintain a reasonable content of humus for any considerable period of cultivation, it is usually necessary to restore the level of soil organic matter by resorting to periodic vegetative fallows which, as we have seen, are also needed to restore soil structure.

Shifting Cultivation

The most widespread farming system whereby annual (and short-term perennial) crop production is alternated with periods of vegetative fallow is that known as shifting cultivation. As this system is practised by many different races in parts of Africa, tropical America, Asia and Oceania, it exhibits many variations, but certain essential features are found wherever it is carried on by traditional methods. Land is cleared from forest or savanna, the debris burned, crops planted on the ash-enriched soil and cropping continued for a period which may vary from one to ten years, but is usually two to four years. After this the land is abandoned to lie fallow under regenerating natural vegetation for some years, and a new area cleared for cultivation. Almost universally, reliance is placed entirely on the bush fallow for the maintenance of fertility; no fertilizers or manures are applied and, even if livestock are kept, little or no use is made of their dung.

In the past this system was operated for centuries by rather sparse populations solely for the purpose of providing subsistence. Practised in this non-intensive manner, it did little harm to the land in the closed forest zones, because fire did not spread beyond the clearings, and the fallow periods were long enough to restore the fertility of the soil nearly to its original level. In the drier woodland and savanna zones fallowing under natural regeneration was less effective in restoring fertility than the forest fallows, and some decline in fertility no doubt occurred. The spread of fires from the clearings of the cultivators also contributed to the degradation of further areas of woodland to savanna, but most of the savanna areas would have been created in any case by the uncontrolled use of fire for purposes connected with stock-keeping and hunting.

In modern times increased population density, combined with the intensive cultivation of cash crops in addition to subsistence crops, has often led to much more intensive shifting cultivation, with shorter fallows and consequent serious decline in fertility, and it is apparent that the traditional system is unsatisfactory under such conditions. However, it is important to understand the traditional system in relation to the environmental and social limitations before attempting to consider what improvements can be made. Consequently in this chapter an account is first given of the indigenous practice of shifting cultivation under the old conditions, followed by a description of its deficiencies under modern conditions. Possible modifications and alternatives are discussed in later chapters.

As shifting cultivation is practised by peoples differing greatly in culture and customs, in different climatic and vegetational zones, and on a wide range of soils, there is naturally great variation in the crops grown, cultural methods adopted, and intensity of cultivation. The principal vegetation zones in which the system is used are: (1) the lowland evergreen and semi-evergreen forest; (2) the semi-deciduous forest; (3) the woodland and savannas; and (4) the montane forest. A brief account of these types of vegetation, and of the predominant kinds of soil associated with them, has already been given in Chapters 2 and 3.

The methods and intensity of shifting cultivation are also influenced to some extent by land tenure customs and the settlement pattern adopted by the people. The settlement patterns influence the local population density and hence the local intensity of cultivation.

There are two quite distinct customary patterns. The more primitive peoples generally build temporary villages, practise shifting cultivation in the immediate vicinity of their dwellings for several years until crop yields have fallen to a certain level, and then the whole community migrates elsewhere to build a new village and open up new land. This, for example, is the custom of the forest-dwelling aboriginal Sakai in Malaya (Dobby, 1942) and used to be the habit of the Kavirondo in Kenya (Humphrey, 1947). Where this kind of movement occurs, land is usually reopened only after a prolonged period of fallow. On the other hand, rather more advanced people often live in permanent villages or towns, with their cultivated and fallow lands covering a fairly large area round about. The prolonged use of a relatively limited area naturally results in a more rapid rotation of the cultivated clearings, and fallow periods tend to become gradually shorter, especially close to the village or town. As the productivity of land in the immediate vicinity of the village declines, the distance from a man's dwelling to his main cultivated area may become considerable, and frequently temporary huts are built on the 'farm' and occupied for periods of days or weeks at a time during the growing season. This settlement pattern is common in southern Nigeria.

People who live in permanent villages often combine limited permanent cropping with shifting cultivation, each family continuously cultivating a small area near the village (maintaining its fertility by applying household refuse, ashes and, perhaps, dung from domestic animals) while farming a larger area, more remote from the village, by shifting methods. A similar combination is also commonly found among people whose staple food is derived from a perennial, such as the banana. For example, the Baganda, in Uganda, customarily live more or less permanently on the same land, each family maintaining near its dwelling a permanent banana garden, which is manured with household refuse and mulched with banana leaves and stems, and with grass that has been spread on the floors of the huts. In addition, each family produces other food and cash crops by shifting cultivation, rotating its clearings around the clan lands (Tothill, 1940).

Most commonly the land is owned communally by the clan or tribe and the control powers are vested in a chief, who exercises them on the advice of a council of elders. Individuals do not own land but have user rights that are permanent so long as they continue to use the land and often, but not always, extend to fallow land. Often the head of a family allocates the use of different parts of his holding permanently to his several wives. User rights are normally heritable, and in polygamous societies when the head of a family dies his sons usually inherit, and divide between them, the use of the land previously cultivated by their mothers. In some places, however, it is, or was, customary for user rights to be periodically reallocated between members of the clan by the chief.

Under customary communal land tenure, especially among more primitive people, some degree of beneficial control of the use of land and natural resources was often exercised by the chiefs and elders, but in many places the extent of these controls has markedly declined in modern times. An individual had to obtain permission before clearing virgin land, and felling might be prohibited in certain areas that were used for hunting, or as a source of building poles, fuel and wild food plants. Cultivation might be forbidden near salt licks, watering places or along river banks. In some tribes the time and extent of burning for hunting purposes was controlled. Where people lived in temporary villages the chief and his advisers decided when the fertility of land in cultivation had declined to a level requiring cessation of cropping, and selected new land to be opened. Almost universally, care was taken in the allocation of land by the tribal authorities or the head of a family to ensure that each holding was representative of land of all types, good, poor and indifferent. In undulating country, for example, this would usually result in each family receiving a strip of land running from the crest of a hill to the bottom of the valley as the main part of its holding and, perhaps, in addition separate portions of any rich alluvium, or of seasonally swampy land.

The Fallow Period

During the resting period fertility is restored to an extent depending on the length of the fallow, the nature of the vegetation and the soil. Nutrients are taken up by the fallow vegetation from a varying depth of soil according to its root range. Part of the nutrients is stored in the vegetation (which may also accumulate a certain amount of nitrogen by symbiotic fixation), and part is returned to the surface soil by rain wash from leaves and twigs, by litter and timber fall, and in the form of dead roots and root exudates. The soil humus is increased during the fallow period, chiefly as a result of the fall of litter. Litter falls continuously throughout the fallow period under regenerating forest vegetation, but there is no big accumulation on the soil surface because it decomposes at a high rate. In the savanna the much smaller amount of litter deposited by the predominantly grass vegetation is mostly destroyed by the annual fires. In the process of decomposition the greater part of the organic matter in the litter is lost through oxidation, but a fraction is broken down to humus and incorporated in the soil. As the humus itself also undergoes decomposition, the amount in the soil does not build up indefinitely during the fallow but tends to rise to an equilibrium level – dependent on the soil, vegetation and climate – at which its rate of decomposition is roughly equivalent to the rate of addition of humified organic matter to the soil from the litter. The total nitrogen content of the surface soil rises with the increase in humus and may also be enhanced by non-symbiotic fixation. During the fallow period there will also be some losses from the soil-vegetation system as a result of run-off and erosion, leaching and denitrification.

The amounts of nutrients stored in the vegetation during the fallow, the quantities returned to the soil, the losses from the soil-vegetation system and the net amounts stored in the surface soil, vary with the soil and the nature and age of the fallow vegetation. In particular, there are marked differences between fallows in the forest and savanna zones. Nye and Greenland (1960) have presented approximate figures for the quantities of nutrients stored in the vegetation, and present in the soil, under various types of fallow.

THE FOREST ZONES

In the forest zones the short cropping period, the native practice of leaving the stumps and roots of trees undisturbed after felling and the prevailing high rainfall and temperatures, all make for very rapid regeneration after cultivation is abandoned, and large amounts of nutrients are quickly accumulated in the secondary forest growth. The rate of accumulation of nutrients in the vegetation is greater in the early years of the fallow than it is later, mainly because the leaves and twigs, which are the parts richest in nutrients, quickly reach their maximum amount, and thereafter storage occurs more slowly in stems and roots. During a fallow of twenty years or more, in the lowland forest areas, a very considerable growth of secondary forest occurs and large amounts of nutrients are stored in the vegetation.

The relative importance of litter fall, timber fall and rainwash in returning nutrients from the fallow vegetation to the soil is indicated by figures obtained by Nye and Greenland (1960) for high forest, forty years old, at Kade, Ghana (Table 11). The major contribution of organic matter, nitrogen, calcium and magnesium to the surface soil comes from the leaf and twig litter-fall, which is estimated at between 9,000 and 10,000 lb per acre per annum in moist evergreen and deciduous forest, this high rate of litter production being attained early in the fallow period. In older forest fallows there is a considerable annual fall of branches and stems, estimated at about 10,000 lb of dry wood per acre per annum at Kade, but compared with litter this contains only small quantities of nutrients, except for calcium and, perhaps, phosphorus. In addition Nye and Greenland estimate that, in mature forest fallows, dead roots, root slough and root exudates also add to the soil about 5,000 lb of dry matter per acre per annum. There is very little information on the amounts of nutrients washed from leaves by rain in inland tropical fallows, but the Kade figures indicate that rainwash contributes remarkably large quantities of potassium and significant amounts of phosphorus.

Table 11. *Nutrients returned to soil from vegetation in high forest fallow at Kade, Ghana*

Nye and Greenland (1960)

Source	lb oven dry material per acre p.a.	Nutrient elements, lb/acre/annum				
		N	P	K	Ca	Mg
Rainwash from leaves		11	3·3	196	26	16
Timber fall	10,000	32	2·6	5	73	7
Litter fall	9,400	178	6·5	61	184	40
Total addition to soil surface		221	12·4	262	283	63

Although a high proportion of the total root weight of most forest trees occurs at shallow depth, soil analyses made at different stages in the crop-fallow cycle have shown that subsoil feeding by the fallow vegetation has an important effect in transferring nutrients from the subsoil, via the vegetation and the litter fall (and in the ash, after burning) to the surface horizons. This counteracts the losses suffered by the surface soil from crop removal and leaching during the cropping period. The transfer of nutrients from subsoil to top-soil probably does not begin until after the first year or two of fallow; during this initial period the top-soil is further depleted by leaching and by the uptake of nutrients by the regenerating vegetation, which has few active roots in the subsoil at this stage. Figures obtained by Popenoe (1959) for the nutrient content of various soil horizons under regenerating forest fallow in Guatemala showed an initial depletion of the surface soil, after which the content of potassium, calcium and magnesium rose in the top 2 inches, and fell in the layers below, as the vegetation developed.

Once a good growth of trees has developed in the fallow it is probable that leaching of nutrients through the soil is fairly well balanced by the uptake of nutrients by plants and their subsequent storage in the vegetation, or return to the soil in rainwash, litter and timber fall, so that a closed cycle of nutrients exists. Nye and Greenland (1960) have pointed out that losses to the soil-vegetation system by leaching will only occur if there is percolation of water through the profile to beyond the root range, and provided that cations are held in the soil solution by balancing anions, the nitrate ion being the quantitatively most important anion in the slightly-acid to acid forest soils. Under fallows in the deciduous and semi-deciduous forest zones (rainfall 45–90 inches), although there will be relatively large amounts of nitrate in the soil, leaching will be restricted both by the uptake of anions by the vegetation, and because the transpiration of a large proportion of the rainfall will reduce through-percolation. However, some through-percolation is likely to occur where the rainfall exceeds 60 inches. In the wetter evergreen forest zone where precipitation (over 80 inches) is considerably in excess of transpiration, through-percolation is bound to occur, but leaching losses will be restricted, not only by the absorption of anions by the vegetation but also by low nitrate levels in the soil, since nitrification will be limited by high acidity. Thus, once a forest fallow has developed, leaching losses are probably not very great in most places, but they are more considerable in the early stages of the fallow, before the tree root systems have developed and penetrated to any considerable depth. Similarly, although there may be some loss of nutrients by erosion in the earliest stages of the fallow, once the regenerating vegetation has developed a full canopy and a litter layer to protect the soil surface, such losses will be slight (except, perhaps, on very steep land).

The above observations refer to the effects of a strong growth of secondary forest, such as develops when a prolonged fallow follows a short period of cropping. If the cropping period is prolonged, or if the land is subjected to frequent cropping with only short intervening fallows, then the vegetation may become more or less permanently degraded to coarse grass with scattered shrubs. A fallow under this kind of regeneration will not have the same effect as a forest fallow but will produce results similar to those of fallowing in the savanna.

THE SAVANNA ZONES

In the woodland and savanna zones the amounts of nutrients stored in the fallow vegetation, and the organic matter and nitrogen content of the soil after fallow, are much less than in the forest zone. Most of the woodland and savanna regions have a comparatively low rainfall and experience a marked dry season. Consequently, after cultivation is abandoned, and to a lesser extent after each annual burning in the dry season, the grass may be slow to form a cover and its annual production of dry matter may be limited by lack of moisture. Except in the tall grass savannas on the margins of the closed forest, grasses and herbs store only relatively small quantities of nutrients, and as a rule the amount of nutrients stored in the fallow vegetation depends largely on the extent to which trees and shrubs are present. The amount of woody vegetation present varies greatly, but is always much less than in a forest fallow, especially as the growth of trees is slow, being limited by the low rainfall, annually checked by fires, and probably retarded by low levels of available nitrogen in the soil induced by the grasses. Almost all the organic matter and nitrogen contained in the leaves and stems of the grasses and in the litter from the trees is destroyed by the annual fires. Nye and Greenland (1960) estimate that in the moist high-grass savanna of Ghana there is an annual contribution to the soil of about 2,700 lb of dry matter per acre from root material, but the amount might be appreciably less in the drier savannas.

The extent to which fallows in the savanna zones transfer nutrients from the subsoil to the surface soil is probably rather variable. Nye and Bertheux (1957) have shown in Ghana that the accumulation of phosphorus in the surface soil is less marked on the savanna than in the forest, and it may be that in areas with only a comparatively slight dry season the savanna grasses are predominantly surface rooting. On the other hand, in parts of East Africa experiencing a pronounced dry season it has been observed that many of the grasses in savannas explore the soil to considerable depth, and under these conditions it would appear that they can effect a considerable transfer of nutrients from subsoil to top-soil. Moreover, the number of trees (whose roots would certainly be expected to explore the subsoil) that occur in savanna fallows is very variable.

Despite the fact that slopes are generally less steep in long-established savanna than they are in forest, erosion losses during the fallow period are probably greater in the savannas. Erosion is slight on well-grassed, moderate slopes, but when land is abandoned after cropping, and again annually after the dry season fires, the grass takes time to form a full cover, so that each year much of the soil is exposed to the direct impact of the early rains. In the drier savannas the grass often fails to cover more than a fraction of the soil surface even when it has attained its peak growth and, despite the low rainfall, a few heavy storms can cause much erosion. Furthermore, steep slopes are by no means uncommon in savanna country, especially where the grass-lands have been recently derived from deciduous forest or from montane evergreen forest.

Leaching losses are probably comparatively slight in most of the savannas, partly on account of the lower rainfall and large amounts of water transpired by the grasses, and partly because the grass cover represses nitrification. The consequence is that, in the absence of any considerable quantities of nitrate anions, the concentration of cations in the soil solution is low.

Clearing and Burning

CLEARING

Under traditional native practice in the forest zone the length of the fallow period is usually between ten and twenty years, but may be longer in places where the population is sparse. Farmers normally prefer to clear good secondary growth rather than virgin forest because the latter requires considerably more labour for felling and clearing, takes longer to dry and is more difficult to burn thoroughly. Native methods of clearing and land preparation in the forest zones have already been described in Chapter 6; they involve little or no soil disturbance by cultivation.

The duration of the fallow in the woodlands and savannas varies greatly. When the vegetation is mainly grass, the land is usually fallowed for less than ten years but in woodland areas, such as the 'miombo', it may be between ten and twenty years. The trees are felled and burned with the grass during the dry season. The stumps and roots of the trees are not eradicated and large trees may be left standing, but usually almost all are felled. In the 'chitemene' system, practised on poor sandy soils in Zambia, branches lopped from trees on surrounding land are piled and burned along with the trees felled on the clearing. After burning the land is usually hoed to get rid of grass roots and sometimes the soil may be scraped into mounds or ridges before planting. Although burning is usual there are some places in predominantly grass savannas where the grass is not burned but is slashed or hoed and buried in mounds or ridges (see p. 167).

BURNING

With the exceptions just mentioned, burning after clearing is the general rule, and is usually essential, if only because it is the only practicable method of freeing the land of debris before cropping. An account of the effects of burning on the soil has already been given in Chapter 6. It will have been noted that, although it has an important value for the shifting cultivator in temporarily enriching the surface soil in phosphate and exchangeable cations, it has an adverse effect on fertility in the long run, since part of the cations added by the ash are lost by leaching and most of the organic carbon, nitrogen and sulphur contained in the vegetation and its litter are lost to the soil.

21 Threshing rice by hand, Malaya

22 Rice fields lying fallow in the off-season, Malaya

23 Vegetable growing in rice fields during the off-season, Malaya

24 Mixed tree culture, Malaya

PHYSICAL CONDITION OF THE SOIL AFTER CLEARING AND BURNING THE FALLOW

In the forest zone, assuming that there has been a long fallow with a strong regeneration of shrubs and trees, the physical condition of the surface soil, and to a lesser extent that of the subsoil, will be good immediately after clearing and burning. The fresh organic matter of the forest litter will have been wholly, or largely, destroyed by the burn, but some of the partially decomposed organic matter immediately below will remain, and no appreciable part of the humus in the top two or three inches of the soil proper will have been lost. The condition of the deeper, less permeable horizons will have been improved by the tree roots, which open up the soil while living and leave channels after their death and decay.

The physical condition of the soil is also improved after a predominantly grass fallow in the savanna zone, but not as much as under the forest. Owing to the nature of the vegetation and the occurrence of annual fires, litter deposition will have been much less, and the humus content of surface soil correspondingly lower, than in the forest. In addition, the improvement of structure will have been annually checked by the exposure of the more or less bare soil to the early rains after the burn during the dry season.

The Cropping Period

CROPS AND CULTURAL METHODS

As might be expected, the crops grown and the cultural methods adopted vary widely, being influenced by a variety of factors, among which the most important are climate – especially the amount and distribution of the annual rainfall – soil, and the customs and dietary habits of the people.

One almost universal feature is mixed cropping. It is rare to see any considerable part of a clearing planted with a single crop. There may, however, be a high proportion of one crop: for example, finger millet (*Eleusine coracana*) as a main crop may be interplanted with lesser amounts of sorghum and beans. The main reason for the practice of mixed cropping is probably that it enables the maximum returns to be obtained for the minimum effort. The more complete cover provided by a mixture of crops shades weeds and checks their growth, thus reducing the labour of weeding. It is quite probable that growing a mixture of several crops, which have different growth habits and root systems and make different demands on the soil, may enable the best use to be made of soil, light and rainfall, and thus provide the maximum return from the land cultivated. Mixed cropping also provides an 'insurance policy' as some crops are likely to give a fair return even if bad weather, pests or diseases cause the partial or total failure of others.

Owing to the excellent physical condition of the soil after clearing and burning a forest fallow, little preparatory cultivation is needed. In some places none is attempted, the seeds merely being scattered on the surface of the friable, ash-enriched soil; but more commonly seeds are dibbled into holes made with a digging-stick or a hoe, without any cultivation of the soil between the planting holes. Less frequently a superficial scratching may be done with hoes and in some areas where roots are planted as the first crop, ridges or mounds may be made.

The commonest practice in the forest zone is to plant a cereal as the first main crop, usually with some admixture of other crops, such as pulses and vegetables; then to interplant with annual, or semi-perennial, roots, and subsequently to interplant again with perennials such as bananas. The cereals and other minor annual crops are harvested in the first year and the roots mainly in the second year, although some, such as eddoes and short-term varieties of cassava, may be dug in the first year and others, such as longer-term cassava varieties, left until the third year or even later. Harvesting of perennials starts in the second year and continues in the third, after which cultivation usually ceases, but perennials and semi-perennial root crops may continue to be harvested for several years after the clearing has been abandoned to natural regeneration.

The above sequence is followed in the forest zone of Ghana as described by Nye and Greenland (1960). Maize is sown as soon as the rains break and is almost immediately interplanted with relatively small amounts of other crops, such as pepper, okra, spinach and other vegetables. Later, either during the growth of the maize or shortly after it has been harvested, the farmer plants root crops such as cassava (*Manihot utilissima*), yams (*Dioscorea* spp), coco-yams, eddoes, or dasheen (*Colocasia antiquorum*), tannia (*Xanthosoma sagittifolium*), and bananas. During the second year some of the bananas and roots are harvested, while the remainder are left to be harvested later as needed, or, if not required, may be abandoned in the regenerating forest. Very similar procedures are followed elsewhere in West Africa. For example, Jurion and Henry (1951) report that, in the Congo, rice or maize or a mixture of the two is first planted, followed by interplanting of cassava and bananas, and the harvesting follows the sequence described above. Again, in Sierra Leone (Waldock *et al.*, 1951), rice is the first crop and is subsequently interplanted with other annuals and cassava.

Similar patterns are found in many parts of South-East Asia. Thus Conklin (1957), describing Hanunoo shifting cultivation in the Philippines, states that hill rice or maize, or both, is first planted at relatively wide spacing, it being common to mix small quantities of seed of other crops (such as pigeon pea, melon, cucumber, millet and sorghum) with the rice seed so that plants of these other species are evenly distributed through the rice. After harvesting the cereals, roots are planted – sweet potatoes, yams, tannia and cassava – again interplanted with pigeon pea and other vegetables. Finally, bananas and, to a lesser extent, fruit trees or other perennials of value (such as bamboo or manilla hemp) are planted, many of which may continue to be utilized for some years after the land has been abandoned to natural regeneration. At some stages during the cultivation of a clearing as many as fifty crops may be found growing together. Similarly, Terra (1958) mentions that in parts of Indonesia where shifting cultivation is practised, rice or maize is usually planted in the first year (but sometimes *Coix lachryma-jobi* or *Eleusine coracana*), followed by root crops and bananas interplanted with pulses and other vegetables, and later by fruit trees, which will be used for some time after cultivation ceases. Similar cropping systems are reported by Barrau (1959) from Pacific Islands.

In South-East Asia, limited and sporadic shifting cultivation is often practised by farmers as an adjunct to the permanent cultivation of swamp rice, which forms their main farming activity. Despite their skill in the more advanced techniques of swamp rice growing, these farmers adopt primitive, and often damaging, methods of shifting cultivation. In Malaya, for example, secondary forest on poor soils and steep slopes is cleared and burned during a dry (or less wet) season and maize planted at wide spacing on the resumption of the rains. The main crop of hill rice is not sown between the maize until a month or six weeks later, because if it were planted earlier it would reach maturity at a time when the weather is too wet for ripening and harvesting. Consequently a good deal of erosion occurs from the bare soil between the maize before the rice is planted. As a rule only one crop of maize and rice is taken, and the land is then abandoned.

It will thus be seen that in most places one cereal crop is taken, followed by roots and perennials, over a cropping period of about three years, during which there is little tillage, since weeds are mainly controlled by slashing. In some places, however, there is a departure from this procedure and another annual crop is taken in the second year, usually, but not invariably, a cereal. This often necessitates a second clearing and burning, which destroys regenerating seedlings and weakens or destroys suckers of trees and shrubs, so that regrowth of the vegetation does not occur rapidly when the land is subsequently fallowed. It also encourages the invasion and development of grasses, which have to be controlled by hoeing during the growth of the second crop. The additional cultivation, the greater exposure of the soil during the cropping period, and the slower regrowth of the fallow vegetation, all tend to increase erosion losses on sloping land.

In the savannas and woodlands there is usually more preparatory cultivation than in the forest zone because the soil is not left in as good a physical condition after clearing and burning the fallow, and on account of the need to get rid of the grass root-stocks. The land is usually dug with a hoe to a depth of about four inches before planting the first crop and this procedure is repeated each season before replanting at the beginning of the rains. Mounds or ridges are often made if a root crop is to be grown, or for the purpose of burying grass if burning has not been done after clearing. Further hoeing is usually carried out during the growth of each crop to control weeds.

In contrast to the forest zone, cereals are by no means invariably the first crop. Where the dominant vegetation is grass, very little nitrate occurs in the soil during the fallow, and after clearing, a low level of available nitrogen persists for a year or more until the rate of mineralization of nitrogen improves. The quick-growing cereals are sensitive to nitrogen deficiency since they need to take up relatively large amounts of this element during a short period of rapid growth. Consequently, where grass has been the main component of the fallow vegetation, and especially where the grass is buried instead of burnt after clearing, legumes or roots are usually the first crop to be planted. On the other hand, a cereal is commonly the first crop where predominantly woody vegetation has been felled and burned, as for example, in the 'chitemene' system mentioned below.

Cropping continues for a period which may be anything from two to ten years, but is usually three or four years, during which a sequence of cereals, legumes and roots is grown, with lesser amounts of interplanted vegetables and other minor crops. Very commonly, pigeon pea or cassava is planted at the end of the cropping period, weeding being discontinued after their establishment, and the crop subsequently harvested during the first year or two of fallow regeneration. The forest zone practice of growing perennial crops and harvesting them after abandoning the land is ruled out by the prevalence of annual fires. Crops and cultural methods vary greatly and only a few illustrative examples can be given here.

In the wetter savanna zones of southern Nigeria and Ghana yams are usually planted first on large, well-prepared ridges or mounds at the beginning of the rains, and a variety of other crops, such as maize, beans and various vegetables, may be sown shortly afterwards on the sides of the ridges. In the second year maize and sorghum are planted on the remains of the yam ridges and interplanted with other, minor crops. In the third year the main crops are commonly groundnuts and millet and thereafter the land may be abandoned, or cassava may be planted in a fourth year.

In the 'chitemene' system practised on poor sandy soils in Zambia (Northern Rhodesia: Trapnell and Clothier, 1937; Trapnell, 1943), in which loppings from trees on a surrounding area are carried in and burned along with the felled vegetation on the cleared land, finger millet (*Eleusine coracana*) is the first crop and is broadcast without any preparatory cultivation, or with only very shallow scratching of the surface soil. A little sorghum and sesame may be mixed with the millet, and small amounts of other crops, including beans, sweet potatoes and cassava, planted around the edge of the millet. Thereafter the land may be abandoned, or cropping may continue for a further two or three years with groundnuts and beans in the second year, sorghum in the third year and perhaps cassava in the fourth year.

In parts of Malawi and southern Tanzania, on land where the fallow vegetation is predominantly grass with only scattered trees, a system is followed in which the grass is not burned after clearing but is hoed up and buried in mounds towards the end of the rainy season. A crop of beans or cowpeas is grown on these mounds on the tail end of the rains. During the dry season the scattered trees are felled and burned. At the beginning of the following rains the mounds are cleaned up and planted with finger millet or with maize and beans, the same crops being sown on the burnt patches. Fairly early in the growing season weeds are hoed out and buried in new mounds between the standing crop, and these mounds are then planted with groundnuts, Bambara groundnuts, beans or cowpeas. These new mounds are not destroyed when the legumes are harvested, and at the beginning of the following rains they are planted with maize or another cereal, and a quick-maturing, climbing variety of beans. Then, as in the previous season, weeding is done and subsidiary mounds containing buried weeds are again formed and are planted with legumes. This procedure is repeated for several years so that a rotation takes place between spots 'A' and 'B' in the field; the 'A' mounds containing buried grass and growing a legume in one year, becoming the 'B' mounds, fertilized with decomposed grass and growing maize in the following year. After several years the whole of the land may be planted to cassava or pigeon pea, after which it is allowed to revert to fallow. Thus, the system (1) minimizes the amount of preparatory cultivation to be done each year in the hot weather shortly before the rains break; (2) recognizes that cereals will not do well on unburnt, newly cleared land; (3) makes use of the manurial value of decomposed weeds.

It will be seen that shifting cultivators, both in the forest and savanna, follow fairly regular crop sequences that are adapted to local conditions. These rotations and the cultural methods accompanying them are primarily determined by the climate, the type of natural vegetation and the dietary habits of the people, but they are also influenced by other considerations.

First, the inherent fertility of the soil will affect the crops grown and the duration of cultivation. For example, on poor, sandy soils cassava will be grown rather than yams, millet may be more satisfactory than maize or sorghum and the land will be cultivated for only one to three years, whereas more fertile soil will grow a greater variety of crops and can be cropped longer. Crop sequences are also related to the declining fertility of the soil during cultivation; certain crops, such as finger millet, being found to benefit especially from first year land while others, such as cassava and pigeon pea, will give a fair crop on relatively impoverished soil and are consequently taken as the last crop. The effect of a deficiency of available nitrogen in savanna land on the selection of the first crop has already been mentioned.

Secondly, the characteristics of certain crops will affect their place in the sequence and their association with other crops in mixtures. Yams, being a long term crop and demanding good preparatory cultivation, are often a first crop, and succeeding crops are planted on the yam ridges with little or no further cultivation. Maize, which responds well to fertile soil, is usually taken early in the rotation; sorghum may be planted at any time but is commonly a second crop. Two consecutive crops of groundnut are rarely taken, since the second crop is almost invariably poor. Cassava and pigeon pea require little weeding once established, as well as being capable of giving a crop on relatively impoverished soil, and are usually grown at the end of the rotation.

Thirdly, crop sequences and mixtures have undoubtedly been chosen in order to facilitate weed control. In the forest, provided that he starts with a clean seed bed as a result of a good burn, the farmer has little difficulty in controlling herbaceous weeds in the first cereal crop. Subsequently the shade of the mixture of roots and semi-perennials reduces weed growth and, as these crops are also less affected by weed competition than cereals, the farmer is able to maintain adequate weed control for a further period of two years simply by slashing. As already mentioned, if the farmer departs from the traditional system to grow a second cereal crop, weed control becomes more difficult. In the savanna, crop sequences and mixtures may also be chosen with a view to minimizing the work of weed control, but here weeds are more troublesome and it often happens that the farmer fails to deal with them early enough to prevent adverse effects on his crops (see p. 164). Some rhizomatous grasses such as *Imperata cylindrica* or *Digitaria scalarum*, are very difficult to eradicate with the hoe and become increasingly troublesome as cropping is prolonged, especially in the wetter savannas. In the drier savannas weeds are generally less troublesome, although in some places *Striga* spp, which are parasitic on some cereals, can be serious.

Fourthly, the customary crop sequences and combinations were chosen with a view to spreading labour requirements evenly through the cropping season, as well as minimizing the total effort required. For example, in the Malawi system already described, the preparation of subsidiary mounds during the previous rains reduces the work of preparatory cultivation during the busy time at the onset of the wet season; early maturing beans are planted after, and harvested before, the maize; new mounds are prepared in conjunction with weeding, and the late-sown legumes on them are harvested after the maize.

While shifting cultivators normally tend to adopt some locally adapted pattern of crop sequence, this does not usually constitute a well-organized rotation that is rigidly followed. Some tribes, it is true, may have a rather highly organized system. For example, de Schlippe (1956) reports that the holdings of the Azande people are composed of a number of 'field types'. Each field type contains a particular mixture, or within-season sequence, of crops suited to its fertility (depending upon the time it has been in cultivation), and each is managed on a standard plan under which field operations are timed and conducted so as to minimize the labour involved and to distribute it as evenly as possible throughout the year. As a rule, however, the operations of the farmer are more in the nature of a traditional procedure which is flexible and subject to variations. Deviations are inevitable because crop sequences, and the areas under different crops, are affected by a number of factors over which the farmer has no control. These include the weather, the incidence of pests and diseases and the family labour supply, which may be reduced by illness, poor food supply, attendance at mourning ceremonies for relatives, or demands from the chief for participation in communal tasks.

The Decline in Yields and Fertility

It is common knowledge that yields fall in successive seasons during the cropping period, but the rate of this decline varies greatly, depending on the soil, climate, topography, crops grown and cultural methods. In the moist evergreen and semi-evergreen forest zones, where rainfall is heavy and most soils are highly leached, yields fall rapidly. The decline is slower on the better soils, and under the lower rainfall, of the semi-deciduous and deciduous forest zones. Provided that erosion is not serious, yields usually fall fairly slowly after highland forest fallows, probably chiefly on account of the high organic matter content and good structure of the soils in this zone, but in some places deficiency of phosphate may lead to a more rapid decline. In the savannas the rate of decline is certainly slower than in the moist evergreen forests, but is very variable, depending on soil and rainfall; the latter severely limits crop growth in many areas, and thus the removal of soil nutrients in the crops.

Under traditional non-intensive methods of shifting cultivation it is unlikely that a build-up of pests and diseases is an important cause of the decline in yields. Where population pressure has led to reduction in the length of the fallow or to the cultivation of the same crop on the same land in successive seasons, and especially where large areas of cash crops, such as maize, cotton or tobacco, are grown in pure stand, a build-up of pests or diseases may certainly occur locally, but it cannot be regarded as an important general cause of decline in yields.

From what has already been said about weed control under native practice in the forest zone it is evidently unlikely that weeds are an important cause of declining yields during the customary short cropping period, although they may become a significant contributory cause if annual crops are grown in two or more successive seasons. In the wetter savanna areas the invasion of cultivated lands by certain grasses that are difficult to control with hand tools (such as *Imperata cylindrica*) is certainly a contributory cause to declining yields. Similarly, increasing weed infestation is also a factor in yield decline in the drier savannas, because of the common failure of farmers to control weeds early enough in the cropping season each year. However, the fall in crop yields during the period of cultivation appears to be mainly due to a decline in soil fertility, which may result from a deterioration in its physical condition, or from the loss of nutrients by erosion, leaching and crop removal, or from a combination of these causes.

THE PHYSICAL CONDITION OF THE SOIL

Under customary native practice in the forest zone there is little preparatory tillage that might break down the good structure obtaining in the surface soil immediately after clearing and burning the fallow vegetation. However, until the first crop provides a cover, the bare soil will be exposed to sun and rain, and some loss of structure will occur as a result of the battering of raindrops and of the rather rapid decomposition of organic matter in the surface layer. Once the first crop is fully established, a protective cover is thereafter maintained by the customary sequence of mixed cropping with roots and semi-perennial crops; as there is little or no tillage for weed control, any further decline in soil structure will be slight. If, however, the soil is cultivated and weeded with the hoe and a succession of annual crops is grown, there will be a more marked deterioration in structure.

The improvement in structure from fallowing in the savanna is less than in the forest and is soon lost, because the land receives much more preparatory tillage and is subsequently repeatedly cultivated for weed control and in preparation for successive annual crops. Pereira *et al.* (1954) showed that the improved structure resulting from three years' rest under vigorous planted and unburned grasses largely disappeared during the first year of cropping. It is unlikely that any greater or less transient benefit to structure would result from most natural savanna fallows. Although a natural fallow may last longer than three years, the grass cover is usually sparser than that of good planted grass, and there are adverse effects from exposure of the soil to sun and rain after each annual burn. Following upon the rapid loss of the improvement conferred by the fallow, there is probably a further slow deterioration in the physical condition of the soil, but there is no good evidence of the extent of this decline on different soil types, nor of its effect, if any, on yields. The general impression of agriculturists is that on most soils yields gradually fall even if fertilizers are applied, and that this is, in part, due to loss of soil structure, but it has not been definitely established that yields cannot be sustained if nutrient supplies are well maintained with fertilizers.

EROSION

In the evergreen and semi-deciduous lowland forest zone some run-off and erosion will occur during the period between burning and the establishment of an effective cover by the first crop, especially on steep land. If the land is subsequently farmed by traditional native methods, the maintenance of a full protective cover of crops, the absence of tillage and the presence of tree roots (which will check gully formation) will all limit erosion. Furthermore, as the plots are usually small and scattered, any soil washed from them is caught and held in surrounding forested areas. Consequently, under traditional practice on moderate slopes, there will be no great amount of erosion during the cropping period. But if the land is steep or more intensive cropping is practised – such as the cultivation of two or more annual crops in succession – then erosion may be much more serious. In the montane forest zone, although the soils are usually inherently of good physical constitution, erosion tends to be greater because of the steeper slopes and the tendency to grow a succession of annual crops.

In general, erosion is probably more serious under native methods of shifting cultivation in the savanna zone than in the lowland forests. The preparatory cultivation before each season's crop and the hoeing to control weeds during the growth of the crop tend to destroy aggregates and seal the soil surface, resulting in run-off. The cropping practice does not maintain a continuous protective cover and, in particular, the common failure to prepare land and plant early lengthens the period during which bare soil is exposed at the beginning of each rainy season. Although large tracts of savanna land possess gentle slopes that are not highly susceptible to erosion, such areas are often characterized by the presence of a hard ironstone layer at shallow depth, so that relatively little erosion can markedly reduce the productivity of the land. Much savanna land, however, has fairly steep slopes, or consists of long moderate slopes surmounted by steep, rocky hill-tops, from which the run-off water flows on to the lower slopes at high velocity.

CHANGES IN THE NUTRIENT STATUS OF THE SOIL

Nye and Greenland (1960) discuss very fully the changes in the nutrient status of the soil in the forest and savanna zones that occur during cropping, and the following brief account draws heavily on their findings.

Humus. There has been a common impression among agriculturists that the humus content of tropical soils declines very rapidly under cultivation, but Nye and Greenland point out that actual measurements from a variety of sources show that the rate of decline is not so great that loss of humus necessarily becomes a serious problem, even after a cropping period of ten years. With increasing time of cultivation the humus content gradually falls to some equilibrium level, dependent on the type of soil and the crops grown, the decline becoming more gradual after the first year or two of cultivation. For forest soils cropped by traditional methods they estimate the rate of decomposition to be about 3 per cent of the total humus content per annum, which is roughly equivalent to the rate of increase under fallow, estimated at between 2 and 5 per cent per annum. Hence a series of short cropping periods, alternating with longer fallows, is unlikely to cause appreciable depletion of humus. The rate of humus decomposition in cropped savanna soils is estimated at 4 per cent per annum, which exceeds the estimated rates of 0·5 to 1·2 per cent per annum increase under fallow, so that, if the land is subjected to approximately equal periods of crop and fallow, it may be expected that humus content will fall markedly.

Nitrogen. In the acid soils of moist evergreen forest zones the rate of release of nitrogen from the soil organic matter is low during the fallow period but is markedly increased by the rise in pH following burning. According to Nye and Greenland, the pH of the top 6 inches of soil does not fall below 5·0 even after eight years under cultivation and it therefore appears unlikely that any significant decline in the rate of mineralization occurs in a two to three year cropping period. In the neutral to slightly acid soils of the moist semi-deciduous forest zone, nitrification proceeds rapidly both under fallow and crop, but these soils have high total nitrogen contents after fallowing. Even after several years of cropping there is no evidence of a significant decline in soil nitrate levels, although Vine (1953) found at Ibadan, in Nigeria, that the levels did fall significantly after eight to ten years. Losses of nitrate occur by leaching during the cropping period, especially in the wetter forest regions, and particularly during the period before a full crop is established. Nevertheless, nitrogen responses are rare on forest soils during the normal cropping period.

In the savannas very little nitrate nitrogen occurs in the soil under a predominantly grass fallow. After clearing and burning the total nitrogen content of the soil is lower than in the forest and the nitrate content is very low, so that nitrogen deficiency occurs in varying degree depending in part on the age and luxuriance of the grass. As a rule there is an improvement in the rate of mineralization during the first one or two years of cultivation, after which there is little change for some years. Compared with the forest zones, leaching losses are limited by the low average nitrate levels and the lower percolation resulting from lower rainfall, but, on the other hand, they are enhanced at the beginning of the rains owing to the flush of nitrate production at this time and the absence of an established crop to protect the soil and absorb the nitrate. Responses to nitrogenous fertilizers are frequent in the savannas, even after the initial depression of nitrate levels by the fallow grasses has disappeared, and it is evident that lack of nitrogen contributes to the decline in yields during cropping.

Phosphorus. A marked increase in available phosphate results from the addition of ash after burning, and although availability declines slowly, soils which rapidly fix soluble phosphate in an unavailable form are in a minority, as evidenced by the residual effects of applied soluble phosphate in many fertilizer trials. Hence, in the forest zone, the effects of the phosphate provided in the ash, estimated by Nye and

Greenland (1960) as 24 lb of P per acre from a ten year fallow, should last for a normal cropping period, but the same may not apply in the savanna, where the ash provides only 4 to 8 lb of P per acre. It is estimated that, in addition, mineralization of the humus during the cropping period releases about 8 lb of P per acre per annum in forest soils and 3 lb in savanna soils. The decline in soil phosphate status during the cropping period is due to removal in crops and changes in availability; losses by leaching are negligible.

Nutrient Cations. Although part of the ash may be washed away by the first rains, the burning of a forest fallow will add considerable quantities of nutrient cations to the exchange complex of the soil. The extent to which cations are lost by leaching during the cropping period will vary with soil and rainfall, and has not been accurately determined. However, over most of the moist forest zone it may be expected that the loss of soluble salts by leaching is considerable during the cropping period, and may well exceed the quantities used by the crops. The reduction in the quantities of available cations by leaching and crop removal will to some extent be compensated for by release from non-exchangeable forms. Nye and Greenland (1960) suggest that the limited evidence indicates that, except for potash, no serious decrease in available cations is likely to occur during a single cropping period of one to three years, but that deficiencies may well arise with more prolonged cropping, or over successive cropping periods with intermittent short fallows. Many forest soils are low in exchangeable potassium, and the consumption of this element is enhanced by growing starchy roots. Consequently a decline in yields due to insufficient potash may occur rather rapidly, and potash responses have been common in fertilizer trials.

In the savannas burning the fallow vegetation adds to the soil appreciably smaller amounts of nutrient cations than in the forest. Except at the beginning of the rains, leaching losses are limited in the manner already described for nitrates. Unless the land is well wooded, crop removal over a period of three years or more could account for a large proportion of the potash added in the ash, but as responses to potash, magnesium or lime are all rare on savanna soils, it would appear that losses by leaching and crop removal are usually adequately replaced by release from the soil reserves.

CONCLUSIONS ON THE TRADITIONAL PRACTICE OF SHIFTING CULTIVATION

In the forest zones, native cultural methods and cropping practice during the short periods of cultivation restrict losses by leaching and erosion, and minimize the work of weed control. A reasonably long fallow leaves the soil in excellent physical condition and restores its humus and nitrogen contents to near their original levels. It also replaces, to a considerable extent, losses of other nutrients from the surface soil that have occurred during the cropping period, but this is only done by transfer from the subsoil, which itself suffers some losses. Consequently a decline in mineral nutrients may result in the rate of regrowth of the fallow vegetation becoming progressively slower after repeated periods of cultivation, necessitating longer fallows if humus and nitrogen are to be maintained. The tendency for this to occur will depend on the total nutrient reserves in the soil, which will vary with soil type. On many forest soils the nutrient reserves would probably enable successive cycles of short cropping periods and long fallows to continue satisfactorily for many years with only a very slow decline in fertility, but this would not necessarily obtain on the fairly widespread soils that are low in potash.

In the savannas traditional shifting cultivation is less satisfactory for the maintenance of fertility. Mainly on account of the predominantly grass vegetation and the annual occurrence of fires, the fallow is much less effective than a forest fallow in improving the physical condition of the soil, building up humus and nitrogen, and transferring nutrients from subsoil to surface soil. During the cropping period, deficiencies of available nitrogen and phosphate often limit yields, while exposure of the soil and tillage for weed control adversely affect soil structure, with consequent increased risk of erosion.

Traditional shifting cultivation worked reasonably well when practised by a relatively sparse population solely for the production of subsistence food crops. In the forest it was the best system that could be devised and resulted in only a slow decline in fertility. In the savanna it was less satisfactory and somewhat wasteful because of the annual burning, but as the population was thinly scattered in most places, it served to support the people at a rather low level of subsistence without causing any disastrous general decline in fertility. Even in the forest, however, the system is one which obviously precludes any long-term improvement in fertility, and if it is intensified as a result of the population exceeding a certain density, then the land deteriorates and yields fall sharply.

SHIFTING CULTIVATION IN PRESENTDAY CIRCUMSTANCES.

During the last sixty years, the practice of shifting cultivation has gradually been intensified, and in many areas (especially in Africa) it is no longer operated by a sparse population merely for the production of subsistence food crops. The abolition of intertribal warfare and slavery, the removal or relief of famine and the greatly improved control of disease have produced a marked increase in human and stock populations. The greater population density, combined with the cultivation of large areas of cash crops in addition to food crops, have increased the pressure on the land, with the result that in many places the fallows have been reduced in duration to such an extent that they are quite inadequate to maintain fertility.

The production of cash crops has also had other effects. The planting of perennial cash crops, or the investment of money derived from the sale of annual crops in better housing, has made people more attached to their dwellings and less inclined to move elsewhere, as they did formerly, when the fertility of the soil in a given locality has declined. Farmers growing cash crops have been encouraged to clear their land of stumps and roots more thoroughly, to plant the cash crops in pure stand, often in widely spaced rows, and to weed them more thoroughly than they did their food crops. Compared with the customary methods of cultivation, all these practices, although they may have advantages in other respects, tend to increase erosion in the absence of proper conservation measures which, unfortunately, are far from being universally adopted. Not infrequently, introduced cash crops have not fitted well into the customary, locally adapted sequence of crops and, in particular, there has been a conflict between the labour demands of cash and food crops. For example, the preparation of the land for, and planting of, cotton may be required at the same time as these operations must be done for finger millet, and the millet may need harvesting during the cotton-picking season. In such circumstances the primary importance of the food crop is likely to lead to late sowing of cotton, late weeding, and delayed picking, which will not only result in low yields but will tend to encourage erosion where planting is delayed into the rains.

In some parts of Africa adverse effects on soil fertility have resulted from a considerable increase in the amount of maize grown, either as a cash crop or for local consumption. Formerly, the main cereals grown were sorghums and millets, and only small areas of low yielding, hard, coloured-grained, maize varieties were grown. With the introduction of the soft, flat, white varieties by Europeans, maize became much more popular as a food crop because these varieties gave higher yields, were easy to grow and easier to pound into flour than the old varieties. Consequently, along with cash crops, large areas of vigorous and relatively high yielding maize were grown. This increased the drain on soil nutrients and, as the maize was usually

The Decline in Yields and Fertility

widely spaced, probably tended to increase erosion, although the practice of interplanting beans and other crops would tend to reduce this.

When the plough has been introduced into areas where cultivation was formerly done entirely with hand tools, its use has often facilitated the opening up and exploitation of larger areas of land without the accompaniment of any proper measures for controlling erosion and maintaining fertility. More thorough removal of stumps and roots must be done when clearing if the plough is to be used; this means that the land is more liable to erosion during the cropping period and that there is a slower regrowth of the natural vegetation during the fallow. Ploughing itself effects greater disturbance of the soil than hoeing, thus resulting in more rapid decline in physical condition and accelerating the decomposition of organic matter. This is liable to reduce the infiltration of rainwater and increase the risk of erosion. If ploughing is done up and down hill, as it often is, erosion is usually severe.

It will thus be seen that, due to higher population densities and the cultivation of cash crops, cropping becomes more intensive and prolonged, fallows are shortened and traditional cultural methods and crop sequences are modified, resulting in deterioration in the physical condition and nutrient status of the soil and increased erosion. Under these conditions, shifting cultivation can become untenable because of a disastrous decline in fertility. As shifting cultivation is essentially suited to the maintenance of the subsistence economy of a sparse population, it is certainly incapable of yielding adequate returns to provide for the higher standard of living nowadays required by the increasing population.

CHAPTER 8

The Maintenance of Fertility under Annual Cropping:
II – Modifications and Alternatives to Shifting Cultivation

Even where there is ample land for a sparse population, the intermittent cropping of traditional shifting cultivation results in a low production per acre from the land as a whole. Under the more intensive shifting cultivation now practised in the more densely populated areas, there is commonly a serious decline in fertility and yields per acre and per man may fall to very low levels. Bearing in mind the rapidly increasing population of the tropics, and the general need for better standards of living, there is clearly a need to modify shifting cultivation, or to replace it by other systems, so that soil fertility can be maintained, or improved, and higher levels of production achieved. In this chapter a number of possible modifications are discussed, but it should be mentioned at the outset that none of these improvements has yet been widely adopted.

IMPROVEMENTS TO THE FALLOW

In the savanna zones, and to a lesser extent in those of deciduous forests, a good deal of effort has been devoted to investigating the value of planted grass fallows as an alternative to natural regeneration. Where the rainfall is adequate for the establishment and maintenance of a good grass cover, and provided that the soil is not initially infertile nor exhausted by intensive cropping, it has usually been found that a rotation of about three years grass and three years cropping has maintained fertility at a moderate level over a long period. But it has also been generally observed that a planted grass fallow has been no more effective than natural regeneration, provided that the latter was not burnt annually. Furthermore, in most experiments continuous cropping has given the highest total yields; although yields increased after a grass rest, these increases were insufficient to make up for the loss in total yield resulting from absence of cropping during the rest period. These results were obtained in trials that ran for five to fifteen years, and it is probable that yields would have declined further on land continuously cropped for longer periods. However, it seems unlikely that the labour involved in establishing grass, instead of merely allowing natural regeneration to occur, will be worth while unless a good return can be obtained from animal production on the grass. It is also improbable that fertility can be maintained at a satisfactory level unless legumes are planted along with the grass, or unless fertilizers and imported feeding-stuffs are used. For these reasons, further discussion of grass rests and leys is deferred until Chapter 13, after the subject of grassland has been covered.

A number of experiments have been made with 'fallows', of three or four years duration, in which the land has been planted with various perennial legumes, such as *Tephrosia candida*, *T. vogellii*, *Cajanus cajan* (pigeon pea), *Centrosema pubescens*, *Pueraria phaseoloides*, *Crotalaria spectabilis*, *Leucaena glauca*, *Glycine javanica*, etc. The general experience has been that the effect of several years under these legumes on soil fertility, as measured by following crop yields, has been very similar to that of an equal period of natural regeneration or of planted grass.

Pigeon pea has been more frequently employed as a restorative crop in experiments of this kind than any other legume. For native farmers, who are averse to planting restorative crops that yield no direct return, it has the great advantage that it will provide human or stock food for at least two years of a three-year 'fallow'. Many varieties will persist for three or four years, providing a good cover and abundant leaf fall, and requiring no weeding after the establishment phase in the first year. The crop is tolerant of a wide range of soil types and climatic conditions and, being deep rooted, it has some effect in enriching the surface soil with nutrients brought up from deeper layers. A number of experiments have shown that three or four years under this crop is equal to, or even slightly better than, a similar period under natural regeneration or planted grass fallow, in its effect in improving soil fertility. The yields given in Table 12 for crops following three-year fallows of pigeon pea, Gamba grass (*Andropogon gayanus*) and natural regeneration in three trials at Yandev in the Southern Guinea savanna zone of Northern Nigeria, show no significant differences between fallow treatments (Dennison, 1959).

Clarke (1962) reports a trial carried out on poor, sandy soil at Matuga on the Kenya coast, in which continuous cultivation was compared with rotations in which three years' cropping followed three years under grazed star grass, pigeon pea or cassava. The pigeon pea and the cassava were only weeded during their first year, and a ground

Table 12. *Yields in lb per acre of crops following different 3 year fallows at Yandev, Northern Nigeria*

Dennison (1959)

	Type of fallow		
	Pigeon Pea	Gamba grass	Natural regeneration
Trial No. 1			
1st year, yams	8,154	8,435	7,830
2nd year, millet and sorghum grain	833	846	791
3rd year, sesame	211	238	210
Trial No. 2			
1st year, yams	8,447	7,610	7,915
2nd year, cowpea	124	120	104
2nd year, millet and sorghum grain	787	773	811
3rd year, sesame	459	411	404
Trial No. 3			
1st year, yams	2,743	2,559	2,709
2nd year, millet and sorghum grain	1,280	1,149	1,370
3rd year, sesame	270	237	238
(Three year fallow)			
7th year, yams	12,659	11,715	11,507
8th year, millet and sorghum grain	691	798	851
9th year, sesame	350	357	343

cover of grass subsequently developed under them. The yields of the sorghum test crops (Table 13) showed that the highest yields were obtained following pigeon pea. The beneficial effect of the weedy cassava and pigeon pea 'fallows' persisted throughout the three years recorded, but that of the admittedly poor cover of grazed star grass

Table 13. *Mean yields of dwarf sorghum, cwt. per acre, following various treatments at Matuga, Kenya*

Clarke (1962)

Treatment, 1953–55	Sorghum yields, cwt per acre			
	1956	1957	1958	Mean
3 years' cultivation	6·79	2·48	9·25	6·17
3 years' cassava—weed fallow	7·91	4·93	11·92	8·25
3 years' pigeon pea—weed fallow	8·41	5·24	13·48	9·04
3 years' grazed star grass	6·64	4·09	8·57	6·43

was less marked. The figures from three trials in Malawi given in Table 14, indicate that yields of maize grown without fertilizer were usually better after a three-year pigeon pea fallow than after a bush fallow, although there was little difference when the maize received

Table 14. *Malawi, maize grain, lb per acre*

Nyasaland (1957–8)

	Chitedze		Tuchila		Mbawa	
Fertilizer	0	100 lb S/A	0	100 lb S/A	0	100 lb S/A
After bush fallow	1933	2844	1215	2139	263	861
After pigeon pea	2862	2920	1545	2068	355	870

100 lb per acre of sulphate of ammonia (Nyasaland, 1957–8). Peat and Brown (1962) report that three-year rests under pigeon pea or grass gave very similar yield responses in following crops at Ukiriguru in Tanzania. Nye (1958) states that in the savanna zone of Ghana a two-year growth of pigeon pea stores three times as much phosphorus, potassium, calcium and magnesium as an established stand of *Andropogon gayanus*, the grass which forms the bulk of the natural regeneration after land has been cropped. In addition, he reports that the pigeon pea had a far better effect on the nitrogen status of the soil than the grass.

It will be noted from Table 13 that in the experiment at Matuga the effect of the weedy cassava was almost as good as that of the pigeon pea and slightly superior to that of the grazed star grass. Three years under weedy cassava was also included as a treatment in a trial at Ukiriguru (Table 15), where its effect on crop yields in the following two years was rather better than that of a three-year, grazed fallow of planted grass. In this trial, however, the grazed 'tumble-down' fallow

Table 15. *Yields, as percentage of control, following restorative treatments at Ukiriguru, Tanzania*

Peat and Brown (1962)

3 year restorative cycle 1948–49 to 1950–51	1951–52		1952–53	
	Cotton	*Millet*	*Cotton*	*Millet*
1. Annual cropping, residues grazed	100	100	100	100
2. Cassava, weeded first season only	167*	140*	104	131*
3. Tumble-down fallow, grazed	186*	150*	113*	120*
4. Planted grass, grazed	154*	107	116*	128*

* Denotes statistically significant difference from control.

(natural regeneration) gave the best results (Peat and Brown, 1962). In many places it is a common practice of shifting cultivators to plant cassava as the last crop before abandoning cultivation, weeding it only in the first year and harvesting the roots as required over the subsequent two or three years. The above experiments show that, apart from its value as a food reserve, weedy cassava has some beneficial effect on soil fertility, at any rate on sandy soils.

ROTATIONS AND MIXED CROPPING

Although fertility clearly cannot be maintained by the rotation of crops alone, a good rotation will usually give better average yields, and result in a slower decline in fertility, than growing the same crop in successive seasons or taking a succession of exhaustive crops. There is probably some advantage in rotating crops of different root habit, requiring different cultivations and making varying demands on the different soil layers. The alternation of legumes and non-legumes is usually advantageous because, although the seeds are harvested, the haulms and roots of such crops as beans, soya beans, groundnuts and pigeon pea are usually incorporated in the soil, and the succeeding crop thus benefits to some extent from the nitrogen fixed by the legumes. For example, in a trial that was continued for thirty years in the Sudan Gezira, a three year rotation of cotton, *Dolichos lablab* and fallow gave much better cotton yields than the rotation cotton, sorghum, fallow, and was also superior to the rotation cotton, fallow, fallow (Dutta-Roy and Kordofani, 1961). A good rotation can also be beneficial by minimizing exposure of the soil and the risk of erosion, by reducing weed infestation and the frequency of cultivation, and by checking the build-up of pests and diseases.

Table 16. *Yields in lb per acre, Nigeria*

Northern Nigeria (1955–6)

Bida	1st cycle 1950–52			2nd cycle 1953–55		
	Sorghum	Groundnut	Cassava	Sorghum	Groundnut	Cassava
Monocropping	315	272	4544	354	128	2790
Rotation	624†	346	5405†	668†	162	4724†
Samaru	Sorghum	Groundnut	Cotton	Sorghum	Groundnut	Cotton
Monocropping	939	717	420	947	781	402
Rotation	1196†	1035†	444	1124†	1027*	489

* Significant increase, $p = 0.05$.
† Significant increase, $p = 0.01$.

An example of the benefit of rotational cropping as compared with monocropping is provided by the results of experiments carried out in Nigeria at Bida, where the crops grown were sorghum, groundnuts and cassava, and at Samaru, where the rotation was sorghum, groundnuts, cotton (Northern Nigeria, 1955–6). The figures of Table 16 show that all crops, except cotton, gave higher yields when grown in rotation than under monocropping. Similar observations of the benefit of rotations, as compared with taking two successive crops of the same kind, were made in a number of trials in Malawi, the results of two of these being given in Table 17. It was concluded that maize did best after groundnuts or tobacco, groundnuts best after cotton or maize and tobacco best after groundnuts, but that cotton could follow satisfactorily after any of the crops, except cotton.

Table 17. *Malawi, effect on crop of preceding crop; yields lb per acre*

	Chitedze			Tuchila			
Preceding crop	Maize	Maize and beans, lb maize	Groundnut	Preceding crop	Maize	Cotton	Groundnut
Maize	2878	2288	618*	Maize	2394	719*	133*
Maize and beans	2915	2295	551*	Cotton	2605	603	215*
Groundnut	3118	2795*	329	Groundnut	3280*	727*	52

* Denotes statistically significant difference.

As already mentioned (p. 168), most peasant farmers follow a flexible, locally adapted sequence of mixtures of crops. Comparisons between pure stands and mixed cropping are affected by many factors, including the crop species, time of sowing, spacing, soil fertility, weather, incidence of pests and diseases, labour requirements and crop prices. This makes it impossible to generalize in comparing the two practices, but the limited experimental evidence available suggests that mixed cropping has advantages with many crops.

Numerous experiments have been made to compare cotton in pure stand and interplanted with other crops. Although the results have been somewhat variable, they have usually shown that, provided planting was done at the best time, a modest reduction in the yield of interplanted cotton was offset by the production from the other crop, so that interplanting gave a higher cash return per acre than pure stands. This was shown in experiments with cotton and maize or groundnuts in Kenya (Kenya, 1955, 1956; Grimes, 1963), and with cotton and maize in Malawi (Munro, 1959). In the Sudan, Anthony and Willimot (1957) obtained similar results when comparing cotton in pure stand and interplanted with groundnuts, soya bean, cowpea or *Phaseolus vulgaris*. With maize and beans at various spacings in Kenya, maximum yields of either crop were obtained in pure stands, but interplanting gave approximately the same cash return per acre (Kenya, 1959, 1960). At several sites in Tanzania a comparison of intercropping and pure stands was made for maize or sorghum with groundnuts, and for castor with groundnuts or soya beans, using various plant populations per acre for each crop (Evans, 1960; Evans and Sreedharan, 1962). In most of these trials the yields of individual crops were less when interplanted than in pure stands, especially at the higher plant populations per acre, but over a wide range of populations intercropping gave the greater overall production per acre.

In the majority of the above experiments more than one acre of crops in pure stand was required to produce the yield of one acre of intercropping. Since the amount of land that can be cultivated by farmers equipped with hand tools is strictly limited, it seems unlikely that there would be any general advantage in replacing the traditional practice of mixed cropping by rotations of pure-stand crops, although in certain cases the latter might prove advantageous by reducing the labour of harvesting, or by making it easier to control pests and diseases.

GREEN MANURING

Growing annual legumes as green manures in rotation with other crops is only likely to be of value where sufficient soil moisture is available, both for the vigorous growth of the green manure crop and for its decomposition to reach an advanced stage during the period between ploughing it in and planting the next crop. The practice is therefore usually ineffective in the drier areas, but under fairly good rainfall, and provided that the soil is neither highly acid nor in very poor physical condition, it can be a useful aid for the maintenance of fertility. As a rule, however, the beneficial effects of green manures are short lived, rarely extending beyond the first, or at most the second, following crop. Consequently green manures must be grown rather frequently in the rotation in order to maintain crop yields at a reasonable level. The effects of green manuring can be illustrated by reference to the results of prolonged experiments carried out under very different conditions at Ibadan in Nigeria and at Salisbury in Rhodesia.

Under an average rainfall of 49 inches per annum at Ibadan, in the lowland deciduous forest zone, there are two cropping seasons a year and it is possible to grow a green manure crop in one of them and a

Green Manuring

food or cash crop in the other. Large-scale replicated experiments on green manuring were carried out for over thirty years (Webster, 1938; Vine, 1953). The earlier trials, which compared rotations having different frequencies of velvet bean (*Mucuna utilis*) green manure crops, showed that the yields of various crops were markedly better when they followed a green manure than after other crops. Although the beneficial effects were short-lived, rotations with three or four green manure crops in four years (eight cropping seasons) appeared to maintain yields. Since other experiments, which were repeated over many years, showed that burning the *Mucuna* and digging in the ashes gave as good yields of the following test crop as did digging in the green *Mucuna*, it was evident that the benefit of the green manure was not due to the organic matter and nitrogen in its tops. It was therefore concluded that the benefit was mainly due to the available mineral nutrients contributed by the green manure, and that the amount of nitrate in the soil (no doubt partly provided by decomposition of the *Mucuna* roots) was adequate for the maize crop. However, Vine (1953) has pointed out that, as the experiments were started on land newly cleared from secondary forest, the gradual decomposition of humus accumulated under the forest must have been largely responsible for providing adequate nutrients, especially nitrate, during the first ten years. Although digging in green manures continued to improve the yield of crops immediately following them well beyond ten years after clearing the forest, this effect appears to have been mainly because of increased production of nitrate following their incorporation in the soil, probably coupled with increased availability of phosphate. This was indicated by the fact that fertilizer experiments consistently showed large responses to nitrogen, less frequently to phosphate and rarely to potash. Thus the earlier reports probably exaggerated the importance of green manures in maintaining yields; although they certainly raised the yields of crops immediately following them, it is uncertain whether three green manures in four years fully maintained fertility.

At Salisbury, which is 5,000 feet above sea level and has one cropping season a year during which an average of 30 inches of rain falls in six months, the alternation of a green manure crop (velvet bean or sunn hemp (*Crotalaria juncea*)) and maize in successive years was shown to maintain maize yields at a good level over a very long period, as indicated in Table 18 (Rattray and Ellis, 1953). The maize yields obtained from alternating with green manure were more than double those from continuous cropping, but if more than one maize crop followed a green manure the yield dropped by nearly 50 per cent (Table 19). For many years the alternation of maize and green manure was recommended in the Salisbury area, but more recent work (Rattray, 1956) has shown that better results are obtained from a rather diversified rotation, with moderate applications of fertilizer and farmyard manure and a green manure crop only once in five or six years.

Table 18. *Comparison of continuous v alternate maize cropping*
Rattray and Ellis (1953)

System	Period	Number of maize crops	Maize yields, bags per acre	
			Total	Mean
Continuous maize	1928–50	22	132·2	6·01
Alternate maize and green manure	1928–50	14	186·9	13·35

Table 19. *Six-year average maize yields, compared with $A = 100$*
Rattray and Ellis (1953)

Successive maize crops after green manure	1	2	3	4
A. Green manure and maize alternating	100			
B. Green manure followed by 2 maize crops	93·5	52·9		
C. Green manure followed by 3 maize crops	95·6	53·5	48·2	
D. Green manure followed by 4 maize crops	87·1	51·5	41·2	36·4

As a rule, the effect of green manures in increasing soil organic matter and total and available nitrogen is probably more important than any increase in the availability of mineral nutrients they may bring about. At Ibadan both factors seem to have been concerned, with nitrate production playing the major part. At Salisbury there is little doubt that nitrogen was the important factor, because green manuring benefited maize crops that received ample dressings of phosphate, and no potash responses were obtained in fertilizer trials. On the other hand, there is some evidence that on certain soils the benefit of green manuring is due to an increase in available phosphate. Orchard and Greenstein (1949) showed that on some South African soils increased maize yields following green manures were largely attributable to the phosphate content of the legumes, and there was no correlation between the maize yields and the amounts of nitrate released. Haylett (1943), again working in South Africa, also attributed the beneficial effects of ploughing in cowpeas, sunn hemp and pearl millet on following maize crops to an increase in available phosphate. The fact that some legumes are able to extract more phosphate and potash from the soil than most other crops is probably of significance in connection with green manuring (Prianishnikoff, 1940; Moser, 1942; Hopkins and Aumer, 1943).

It is generally considered desirable to plough in a green manure crop when it is in the growing green stage rather than to allow it to mature and set seed, or to harvest seed and plough in the senescent plants. As the plants approach maturity their percentage nitrogen content falls and their C:N ratio widens. Consequently, burying a mature crop, especially if it has ripened seed, may result in little if any immediate increase in the available nitrogen content of the soil. It is also desirable to plough in a green crop at a time when there will be adequate soil moisture to permit of decomposition getting well started before the next crop is planted. On the other hand, if ploughing in is done too early and is followed by a lengthy wet period before planting the next crop, much of the nitrate released by the decomposition of the green manure may be leached before the young crop is able to absorb it. At Salisbury (Rattray, 1956) it was found that if the green crop was allowed to mature and set seed before ploughing it in its effect on the yield of the following maize crop was appreciably reduced (Table 20). Furthermore, the time of ploughing in the green crop was important; burying it in April or early May usually being better than doing so in February or March.

Table 20. *Maize yields in bags per acre*
Rattray (1956)

Preceding crop	Ploughed in green	Seed taken
Sunn hemp	24·53	19·37
Velvet beans	24·30	15·83
Soya bean	22·19	18·20
Sunflower	18·01	11·96

It is not likely to be practicable to maintain fertility by green manuring alone. The effect of ploughing in annual legumes on crop yields is too short-lived, and they do not produce any lasting increase in the amount of humus, total nitrogen and other nutrients in the soil. Even alternating green manures and other crops in successive seasons, although it may maintain yields at a fair level over a long period, probably does not serve to prevent a gradual decline in soil organic matter. On the other hand, less frequent green manuring may be useful if it is employed in conjunction with other measures for the maintenance of fertility, as it is, for example, by peasant farmers in parts of India and China. At present, however, green manures are not used to any extent in small-scale farming in the majority of tropical countries. Most peasant farmers are naturally averse to growing a crop which occupies the land for the whole, or the greater part, of the rainy season without giving any direct return, and which demands labour for planting and for turning in at times which coincide with the busy periods of planting and harvesting food crops. Green manures are most likely to be used where they can be grown as short-term crops, occupying the land for only a part of the rainy season, either before or after a main crop.

FARMYARD MANURE AND 'KRAAL MANURE'

Farmyard manure, or pen manure, consists of a rotted mixture of the excreta of animals and the straw provided for their bedding, which should be sufficient to absorb the dung and urine. Even when made under cover – which should always be provided in order to avoid losses of nitrogen and soluble mineral nutrients by exposure to sun, wind and rain – its composition is very variable. For example, analyses of pen manure made under cover at various places in Nigeria, given by Phillips (1956), show that moisture content varied from 24·4 to 70·8 per cent, and the percentage of nutrients on the dry matter ranged from 1·03 to 2·12 for N, 0·49 to 1·30 for P_2O_5, and 2·33 to 5·54 for K_2O. The manure customarily available from the herds and flocks of most African farmers and pastoralists is usually referred to as 'Kraal manure', 'boma manure' or just 'cattle manure', and is of poor quality, since it consists of the weathered excreta of animals kept in uncovered pens, usually without bedding, and is commonly mixed with a considerable amount of soil. In many parts of Asia animals are normally kept under cover at night and provided with limited bedding, so that the manure is of fair quality.

Numerous experiments, in many parts of the tropics, have shown that the majority of annual crops respond well to applications of farmyard or kraal manure. There is no doubt that, on the majority of soils, adequate dressings of farmyard manure can maintain crop yields under continuous cultivation, provided that a suitable rotation is followed. The rates and frequencies of application of manure required to maintain yields vary with the climate, soil, crops grown and the quality of the manure.

Russell (1958) has suggested that in East Africa; 'dressings of the order of one ton of farmyard manure per acre per year, or three tons every three years, or five tons every five years, are probably necessary to maintain yields in the absence of fertilizers.' This is probably a fair generalization for the soils of a large part of East Africa. It is supported by the results of a large, long-term fertility experiment at Serere in Uganda, which indicated that fertility could be maintained by applying 5 tons of farmyard manure per acre once in five years to a rotation of four years' cropping and one year's fallow. (Jameson and Kerkham, 1960.) However, on some soils larger dressings of manure are likely to be needed in order to obtain satisfactory yields, or even to maintain fertility. For example, trials on poor, sandy soils in the Kenya Coast Province (Kenya, 1957; Grimes and Clarke, 1962) suggest that under these conditions dressings of the order of 3 tons of manure per acre per annum are needed to maintain yields and fertility at a modest level under continuous cultivation. Similarly, in Northern Nigeria, trials at Yandev showed that at least 2 tons of manure per acre per annum was required to maintain yields over a period of nine years when applied to a rotation of yams, sorghum and sesame (Dennison, 1959), while in experiments at Bida as much as 5 tons per acre per annum was needed to maintain the fertility of land continuously cropped on a three-year rotation (Northern Nigeria, 1955–6).

In some places remarkably large yield responses have been obtained from quite small applications of farmyard manure, but this probably only occurs where the manure supplies a mineral nutrient in which the soil is markedly deficient. For example, at some sites in Northern Nigeria one ton of farmyard manure per acre produced large increases in the yield of sorghum, but it was shown that the same result could be obtained with a dressing of fertilizer of equivalent phosphate content (Hartley, 1937). Prolonged residual effects of farmyard manure have been observed in a number of places. The Serere experiment, already mentioned, provided an example of this, for the effect of 5 tons of manure lasted until the fifth year after its application. More remarkable results were obtained at Ukiriguru in Tanzania where residual responses to 7 tons of farmyard manure or compost were still evident thirteen years after application, (Peat and Brown, 1962.) The reasons for these very prolonged residual effects have not been clearly established. They may be partly due to the fact that much of the organic matter of the manure is incorporated in the soil humus which only decomposes slowly, largely in periodic flushes when the soil is wetted after drying (Birch, 1958). However, it is unlikely that this can account for the very prolonged residual effects sometimes observed, which are more likely to be due to non-mobile nutrients, such as phosphorus or trace elements, contained in the manure.

Well-rotted farmyard manure is usually best applied very shortly before planting a crop. Applying it earlier in a dry season, especially if it is left partly exposed, is liable to lead to some loss of nitrogen by volatilization and, possibly, to loss of soluble nutrients by leaching at the beginning of the rains. It is usually impracticable to apply it satisfactorily after planting an annual crop, and in any case it does not decompose speedily enough to exert its full effect on a standing crop. If the manure contains much undecomposed straw, its application shortly before planting may result in the crop temporarily suffering from shortage of available nitrogen. Consequently, it may be desirable to plough in such manure at the end of a rainy season, so that further decomposition can occur before a crop is planted at the break of the next rains.

When dressings of farmyard manure are applied at intervals of several years it may be advantageous to apply them to a particular crop in the rotation, either because that crop is of high value, or because it responds better to the manure than other crops. However, there is not much information on the relative responses of different crops. Large yield increases are usually obtained with yams and sweet potatoes, but cassava may not respond so well. Cereals give big responses, but those of cotton and legumes are smaller. Dennison (1961) reports that sorghum gave the best response of all the major crops grown in Northern Nigeria; sweet potatoes and millet also benefited markedly, but cotton, sesame and groundnuts gave comparatively poor responses. Experiments have shown little difference in overall yields between applying manure annually and applying an equivalent amount in larger dressings at intervals of several years. According to Dennison (1961) trials in Northern Nigeria showed no significant difference between applying 3 tons per acre every third year, or one ton per acre per annum. In a long term trial at Matuga, in Kenya, which compared 3 tons per acre per annum and 9 tons per acre every third year (Table 21), there were no significant differences between the two treatments in the average yields of the crops of maize, sorghum, cassava and sweet potatoes grown in the rotation (Grimes and Clarke, 1962).

Most of the effect of farmyard manure is due to the nutrients it contains, rather than to any special effect associated with its organic matter content. In a number of experiments fertilizer dressings of equivalent NPK content have proved just as effective as farmyard manure. For example, in the experiment at Matuga, mentioned above, there was no significant difference between the yields of plots receiving 3 tons of farmyard manure per acre per annum and those which annually received an equivalent amount of N, P, and K as inorganic fertilizer (Table 21).

On the other hand, in some experiments, especially long-term ones, farmyard manure has given better results than equivalent amounts of N, P and K in inorganic fertilizers. Dennison (1961) states that farm-

Table 21. *Mean yields in cwt per acre, Matuga, Kenya*
Grimes and Clark (1962)

	No Manure	Fertilizer annually	3 tons f.y.m. per acre annually	9 tons f.y.m. per acre every 3rd year
Maize grain, average of 6 crops	4·65	7·73	6·97	7·52
Sorghum grain, average of 6 crops	8·43	15·37	13·14	12·64
Cassava roots, average of 6 crops	50·25	81·50	76·50	73·87
Sweet potato, average of 4 crops	20·60	40·78	43·28	37·35

yard manure gave better results than equivalent amounts of NPK fertilizer in long-term, continuously cropped, experiments in Northern Nigeria. Doughty (1953) found that all the crops he tested in a series of experiments in the Lake Victoria area responded better to dung than to fertilizers. Cattle manure was included in a number of factorial trials in Ghana, along with sulphate of ammonia, single superphosphate, muriate of potash and lime, and it was found that the manure nearly always gave better results than the inorganic fertilizers, especially after some years of cropping (Djokoto and Stephens, 1961). The nutrient contents of the manure and the fertilizer dressings were not equivalent in these experiments, but it was concluded that part of the effect of the manure could not be accounted for by its content of major nutrients. The superiority of manure in trials where the inorganic nitrogen is supplied as sulphate of ammonia may be partly because this fertilizer tends to lower soil pH and exchangeable base content, whereas farmyard manure usually slightly raises both. It is also quite probable that a small part of the effect of farmyard manure is due to a beneficial effect of its organic matter on the physical condition of the soil and possibly also on micro-biological activity, such as nitrogen fixation. In addition, the relatively slow decomposition of the organic matter should result in a steady supply of balanced available nutrients, whereas the application of fertilizers may provide a somewhat unbalanced supply of nutrients, especially in soils of low nutrient status.

Farmyard manure is unquestionably a most useful aid in maintaining fertility and should be used wherever it is practicable and economic to do so. There are, however, several limitations to its extensive use in peasant agriculture. First, there are large areas, especially those infested with tsetse fly in Africa, where no cattle are kept. Secondly, in most places where cattle are kept they are not integrated with crop husbandry, and manure is not made and used, so that it is first necessary to persuade native farmers to improve their methods of husbandry and to make and use good manure. A third difficulty is that where peasant farmers do not possess any wheeled transport, a very great deal of labour is involved in carrying crop residues, etc. to the cattle sheds, and in the transport and application of the manure to the field. Finally, it should be noted that, since there is very little use of imported cattle foods in the tropics, any system of agriculture which relies mainly on the use of farmyard manure to maintain crop yields, at present necessarily demands a relatively large area of grazing land which is robbed of its nutrients for the benefit of the arable land. Such a system must become increasingly difficult to maintain as population pressure on the land increases.

COMPOSTS

Compost consists of a partially decomposed mixture of household refuse, crop residues, weeds and other waste vegetable material, either with or without the addition of some animal or human excreta. In some places where stock are kept, the preparation of composts containing a small proportion of animal excreta is advocated as a means of making available larger amounts of organic manure than could be obtained by the ordinary method of making farmyard manure. Peat and Brown (1962) report that at Ukiriguru, in Tanzania, it was possible to make about 3 tons of farmyard manure (averaging 1·5 per cent N and 0·69 per cent P_2O_5 on the dry matter) per beast per annum, but as much as 15 tons per annum of compost, averaging 0·93 per cent N and 0·52 per cent P_2O_5 on the dry matter. The nutrient value of composts varies with the nature of the materials used in their preparation, but even those made without any animal excreta may contain useful amounts of the main plant nutrients. For example, analyses of five composts made from grasses, weeds and household refuse in Nigeria showed that moisture content varied from 21·9 to 50·4 per cent, and the percentage nutrients on the dry matter from 0·43 to 0·91 for N, 0·16 to 0·56 for P_2O_5 and 0·42 to 2·7 for K_2O (Phillips, 1956).

The advantage of composting as opposed to burying raw crop residues and weeds, is that it enables partially decomposed organic matter to be applied to the land at the best time so that available plant nutrients are rapidly released for the growth of a crop. Burying raw vegetable wastes of wide C:N ratio may easily result in a temporary shortage of available nitrogen for the following crop. On the other hand, burying fresh vegetable material of narrow C:N ratio is liable to result in decomposition reaching an advanced stage, and some nitrogen being lost by leaching before a crop can utilize it, especially in areas of monsoon climate. Some loss of nutrients, especially nitrogen, is involved in composting, but this can be minimized by appropriate methods of preparation and by composting a mixture of materials of suitable average C:N ratio. Extensive use of composts is made by industrious and painstaking farmers in some of the more densely populated parts of Asia, notably in China and India, but their use is small in other parts of the tropics, especially in Africa.

Crops usually respond to dressings of good compost very much in the same way as they do to farmyard manure, although larger dressings of compost are required to produce equivalent results. The use of compost is therefore to be encouraged wherever it is practicable, but there are several limitations which make it unlikely that, in most countries, it will make any great contribution to the maintenance of fertility. First, although the preparation of compost from household and township refuse is usually practicable and worthwhile, the amounts of such materials which are available are insignificant in relation to the very large areas of cultivated land. Secondly, the composting of other materials such as crop residues and weeds involves a very high labour requirement, especially for farmers who have no wheeled transport, in carrying the residues to heaps, turning the heaps, and conveying and applying the finished product to the field. Thirdly, lack of water is often a problem in composting. In humid areas the rainfall may supply sufficient moisture to the heaps, but in many regions crop residues become available at the beginning of a dry season, and composting them before the next planting season is impossible unless water is carried to the heaps. Even if water is available, carrying it to water the heaps is a very laborious procedure. Finally, it should be noted that the amount of compost that can be made from crop residues is quite inadequate to maintain fertility by itself, and that reliance on compost materials from outside the arable area is not only extremely laborious, but also becomes increasingly impracticable as population pressure increases, for the same reason as that given in the section on farmyard manure. It is robbing Peter to pay Paul, and demands a relatively large area of waste land. Moreover, as nutrients are steadily transferred from this land to the arable, progressively less and less regrowth of vegetation is produced for composting.

The essence of good composting is to make the best use of available materials and conditions; consequently appropriate methods of preparation will vary in different places, but the following general points may be noted:

1. Composting materials of low C:N ratio by themselves is wasteful, since much organic matter and nitrogen will be lost in the process.

Materials of wide C:N ratio, such as most cereal straws, will decompose only slowly by themselves; if they have to be composted alone, then addition of nitrogen is desirable, at the rate of $\frac{1}{2}$ to $\frac{3}{4}$ cwt of sulphate of ammonia per ton of straw. Wherever possible, however, composting of mixed residues is desirable, both as a safeguard against unsatisfactory nitrogen status, and to give better structure and aeration to the heap.

2. An approximately neutral reaction is needed in the heap: hence if sulphate of ammonia is added it may also be necessary to add $\frac{1}{2}$ to 1 cwt of ground limestone per ton of straw.

3. The microorganisms responsible for decomposition require a supply of phosphate, and it may be desirable to ensure this by adding 10 to 20 lb of superphosphate per ton of material. Alternatively, the addition of wood ashes may serve both this purpose and (2) above.

4. In order to obtain satisfactory conditions of aeration and moisture within the heap a balance must be struck between making heaps so small and open that they dry out readily, or so high and compact that anaerobic conditions are produced. Generally 3 or 4 feet is a suitable height.

5. Initial inoculation of the heaps with some well-rotted compost is desirable to provide the necessary microorganisms, if no animal excreta is included in the material to be composted.

6. Periodic turning and rebuilding of the heaps, by improving aeration in the earlier stages, accelerates decomposition and makes it more uniform.

Using mixed residues and turning the heaps and under good conditions of moisture and aeration, the material should break down to a structureless compost with a C:N ratio of about 10–15 in about three months. If conditions are less favourable, or if a large proportion of less readily decomposable materials, such as straw, is used the process will take longer, but provided there is adequate moisture will usually be completed in not more than six months.

OTHER ORGANIC MANURES

Except for night soil, other organic manures are little used on annual crops. They are not often available in quantity locally and are bulky and costly to transport. The majority of oilseeds produced in the tropics are exported overseas for oil extraction, so that the residual cakes and meals are not widely available. Even where oil extraction is done locally, the cost of cake is commonly high because of its value as a cattle feed. Moreover, when applied to annual crops such materials have often failed to give greater yield increases than an equivalent quantity of NPK applied more cheaply as artificials, or even than a dressing of sulphate of ammonia of equivalent nitrogen content. Exceptions to the above general statements are, however, found in some areas; for example, where a locally available organic manure can be applied to high value crops, as in the use of prawn dust by Chinese market gardeners in Malaysia.

Night soil, which is solid and liquid human excrement, has been used on a very large scale for centuries in China, both for direct application to the land and in the preparation of composts. It is also extensively used in parts of India, mainly in composts, but to a lesser extent for direct application. Before being used for direct application, night soil is normally diluted and stored for a variable period, partly because this is necessary to avoid damage to plants, and partly because the main need for night soil in the fields is seasonal. During storage, which may be in pits, concrete tanks, wooden barrels or stone jars, appreciable losses of organic matter and nitrogen occur. Although the use of night soil has been a major factor in the maintenance of fertility in China and some other areas, it has undoubtedly been the cause of a tremendous amount of human disease and ill-health. For a discussion of improved methods of handling night soil, and of sewage disposal, reference may be made to Ignatieff and Page (1958).

Dependence on farmyard manure, composts and other locally available organic manures is characteristic of some of the more advanced traditional systems of dry-land farming in the tropics. Although such systems are an improvement on shifting cultivation they work in a closed cycle, which makes it impossible fully to correct the inherent nutrient deficiencies displayed by many soils, or to replace nutrients lost by leaching or removed in crops and stock products. Except over limited areas these systems are unable to maintain fertility and yields beyond a moderate level. They are thus in contrast to modern intensive alternate husbandry in the temperate zone which, by making skilful use of grazing animals and grass-legume leys and by importing fertilizers and feedingstuffs on to the farm, enables fertility not merely to be maintained but to be steadily improved and makes high yields per acre possible.

FERTILIZERS

It has not been clearly established whether it is practicable to maintain fertility by the use of fertilizers alone on land that is kept under continuous rotational cropping and where suitable measures for soil and water conservation are adopted. In a number of experiments fertilizers have maintained the yields of a rotation of crops grown continuously for some years, but it remains doubtful whether reliance on fertilizers alone is possible, and very doubtful whether it would be generally practicable and economic. It seems possible that in most places farmers would find it necessary to resort to the periodic use of a fallow or some kind of restorative crop, or to the application of organic manures, in addition to fertilizers, in order to maintain the soil in a satisfactory physical condition.

It is clear, however, that fertilizers can raise the yields of crops grown by shifting cultivation and can reduce the length or frequency of the fallows required. Nye and Greenland (1960) have reviewed a large number of fertilizer trials carried out in many parts of Africa with the main crops grown by shifting cultivators and have shown that the results conform to a fairly consistent pattern. On forest soils, whether oxysols or ochrosols, responses to nitrogen have been small after fallows lasting ten years or more, but large on land more intensively cropped with only short fallows. The effect of phosphate has varied with soil type, but in many places small responses have been obtained in the first year of cropping and larger responses in subsequent years. Response to potash has very commonly been obtained after short fallows or on land that has been cropped for a number of years. On savanna soils moderate to large responses to nitrogen have been very consistently observed, even where the land is cropped only infrequently; in fact, very marked responses are usual immediately following a long grass fallow. Yield increases from phosphate have also been common, tending to be small on newly cleared land but increasing with cropping. In contrast to forest soils, response to potash has been remarkably rare, even on land intensively cropped.

Formerly there was a rather general impression that the use of fertilizers in peasant farming would be uneconomic, but many post-war experiments have demonstrated that fertilizing a variety of peasant-grown crops can be profitable. For example, in various parts of Ghana it has been shown that, without any change in farming practice, fertilizers can economically increase yields of yams, upland rice, maize, millet and sorghum (Nye and Stephens, 1962). Phosphate application to groundnuts, sorghum and cotton has been found profitable over large areas in Northern Nigeria (Greenwood, 1951), and the application of nitrogen to yams, maize and cassava has proved economic in densely populated parts of south-eastern Nigeria (Irving, 1954). In the Lake Province of Tanzania, nitrogen and phosphate have given large and profitable increases in cotton yields (Peat and Brown, 1962). Phosphate, and to a lesser extent nitrogen, have given economic responses with a variety of crops on peasant farms on certain widespread soil types in Kenya and Uganda (Doughty, 1953).

Response to fertilizers may, of course, be limited by lack of soil moisture. Where the rainfall is marginal for crop production no benefit from fertilizers may be obtained, and in other areas, where cropping intensity and yield potential are limited by low or poorly distributed rainfall, only relatively small dressings may prove economic. Where drought is likely to occur during the growing season it is possible for the ill-advised use of fertilizers to have an adverse effect by stimulating a strong early growth of plants which more rapidly exhaust the moisture supply. Too much nitrogen is liable to do this by encouraging vegetative development and delaying ripening, but phosphate is usually helpful by hastening maturity. In many areas of moderate rainfall satisfactory responses to fertilizers will only be obtained if measures are adopted to ensure a maximum infiltration of rainwater into the soil; this has already been illustrated by reference to experiments with cotton in Northern Nigeria (p. 124).

In addition to adopting sound measures for the conservation of soil and water, it is necessary to follow other good husbandry practices if the best results are to be obtained from fertilizers. These include the use of a suitable crop rotation, planting at the best time, good weed control, optimum spacing, the use of improved varieties, pest and disease control and the return of all possible crop residues to the soil. Improvements in these matters are needed on peasant holdings in order to maximize the profits from fertilizers. In particular, most peasant farmers tend to plant too late and fail to weed thoroughly.

Nitrogen is probably the most commonly needed element for annual and short-term perennial crops. Responses from a variety of crops are commonly obtained on both savanna and lowland forest soils, but are less frequent in the montane forest areas, where the soils usually have a higher organic matter content. The crops that are most responsive to nitrogen are the cereals, sugar-cane and grasses for all of which good returns are usually obtainable from relatively high rates of application of nitrogenous fertilizers, but good responses are often obtained with many other crops, especially cotton, sisal, pineapple, yams and sweet potatoes. Legumes usually show little or no response, but groundnuts have sometimes benefited from application of nitrogen. Sulphate of ammonia (20·6 per cent N) is by far the commonest form in which nitrogen is applied. The use of this fertilizer lowers the pH and the available potassium, calcium and magnesium contents of the soil, and unless this tendency is corrected by the application of lime, it would seem wise to consider other types of nitrogenous fertilizer, such as nitro-chalk (15·5 per cent N) or ammonium nitrate (35 per cent N), as alternatives to the long-continued use of sulphate of ammonia. Some of these alternatives also have the advantage of a higher nitrogen content, and hence a lower transport cost per unit of nitrogen, which may make them more economic for the many farmers in the tropics who are remote from sources of supply. Since nearly all nitrogenous fertilizers are either readily soluble in water or are speedily converted to soluble forms in the soil, it is desirable to apply them in a manner which will avoid leaching as far as possible, and will ensure that the nitrogen is available in the root zone at the time when crop demands are highest. It has often been found advantageous with cotton, maize, sorghum and millet to apply part of the fertilizer at planting time and part at or slightly before the flowering time of the crop.

The majority of tropical soils contain too little available phosphate for optimum crop growth. Consequently widespread responses to phosphatic fertilizers have been obtained with cereals, yams, sweet potato, cassava, cotton, and legumes, which have a high demand for phosphorus. Superphosphates, containing water-soluble phosphate, are the commonest phosphatic fertilizers applied to annual crops. Single superphosphate, with 18 per cent P_2O_5 (and also containing calcium and sulphur) is more expensive to transport than double or triple superphosphates, which contain 42–49 per cent P_2O_5 but no sulphur, although the initial cost of the P_2O_5 unit is slightly higher in the latter. Good quality, finely-ground rock phosphates are probably generally more economic than superphosphates on acid soils. As they contain phosphate in a water-insoluble form which only becomes available relatively slowly, following reaction of the fertilizer with the soil, they do not have the same effect as an equivalent quantity of superphosphate in the season of application, but continue to exert a residual effect in subsequent seasons. Phosphate is normally incorporated in the surface soil at planting time, because most crops have a particular need for it during the early stages of growth, and it is not subject to leaching. Later top dressing of growing crops is not usually effective.

In some acid soils applied water-soluble phosphate is rapidly 'fixed', or converted into unavailable forms by combination with soil iron and aluminium, so that the amount recovered in crops is usually less than 30 per cent of that contained in the fertilizer, and residual effects are shortlived. On such soils it is advantageous to apply water-soluble phosphate in small annual dressings rather than in larger amounts less frequently, and to place the fertilizer in bands or pockets near the seed at sowing time, although with most crops it should not be placed within half an inch of the seed in order to avoid reducing germination. By placing the phosphate in this manner, its speedy absorption is facilitated by the presence of a high concentration near the plants, and fixation is reduced because the fertilizer reacts with a much smaller amount of soil than it would if it were broadcast. However, in the majority of fertilizer trials water-soluble phosphatic fertilizers have given marked residual effects extending over several seasons, and it would therefore appear that soils which fix phosphate severely are in a minority. Rock phosphates, which do not contain water-soluble phosphate, should not be placed, but preferably thoroughly mixed with the soil in order to promote reaction of the fertilizer with the soil to increase its availability.

Annual and short-term perennial crops respond to potash fertilizers far less commonly than they do to nitrogen and phosphate. Potash responses are usually confined to limited soil types, usually light sandy soils that are less well supplied with potassium than the heavier soils. On such soils tobacco, yams, sweet potatoes, cassava, and possibly cotton are more liable to suffer from potash deficiency than most other crops, although some legumes also have a rather high demand for this element. Potash is usually applied at planting time as sulphate or muriate of potash, the former being preferable for some crops, especially tobacco. On some highly weathered, potash-deficient sandy soils that are much subject to leaching, it may be desirable to apply part of the potash as a top dressing. Potassium is fairly mobile in most soils and although certain clay minerals fix potassium, fixation is rarely a problem.

The use of modern compound NP, or NPK, fertilizers of high analysis may have certain advantages for farmers in remote areas, since these concentrated fertilizers incur lower costs per unit of nutrient for bagging, handling and transport. They are also usually granulated, which greatly improves their handling and storage qualities. However, considerable care is needed in the selection and use of such materials. The use of compound fertilizers is only justified when the land requires two or three nutrients in some well defined ratio. Hence it is important to be sure that a selected compound fertilizer meets local nutrient requirements, especially as many of those available may have an unsatisfactory potash content for tropical crops and soils. Furthermore, although granular fertilizers are entirely satisfactory if all their ingredients are soluble, their value is more doubtful if some of the ingredients are insoluble. Some granulated compound fertilizers contain a large part of their phosphate in a water-insoluble form and this is likely to be useless unless the granules break down readily and completely in moist soil.

A number of experiments have shown that responses to lime on both forest and savanna soils are small, or absent, unless the pH is below 5·0 (Nye and Greenland, 1960). Very acid soils, on which responses might be expected, are rare in the savanna zones but occur to some extent in the lowland forest areas. For example, on soils of pH about 4·1 in the Umuahia area of Nigeria maize showed large responses when sufficient lime was applied to raise the pH to 5·0, but further addition of lime had no effect. Lime has been very little used on annual crops and in many places remote from a source of supply its use to correct acidity would be very expensive, although small dressings to correct a deficiency of calcium might be economic. Reports of deficiencies of other secondary nutrients in annual crops are rare. Sulphur deficiency in groundnuts has been established over wide areas in Northern Nigeria and Ghana (Greenwood, 1951), and is suspected in certain crops in parts of East Africa. Soils with low levels of magnesium are not uncommon in the humid tropics but deficiencies of this element do not appear to cause trouble with annual crops. Trace element deficiencies, although well known in perennials, are rare in annual crops grown on peasant farms, even where the land has been intensively cropped.

It is difficult to see how fertility can be maintained and at the same time more and better food produced for the rapidly increasing populations of the tropics, without a greatly increased use of fertilizers. As population pressure increases, it will become increasingly difficult to avoid a decline in yield and fertility under any of the traditional systems of farming, which rely on fallows and locally available organic manures, and thus operate within a closed cycle of nutrients. At present only very small amounts of fertilizers are used by peasant farmers, and their employment on a much larger scale would seem to be an essential and relatively simple means of both increasing production and maintaining fertility.

However, it is unlikely that the problem can be speedily solved simply by a rapid and massive increase in the use of fertilizers. In the first place, the majority of peasant farmers can, at best, only afford to buy small quantities of fertilizer at present, and the amounts used can only be expected to increase slowly as the economy of tropical countries develops, and governments are able to help with subsidies and the establishment of local fertilizer factories. In the second place, using more fertilizers is not likely to be highly effective in most parts of the tropics unless it is accompanied by improvements in husbandry and in soil and water conservation. Indeed, there is a risk that a greater availability of fertilizers might make matters worse by enabling farmers temporarily to make more intensive use of land without proper conservation measures, thus leading to increased denudation and erosion. Furthermore, as our knowledge of the overall nutrient requirements of different crops on different soils is at present limited, there is a danger that a large use of fertilizers in peasant farming could induce conditions of nutrient imbalance in the soils. Bush fallowing, or the application of most organic manures, maintains a reasonable balance between the nutrients added to the soil, but the widespread soils of low nutrient status could easily become unbalanced by the excessive use of a fertilizer supplying only one or two nutrients, or through failure to provide a particular nutrient.

It may be concluded that, although increased use of fertilizers will be necessary, the problem of increasing production and maintaining fertility is not likely to be resolved by fertilizers alone nor, indeed, by reliance on any other single measure for maintaining fertility, but only by evolving better farming systems employing a combination of fertility-building practices appropriate to local conditions.

CHAPTER 9

The Maintenance of Fertility under Annual Cropping:

III – Permanent Farming Systems Associated with the Production of Swamp Rice

In many parts of Asia, swamp rice has been continuously grown on the same level and bunded fields for many centuries. The technique of permanent wet rice production was probably originally introduced from China, but the practice has long been widespread in many tropical Asian countries, where it has formed the backbone of permanent farming systems capable of supporting relatively high population densities, much higher than those that can be maintained by shifting cultivation. Although these systems invariably include permanent rice fields as an essential, and often as the principal, feature, they usually embrace one or more other enterprises, some of which may be closely integrated with the rice production, while others may be unrelated ancillaries. The other elements which may form components of a system are as follows:

1. The rotation of other annual crops with rice in the wet padi fields.
2. Permanent dry-land cultivation of annual crops, usually on only a small area.
3. Perennial crops on dry land, commonly only a small plot containing a mixture of plants of value in a subsistence economy, but sometimes plantations of cash crops, such as rubber, in pure stand.
4. Limited shifting cultivation of annual and semi-perennial crops.
5. Livestock.
6. Fish culture.

These components are used in varying numbers, combinations and proportions to form a variety of farming systems differing widely in complexity and intensity. For example, in the extensive, alluvial plains of the north-west of Malaya, the farmer's principal enterprise is the production of rice but, owing to the occurrence of a dry season and the limited availability of irrigation water in most places, the padi fields annually produce only one rice crop, in a growing season of about seven months, and lie fallow for the rest of the year. Other enterprises are of much less importance. Cattle or buffaloes are kept for work and graze the padi fields in the off-season, but are not otherwise integrated with crop husbandry. Fish culture is practised to a limited extent in the padi fields, and the farmer will usually have a small area of slightly higher land near his house on which he grows fruit and other trees and a few vegetables and spices. If his padi holding is small, he may also practise limited shifting cultivation of annual crops on hilly land at some distance from his dwelling.

On the other hand, in small inland valleys in other parts of the country, where the amount of land that can easily be rendered suitable for wet rice production is limited, the Malay farmer will only have a comparatively small padi area, again producing one rice crop a year, but a larger dry-land area, where there may be a rubber plantation as his principal enterprise, together with a plot of mixed trees and a small proportion of annual crops. He may also do a little shifting cultivation of annual crops on hilly land. Livestock normally play only a small part in the system; there will usually be a few goats and poultry, but cattle, if present at all, are usually kept only for work.

As a third example, Chinese peasant farming under favourable conditions in Malaya is usually characterized by more intensive permanent cultivation and more productive use of livestock on a system in which the various components are closely integrated. Where moisture supply and drainage permit, other annuals, especially vegetables, are rotated with rice in the padi fields. Livestock, particularly pigs, are grown for meat, being fed partly on crop by-products, and full use is made of their manure. Part of the animal excreta may be used to fertilize fish ponds, and mud from the ponds may periodically be applied to an orchard of fruit trees.

Similar systems occur in many other Asian countries. For example, Terra (1958) describes a Javanese system in which the components are: (1) wet rice; (2) permanent cropping of non-irrigable land with cassava, sweet potato, maize and groundnuts; (3) mixed trees, other perennials and some vegetables, near the dwelling; (4) cattle, goats, sheep and poultry; (5) fish culture in fresh-water ponds. A less highly developed Sundanese system, which he mentions, includes wet rice; trees and other perennials around the dwelling; shifting cultivation (principally of hill rice), and livestock.

Since these rather intensive permanent farming systems include swamp rice growing as an essential feature, they only occur where conditions are suitable for this crop. Swamp rice requires level fields, medium to heavy soils that are not too freely drained and an ample water supply, provided either by rather high rainfall, by flooding, or by irrigation. Originally such systems were developed in the fertile flood-plains of the great rivers of the East, in extensive areas of coastal alluvium and on more limited flat land in inland valleys. Subsequently there was an extension of wet rice growing to areas which were less suitable in their natural state. Swampy or seasonally heavily flooded areas were reclaimed by drainage and flood control measures, some of these being coastal mangrove swamps, where protection from the sea was also necessary. Irrigation works were constructed to conserve and convey water for drier areas. The gentle slopes of inland valleys were made into broad terraces and, in places where the population reached a high density, such as parts of Ceylon, Java and the Philippines, terraced irrigable rice fields were laboriously constructed even on steep hillsides. (Plate V.)

Although these systems are so far almost entirely confined to Asia, relatively large areas that are (or could be rendered) suitable for wet rice growing exist in other parts of the tropics, and in such places there is the possibility of replacing the predominant shifting cultivation with more permanent systems based on rice. Wet rice growing has already been introduced into some of these places; for example, on coastal plains protected from the sea in Surinam and British Guiana, on land reclaimed from mangrove swamps in Sierra Leone, on flood lands protected by bunds along the Gambia River and in inland valleys in Madagascar. There are undoubtedly further extensive areas in Africa and South America where permanent farming systems based on rice could be developed. However, for a variety of reasons – amongst which are the need for surveys to determine the suitability of land, the high capital cost of irrigation, drainage and flood control, the difficulty of finding profitable crops for rotation with rice and the necessity of training native farmers in techniques that are entirely new to them – such projects require thorough preliminary investigation and careful planning and execution.

In the following brief discussion of the various components of these systems no attempt is made to go into detail on the subject of rice production, for which reference may be made to standard works on this crop, such as those of Ghose *et al.* (1956) and Grist (1965). The object here is to mention only those features of rice growing that have an important influence on the efficiency of the farming systems under discussion.

Swamp Rice and Associated Rotation Crops

There is great variation in the efficiency with which the necessary water for the rice crop is provided and controlled. In some less developed areas it may be obtained merely by impounding rain water in bunded fields, or by relying on the seasonal flooding of rivers. Both these methods are chancy, especially the latter. If the floods come earlier than expected the crop may not be sufficiently advanced to withstand inundation without damage; if too late, the rice may have already suffered a severe check and be unable to take full advantage of ample water when it becomes available; and if there are unusually late floods they may interfere with ripening and harvesting. Some form of constructed irrigation system is therefore usual, but in many places this has hitherto comprised small dams and channels built by the farmers themselves, providing very imperfect water control. Only in relatively recent times has there been any considerable development of large, well-constructed and efficient irrigation facilities. On the other hand, in other countries efficient irrigation was long ago established on a large scale, as for example, in parts of China, where a gigantic canal system has been in use for centuries. Similarly, in the drier parts of Ceylon there was formerly an extensive and well-organized system of irrigation by means of water stored in tanks, or reservoirs, which is believed to date from about 400 B.C. (Grist, 1965).

Although broadcasting seed in the field is customary in some places (for example in Ceylon), as a general rule rice seed is sown in nurseries. Seedlings about six weeks old are transplanted into fields that have already been flooded and thoroughly cultivated, so that the soil is reduced to the consistency of creamy mud, and the seedlings can be readily thrust into holes made with a stick or the hand. The main reason for this laborious procedure is that when land grows rice continuously, weeds can only be effectively controlled if the fields are flooded and thoroughly cultivated before planting. Such flooding can usually only be done when sufficient water becomes available early in the rainy season, and thereafter the preparatory cultivation takes some time. The late sowing of seed would then reduce yields, or might be impractical if the standing water cannot be removed from the fields at planting time, but these difficulties are overcome by having plants of a suitable size ready in a nursery for transplanting. (Plates 19 and 20.)

The preparatory cultivation following the initial flooding may be done entirely by hand, the weeds being slashed or hoed, and either piled on the bunds or trampled into the mud after the land has been hoed. When the soil is deep and soft, so that animals sink into it, there may be no alternative to hand cultivation. More commonly, whenever the soil is shallow or possesses a firm layer within a few inches of the surface, preparation is done with cattle or buffaloes, using implements which vary in number and kind in different localities, but usually include ploughs and harrows, and often rakes and rollers as well. In order to obtain good weed control and to puddle the soil so that the seedlings are easily transplanted, preparatory cultivation is usually very thorough. For example, as many as eight ploughings, followed by some form of harrowing, are not uncommonly given in parts of India (Pillai, 1958). (Plates 17 and 18.)

Harvesting of rice is almost universally done by hand and in many places each ear is cut off separately with a small knife, partly because of the uneven ripening of many of the older indigenous varieties, and partly on account of superstitions. However, the use of sickles is becoming much more widespread with the introduction of more evenly ripening varieties, and has always been customary among some people, for example the Chinese. Threshing and winnowing may be done by hand, or the grain may be trodden out by animals. (Plate 21.)

In most places only one rice crop is taken annually, the locally adapted varieties grown in different places varying in their period of maturation from five to as much as eight months. For the remainder of the year the land lies fallow and may be grazed by domestic animals, where these are kept. (Plate 22.) On the other hand, if the population is dense and provided water supply and drainage permit, a second crop of rice may be taken during the same year. Growing two rice crops annually on the same land demands an ample water supply throughout most of the year, which can usually only be provided by an efficient irrigation system with good water control and drainage. Many existing irrigation schemes, which are adequate for one rice crop, would need to be modified and improved to allow of growing two rice crops in the year. Double cropping is also only possible if short-term, non-photoperiodic varieties, adapted to local conditions, are available and in many places such varieties have not yet been established. Furthermore, the preparation of the land, planting and harvesting can only be accomplished twice within a year if the cultivators are prepared to work hard and to observe a rather strict discipline in regard to planting, irrigating and harvesting times. Failing this it becomes impossible to organize satisfactorily the provision of adequate water to all the farmers participating in an irrigation scheme. For one or more of these reasons double cropping with rice is only practised to a limited extent in most countries, but in some areas, especially where the population is dense, it has been customary for centuries. For example, Gourou (1953), mentions that on the densely populated alluvial plains of the Ton

King delta, where rice has been grown for 2,000 years, 50 per cent of the land grows two rice crops a year.

Following a main rice crop with other annuals in the same year is more common than double cropping with rice, but is limited to areas where adequate moisture is available from rainfall and irrigation, and where the padi fields can be sufficiently well drained for crops that require good aeration of the soil. Where such conditions prevail, the extent to which other crops are grown in the off-season for rice depends mainly on the population pressure on the land and the energy and skill of the people. (Plate 23.)

Highly intensive use of the padi fields for this purpose is made in very densely populated countries, such as China, where people are both extremely hard working and possessed of great agricultural knowledge and skill. Wherever water supply is adequate other crops are grown in rotation with rice, and where drainage is poor it is improved by constructing ridges and channels. Mixed and successional cropping is practised with skill in order to maximize production, minimize labour requirements, and make the utmost use of the land and the growing season. Crops are usually planted in carefully spaced rows, often with recurrent rows of crops of different ages, a second crop being planted as the first approaches maturity. Planting alternate rows of cereals and legumes is common, as it is thought that this permits the soil to be more completely foraged by the root systems and provides an even distribution of the nitrogen-fixing nodules of the legumes throughout the field. Well-rotted composts and other manures are applied to the land to release nutrients rapidly for the growing crop. These and other practices are well described by King (1927) and it will be seen that, combined with the transplanting of rice, they are designed to economize in time so that the maximum possible use of the land can be made throughout the entire growing season.

The conditions obtaining in the padi fields are clearly very different to those experienced in the dry-land cultivation of annual crops and, in general, they are more conducive to the maintenance of fertility. There are no losses by erosion from the levelled, bunded fields, but an inflow of soil and nutrients in irrigation water, or by wash from higher land. The quantities of nutrients in the irrigation water vary greatly with locality. In some places they may be adequate to meet all the major element requirements of a good crop of rice annually (Grist, 1965), and often the amounts of some, but not all, nutrients will be sufficient to provide at least a large proportion of the needs of the rice. For example, in parts of Malaya the calcium and magnesium in the irrigation water are adequate for a good crop of rice, the potash sufficient to meet a large part of the requirement, the nitrogen supplied is appreciable, but the phosphate low (Kanapathy, 1962). Losses of nutrients by leaching are commonly small. Although large amounts of water may be applied to the land, much of it is disposed of by evaporation and by the transpiration of the growing crop. The soils are not usually free draining and the percolation of water through them is further reduced by the destruction of structure resulting from preparatory cultivation and puddling. Loss of structure, which may seriously reduce the productivity of dry land under annual cropping, is of much less importance for the rice crop, although it may have some adverse effect on the yields of other crops grown in rotation with rice. There is no loss of organic matter due to burning, as there is in shifting cultivation, but a regular addition of organic matter to the soil by ploughing in crop residues and weeds. The contribution made by the latter may be considerable where only one rice crop is taken annually and weeds grow strongly during the off-season fallow.

The padi fields are waterlogged for part of the year, which means that essentially anaerobic conditions prevail in the soil below the top one or two centimetres. Rice is adapted to grow in soil containing very little oxygen. It possesses large air spaces in the leaf sheaths and root cortex which enable oxygen to be readily transferred from the shoot to the roots, at any rate up to the tillering stage (van Raalte, 1940, 1944). Subsequently, as shoot elongation occurs and nodes develop, thus increasing the distance over which oxygen must be moved and the resistance to its passage, it may be less readily transferred. But at this stage the plant develops a mass of very fine much-branched roots that grow horizontally on the soil surface and in the irrigation water, or upwards into the air, and are able to absorb oxygen (probably mainly that dissolved in the irrigation water) and to transport it to other parts of the root system (Sethi, 1930; Alberda, 1953). There is also some evidence that rice possesses an enzyme system that enables its tissues to function anaerobically for a limited time; thus rice is able to germinate and grow under very low oxygen concentrations at which wheat and barley fail to do so (Cannon, 1925; Taylor, 1942; Vlamis and Davis, 1943, 1944).

Important effects on soil nutrients result from the fact that while the soil is waterlogged it develops reducing conditions below the shallow surface layer that remains in contact with air, or with oxygen-containing irrigation water. The anaerobic decomposition of organic matter proceeds more slowly than the oxidative decomposition occurring in dry-land soils and results in the production of methane instead of carbon dioxide, ammonia instead of nitrates, and sulphides instead of sulphates. The availability of certain cations in the soil solution is increased; this applies especially to iron and manganese, but also to calcium and magnesium, which are freed in the soil solution by base exchange following the formation of ammonia (Pearsall, 1950). There is also some evidence of increased availability of phosphate (Pearsall, 1950; Islam and Elahi, 1954; Kanapathy, 1957; Mitsui, 1960). Thus a slower rate of decomposition of organic matter, and greater availability of many of the nutrient elements, while the soil is waterlogged, may favour the maintenance of fertility in padi fields in comparison with cultivated, dry-land fields.

There is evidence that significant amounts of nitrogen are fixed in waterlogged padi fields. Details of the process have not been fully elucidated but probably most of the nitrogen is fixed by blue-green algae growing on the soil surface or floating on the surface of the irrigation water. De (1939) showed that algae fix nitrogen in rice soils, and Watanabe *et al.* (1951) demonstrated that thirteen species of blue-green algae, all common in south-east Asia, were capable of fixing nitrogen, and that their presence significantly increased the growth and yield of rice in water culture, pot and field trials.

Some workers have suggested that the principal organism concerned in nitrogen fixation is *Azotobacter*, which uses algal bodies as a source of food, but De (1939) showed that algae alone fix as much nitrogen as algae and *Azotobacter* together, and Sulaiman (1944) found that dead algal bodies were ineffective as a source of nutrients for *Azotobacter*, and that the latter failed to multiply, or to fix appreciable quantities of nitrogen, in waterlogged soil. However, *Azotobacter* may be capable of fixing a certain amount of nitrogen in waterlogged soils if it is in association with living rice roots. Sahasrabudde (1936) showed that nitrogen fixation was helped by the presence of growing rice roots, and Uppal *et al.* (1939) found that total nitrogen and numbers of *Azotobacter* were almost invariably higher in soils carrying rice than in similar soils uncropped. Encrustations of ferric compounds observed around rice roots indicate that oxygen diffuses outwards from the roots to oxidise ferrous to ferric compounds (Sturgis, 1936; van Raalte, 1944); consequently it may be that changes in oxygen concentration or reaction near the roots may favour fixation of nitrogen by *Azotobacter* or other organisms.

There appear to be no reliable figures for the amount of nitrogen fixed in the field, but some estimates have been made from pot experiments. Watanabe *et al.* (1951) estimated that fixation at the rate of 20 lb of N per acre occurred in an unspecified period during the growth of one rice crop. De and Sulaiman (1950), who grew rice for five years in presence or absence of algae, found that the nitrogen content of the soil increased from 895 to 1014 p.p.m. with algae present, but declined to 814 p.p.m. in their absence. During the first two or three years there was little difference in rice yields from the two treatments but thereafter there was a progressive increase in yield in presence of algae and a decline in their absence. De and Mandal (1956) used gas analysis to measure the amount of nitrogen fixed when rice was grown in closed illuminated jars, and obtained figures ranging from 13·8 to 44·8 lb of N in five or six weeks. Willis and Green (1948) concluded that the amount of nitrogen fixed was more than enough to support a good rice crop.

Some nitrogen may be lost from waterlogged soils in the form of gas. This is due to the fact that, if nitrate derived from the surface application of fertilizers and manures, or from the decomposition of soil organic matter in the shallow oxidizing layer, passes down into the reducing zone, it is reduced to nitrite and then to nitrogen. For this reason, sulphate of ammonia is sometimes placed into the reducing layer. The ammonium ions can then remain unchanged and be absorbed as such by the rice roots, instead of being applied as a surface dressing to the oxidising layer, where ammonia is converted to nitrate, which may move downward and be reduced to gaseous nitrogen. However, if there is a vigorously growing rice crop on the land to absorb nitrate or ammonia then losses of nitrogen, either as gas or by leaching, are likely to be slight.

Farming practice, although very variable, is generally calculated to maintain the fertility of the rice fields at least at a moderate level. As a rule, where land is used non-intensively and only one rice crop a year is grown, no very positive measures for the maintenance of fertility are taken by the farmer. The annual fallow, with the dung dropped by animals grazing it and the ploughing in of weeds and stubble, will serve to maintain fertility at a level adequate for the production of one crop a year. The more intensively the land is used the more energetic and skilful are the farmers in adopting a variety of methods of maintaining fertility. Practices described by King (1927) for parts of China, where land has long been intensively cropped with rice and other annuals in rotation, provide an outstanding example. All animal manure, household and crop wastes, and plant ashes derived from fuel, are carefully collected and applied to the land, usually in the form of composts. Night soil is widely used, commonly being stored in earthenware receptacles, diluted with water and thriftily applied to individual plants at appropriate intervals. Mud, rich in organic matter, is periodically dug out from canals, reservoirs and fish ponds and spread on the land, or used in composts. Green manures may be included in the rotation, sometimes for direct ploughing in, but more often for composting. Hillsides bearing trees and shrubs are regularly cut over to provide material for trampling into the mud of the rice fields, or for the preparation of composts. Great use is made of composts, care being taken to see that they are well rotted, with much of their organic matter broken down, so that they rapidly release nutrients to a growing crop. It should be mentioned, however, that although the fertility of the densely populated alluvial plains of China has been maintained at a relatively high level by deposition of silt from irrigation or floodwater, by application of night soil, and, to a lesser extent, by green manuring with vegetation imported from hilly land, this has been done partly at the expense of other land. Much hilly land in China has been depleted of fertility and the general level of fertility of lands other than the alluvial plains is only moderate. Furthermore, the widespread use of night soil has resulted in a high death and disease rate from soil and water pollution (Richardson, 1946).

Where drainage and water supply during the 'off-season' permit the growing of other crops in rotation with rice, the tillage and the resulting aeration of the soil have beneficial effects, provided that cultivation is not done too deeply, nor when the soil is too wet. The physical and biological condition of the soil is improved, organic matter is oxidized to release nutrients, especially nitrate, and weed control in the following rice crop is made easier. Compared with fallowing, the rotation of other crops with rice does not necessarily reduce rice yields, and may even improve them, although the effects vary with soil, intensity of cropping, crops grown and whether or not manuring is done. For example, on the highly fertile soils of the Tanjong Karang area of Malaya, farmers have for some years practised intensive off-season cropping, with a variety of crops but with little or no manuring, and there is evidence that this has increased rice yields, which are very high. Similarly in Province Wellesley, prewar experiments (Hartley, 1947) showed that off-season cropping with groundnuts, sweet potato, okra, brinjal or cowpea enhanced padi yields, at least over a few years, while in a postwar experiment (Allen, 1956) off-season cropping with sweet potatoes for eight successive seasons gave higher yields of rice than those following fallow, even when no manure was used. On the other hand, the results of experiments in various parts of India, reported by Pillai (1958), seem to indicate that rice yields following non-leguminous crops are lower than those after fallow, but that growing and harvesting a legume in the off-season does not reduce rice yields, and may even improve them. In general, prolonged and intensive off-season cultivation of non-leguminous crops results in declining yields, unless manuring is done. The use of fertilizers and manures is therefore usual where such intensive cropping is practised, although in most parts of Asia their use on the rice crop is as yet on a small scale.

Growing green manures in preparation for rice, although well established in some places, is impracticable in others, where the land is either too dry or too ill-drained during the off-season. Where it is practicable it is usually beneficial. Grist (1965) reports that, in general, experimental evidence indicates that green manuring improves rice yields, and Pillai (1958) reaches a similar conclusion for India. Ploughing in is normally done only ten to fourteen days before transplanting. Earlier incorporation into semi-dry soil results in the decomposing green manure producing considerable quantities of nitrate which, upon subsequent flooding, are liable to be lost as gaseous nitrogen, or by leaching. Where it is impracticable to grow green manures in the padi fields, leafy material from waste land or from the bunds is often trodden into the mud after flooding and a few days before transplanting.

Livestock

Where livestock are kept, the part they play in the farming system varies greatly. For example, the majority of Malay farmers keep cattle and buffaloes primarily for work, and do not attempt any appreciable production of meat or milk, nor make use of the dung to manure their fields. During the rice-growing season their animals subsist on rough grazing of poor quality, or on grass cut from waste land for stall-feeding. In the off-season they graze the padi fields, where the grazing is often poor, with the result that they are in poor condition to tackle the heavy job of land preparation at the beginning of the wet season.

On the other hand, the animals, principally pigs, kept by Chinese farmers in Malaya, are well integrated into the farming system. They are largely fed on cassava and on crop residues, such as unsaleable parts of fresh vegetables and rice by-products, and their manure is carefully conserved and applied either to crops or fish ponds. Similarly, where stock are kept in China they play a useful part in the farming system, and in many parts of India and Ceylon farmers make and apply manure, and fold animals systematically on the padi fields during the off-season.

Permanent Cropping with Annuals and Perennials on Dry Land

A common component of systems associated with swamp rice in the wet tropics is a plot of unirrigated higher land, around the dwelling house, on which the farmer grows both annual and perennial crops. A great variety of annual crops may be grown, including vegetables, condiments, spices, tobacco, medicinal herbs and, sometimes, a little maize and several kinds of roots; but usually the total area devoted to annuals is small, and some of them are grown under the shade of trees. Most of this plot is usually devoted to an unsystematic, confused mixture of many kinds of trees and shrubs, partly self-sown, which provides the family with firewood, timber, building poles and materials for making baskets, mats, hats, roofing etc., as well as a variety of fruits and certain other foodstuffs. In Malaya, for example, the mixture will commonly include citrus, rambutan, durian, mango, and other fruit trees, coconut palms, coffee, bananas, pineapples, bamboos and a number of less well-known species. (Plate 24.)

This mixed tree culture is by no means confined to Asia, but is found in other parts of the wet tropics, such as the West Indies and parts of Africa, often in association with shifting cultivation. It produces conditions similar to the forest, the soil being protected from the direct impact of rainfall by a closed canopy and by the deposition of litter, so that erosion losses are slight, and a more or less closed cycle of nutrients is established. Consequently the land can be used permanently in this manner with only very slight, if any, decline in fertility. For this reason, and also because it is possible that some species benefit from the shade of others, various authorities have suggested that a mixed tree culture forms an ideal basis for peasant land use in the wet tropics. It has further been suggested that mixed tree culture might be accompanied by the short-term cultivation of a small plot of annual crops, which could rotate round the perennial area, the tree crops being replanted at the end of the period of annual cropping. However, although no figures are available for the productivity of such mixed plots of trees, the indications are that it is not high. It seems fairly certain that greater returns can be obtained from tree crops by planting them in pure stands, where the particular cultural requirements of a single crop can be met without the risk of creating an environment that might be unsuitable for other crops, as might happen in a mixed stand.

Shifting Cultivation

The limited, and often sporadic, shifting cultivation of hill rice and other annuals that is frequently practised by farmers whose main concern is the permanent cultivation of swamp rice, has already been described in Chapter 7. It should be noted that, even where the population is fairly dense, this shifting cultivation will usually do far less extensive damage to the land than that which occurs in the savannas and forest margins of Africa. The reasons for this are that smaller plots are cultivated for only short periods, fallow regeneration is rapid and vigorous, and there is no extensive damage from fires. Only where cultivation is prolonged, or is several times repeated after only short fallows, will soil fertility be severely depleted and the natural vegetation degraded to a fire subclimax of coarse grass and shrubs.

Fish Culture

Fish culture is not infrequently undertaken in conjunction with the farming systems under discussion. In some places a not inconsiderable production is obtained from the padi fields themselves, which can be easily populated by introducing a few pairs of indigenous air-breathing fishes that are easily transported and prolific. When the water is drained off to ripen the rice, the fish move to troughs or tanks dug in the corners of the fields, and are then harvested. More important is the pond culture of fish which is linked with farming operations by the use of the excreta of domestic animals, especially pigs, to fertilize the pond water, and the periodic manuring of the fields with mud scooped from the ponds (Hickling, 1961).

The Low Standard of Living Associated with Farming Systems Dependent on Swamp Rice

Although farming systems that are largely dependent on swamp rice production are generally effective in maintaining soil fertility and crop yields at a satisfactory level, and have been operated successfully for many centuries in Asia, they usually procure only a low standard of living for the farmers. There are two main reasons for this: (1) problems arising from population density and land tenure in the rice growing areas; (2) the low returns per man obtainable from traditional methods of rice growing.

Population density is usually high in relation to the areas available for wet rice growing, and land is usually individually owned and inherited. The combination of increasing population with individual land ownership, and laws of inheritance that commonly involve subdivision among heirs, naturally tends to result in many people having holdings which are too small and fragmented to yield a satisfactory return. In addition, although the ability to borrow money on the security of his own land should be advantageous to the farmer, it is frequently not so because proper credit facilities are not available and the small farmer becomes chronically indebted by borrowing at high rates of interest from large land-owners and traders. The position of tenant farmers, of whom there are many, is frequently still more unsatisfactory. As land becomes scarcer, so landlords raise rents and often, in order to profit from rising rents, they will only lease land on short tenancies. Furthermore, the tenant is frequently forced to sell his surplus crops to the landlord, who provides advances against the crop, throughout the growing season, at high rates of interest. Thus tenant farmers can often only afford to rent small plots of rice land, have no security of tenure, and are chronically indebted. These problems can be dealt with by legislation to control rents, to provide security of tenure and to prevent fragmentation, by the establishment of cooperatives and the provision of credit and by improved marketing arrangements, but in many places progress on these lines has so far been limited.

Bearing in mind that rice is not a high-priced crop, the returns obtainable from its production by the laborious traditional methods are low. Even with the aid of animals, land preparation on heavy, wet soils is slow and arduous. Both transplanting and harvesting make heavy demands on labour, and outside help is often required with these tasks on family holdings. Threshing and winnowing, although not heavy work, are time consuming. Thus, even where there is no serious shortage of padi land, the area which a man and his family can cultivate is limited, the input of labour is high, the surplus rice available for sale after providing for the needs of the family is not large and it yields only a small cash return. Where it is possible to grow other crops on the same land during the off-season, these will provide further income, but in many places this is impracticable.

In many countries much could be done to mitigate these difficulties by various measures to increase the yield per acre of rice. Probably of paramount importance is more work on the selection and breeding of better varieties, particularly the production of hybrids combining the high yield potential of *indica* varieties with the enhanced fertilizer response of *japonicas*. In many places improved irrigation, drainage and water control could markedly increase production. If accompanied by the introduction of improved varieties and good water control, a greater use of fertilizers would be profitable. Yields could also be raised by better control of pests and diseases and by general agronomic improvements, such as better nursery techniques and more timely planting.

However, even where really high yields are obtained by the most efficient application of traditional methods, the return per man remains relatively low. Consequently in recent years there has been much interest in mechanization, which might make a higher output per man possible by reducing the labour required for some of the more laborious operations, permitting more timely planting, and enabling a family to handle a larger holding. Tractor preparation of land and mechanical threshing are already extensively employed in some countries, but economic methods for fully-mechanized rice production, including the elimination of hand transplanting, have been successfully operated only in limited areas, usually where very good control of irrigation water is possible, or where most of the preparatory cultivation can be done before flooding. A major problem in long-term mechanized rice production is weed control, which is effected under the traditional system by preparatory cultivation after flooding combined with the practice of transplanting. The use of modern herbicides will probably enable this problem to be overcome to an increasing extent. However, both for technical reasons and on account of the capital and operating costs of machinery, mechanization is unlikely to be extensively introduced in the great swamp rice growing areas in the foreseeable future.

CHAPTER 10

The Culture of Tree and Bush Crops

For several reasons the productivity of many of the older plantations of perennial crops, both on estates and smallholdings, is low by modern standards. In the first place, these areas were mostly planted with low-yielding, unselected seedlings. Secondly, erosion and loss of soil fertility often resulted from failure to adopt adequate measures for soil and water conservation during the early years of the plantation. Thirdly, fertilizers have only been used on a modest scale, if at all, often because they failed to produce profitable responses from planting material of low yield potential. Fourthly, owing to lack of knowledge and experience, practice in respect of such matters as planting density, transplanting, pruning and shade management has not been of the best. The replanting of these older plantations with improved planting material, coupled with the adoption of more efficient and intensive cultural methods, has been proceeding for some years at a pace which varies with the crop and the locality, but large areas of most of the main tree crops still await replanting and the introduction of modern methods.

The most important factors that make for high productivity in modern plantings, or replantings, are as follows:

1. The use of improved planting material, selected to give high yields of good quality produce, possessing good secondary characters.
2. Cheap and practicable methods of propagating the improved planting material on a large scale.
3. Efficient transplanting and early care of the young trees, in order to obtain a full stand of plants that grow vigorously and reach the productive stage as quickly as possible.
4. Good horticultural practice in respect of such matters as spacing, shade management, windbreaks, pruning, etc.
5. Methods of soil management, including soil conservation works, cover crops, weed control, mulching and manuring, which will maintain soil fertility and sustain high yields.
6. Control of pests and diseases.
7. Improved methods for the exploitation of the crop, such as modern techniques of tapping and yield-stimulation in rubber, or the use of synthetic growth substances to induce early and uniform fruiting in pineapples.

Production of Improved Planting Material

In plantations of unselected seedlings there is usually great variation in yields of individual trees, the majority being poor yielders while a small proportion are outstandingly productive and produce a large part of the total crop. For example, studies of seedling rubber in Malaya showed that 9·6 per cent of the trees produced 28 per cent of the total crop; and in a seedling plantation of tung oil trees (*Aleurites montana*) Webster (1950) found that 70 per cent of the trees yielded less than 20 lb seed per tree but 3·8 per cent gave over 50 lb per tree. The first step in improvement is therefore to pick out high-yielding individuals which also possess other desirable characters, such as vigour, early cropping, disease resistance, or wind resistance. Selection will also usually be made for features contributing to the easy exploitation of the tree or to good quality in the final product, such as good bark and latex quality in rubber, or thin-shelled pods, large bean size and good flavour in cocoa.

With some crops considerable improvement can readily be achieved by multiplying selected mother trees by vegetative propagation. With tea and cocoa, selections can be multiplied as rooted cuttings, thus producing clonal plants that are genetically identical with the mother tree. With other species which have so far proved difficult to propagate on their own roots on a large scale (e.g. rubber), multiplication has been by budding or grafting on to seedling rootstocks. By this method only the scion is genetically identical with the selected parent, and its performance may be modified by the effect of the seedling rootstock. With a few crops that are mainly self-pollinated and highly self-fertile, multiplication of primary selections can be by seed, since the open-pollinated seedling progenies of selected mother trees are fairly uniform and true to type. Thus with *Coffea arabica*, in which 90 per cent of the seed results from natural self-pollination (Krug, 1958), trials have usually shown no significant differences in yield between open-pollinated seedling progenies and vegetatively propagated clones derived from the same mother tree (Sanders, 1951). Open-pollinated seed can also be used for propagating selected mother trees of crops which exhibit a high degree of polyembryony, such as some varieties of mango and citrus. On the other hand, the use of open-pollinated seed from selected mother trees of cross-pollinated species is likely to be of little value, since unknown male parents are involved and the resulting progenies are usually highly variable and give average yields little, if any, better than those of unselected seedlings. This is illustrated by the yields of an experiment with coconuts in Ceylon (Table 22) which compared the preformance of seedling progenies derived from nuts taken from: (a) selected high-yielding palms, (b) selected low-yielding palms, and (c) a heap collected from a block of palms known to give a high total yield (Coconut Res. Inst. Ceylon, 1953). It will be seen that there are no significant differences in yield between the three types of material. With cross-pollinated species that cannot be vegetatively propagated, such as coconut and oil palm, improvement can only be effected by the controlled crossing of selected trees to produce legitimate seedling families.

Table 22. *Average yields of coconuts per acre per annum for three years* 1951–3
Coconut Res. Inst. Ceylon (1953)

Source	Nuts	Copra (lb)
Selected seedlings from		
A. High-yielding palms	2,677	1,249
B. Low-yielding palms	2,573	1,309
C. Heap nuts	2,689	1,261
Selection in nursery	2,646	1,273
No selection in nursery	2,358	1,136
Significant difference ($p = 0.05$) between selection and no selection in nursery	142	127

In the long term, a breeding programme is necessary for the continued improvement of all crops, since vegetative propagation, or the use of open-pollinated seed of self-pollinated crops, results in clones or seedling progenies which, at best, reproduce the qualities of a selected mother tree and usually fall short of this. Further improvement is dependent on a breeding programme, which ultimately aims at combining desirable qualities from a number of parents to produce planting material possessing high yield potential and all other desirable commercial qualities.

Breeding programmes have been operated with most of the major plantation crops for varying periods, and variable progress has been achieved. With coconuts no bred material yet appears to have been planted on a commercial scale. Legitimate seedlings resulting from controlled crosses between high-yielding palms have been planted in isolated gardens in Ceylon and elsewhere for the production of elite seed (Liyanage, 1953), but no records of palms raised from such seed yet appear to be available. Long-continued breeding in the banana has not yet produced a variety that is cultivated on any considerable scale (Simmonds, 1959). Improved planting material of cocoa is still mainly vegetatively propagated from primary selections, but recently exceptionally good vigour and yields have been obtained with hybrid seedling progenies derived from crossing selected clones (Cope and Bartley, 1957). Improved varieties of *Coffea arabica* are in use as a result of breeding in Brazil and elsewhere (Krug, 1958), but less progress has been made with the cross-pollinated, self-sterile *Coffea canephora* (Ferwerda, 1958). Large increases in commercial yields have been obtained by breeding oil palms (Hartley, 1957; Haddon and Tong, 1959). Unselected seedling rubber in Malaya yields at most about 500 lb per acre, but selection and breeding has produced clones that are already giving 2,000 lb on a commercial scale, and newer clones, still under large-scale experimental trial, have yielded at the rate of over 2,500 lb per acre.

Propagation by Seed

Seed is used for the propagation of coconuts, oil palms, all three species of coffee and rubber, the last being planted as clonal seedlings or buddings on seedling rootstocks. A good deal of tea and cocoa is still planted as seed, and citrus is usually budded on seedling stocks. It is generally desirable to sow seed as soon as possible after it is harvested; many seeds, including those of rubber, oil palm, tea, cocoa and coffee, lose viability fairly rapidly on storage. With some seeds viability can be maintained for considerable periods under appropriate conditions of storage; for example, stratification in moist sand or charcoal, in a cool place, maintains the viability of tea and tung seeds for some months (Leach, 1936; Webster, 1948), and storage at room temperature in charcoal or sawdust of 20–30 per cent moisture content maintains the viability of rubber seed for at least four months (Rub. Res. Inst. Malaya, 1962). Sound, fresh seed of most of the plantation crops germinates fairly rapidly on planting in ordinary nursery beds, but oil palm seed is an exception and gives slow and erratic germination unless special methods are adopted. The technique now recommended is to place seed of saturated moisture content (20–21 per cent), contained in polythene bags, in heated chambers, where it is maintained at a constant temperature of 38–40°C for seventy to eighty days. It is then cooled to ambient temperatures (25–28°C), when a rapid flush of germination occurs, lasting twelve to fifteen days and resulting in over 80 per cent germination with good seed (Rees, 1959a and b).

Although seed can be sown 'at stake' in the field, it is usually better to raise seedlings in nurseries or plastic bags, and then to transplant to the field. Nurseries should be sited near water, in a sheltered position free from risk of flooding, on reasonably fertile soil possessing adequate depth and good drainage. Preparation should include thorough and fairly deep cultivation, eradication of weeds and removal of tree roots and stumps. It is important to select a row spacing and plant density which will give good growth combined with economy and convenience in working, but the best layout naturally varies with the crop and the type of planting material to be raised.

Rigid selection of plants in the nursery before transplanting is of the greatest importance. All poorly grown, off-type, or diseased seedlings, or those with twisted or bent roots, should be discarded and only the best and most vigorous plants transplanted to the field. The importance of selection of coconut seedlings for early germination and vigour in the nursery is illustrated in Table 22, from which it will be seen that, irrespective of the origin of the seed, selection in the nursery resulted in marked and significant increases in yield. With this crop selection in the nursery is of far greater value than planting open-pollinated seed from selected mother trees.

Vegetative Propagation

The main purposes for which vegetative propagation of perennial crops is used are as follows:

1. To propagate plants which do not produce viable, or sufficient, seed; examples being the banana and the navel orange.
2. To establish clones from selected high-yielding mother trees.
3. To select for sex in dioecious trees, such as the nutmeg, or to eliminate a large proportion of low-yielding, predominantly male trees in a monoecious species such as *Aleurites montana*, the tung oil tree (Webster, 1950).
4. To provide resistance, or immunity, to pests or disease by a suitable combination of stock and scion.
5. To obtain earlier fruiting; plants propagated vegetatively generally come into bearing more quickly than corresponding seedling types.

The chief drawback to vegetative propagation is that it requires more skill and is more expensive than using seed. Unless really efficient techniques are evolved, the use of vegetative propagation on a large-scale is likely to be slow and expensive.

Information on the general principles and practice of vegetative propagation can be obtained by reference to such standard books as those of Adriance and Brison (1955) and Garner (1958). Details of the techniques used for the vegetative propagation of a wide range of tropical and subtropical plantation crops and fruit trees have been published by Fielden and Garner (1936, 1940); although new techniques have been developed for some crops, the majority of the methods they described are still standard, or have only been slightly modified. The four main methods of propagating tropical perennial crops are by cuttings, by layering (particularly aerial layering, or marcotting), by budding and by grafting.

CUTTINGS

Satisfactory methods for the propagation of tea and cocoa by stem cuttings have long been available and are used commercially. Large-scale propagation of stem cuttings of citrus, coffee and rubber is also practicable, but has not hitherto been extensively employed. Leafy, semi-hardwood, or softwood, cuttings have generally been found preferable to leafless hardwood cuttings. The latter have proved unsatisfactory, or inferior to leafy cuttings, with cocoa (Evans, 1953), tea (Eden, 1958), *robusta* and *arabica* coffee (Cramer, 1957; Fernie, 1958) and rubber (Tinley, 1961). Leafless hardwood cuttings are dependent upon their reserves of carbohydrates to maintain them until they root and shoot, and the unsatisfactory results obtained with them are probably mainly because there is insufficient accumulation of such reserves in the stems of evergreen tropical tree crops. The presence of leaves on cuttings is advantageous partly due to their photosynthetic activity and partly because they supply auxins which stimulate rooting. Evans (1953) demonstrated that the former was by far the more important function of leaves on cocoa cuttings, but with citrus their action in supplying auxins is important (Adriance and Brison, 1953).

The best results are obtained when cuttings are made from young stems possessing relatively good reserves of nutrients. With cocoa (Evans, 1953) and rubber (Tinley and Garner, 1960) such material is obtained by taking stem pieces from vigorous flushes that have just matured on well-fertilized nursery plants, which are frequently pruned. Coffee cuttings are best made from the upper parts of vigorously growing young suckers (Cramer, 1957; Fernie, 1958), and it has been shown that cuttings from new growth on pruned tea are superior to those from bearing bushes (Tubbs, 1947; Wight, 1942). It is important to conserve the moisture content of leafy cuttings before setting by taking them in the early morning and adopting precautions to avoid water loss during their preparation.

To obtain good results with some species, including cocoa, coffee and rubber, it is necessary to maintain a high atmospheric humidity, and to control light intensity and temperature carefully, during the rooting period. Too low a light intensity may reduce the photosynthetic activity of the leaves to a level inadequate to supply the carbohydrate requirements of the developing roots, the formation of which may be prevented or delayed. On the other hand, too much light may raise the temperature and reduce humidity sufficiently to cause wilting and death of the leaves. Formerly, cuttings of some species were rooted in shaded concrete or brick bins with movable glass tops; constant skilled attention, careful adjustment of shade and frequent watering were needed to maintain satisfactory conditions within the bins. Modern mist-spray systems, which can be used on shaded beds as well as in glasshouses or plastic tents, have simplified matters since the cool spray of water falling continuously on the leaves maintains humidity and prevents undue temperature-rise, while permitting higher light intensities than could be used in the old bins. Cocoa cuttings are rooted under mist-spray in brick-sided beds with overhead shade to keep the light intensity at 20 per cent of full daylight (Evans, 1953), and rubber cuttings are rooted under similar conditions, but using a light intensity of about 50 per cent (Tinley and Garner, 1960). (Plate 25.)

The rooting medium in which cuttings are set should be fairly retentive of moisture, yet well-drained and aerated. Sand, peat, coir fibre, composted sawdust and various synthetic media, such as vermiculite, are all used. Porous media are probably better than sand because they retain air within, as well as between, the particles and provide a satisfactory air: moisture relationship over a wider range of moisture contents.

Some species are more easily rooted, and are less exacting in their requirements in respect of atmospheric humidity, light, temperature and air: moisture relations in the rooting medium, than others. For example, tea is successfully rooted using open beds of loamy soil in which the cuttings are shaded by means of bracken fronds stuck into the soil (Eden, 1958).

After rooting has occurred cuttings require a 'hardening-off' period before being fully exposed to the outside atmosphere and finally transplanted to the field. Conditions favourable to maximum root growth are required during this period, so that the newly formed roots can speedily develop an absorbing surface capable of meeting the water requirements of the growing shoot in full daylight. Cuttings raised under mist-spray are usually transferred to polythene bags after rooting, left for about a week in the drift of the spray beside the beds, and then kept for a further period under less humid and shady conditions before planting out. (Plates 26, 27.)

LAYERING AND MARCOTTING

Layering is a term which covers all procedures designed to cause stems to root before separation from the parent plant, but the only method that is at all widely used in the tropics is aerial layering, or marcotting, which is employed with some species of fruit trees that are not readily rooted from cuttings. The technique consists in partial or complete ring-barking of a suitable stem, covering the damaged area in wet moss wrapped in a polythene sleeve, and leaving it until roots are visible through the polythene before severance from the tree. Although a rather laborious procedure, it is still widely used to propagate various fruit trees, including the mango and the nutmeg.

GRAFTING

Of the numerous methods of stem grafting, none is normally employed with any of the major plantation crops and the most widely used method is that of approach-grafting, or inarching, which is used for the propagation of some fruit trees, especially mangoes. Usually this involves taking seedling rootstocks, aged twelve to eighteen months and contained in baskets or bamboo pots, supporting them on scaffolding or platforms to bring them adjacent to terminal scion shoots of similar diameter, and grafting the two together. As about three months is required before a union is formed and the graft can be severed from the parent plant, followed by a month's hardening-off, the whole process requires a minimum of eighteen months. Bharath (1958) has described a simpler technique of approach-grafting mangoes, which might be used for some other species. This depends on the use of seedling stocks only six weeks old, with the roots and seed surrounded by moist moss wrapped in polythene, which can be tied, and grafted, on to a selected scion on the parent tree. As a union is formed in about four weeks, this method produces grafted plants in six months and more cheaply than the older technique.

BUD-GRAFTING

Budding is only employed extensively for two major crops, rubber and citrus, but it is also fairly widely used for a number of fruit trees, including avocado, mango, durian, litchi and rambutan. The two commonest methods are shield, or T, budding, and various forms of patch-budding, including the modified Forkert method, which is used for rubber. All the common budding techniques have been repeatedly described in the literature on various crops, and most of them are given by Fielden and Garner (1940) and by Garner (1958). The standard practice with rubber has been to bud seedling rootstocks at ten to twelve months after planting in either the field or the nursery, using budwood from shoots one to two years old. (Plates 28–32.) A recent innovation is 'green strip budding', whereby seedlings three or four months old are budded with small buds taken from very young shoots (Hurov, 1961; Tinley, 1962). This procedure has the advantage of being simpler and quicker than conventional budding, and it reduces the period of immaturity in the field. 'Green strip budding' might well be used with advantage for other species.

STOCK: SCION RELATIONS

The use of genetically variable seedling rootstocks for grafting or budding introduces some variability into the resulting clone. Selected clonal stocks, raised as cuttings, or as nucellar seedlings from polyembryonic seeds, would not only eliminate, or reduce, variability due to rootstock, but might also offer the possibility of using favourable combinations of stock and scion that might have advantages in enhanced vigour, disease resistance, etc. So far, however, clonal stocks have only been used commercially for certain varieties of citrus, where they may confer resistance to certain root diseases, as well as greater vigour and increased yields of better quality fruit. There is very little information on stock: scion relationship for other tropical crops. Experiments with cocoa in Trinidad indicated that stock: scion interaction is negligible, and hence that the use of selected stocks is unlikely to have any advantage over propagating clonal material on its own roots (Murray and Cope, 1953, 1955, 1959). With rubber, investigations with clonal seedling stocks of *Hevea brasiliensis* have been inconclusive (Buttery, 1961), but it is generally considered that only the vigour of the stock is important and the universal practice is to use ordinary or clonal seedling stocks selected for vigour and uniformity.

Transplanting

The planting material for transplanting may be either entire leafy plants (seedlings, rooted cuttings, or small budded plants) or pruned plants that are devoid of leaves such as seedling and budded stumps.

Stumps are prepared from seedlings in the nursery by cutting back the main stem to 4 to 18 inches of brown wood, and pruning the taproots to 12 to 18 inches and the lateral roots to 3 to 9 inches, depending on the species. Seedling stumps are commonly used for tea and coffee, and some rubber is planted as clonal seedling stumps. Budded stumps are prepared by budding seedlings about twelve months old in the nursery, cutting back immediately above the dormant bud-patch and pruning the roots. Stumps are normally planted with bare roots and, being leafless and devoid of absorbing rootlets at the time of transplanting, they have to make their initial growth in the field at the expense of their stored nutrients. They should therefore be sufficiently well grown to have fair nutrient reserves. On the other hand, excessively large stumps cost more to dig out, transport and plant, and are slower in starting growth after transplanting than smaller, younger stumps. There is probably no advantage in leaving the lateral roots longer than about 6 inches; experiments with rubber (Morris, 1934) and tung oil trees (Webster, 1942) showed no benefit from transplanting budded stumps with longer laterals. With budded stumps the stem is usually cut back above the bud-patch seven to ten days before lifting from the nursery, in order to break the dormancy of the buds and cause them to swell before transplanting. This gives quicker initial growth in the field.

Larger budded plants, on which the scion has been allowed to grow for a year or more in the nursery and then pruned back before transplanting, are used for some species. With citrus and some other fruit trees such plants are transplanted with the primary branches formed but pruned back. Rubber is sometimes transplanted as 'stumped buddings', i.e. with the scion cut back to leave 6 or 7 feet of unbranched stem. The use of such planting material reduces the period of immaturity in the field, but requires a longer period in the nursery and increases the cost of lifting, transporting and planting.

Plants may be transplanted with bare roots or with a ball of earth about the roots, or they may be raised and transplanted in baskets or polythene bags, thus avoiding damage or disturbance to the roots. Bare root planting naturally delays regrowth after transplanting but it is cheaper than the ball of earth method and permits inspection of the roots for disease or malformation. Bare root planting is used for stumps but is not satisfactory for most leafy plants, although it is commonly used for coconuts and oil palm seedlings in smallholdings, since the weight of such seedlings plus a ball of earth makes it impracticable for peasants to carry them any considerable distance. When planting oil palm seedlings with naked roots, improved survival and growth is obtained by cutting the roots 6 to 8 inches from the plant one month before removal from the nursery, by dipping the roots in a clay slurry after lifting, and by pruning off some of the foliage (Sparnaaij and Gunn, 1959).

In 'ball of earth' planting, plants are lifted with a ball or cube of earth which may be secured about the roots by wrapping in hessian or large leaves. Compared with bare root planting, this method reduces

I Lowland evergreen rain forest, Sabah

II 'Miombo' woodland, Tanzania

III *Acacia-Commiphora* thicket, Tanzania

IV Grassland following clearing of *Acacia-Commiphora* thicket, Tanzania

losses and gives speedier regrowth in the field, but is more expensive, does not permit inspection of the root system, and results in a big loss of top-soil from permanent nursery sites. Raising and transplanting plants in baskets or polythene bags is generally the most satisfactory method with leafy plants. Although it may be more expensive than other methods it is usually well worth while, since losses should be negligible and plants grow vigorously in the field with little or no check from transplanting.

Minimum sizes of planting holes are often prescribed for different crops and it is sometimes suggested that large holes are advantageous in promoting more rapid and extensive development of the root system. However, unless there is some definite impediment to root penetration, such as a very stiff clay soil or a hardpan, there appears to be no advantage in large holes, unless they are filled with rich surface soil, when there may be some benefit from the greater volume of surface soil adjacent to the plant. Apart from this, the size of the hole in itself is unimportant, so long as it is big enough to accommodate the root system of the transplant and to permit of reasonable consolidation of soil around the roots without misplacing or damaging them.

Plant Density and Spacing

The optimum plant density and spacing naturally vary with the crop, and to a lesser extent with soil, climate and economic factors, but one or two general points may be noted. Trees that bear fruit mainly at or near the extremities of the branches, such as tung, cashew and the majority of edible fruit trees, must be spaced sufficiently widely to prevent yields being reduced by their branches rubbing against or crowding each other. Similarly, mature oil palms and coconuts, which are sun-loving plants, must be spaced wide enough to prevent their crowns shading each other. For such crops, initial planting at a spacing appropriate to mature trees involves some waste of space during the early years. This may be countered by planting at a higher density and subsequently thinning, or by intercropping during the early years. On the other hand, with cocoa (which bears its pods mainly on the older branches) and also with tea and rubber, yields per acre tend to increase with increasing plant density up to a fairly high limit, and fairly close initial spacing has the further advantage that a continuous canopy, which protects the soil and shades out weeds, is formed as early as possible.

Careful consideration must be given to factors other than optimum yield in deciding upon the best spacing to adopt. One of these is the balance between the value of the enhanced early yields obtained from close planting and the cost of the extra plants, which is important where the plant population per acre is in any case high, as with tea. Other factors which may influence a decision on spacing are the profitability of early intercropping, the incidence of disease (which may be increased by high plant density) and the ease and economy of mechanical cultivation, spraying and harvesting. With rubber, yields per acre increase with increasing plant population up to about 400 per acre, but at the higher densities bark thickness, bark renewal and yield per tree are all reduced and a proportion of the trees never reach tappable size. As these factors increase tapping costs, it is more profitable to plant at medium densities which, though giving lower yields per acre, give higher yields per tree and per tapper.

The commonest planting systems in use are the square, equilateral triangle, avenue and contour systems. The equilateral triangle system has some advantage over square planting as it permits a fuller use of the land, accommodates about 20 per cent more trees per acre at the same planting distance, and allows of mechanical cultivation in three directions instead of two. Avenue planting facilitates both early intercropping and mechanical cultivation, but the closure of the canopy and shading out of weeds between the rows is delayed as compared with square or triangular planting. On slopes liable to erosion, contour planting is advisable in order to permit of contour cultivation and to make use of the tree rows as an aid to soil conservation.

Soil Management after Planting

At any rate for the first few years after planting tree or bush crops, some form of protective ground cover is usually needed to shield the soil from direct sunlight, which raises soil temperatures and increases the rate of decomposition of organic matter, and to afford protection from the direct impact of rainfall, which breaks down soil aggregates and seals the surface, thus reducing or preventing the infiltration of rain water. Even on flat land reduced infiltration of rainfall is undesirable, because part of the water is lost by evaporation while standing on the surface and the entry of air into the soil is also reduced, both circumstances rendering conditions less favourable for root growth. On sloping land, water in excess of that which can readily percolate into the soil runs off the surface and causes erosion. Ground cover also helps to check erosion by impeding the movement of run-off water by physical obstruction, but this effect is much less important than that of protecting the soil from rainfall impact. With some crops, such as cocoa and rubber, the provision of a protective ground cover may not be essential once the trees themselves have formed a complete canopy

Soil Management after Planting

and deposited leaf litter, unless the land is on a fair slope. On the other hand, crops that do not form a full canopy, such as citrus and coffee, usually require permanent ground cover. Tea, which is encouraged to spread by management and pruning, soon provides adequate soil protection on its own.

Clean weeding, without any anti-erosion measures, was widely practiced in the early days of the plantation industry in the tropics when the need for soil protection and conservation was not appreciated, and on sloping land it commonly resulted in severe erosion. Even on flat land, or on moderate slopes where adequate earthworks have been constructed to control erosion, clean weeding is not usually advisable; by removing all weed competition it may temporarily enhance growth and yields, but it accelerates decomposition of soil organic matter, may increase leaching losses, and leads to deterioration in the physical condition of the soil. In certain circumstances the last effect may take some time before it reaches the stage of reducing yields. For example, in a tillage experiment with coffee in the East Rift area of Kenya, clean weeding during the rainy season (by means of a hand fork-hoe, disc harrow, or tractor-mounted three-furrow plough) was compared with periodic slashing of the weeds, or allowing them to grow unrestricted until the onset of the dry season and then controlling them with the same implements. Compared with the other two treatments, clean weeding had an adverse effect on the physical condition of the soil and reduced its ability to accept infiltration of rain, the hand fork-hoe causing the least deterioration. Yet clean weeding consistently gave the highest yields throughout a period of fifteen years, though its advantage in this respect decreased during the last five years of the trial (Jones and Wallis, 1963; Pereira, Dagg and Hosegood, 1964). However, this experiment was conducted on a soil with a rather stable structure and under somewhat low rainfall (30–40 inches), where weeds readily reduced coffee yields by competing for water. In most circumstances the adverse effects of clean weeding are more speedily apparent in reduced growth and yields. Consequently clean weeding is now usually restricted to circles around the trees, or to strips along the tree rows, and the soil in the interrows is protected by intercropping, by planting cover crops, by developing a suitable cover of natural regeneration through selective weeding and cutting, or by mulching.

INTERCROPPING

Intercropping with annuals or short-term perennials is practicable for the first few years and is usually practiced by smallholders in order to provide food or a cash return while waiting for the tree crop to come into bearing. On estates it is rarely done, presumably because it is not usually profitable. Unless it is practiced circumspectly, intercropping can easily produce adverse effects on the tree crop by shading, by competition for nutrients and water, and by deterioration in soil structure resulting from excessive or ill-timed cultivation. Such adverse effects during the early years of the tree crop may have a prolonged influence on subsequent growth and yields. On the other hand, interplanting with suitable crops, in a manner which maintains fertility and avoids competition, may be profitable and do no harm to the tree crop, or may even have beneficial effects. For example, Webster (1950), in an experiment with tung trees (*A. montana*), which compared intercropping for the first four years with soya bean, maize, velvet beans as green manure and *Calapogonium mucunoides* as a cover crop, found that significant yield increases from the soya bean treatment persisted at least until the eighth year, four years after intercropping had ceased. Leguminous crops are very suitable for intercropping but most other crops should not be grown without the application of fertilizers, unless the soil is very fertile. In practice this normally means that the most suitable intercrops (other than legumes) are those which will pay for a good dressing of fertilizer, such as tobacco, pineapples or bananas. The short-term crops should not be planted too near the trees, nor should their cultivation be continued when the tree roots spread into the interrows. If rainfall is limiting for the tree crop, then an intercrop is clearly undesirable on account of competition for soil moisture.

COVER PLANTS

Cover plants may be provided either by controlling the natural regeneration of indigenous plants by selective weeding and cutting, or by planting suitable herbaceous or shrubby species, which are usually legumes. Apart from protecting the soil, cover plants may have other advantages. By the deposition of leaf litter and the death of their roots they add organic matter to the soil, which improves its physical condition and raises its base exchange capacity, thus increasing the proportion of potassium, calcium and magnesium that is held in a form available to plants. As the added organic matter decomposes it gradually releases available plant nutrients. In addition to extracting nutrients from the soil and returning them in a more available form in the deposited organic matter, cover plants may also make nutrients more accessible to relatively shallow-rooted tree crops by checking leaching and, in the case of deep-rooted cover plants, by drawing up nutrients from the deeper parts of the soil profile and depositing them

Soil Management after Planting

in the surface layers. Good cover plants produce excellent physical and nutrient conditions in the surface soil, and this encourages the development and spread of the feeding roots of most tree crops in the surface layers. By contrast, tree crop roots appear less able to compete with many weeds which restrict their development in the surface soil and result in a poorer root system in deeper layers that are less rich in nutrients.

SOWN COVER PLANTS

Sown cover plants are usually leguminous creepers, such as *Pueraria phaseoloides, Centrosema pubescens, Calapogonium mucunoides, Glycine javanica, Stylosanthes gracilis, Dolichos hosei, Indigofera spicata*, or shrubs, as for example, *Desmodium ovalifolium, Flemingia congesta, Indigofera arrecta*, and *Crotalaria* spp. The principal advantage of leguminous cover plants is almost certainly the considerable amounts of nitrogen which they mobilize and subsequently release, but there is also evidence that they may increase the amounts of other available nutrients, especially phosphate. It has not been definitely established that leguminous covers generally fix nitrogen under field conditions, but evidence from pot experiments, and from soil and leaf analyses, makes it highly probable that they do. For example, Watson (1957) demonstrated that *Centrosema pubescens* fixed considerable amounts of nitrogen in pot trials, and in a number of experiments in Malaya it has been shown that growing a mixture of *Pueraria phaseoloides, Centrosema pubescens* and *Calapogonium mucunoides* between rubber significantly increases the nitrogen content both of the soil and of the leaves of the trees as compared with the effect of covers of controlled natural regeneration, or of grasses (Watson *et al.*, 1963). In the same experiments it was evident that the leguminous covers had increased the magnesium content, and on some soils the phosphorus content, of the rubber leaves. As another example, Gethin Jones (1942) showed that growing a cover of *Glycine javanica* for nine years raised the nitrogen content of the top 9 inches of soil from 0·206 per cent to 0·270 per cent, the phosphate content from 9·9 to 19·3 p.p.m. and the potash from 0·054 per cent to 0·096 per cent. In addition, the cover crop and its litter contained considerable quantities of all three nutrients (Table 23).

The chief disadvantage of all cover plants, whether sown or naturally regenerated, is that they compete to some extent with the tree crop, at any rate while they are growing vigorously. For example, during the first two years after planting, the leguminous creepers in the Malayan experiments mentioned above grew vigorously and took up considerable quantities of nutrients and water from the soil, and some

Table 23. *Dry matter and nutrients (lb per acre) in* Glycine javanica *cover*
Gethin Jones (1942)

Source	Earth-free dry matter	N	P_2O_5	K_2O
Cover crop	5,700	130	23	113
Litter	12,000	280	88	197
Total	17,700	410	111	310

competition must have developed as the rubber roots began to spread from the clean weeded row strips into the interrows. From the third to the fifth years, however, the legumes gradually died out owing to the effect of the heavy shade of the developing rubber canopy, depositing much organic matter in their litter and dead roots. The nutrients released from this organic matter enhanced the growth of the trees and increased early yields in comparison with the effects of natural covers and planted grasses (Watson *et al.*, 1963; Mainstone, 1963). The fact that these creeping legumes die out in the shade of tree crops that develop a full canopy has the disadvantage that it tends to leave the soil bare, or to permit the development of undesirable weeds. Consequently it is probably advantageous to plant a mixture of creepers and shade-tolerant shrubs, so that as the former die out the latter develop to form a persistent ground cover under the trees (Wycherley, 1963). Such mixtures seem likely to confer long-term benefit to the soil from the varied effects of the different types of plants, the shallow-rooted short-term creepers depositing much organic matter as they die out in the early years, and the deeper-rooted persistent shrubs continuing to provide organic matter (including nutrients derived from the subsoil) over a much longer period.

The severity and duration of the competition afforded by cover plants is affected by rainfall, soil fertility, the species of crop and cover plants that are grown and the management of the cover. Consequently the effects of cover plants on the growth and yields of tree crops have been found to be somewhat variable, and it is difficult to define precisely conditions under which the benefits of sown covers are likely to provide an adequate return for the cost of establishing and maintaining them. Generally speaking, it would seem that on fertile soils sown cover plants are likely to have little advantage over the selective weeding of natural regeneration, but on poor soils, provided that rainfall is good, their advantages are likely to be sufficient to justify their cost.

If cover plants are sown it is clearly desirable that they should speedily become established and spread throughout the interrows. One factor which affects speed of establishment is that leguminous cover crop seed usually contains a proportion of seeds having hard, impermeable testas, and unless it is pretreated germination is often low and establishment slow and uneven. Such seed may be treated in three ways:

1. By soaking in sufficient hot water at 75°C to cover the seed, adding an excess of cold water after cooling, and leaving to soak overnight.
2. By soaking for fifteen minutes in concentrated sulphuric acid, draining off as much acid as possible, washing the seed several times in water to remove all traces of acid and then leaving it to soak overnight in water.
3. By mechanical scarification in commercial machines designed for the purpose.

The third method is the best, since it avoids danger to seed and personnel from careless use of acid, obviates difficulties in drying and storing seed after treatment, and does not make success dependent on favourable weather immediately following sowing, as it is with the other two methods. Other factors that affect speed of establishment are seed rate, depth of sowing and manuring. Seed rates will vary with the species of cover plant and the spacing of the tree crop. Seed of most species should not be sown deeper than half an inch. Application of phosphate at or before sowing usually aids establishment and on some soils other nutrients may be beneficial.

CONTROLLED NATURAL REGENERATION

This method of soil management aims at establishing and maintaining, by slashing and selective weeding of the natural regeneration, a cover of the more desirable indigenous plants, which do not compete strongly with the crop by reason of the vigour of their growth above ground, or by possessing a vigorous, shallow root system. This means endeavouring to eliminate all large shrubs, sapling trees, climbers, vigorous grasses and bracken, and encouraging less vigorous species with soft tissues that decompose rapidly when deposited as litter. A cover of desirable species is most readily developed when plantations are established on land newly cleared from forest, with only light burning after felling. If thorough burning is done after clearing forest, or if planting follows the clearing of poor, secondary forest or grassland, the natural regeneration tends to be rather unsatisfactory, consisting mainly of vigorous grasses and other competitive species. In recent experiments in Malaya, with both new-planted and replanted rubber, selectively weeded natural regeneration retarded the growth of the trees compared with leguminous cover crops, but the difference between the two treatments was not large on fertile coastal alluvial soil or where rubber was planted after clearing good secondary forest on inland soils (Watson et al., 1963).

Non-selective cutting of weeds may also be regarded as a form of controlled natural regeneration. Where rainfall is good, a moderate growth of weeds may not compete seriously with the crop for water, and periodic cutting of weeds with tractor-drawn brush cutters, together with the application of fertilizers to compensate for nutrient competition, may prove an economic and satisfactory procedure. Periodic cutting near ground level tends to eliminate shrubs and herbs and to encourage the dominance of grasses, which are more resistant to repeated defoliation. If the grass species that become dominant are only of moderate vigour, then relatively infrequent cutting and modest compensatory applications of fertilizer will probably ensure satisfactory growth of the tree crop. On the other hand, if vigorous grasses develop, more frequent cutting and liberal applications of fertilizer, particularly of nitrogen, will be necessary. In places where inadequate rainfall is liable to limit crop performance, periodic cutting of weeds cannot be recommended, owing to competition for water. For example, in the Kenya coffee tillage experiment mentioned above, Pereira and Jones (1954b) found that slashing weeds reduced the average yield by 27 per cent over the first five years, as compared with clean cultivation. Since the latter cannot be recommended as a long-term policy because of its ill-effects on the soil, the best procedure in drier areas is likely to be mulching.

MULCHING

Mulching with grass, weeds, brushwood, or other vegetable trash, protects the soil from sun and rain without the competition for moisture and nutrients which accompanies the use of cover plants. Mulching also markedly reduces weed growth, presumably mainly by reducing light intensity. Compared with soil that is bare, or under the majority of crops, a mulched soil has a lower average temperature, and a restricted diurnal temperature range. This effect is naturally of minor importance under tree crops that provide heavy shade, but even under cocoa Smith (1954) found that mulching reduced average soil temperature slightly and the diurnal range considerably. By restricting weed growth, lowering soil temperatures and protecting the soil from wind, mulches reduce evapo-transpiration losses. The infiltration of rain water is also improved because the mulch prevents the breakdown of soil structure by the impact of rainfall, and because

Soil Management after Planting

increased earthworm and termite activity under a mulch provide many channels for the percolation of water. The improvement in infiltration has a much greater beneficial effect on soil moisture status than the reduction in evaporation. For this reason it is desirable to apply mulch at the beginning of a rainy season, in order to aid infiltration, rather than at the end of the rains when it can only reduce the rate of drying out during a following dry season. For example, Pereira and Jones (1954a) showed in Kenya that mulching before the rains had a much greater effect on coffee yields than mulching immediately after the rains (Table 24). In fact, mulching alternate rows before the rains gave better yields than mulching all rows after the rains.

Table 24. *Yields of clean coffee, cwt per acre*
Pereira and Jones (1954a)

Year	Total rainfall (inches)	No mulch	Mulched all rows before rains	Mulched alternate rows before rains	Mulched all rows after rains
1950	24	0·76	1·52	1·09	0·71
1951	54	6·81	12·08	10·58	9·82

The organic matter added to the soil by the decomposition of a vegetable mulch improves soil structure and gradually releases nutrients in an available form. Some nutrients may also be contributed to the soil by leaching of the undecomposed mulch. The extent to which nutrients derived from the mulch benefit crop performance naturally varies with the initial fertility of the soil and the composition of the mulch, but the contribution made by some mulches may be considerable. For example, the standard, alternate-row mulch of 10 tons (dry weight) of elephant grass per acre applied to coffee in Kenya contains 900 to 1,200 lb of potash and 500 to 600 lb of phosphate. However, the addition of organic matter of wide C:N ratio, probably coupled with a lower rate of nitrification and more leaching under the mulch during the wet season, not infrequently results in mulched crops suffering from nitrogen deficiency unless compensating fertilizer applications are given.

Mulching has proved an effective means of improving the yields of many tropical crops and is extensively used in various places for coffee, citrus, bananas, cocoa, pineapples and a variety of fruit trees (Jacks *et al.*, 1955). On account of its marked effect in improving soil moisture status, the greatest benefits from mulching are usually obtained in areas of relatively low rainfall, although it has also frequently increased yields in places where lack of rain is not a serious limiting factor. In the rather dry East Rift area of Kenya mulching has a greater effect on coffee yields than any other cultural or manurial treatment; coffee is almost universally mulched, many planters devoting part of their land to the production of elephant grass for mulching.

The chief disadvantage of mulching is the high cost of cutting, transporting and applying large quantities of material. With some crops, especially when they are young, costs can be reduced by cutting plant material grown in the interrows to mulch limited areas near the trees. More commonly it is necessary to carry in material from outside the plantation, in which case the practice is only likely to be economic if the crop is of relatively high value, and it is desirable to reduce costs by mechanizing the cutting, transport and application of the mulch. It is desirable that the mulch shall have as long a life as possible; consequently materials that do not decompose too rapidly are to be preferred, and cultivation to control weeds growing through the mulch may need to be minimized, chemical weed control being useful in this connection. In some circumstances it may pay to combine mulching with periodic tillage by an implement which both controls weeds and gradually incorporates the mulching material into the top soil. In a trial with coffee in Kenya, clean weeding of mulched plots with a hand fork-hoe during the rains did not improve yields compared with allowing the weeds to grow unchecked until the end of the rains. But weeding the mulched plots by tillage with a modified rotary hoe (designed to simulate the action of the hand fork in leaving the soil in a cloddy condition) did significantly increase yields. In addition to controlling weeds, the rotary hoe incorporated the coarse mulching material in the top-soil, thus maintaining the soil in a better physical condition than when the hand fork-hoe was used. The benefits of mulching and of frequent weeding with the rotary hoe were additive, and both gave similar increases in profit (Jones and Wallis, 1963; Pereira, *et al*. 1964).

Manuring

It is especially important to ensure that perennial crops are adequately supplied with nutrients during establishment and immaturity. Failure to do so is liable to result in poor early growth, which is not easily rectified later and may have a prolonged adverse effect on yield. On most soils the application of nitrogen and phosphate at planting, and for several years thereafter, results in improved vigour which is commonly reflected in enhanced yields when the trees come into bearing. Potash and magnesium are not so generally beneficial, but may be on some soils.

On the other hand, the response of mature perennial crops to manuring is much more variable, being dependant on the species, type of planting material and cultural methods, as well as on soil and rainfall. For example, fertilizers applied to low-yielding, unselected seedling rubber and coconuts, or to seedling cocoa under heavy shade, will often fail to produce economic increases in yield. Provided that it has been manured in immaturity, even high-yielding mature rubber commonly gives small responses to fertilizers, probably because only relatively small quantities of nutrients are removed in the latex. By contrast, provided that rainfall is adequate, large responses are usually obtained from high-yielding tea, oil palm, citrus and unshaded coffee, where the crop harvested makes a considerable drain on soil nutrients. Mature tree and bush crops often fail to benefit from potash, phosphate and farmyard manure on soils where annual crops, especially cereals, show marked responses to these manures. Annual crops have to make rapid growth to maturity within a short season, and must be able to extract available nutrients speedily with a relatively restricted root system, especially in the early stages of growth. Established perennial plants have much bigger root systems, which explore a greater volume of soil over a long period, and are thus more efficient in extracting the nutrients they require.

Perennial crops respond to nitrogen more commonly than any other nutrient. This is especially true of tea, where the harvested leaf is of relatively high nitrogen content; plucking provides a continuous stimulus to growth, and a linear response is often obtained up to 80 lb of nitrogen per acre per annum. Bananas and coffee also make big demands for nitrogen, and though less may be required by some other crops, nevertheless nitrogen is the nutrient to which responses are most widely obtained. On most soils phosphate is beneficial in encouraging early growth, and its application at planting and during immaturity is usually to be recommended, but the response of mature crops to phosphate is much less general. Bananas are well known to have a high potash requirement, and citrus and coffee also make relatively heavy demands for this element because of the amounts contained in their fruits. There is a fairly heavy removal of potash in the leaf and prunings of tea and a need to apply potash not infrequently develops where heavy nitrogen fertilizing is practised. With other crops potash responses are usually restricted to certain soils. The need to correct deficiencies of subsidiary and trace elements is experienced more commonly with perennial crops than with annuals. For example, in various places it has been found necessary to correct deficiencies of magnesium in citrus, oil palms, rubber and cocoa, of sulphur in tea, of iron and zinc in coffee and citrus, of boron in coffee and oil palm, and of manganese in coffee and rubber.

Organic manures, such as oil cakes, farmyard manure and composts, are nowadays little used on plantation crops, partly on account of their high initial cost per unit of nutrient and the expense of handling, transporting and applying these rather bulky materials, but also because experiments have often shown them to be no more, or even less, effective than dressings of inorganic fertilizers of equivalent nutrient content.

In places experiencing an alternation of wet and dry seasons it is of little value to apply fertilizers in a dry period, and they are generally applied at, or shortly after, the onset of the rains. If high rates of application of nitrogen and potash are needed, it is desirable to apply these fertilizers in several doses, one at the beginning of the rains and the others later in the wet season, but there appears to be no advantage in applying phosphate more than once a year. For some crops the time, or frequency, of application of fertilizers may be affected by the occurrence of periods of peak nutrient demand associated with times of maximum growth, flowering or fruiting. For example, coffee has particularly high nutrient requirements during the first three or four months of the rainy season when it is making rapid shoot growth at the same time as setting and developing fruit. The nutrient uptake of cocoa is increased at a time of simultaneous flowering and flushing. Fertilizers are probably best applied to rubber immediately before the refoliation which follows the annual 'wintering', or shedding of foliage. After pruning tea the bushes are not only temporarily leafless but the absorbing elements of the root system have also to be largely renewed; consequently it is inadvisable to apply fertilizer during a period extending from shortly before pruning until 4 to 6 inches of new shoot growth has been made after pruning.

Fertilizers are usually applied to young trees in the first few years after planting by broadcasting over a circular area roughly corresponding with the root spread, usually starting with a diameter of about 18 inches and gradually widening the circle. With mature trees at normal spacings the root systems will usually permeate most of the interrow area and as a rule no method of placement has any marked advantage compared with broadcasting over the whole area, or over a broad band between the rows. Placing the fertilizer in a band $1\frac{1}{2}$ to 2 feet wide around the trees under the extremities of the branches is quite frequently recommended, and placement in short trenches or pockets at intervals around the trees has also been tried. Bearing in mind that application by such methods is more expensive than broadcasting, the evidence is that the latter procedure is just as good.

Shade

The most important perennial crops for which shade may be considered necessary are cocoa, coffee and tea. Shade for these crops takes two main forms: (1) temporary, provided for a limited period after planting, either by artificial means or by interplanting; (2) permanent, provided by interplanting larger trees at relatively wide spacing and maintaining them throughout the life of the crop.

TEMPORARY SHADE

In the absence of both shade and mulch, young coffee and cocoa plants tend to make rapid initial vegetative growth, to bear one or two relatively heavy crops and then to suffer from unthriftiness and dieback. Temporary shade is generally regarded as essential for young cocoa and is usually provided by interplanting with food crops, although leguminous trees and shrubs may also be used. For example, dasheen, eddoes (*Colocasia* spp), tannia (*Xanthosoma*), cassava and bananas are commonly used in the West Indies, but *Leucaena glauca* and *Gliricidia maculata* have been successfully employed in New Guinea. The function of these plants is to protect and shade the soil and the young cocoa until the latter has had time to develop a fairly strong root system and to achieve some degree of self-shading. Temporary shade is not essential for young coffee. Under good rainfall, shading of young coffee (planted at normal spacing) by interplanting rows of leguminous shrubs, such as *Tephrosia* spp and *Crotalaria* spp, is usually beneficial, but under drier conditions these shrubs compete too much for moisture and better results are obtained by mulching, without shade. In modern methods of 'sun-hedge' culture (Cowgill, 1958) coffee is planted closely in single or double rows so as to provide a relatively high degree of shading of the soil that is being explored by the young root system, and of self-shading of the bushes themselves. Mulching is not essential when planting is done in this way, but it is always an advantage. Interplanted leguminous shrubs are sometimes used as temporary shade for tea, but they are generally considered to be inferior to mulching combined with the use of grass cones or earthenware cylinders to protect the collars of young plants for a few weeks after planting.

PERMANENT SHADE

Apart from reducing light intensity, shade trees may markedly influence the moisture and nutrient relations of the crop, may have chemical and physical effects on the soil by the deposition of litter and the action of their roots, and may affect the incidence of pests and diseases and of damage by wind and hail. Their overall effect, resulting from the interaction of these factors, necessarily varies with local climate and soils. Consequently it is not surprising that permanent shade trees have traditionally been planted in cocoa, coffee and tea plantations in some areas, whereas in others they have been thought unnecessary, or their value has been disputed.

Light intensity and nutrition

In the past the protagonists of shade were greatly influenced by the belief that cocoa, coffee and tea, being indigenous to forest areas, could not flourish in full daylight, but recent investigations have shown that this is not necessarily true. For example, in an experiment conducted in Trinidad, in which cocoa was grown with and without NPK fertilizer, under light intensities of 15, 25, 50, 75 and 100 per cent full daylight, using artificial shade of bamboo slats, it was found that the growth and yield of fertilized cocoa increased with increasing light intensity up to 75 per cent, and its performance was only slightly reduced in full daylight. On the other hand, without fertilizer, maximum yields were obtained at 50 per cent light intensity and fell markedly at higher intensities. It was concluded that the yield of cocoa is linearly related to the logarithm of light intensity but that this relation is conditioned by nutrient availability, and when nutrient availability is low the linearity breaks down at a lower light intensity than when nutrient supply is high (Evans and Murray, 1953; Murray, 1953-5). Similar evidence was obtained in Ghana, where cocoa was grown with and without NPKMg fertilizer and either under *Gliricidia maculata* shade or unshaded; the unshaded cocoa gave the best yields, whether fertilized or not, and the response to fertilizer was much less under shade (Cunningham and Lamb, 1958).

The work of Tanada (1946), Sylvain (1952) and Alvim (1958) has shown that *arabica* coffee behaves in much the same way as cocoa, its net assimilation rate, growth and yields increasing with increased light intensity up to 100 per cent full daylight. The majority of the world's *arabica* coffee is, in fact, grown without shade, and experience in Hawaii and elsewhere has shown that, provided conditions are favourable and ample fertilizer is applied, yields far in excess of those obtainable under shade can be achieved (Cowgill, 1958).

With tea, nitrogen appears to be the most important element in relation to light intensity. In a trial in Assam, in which shade was provided by bamboo screens, average yields over seven years showed that in the absence of nitrogen shaded tea gave the highest yields, but tea receiving 80 lb per acre per annum of fertilizer nitrogen gave significantly greater yields without shade (Dutta et al., 1958; Wight, 1959). Van Dierendonck (1959) quotes data from an experiment in Java, in which tea was grown under light intensities of 25, 50, 70 and 100 per cent full daylight, which indicates that yield was almost directly proportional to light intensity. Laycock and Wood (1963), after taking into account the effect of shade trees on moisture and nutrient relations in a field experiment in Malawi, concluded that the shade trees reduced yields by a shade effect *per se*.

There is thus evidence for all three crops that there is a strong interaction between light intensity and nutrition. However, provided nutrient uptake is not limiting, photosynthetic activity, net assimilation rate, growth and yields increase with increasing light intensity up to, or closely approaching, full daylight. On the other hand where nutrient uptake is below optimum, shade may be helpful by reducing light intensity and thus limiting the photosynthetic activity to a level that can be sustained by the available nutrient supply.

The effect of shade trees on soil fertility

Leguminous shade trees may be expected to increase total soil nitrogen by symbiotic fixation, although the amount of nitrogen that they fix does not appear to have been accurately determined. Wight (1959) equated the value of twenty-seven *Albizia chinensis* trees per acre in a tea plantation to that of 80 lb per acre per annum of nitrogen applied as sulphate of ammonia, but this does not necessarily represent the amount of nitrogen fixed, if only because shade *per se* has been shown to reduce the response of tea to applied nitrogen.

Deep-rooted shade trees may enhance the fertility of the surface soil, and make nutrients more available and accessible to shallow-rooted crops, by extracting nutrients from the deeper soil layers and depositing them on the surface in the organic matter of their leaf litter. This effect may be increased by pruning the shade trees and leaving the loppings on the surface, as is commonly done in tea gardens. On the other hand, the shade trees necessarily compete to some extent with the crop for nutrients, and a large part of the nutrients that they extract from the soil will not be returned in the litter, or loppings, but locked up in their trunks and branches. Furthermore, though a considerable proportion of the roots of tea, cocoa and coffee occur near the surface, it does not follow that there are not sufficient deeper roots to explore the soil adequately, and it is certain that the distribution of these roots is affected by the environment. In a deep, friable, well-drained soil, *arabica* coffee can have a well-developed root system extending to a depth of 15 feet or more (Nutman, 1933). From several places it has been reported that as much as 70 per cent of the fibrous roots of tea occurs in the top 30 cm of soil, (Eden, 1940; Thomas, 1944; Barua and Dutta, 1961), but elsewhere Kerfoot (1962) found that tea developed an extensive system of deep secondary roots and considered that it was quite capable of adequately exploring the soil without any assistance from shade trees.

The mulch of litter deposited by the shade trees will have beneficial physical effects on the soil, but this may not be very important if the soil is also protected by the canopy and litter provided by a full stand of tea or cocoa. It is also possible that the deep roots of shade trees may open up the subsoil, thus improving drainage, and that the death of these roots will leave channels to facilitate the deeper penetration of crop roots. But the latter effect has not been established, and it may be doubted whether the shade tree roots are significantly beneficial unless there is some definite impediment to drainage or crop root development, such as a hardpan.

The effect of shade trees on soil moisture

The shade trees and their mulch of litter will help to conserve moisture by lowering the temperature of the ambient air and of the soil, and by reducing wind speed and air movement, thus reducing evaporation from the soil and transpiration from the crop. But against this must be set the water transpired or evaporated after interception from the shade trees themselves, which will depend in part on the species, spacing, age and management of the trees. Where rainfall is good and well-distributed, a favourable balance may be achieved between these two effects, especially if the shade is well managed, but under such circumstances moisture conservation is not likely to be highly important. On the other hand, it is unlikely that shade trees assist in conserving moisture in areas subject to dry seasons or droughty spells, and where rainfall is marginal for a crop. For example, with coffee in the East Rift of Kenya, shade reduces yields by competing for moisture. Similarly in Malawi, Foster and Wood (1963) showed that over a period of four years there was less available moisture in the top 7 feet of soil under mulched coffee when it was shaded with *Grevillea robusta* and *Albizia lebbek* than when it was unshaded. In the same country Laycock and Wood (1963) found that shade trees in tea reduced available soil moisture in the soil at 5 and 10 feet, although they had little effect on the moisture content of the top foot.

Subsidiary effects of shade trees

Among the subsidiary effects of shade probably that of most general value is that they reduce weed growth. This is certainly an advantage

V Terraced rice fields, Ceylon

VI Rice field with mixed tree culture in background, Malaya

VII East African goats at Entebbe Livestock Station, Uganda. Note the wide variation in colour and conformation. Generally speaking, flocks of sheep and goats are less uniform in the tropics than herds of cattle, and they are very difficult to classify with precision

VIII Jamaica Hope cow from the Government Herd maintained at the Bodles Old Harbour Livestock Station, Jamaica. This magnificent cow is very near to the 'ideal' animal aimed at in the evolution of this new tropical breed. The Jamaica Hope cattle have been evolved from Sahiwal cattle from India, Jersey cattle from Europe and, more recently, crossbred Holstein cattle from various sources. Average milk yields of the better herds of Jamaica Hope cattle exceed 1000 gallons per annum

with coffee in some places, but is unlikely to be of much importance in established tea or cocoa, where the crop canopy limits weed growth. Shade may also reduce damage caused by high winds, hail or sunscorch. Tea is liable to sun-scorch after pruning, and hail can cause damage to it by direct wounding and by encouraging the incidence of brown blight fungus. Shade reduces the incidence of certain insect pests, such as thrips in coffee and cocoa, but it tends to encourage the development of certain fungus diseases, notably that of blister blight in tea.

The overall effect of shade trees

From the above discussion it will be seen that, apart from the likely addition of nitrogen by legumes, it is not certain whether shade trees generally exert any effect on soil fertility that is beneficial to the crop. They have adverse effects on moisture conservation during dry seasons or spells, although they may be of some benefit under wetter conditions when, however, moisture conservation is less important. By reducing light intensity, and hence the level of photosynthetic activity, they may lower the yield potential of crops that are amply supplied with nutrients. Leaving aside their influence on the incidence of pests, diseases and wind damage, it would seem that, in general, shade trees are only advantageous where they are needed to permit modest yields to be obtained by ameliorating unfavourable climate or soil. Provided that the soil structure, aeration and drainage are such as to permit the development of a good root system, and provided that adequate moisture is available from rainfall or irrigation, and adequate nutrients from the soil or applied fertilizers, then the crop can cope with a high rate of photosynthesis induced by high light intensity and will give the best yields with little or no shade. On the other hand, if unfavourable conditions limit the uptake of water and nutrients, the crop will be unable to sustain a high rate of photosynthesis under high light intensity and will become unthrifty and low-yielding. In these circumstances, if it is impracticable to modify the environment so as to raise the nutrient uptake near to the optimum, then it is probably necessary to accept the need for shade as a buffer against the effects of unfavourable conditions and to be content with lower, but continuing yields.

Shade management

Where shade is used it is important to minimize its competition with the crop for nutrients and moisture, and to regulate light intensity, by planting shade trees at a suitable density and by proper shade management. It will not usually be satisfactory merely to plant shade trees and allow them to grow unchecked, as this is likely to lead to strong competition with the crop, and to shade which is too dense, or of an unsatisfactory pattern. Regular pollarding, or lopping, of shade trees will commonly be needed to control the rate of growth and the density of the canopy, and to maintain a favourable proportion of useful foliage to unwanted wood. It may also be necessary to undertake rotational planting of shade by planting initially at a high density to provide adequate shade quickly, subsequently thinning by stages, and replanting at some of the thinned points in order to grow new trees to replace older ones when they get too large and have to be removed.

Windbreaks

Shelter belts, or windbreaks, are needed for a number of crops (including cocoa, coffee, tea, citrus and bananas) in areas exposed to wind. The main effect of a windbreak is reduction in wind speed, which has the following effects:

1. Reduction in mechanical damage to crops, such as breakage of branches, tearing of leaves, and blowing off of flowers and young fruit.
2. Improved growth and yields, irrespective of the reduction of mechanical damage.
3. Reduced evaporation from the soil and transpiration from the crop, both of which are functions of absolute wind speed.
4. Slight increases in daytime air temperatures and atmospheric humidity.
5. Possibly, improved pollination of some species, where winds interfere with insects visiting the flowers.

The amount and extent of reduction in wind speed depends on the height of the wind break and its density, or permeability to wind flow. Very dense or solid wind breaks are undesirable (Caborn, 1957; Hogg and Ibbet, 1955). They reduce wind speed for a distance of 2 to 3 H to windward (where H = the height of the belt), and produce a marked reduction in wind velocity immediately to leeward, but the force of deflection of the wind over the top of such belts produces vigorous eddying between 5 and 10 H to leeward, which may damage crops (Fig. 8). Furthermore, there is a rapid recovery of wind speed which reaches about 80 per cent of full value at about 10 H to leeward. With moderately permeable belts part of the air stream passes through the trees with reduced velocity, producing a cushioning effect on the leeward side which minimizes eddying. Subsequently the wind speed increases more slowly than behind a dense belt, only reaching 80 per cent of full velocity at between 15 and 20 *H* to leeward. The optimum density for windbreaks is considered to be about 50 per cent, with porosity distributed as evenly as possible throughout the height. Such belts reduce wind speed below 80 per cent for about 2 *H* to windward and 15 to 20 *H* to leeward and exert some reducing effect as far as 30 to 50 *H* to leeward. With successive belts of this kind at intervals of 20 to 25 *H* an adequate reduction in wind speed should be effected over the whole area. Where the direction of the prevailing wind varies at different times of the year a grid of belts may be needed.

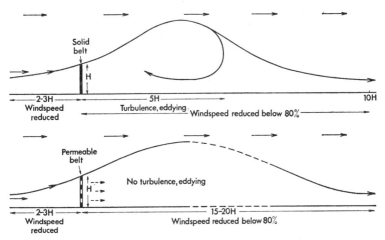

FIG. 8. Effect of 'solid' and permeable shelter belts

The width of shelter belts is not an important factor except in so far as it affects permeability, which is gradually reduced with increased width, but the value and availability of land is obviously a major consideration in this connection. Usually two or three rows of trees will be sufficient. Most mature trees shed their lower branches and thus cease to provide an effective barrier near the ground. Consequently it is commonly desirable to have one or two rows of shrubs planted on the windward side of the belt. Trees for windbreaks must obviously be species which will not blow over, or suffer mechanical damage from wind. It is also desirable that they should be quick growers, and that neither their branches nor their roots should spread widely to shade the crop or compete with it. Some crop reduction due to root competition near the belt is, however, inevitable.

Pests and Diseases

The year-round, or seasonal, high temperatures and humidities characteristic of most parts of the tropics favour the rapid multiplication of organisms causing plant diseases, and of insect pests that attack crops both in the field and in storage after harvest. Consequently heavy losses are often suffered with both annual and perennial crops, or else considerable expense is incurred on control measures. In the past severe epidemics have eliminated or seriously reduced the production of certain crops in some tropical countries, and even today pests or diseases may make it impracticable or uneconomic to grow specific crops in certain areas. For example, when coffee leaf rust (*Hemeleia vastatrix*) first became epidemic in the latter half of the last century, it wiped out the *arabica* coffee industry of Ceylon and some other parts of Asia. Nowadays it is possible to control this disease by spraying but, even so, its incidence renders coffee growing uneconomic in some places, for example, below certain altitudes in East Africa. Similarly, estate production of *Hevea* rubber is impracticable over most of tropical America because of the prevalence of South American leaf blight (*Dothidella ulei*).

No doubt such limitations and losses will be progressively reduced as better control measures are devised or resistant varieties are produced. Great improvements in the control of pests and diseases have already been made. Better formulations of the older insecticides and fungicides have been devised and new, more effective chemicals have been brought into use. Improved wetters or stickers have become available, which enable a uniform deposit of a chemical to be maintained on leaves despite heavy tropical rainstorms. Improvements have been made in spraying equipment and techniques, notably the introduction of low-volume spraying. But there are many diseases and pests which cannot yet be fully controlled, or against which chemical control measures are expensive. Moreover, while well-financed and organized estates may find it practicable and economic to adopt chemical control measures, such measures are as yet little adopted by the majority of peasant farmers, partly because they do not have the technical knowledge necessary for the proper use of chemicals but largely on account of lack of money to buy materials and equipment. Absence of readily accessible water supplies may also make spraying difficult. Consequently very heavy losses from pests and diseases are often experienced on peasant holdings, although they are probably more serious with annual crops than with perennials.

Considerable progress has been made in the selection or breeding of resistant varieties, but in many instances the resistance falls short of immunity and is not effective under all environmental conditions, nor against all the physiological races of a plant parasite. For example, *arabica* coffee varieties that are resistant to one or more of the races of leaf rust are satisfactory in some areas, but not in other places where other races of the fungus to which they are susceptible occur. Furthermore, there are many important diseases against which crop varieties combining good commercial qualities with satisfactory resistance have not yet been produced. This is true, for example, for *Hevea* rubber in respect of South American leaf blight (*Dothidella ulei*) and the two major root diseases caused by *Fomes lignosus* and *Ganoderma pseudoferreum*, and resistance to other important leaf diseases of rubber, caused by *Phytophthora*, *Oidium* and *Gloeosporium*, is inadequate in most commercial varieties in areas where these diseases are prevalent. For most tropical plantation crops similar examples could be given of important diseases against which resistance is inadequate, or non-existent, in commercial varieties.

CHAPTER 11

Natural Grasslands and their Management

Good management of natural pastures requires an understanding of the ecology of the grassland community with which the grazier has to deal. This is especially true of the natural grasslands of the tropics, since most of these occur in regions where the natural climax vegetation is forest or woodland, and thus form relatively unstable subclimaxes, in which the maintenance of grass as an important component of the association (in competition with trees and shrubs) is largely dependent on the biotic factors of cultivation, grazing and burning. Hence a first essential is to understand the status of the grass cover in relation to an existing, or former, cover of trees and shrubs and to appreciate the influence of climatic and soil conditions, as well as management, on the balance between the two types of vegetation. It is also desirable to know the flora of the grass cover itself, to appreciate the agronomic characteristics of its more important components, and to understand the significance of changes in its botanical composition. These changes may arise from local variations in soil and other natural conditions, or may indicate trends towards improvement or deterioration resulting from management.

In this short survey it will be impossible to describe in detail the many types of natural grassland, or to give a full account of diverse and complex management problems which arise therefrom. All that can be attempted is a broad classification of the main types of pastures, indicating their physiognomy and their relation to climatic conditions and to the climax vegetation, together with an account of some principles of management that are generally applicable to a number of important grassland types occurring in regions of seasonal rainfall. At present there is insufficient information to allow of a satisfactory floristic classification of tropical grasslands, although details of the composition and ecology of grasslands in various parts of the tropics is steadily accumulating as a result of past and present surveys. Examples of such studies are the FAO grassland map of Africa, the grassland surveys of India and Latin America (Whyte, 1959), and the more limited surveys made by Edwards (1956) in Kenya, by Meredith

(1947) and Acocks (1953) in South Africa and by Rattray (1957) in Rhodesia.

Types of Natural Grassland

In the following summary some of the more important types of grassland are classified by reference to their relationship to the major climax formations already described in Chapter 3.

GRASSLANDS DERIVED FROM FOREST AT LOW AND MEDIUM ALTITUDES

In the regions of lowland evergreen, semi-evergreen and deciduous forests, prolonged or repeated cultivation, followed by grazing and burning, may result in the vegetation becoming degraded to a fairly stable association of coarse grass and scattered shrubs. Large areas of such grassland, dominated by *Imperata cylindrica*, occur in Malaya and parts of Africa. The lowland 'Talawa' grasslands and 'Savanna Forest', and probably some of the lower elevation 'patanas' of Ceylon, are similarly derived (Holmes, 1951). These grasslands readily result from cultivation or burning at the drier margins of areas of deciduous seasonal forest. Examples of this are the 'tall grass' areas of Uganda (dominated by elephant grass, *Pennisetum purpureum*), and the considerable stretches of former forest land in West Africa that have become dominated by *Imperata cylindrica* and species of *Ctenium* and *Andropogon* (Keay, 1951).

With the possible exception of the elephant grass areas, these grasslands are of poor productivity, being coarse, unpalatable and of low nutrient value. They are not capable of much improvement by management; more intensive land use demands their replacement by planting improved pastures, or other crops.

GRASSLANDS ASSOCIATED WITH, OR DERIVED FROM, BROADLEAVED WOODLANDS

In the broadleaved woodlands, which occur under seasonal rainfall in the range of 25 to 55 inches, the presence of a continuous ground cover (mainly of grass) under the widely spaced trees readily permits the passage of fire during the dry season. As a result of annual fire, large areas of these woodlands have been converted to subclimax savannas of grass with scattered, fire-resistant trees. An indication of the nature of these grasslands can be given by brief consideration of two distinct types that are widespread in Africa.

The first, occurring in regions with roughly 35 to 50 inches of rain per annum and with three to five dry months, includes the types described for West Africa by Keay (1959) as 'undifferentiated: relatively moist types', and called by Edwards (1956) in East Africa 'scattered tree grassland: low tree-high grass'. Both consist of woodlands and savannas characterized by a tall, relatively dense growth of tussocky grass, in which species of *Hyparrhenia* are dominant, and by a varying density of comparatively small trees, mostly 10 to 30 feet high, but with isolated patches of taller trees, usually associated with exceptional ground water conditions.

The second type, known as 'miombo' in East Africa, consists of woodlands and savannas characterized by trees of the genera *Brachystegia*, *Julbernardia* and *Isoberlinia*, and again the ground cover is of tall tussocky grasses with *Hyparrhenia* spp. dominant, but the clumps of grass are usually comparatively widely spaced. This type occurs in regions with one severe dry season annually, usually of not less than six months duration, and although the average rainfall is as high as 50 inches in some places it is usually between 25 and 35 inches. It is found in parts of West Africa (for example in Northern Nigeria) and covers huge areas in Tanzania, Portuguese East Africa, Malawi, Zambia and Rhodesia, being referred to by Rattray (1957) in Rhodesia as 'tall grass Hyparrhenia veld'. (Plate II.)

In both these types the pasturage is what is commonly known in Africa as 'sour veld': meaning that the young grass is palatable and readily eaten by cattle, but soon becomes rank, unpalatable and of low nutrient value. At the onset of the dry season growth ceases abruptly and the tall grass dries out to form coarse, standing hay of poor quality. There is, however, a small proportion of finer, creeping and stoloniferous grasses of better nutrient value than the taller species, and a rather sparse occurrence of legumes. The composition and productivity of the herbage can be improved to some extent by the controlled use of fire and correct, intensive grazing management which will, in some degree, repress the taller grasses in favour of the better creeping grasses. Thinning out the trees to eliminate tsetse fly and encourage the grass is not highly expensive and bush regeneration in pastures is not a serious problem. Nevertheless, these grasslands are at best only moderately productive and in the better rainfall areas their replacement by sown pastures of improved grasses will usually give a better return. Over much of the less productive 'miombo' country, however, the length and severity of the dry season prevents the use of sown pastures, and the farmer must be content with relatively extensive ranching of the natural grassland.

GRASSLANDS ASSOCIATED WITH, OR DERIVED FROM, THORN WOODLANDS AND THICKETS

The vegetation occurring in these regions of 15 to 30 inches of seasonal rainfall per annum, varies from woodlands with a continuous cover of grass under widely spaced trees to thickets of small trees beneath which grass is very sparse. Grasslands in this category almost certainly form the most widespread type found in the tropics. In West Africa they occur in the Northern Guinea, Sudan and Sakel savanna zones (Keay, 1951). In East Africa they are extensive from just south of the Gulf of Aden through Somalia to parts of Uganda, Kenya, Tanzania, Malawi, Zambia, Rhodesia and South Africa. In this region they include the types referred to by Edwards (1956) as 'scattered tree grassland and open grassland – *Acacia-Themeda*' and 'dry bush with trees: *Commiphora – Acacia* – Desert grass'; and several of the types described by Rattray (1957) for Rhodesia, namely: (1) *Eragrostis* – other species veld, (2) *Aristida* – other species veld, and (3) *Cenchrus* – other species veld. Similar grasslands also occur in India, and in parts of Central and South America (Roseveare, 1948).

The grasslands grouped together under this heading all have certain common characteristics of agricultural importance. They are all unstable subclimaxes and there is a constant tendency for the spread or encroachment of trees and shrubs which has to be combated by management, although some of the woody species make an important contribution to stockfeed during the dry season. The herbage consists predominantly of perennial grasses less than three feet high and is usually 'sweet veld', though in some types the proportion of species that are palatable only when young justifies the term 'mixed veld'. Legumes are scarce. Owing to the low and unreliable rainfall, falling in only a short wet season, carrying capacity is low and seasonally very variable. Natural water supplies are usually sparse and unreliable and the provision of adequate and well-distributed artificial supplies is disproportionately expensive in relation to the returns obtainable from the land. Despite these common features, there are important differences between the grassland types in this group. Some of these can be illustrated by reference to two types of grassland which occur towards the extremes of the rainfall range mentioned above.

The 'scattered tree grasslands: *Acacia-Themeda*', described by Edwards (1956) and occurring at medium altitudes in Kenya under an average rainfall of 20 to 30 inches, form one of the most productive types. The flat-topped *Acacia* trees are widely spaced and the grass cover is relatively good. Tsetse fly is either absent or can be easily eliminated by limited, inexpensive clearing of trees and shrubs, and, though bush regeneration is a constant threat, it can be checked by a combination of grazing management and controlled burning. The maintenance of the dominant grass, *Themeda triandra*, which is a good and palatable perennial species, requires light rotational grazing and the periodic use of fire, in the absence of which it is replaced by inferior grasses. The carrying capacity of the unimproved pasture is about a beast to 20 to 25 acres, but under good management this can be doubled. (Plate 2.)

The less productive grasslands associated with *Commiphora-Acacia* bush and thicket are very extensive in East Africa. These consist mostly of deciduous bushes 10 to 15 feet high with more widely scattered taller trees and a somewhat sparse ground cover of scattered tufts of perennial grasses, augmented for short periods after rain by the growth of annual grasses and herbs. In many places, discriminative clearing to eliminate tsetse, and further clearing to promote grass development, are needed. Such clearing is expensive, owing to the strong root systems of the trees and their capacity to regenerate. At best the pasturage is of low productivity, and bush regeneration has to be constantly combated by grazing management, burning or, possibly, by the use of arboricides. The grass cover is sensitive to grazing pressure and anything more than light utilization with cattle can easily lead to a marked reduction in the perennial grasses, while at the same time encouraging the spread of bush. Periodic severe droughts also tend to kill out perennial grasses which may take several years to develop again. Carrying capacity will rarely exceed a beast to 15 acres and is more commonly of the order of a beast to 25 to 30 acres. (Plates III and IV.)

SUBDESERT SCRUB AND GRASS

This type, which usually occurs under a rainfall of around 10 inches per annum and is only capable of supporting a sparse population of nomadic pastoralists, has already been described in Chapter 3.

GRASSLANDS DERIVED FROM MONTANE EVERGREEN FORESTS

Grasslands resulting from the clearing of montane evergreen forests occur at elevations of 3,500 to 10,000 feet but mostly within the range 5,000 to 9,000 feet. Annual rainfall may be anything from 40 to 200 inches but is usually between 50 and 100 inches. The dry season is usually of only moderate intensity but in some places it may be virtually absent, as in the highlands of Ceylon, or more pronounced, as in Rhodesia.

In East Africa this type of vegetation has been called 'highland grassland and forest' by Edwards (1956), and described as consisting 'mainly of a patchy distribution of comparatively small areas of forest in extensive areas of undulating, open grassland'. The grass species are of short to medium height. In the wetter and more fertile parts of the Kenya Highlands the dominant species is Kikuyu grass (*Pennisetum clandestinum*), which makes a productive sward in association with the indigenous clover, *Trifolium semipilosum*. Elsewhere *Themeda triandra* is maintained as a dominant by periodic fire, but tends to give way to coarse species, such as *Pennisetum schimperi* and *Eleusine jaegeri*, if grazing is intensified and fire prevented. In Rhodesia the type called 'short grass mountain grassland' by Rattray (1957) is usually dominated by *Themeda triandra* on more fertile soils and elsewhere by *Loudetia simplex*. Similar grasslands exist in Ceylon, where they are known as 'wet patanas' and are usually dominated by *Chrysopogon zeylanicus*, in parts of India and Burma (Champion, 1936), and at higher altitudes in Peru, Ecuador, Venzuela and Costa Rica (Roseveare, 1948). (Plate 1.)

The productivity of these grasslands is generally relatively high but naturally varies considerably with rainfall, soil and botanical composition of the herbage. Dense swards of a nutritious grass, such as Kikuyu grass, growing in good soil and under moderate rainfall without a severe dry season, will carry a beast to the acre. On the other hand the mountain grasslands of Rhodesia are less productive owing to the more severe dry season, to the dominance of less valuable grass species and to a tendency for shrubs and bracken to invade the pasture; at best they will carry a beast to about seven acres. High rainfall, by leaching nutrients from the soil and increasing its acidity, can also reduce the value of these grasslands. This appears to apply to the highland wet patanas of Ceylon where a productive sward can only be maintained by regular and liberal applications of lime and fertilizers (Anker-Ladefoged, 1955).

HIGH-ALTITUDE GRASSLANDS

On the higher slopes of mountains, usually above 9,000 feet, and lying between the limits of the forest and the snowline, there are natural climaxes composed mainly of low-growing herbs and short grasses. Comparatively large areas of these 'alpine meadows' or 'paranas' exist in the mountainous parts of Peru, Ecuador, Columbia, Venezuela and Bolivia, where they are grazed by sheep, llamas and alpacas and, to a lesser extent, by cattle (Roseveare, 1948). Elsewhere in the tropics such grasslands are limited in extent and, since they are also rather inaccessible and exposed, they are little used.

SEASONAL SWAMP GRASSLANDS

Natural climax grasslands and savannas occur in areas of seasonal rainfall where the nature of the soil or the topography results in annual alternation of excessively wet and dry soil conditions, thus preventing or restricting the growth of trees. The lower regions of the Venezuelan llanos and the Beni region of Bolivia, situated in the upper reaches of the Amazon, form the best known examples of this type. These grasslands, parts of which contain scattered trees and shrubs, occupy a vast area of plains and low hills where the annual rainfall of 40–70 inches falls in one wet season of six or seven months duration, during which large tracts are also inundated by the flooding of the Rivers Orinoco and Amazon and their tributaries. During the severe dry season the surface layers of the soil become parched, the grass rapidly dries out, and is regularly burned towards the end of the dry season to promote a transient growth of young herbage for hungry cattle. Similar extensive llanos occur in parts of Colombia and Brazil.

The utilization of such grasslands is clearly largely conditioned by the alternation of floods and droughts. During the rainy season only islands of higher land are available to the cattle, which often swim from one island to another. Abundant feed is available early in the dry season but the tall grass quickly dries out, becomes unpalatable and is burnt off. The continued use of fire tends to eliminate any better species and leave only those which are resistant to flood, drought and fire (Roseveare, 1948). Any improvement in land use would appear to depend upon the practicability of major flood control measures, possibly accompanied by irrigation. The feasibility of the combined use of an extensive area of seasonally flooded llanos and a smaller, more productive, area in the foothills of the Andean mountains should also be considered. Cattle would be bred and fattened on the drier pastures, grazing improved grasses under fairly intensive systems of management, and could be run as store cattle on the less productive llanos using an extensive system of ranching.

Brief reference should be made to the numerous low-lying, fertile areas of seasonally waterlogged or poorly drained land known as 'vleis' or 'mbugas' etc., which have been mentioned in Chapter 3. These are customarily used by Africans for dry-season grazing but their value could be improved by drainage and by planting better species, since coarse grasses and unpalatable sedges usually form a large proportion of the herbage.

Management and Utilization

LOW PRODUCTIVITY OF THE MAJORITY OF TROPICAL GRASSLANDS

Although the relatively high-quality grasslands derived from montane evergreen forests are of considerable importance in some mountainous countries, such as Kenya, their total extent in the tropics is comparatively small and they are increasingly being replaced by crops. By far the greater part of the natural grasslands are those of the drier broadleaved woodland, thorn woodland and thicket zones, where climatic conditions limit productivity and pose management problems. Over most of these areas the low and unreliable rainfall, and the occurrence of severe dry seasons, result in a sparse sward of discrete tufts of grass which, in the drier areas, may cover as little as 5 per cent of the ground (West, 1958). Many of the pastures are dominated by coarse grasses that are palatable and nutritious only when young, and legumes are scarce. The carrying capacity is, therefore, at all times rather low but it is considerably greater during the warm, wet, growing season than it is in the dry season, when growth stops more or less abruptly and the grass dries out rapidly to form very poor quality standing hay. This marked seasonal variation in the carrying capacity, or lack of dry-season feed, poses a major management problem.

Furthermore, all these grasslands are more or less unstable subclimaxes, and there is a natural tendency for the grass to be invaded by trees and shrubs that are better adapted to the climatic and edaphic conditions. This tendency is, in general, greater in the drier areas but it is present in all these grassland types and is encouraged by the exclusion of fire, or by anything that weakens the competitive power of the grass, such as overgrazing. Bush encroachment must constantly be checked by management, because it results in lower carrying capacity, less satisfactory distribution of feed throughout the year, increased bare ground and consequent risk of erosion and, in places where the pest occurs, entry of tsetse fly.

It is against the background of the above limitations, which apply to the majority of the more extensive grassland types, that the management and utilization of natural pastures is considered in the following paragraphs. The grazier has to work within the limitations imposed by climate. The low rainfall usually precludes the possibility of his radically changing, or replacing, the natural herbage, which consists of species adapted to the conditions and includes a fair proportion of perennial species. His aim must be to improve the natural sward by encouraging the development and spread of its more valuable components.

AIMS OF MANAGEMENT AND FACTORS INVOLVED

The main aims of pasture management can be summarized as follows:

1. To provide, as far as possible, a uniform and year-round supply of herbage for a maximum number of stock.
2. To utilize the herbage at a stage which combines good nutrient quality with high yield.
3. To maintain the pasture in its most productive condition by encouraging its best species and by promoting as full a ground cover as possible. This will protect the soil from insolation and the beating action of rainfall, thus preventing run-off and erosion.

In evolving methods of management which will attempt to achieve these aims, consideration must be given to the following factors:

a. The influence of seasonal growth and of grazing on the maintenance of the sward.
b. The variation in the composition and feeding value of the herbage with stage of growth.
c. The value of certain trees and shrubs as browse plants.
d. The need for bush control.

The influence of seasonal growth and of grazing on the maintenance of the sward

It is well known that too early, too heavy, or too frequent grazing can result in reduction in the yield and vigour of a pasture, death of some of the plants, and consequent development of bare or weedy patches. This weakening of the herbage is generally considered to be due to depletion or exhaustion of its nutrient reserves. In seasonal rainfall areas the perennial grasses, which form the most important component of the pastures, make very rapid growth during the early part of the rains and quickly reach the flowering and seeding stage. Translocation of carbohydrates for storage in the roots and stem bases begins as soon as a certain amount of leaf has been formed and continues until the production of inflorescences, when it ceases or is reduced. At the onset of the dry season growth stops and the aerial parts of the grass dry out. At the beginning of the next rains new growth initially takes place at the expense of the stored food reserves,

Management and Utilization

which are not replenished until a certain amount of new leaf has been produced. Similarly, defoliation by grazing or cutting necessitates translocation of reserve nutrients to the remaining meristems for the intial growth of new leaves and tillers.

Evidence has been produced by many workers – notably by Coetzee *et al.* (1946) and Weinmann (1946, 1948b) for tropical pastures – that defoliation by cutting or grazing brings about a reduction in the weight and carbohydrate content of the roots which is roughly proportional to the intensity or frequency of the defoliation. This does not necessarily imply a causal role for the food reserves in initiating or promoting new growth (May, 1961). However, though work in the temperate zone suggests that under favourable conditions for regrowth the reserves may be of less significance than was formerly supposed, they are evidently important when the environment limits herbage production (Baker, 1961). As the latter circumstances certainly apply to the tropical pastures now being discussed, it may be assumed that depletion of carbohydrate reserves is responsible for the bad effects of too early or too intensive grazing.

Two other factors should be mentioned in connection with overgrazing. The first is that stoloniferous, rhizomatous and prostrate plants, with a high proportion of basal leaves below the grazing level, are more resistant to intensive grazing than taller, tufted species. Such grasses predominate in some tropical pastures – for example Kikuyu grass (*Pennisetum clandestinum*) in parts of the Kenya Highlands – but they are not important components of the grasslands now under consideration, which are therefore somewhat sensitive to overgrazing. Secondly, since cattle are selective in their grazing, it is the more palatable species that are most adversely affected by overgrazing and hence one of the bad effects is a change for the worse in the botanical composition of the herbage.

These ill-effects can be avoided by adjusting stocking rates to the carrying capacity, by refraining from grazing *immediately* regrowth starts after the dry season, and by allowing the pasture to have a periodic rest during the growing season in order to replenish root reserves. Adequate rest cannot normally be provided by the intervals between rotational grazing, as it can with temperate grasslands, but necessitates periodically closing the pasture to grazing for a considerable part of the growing season. As a rule it is unsatisfactory to rest land at the same time in each resting year, as this tends to encourage undesirable species which benefit from resting at that particular time. For example, in East Africa regular resting during the first half of the rains encourages annuals at the expense of perennials, hence, it may be desirable to alternate between resting from mid-wet to mid-dry season and resting from mid-dry to mid-wet season. Grazing during the dry season does no harm provided that it is not so hard as to leave virtually no vegetation to protect the soil at the onset of the next rains.

There are three 'critical periods' in the growth of grasses which spread and reproduce by seeding and not by vegetative means. These are: (1) the flowering period; (2) the period of change from a seminal rooting system to a coronal rooting system (generally occurring about a month or so after the seed has germinated); (3) the period of major transference of carbohydrates from the aerial part to the underground part of the plant, which usually occurs at the onset of the dry season. Natural grasslands containing a considerable proportion of such grasses should be more lightly grazed, and preferably rested, when the more important components of the sward are in one of the other of these three critical periods.

Variation in composition and feeding value of herbage with stage of growth

Data from various parts of the tropics, most of which are reviewed by French (1957), all indicate that the changes in the composition and feeding value of tropical pasture grasses with age are similar to those that have long been known to occur in temperate grasses (see Watson, 1951). However, most young tropical grasses are usually less nutritious than temperate grasses of the same age, and their feeding value declines more rapidly with advancing age. In the tropics young grass is usually lower in crude protein and higher in crude fibre than in the temperate zone. As it grows to maturity the crude protein, phosphate and potash contents fall, the crude fibre rises, and the carbohydrate and lime contents remain fairly constant. Very low contents of phosphorus and sodium are often found in the dry herbage on which animals have to subsist in the dry season. The total yield of herbage and the percentage dry matter content increase up to a fairly advanced stage of maturity.

This is illustrated by the figures in Table 25, showing changes in the composition of the grass herbage of Rhodesian sandveld, obtained by cutting samples at monthly intervals from just after the onset of the rains in December until the beginning of the dry season in May. (Weinmann, 1948a). The crude protein contents of some of the more important grass species in the same pasture are shown in Table 26. The crude protein content fell, and the crude fibre rose, throughout the growing season. Dry matter yield increased up to early April after which it fell, due to the death of some of the plants and the shedding of leaves early in the dry season.

Table 25. *Seasonal changes in composition of grass herbage of Rhodesian veld (Marandellas)*

Weinmann (1948a)

	Dec.	Jan.	Feb.	March	April	May
Composition, percentage on dry matter						
Crude protein	8.52	7.09	4.95	5.05	4.05	3.39
Crude fibre	33.3	35.2	38.4	37.4	39.5	38.4
Carbohydrates	51.1	50.0	49.5	50.2	50.0	51.5
P_2O_5	0.30	0.33	0.31	0.25	0.26	0.19
K_2O	2.28	2.01	1.79	1.49	1.44	1.05
CaO	1.14	1.19	1.29	0.94	1.37	1.11
Yields						
lb dry matter per acre	204	510	908	1214	1305	940
lb crude protein per acre	17.4	36.0	45.0	61.4	52.7	31.9

Table 26. *Seasonal changes in crude protein content (percentage on dry matter) of individual grass species in Rhodesian veld (Marandellas)*

Weinmann (1948a)

	Dec.	Jan.	Feb.	March	April	May	June
Brachiaria brizantha	10.35	7.65	6.70	5.37	4.79	3.54	2.61
Cynodon dactylon	13.58	11.27	7.44	8.46	6.43	5.25	4.76
Digitaria brazzae	9.13	6.11	3.98	4.27	4.03	3.71	2.69
Eragostis chalcantha	8.87	6.04	4.92	4.84	4.30	3.76	2.99
Heteropogon contortus	7.38	5.58	4.86	3.69	3.74	3.69	2.86
Hyparrhenia rufa	8.98	6.43	4.17	3.11	2.76	2.13	1.54

The crude protein content of tropical grasses usually falls very rapidly in the first few weeks of growth, after which the decline is slower until the late flowering, or seeding, stage. After seeding, and with the onset of the dry season, there is a further decline, partly due to the shedding of seed and desiccated leaves. The digestibility of the crude protein also declines in proportion to the percentage protein. Glover and French (1957) have established the relationship: Digestibility coefficient = 70 (log % protein in dry matter) − 15. In general, the variation in protein content between species is less than that between different stages of growth of one species but there are, nevertheless, important differences between species. This can be seen from a comparison of the figures in Table 26 for *Brachiaria brizantha* and *Cynodon dactylon* with those for *Heteropogon contortus* and *Hyparrhenia rufa*. There is usually a preponderance of species of low protein content. For example, Dougall and Bogdan (1958a) found that of fifty-eight species sampled at early flowering stage, thirty-four had crude protein contents of less than 10 per cent and only seven exceeded 15 per cent.

The increase in crude fibre is probably slight after the first six weeks of the growing season but there is generally a further rise in the dry season. Its digestibility decreases with advancing age but the decline is not as marked as it is with the crude protein. It has generally been considered that the digestibility of the crude fibre of tropical grasses is higher than that for temperate grasses of comparable fibre content, but it now seems probable that the apparent difference is associated with a difference in the chemical properties and alkali solubility of the lignins between the two zones, and is thus an artefact introduced by the techniques chosen to isolate the crude fibre fractions in the laboratory (Quarterman, 1961).

As with temperate grasses, after the earliest stages of growth the feeding value of the leaves of tropical grasses is higher than that of the stems. Juko and Bredon (1961) have produced figures for six species in Uganda which show that the leaves contain about 50 per cent more crude protein and 20 per cent less crude fibre than the stems, and that even in the dry season the crude protein and starch equivalent in the leaves is adequate for maintenance.

Ideally, the grazier's aim for as much of the year as possible would be to use the grass at a moderately young stage (but not at a critical period, see page 258) which achieved a good compromise between feeding value and yield. This is done on highly productive temperate pastures by frequent rotational, zero or strip grazing, but such intensive utilization is impracticable for the much less productive natural pastures of the tropics. Efforts to use the grass at the optimum stage can only meet with limited success. Grazing must be relatively light because stock numbers must be adjusted to the low, dry-season carrying capacity. Hence, in the rains the grass grows away from the stock and soon reaches the more mature stage of poorer quality while in the dry season the protein content of the standing hay is insufficient to provide even for maintenance.

Fodder from trees and shrubs

In most of the grasslands of seasonal rainfall areas, trees and shrubs make a considerable contribution to stock feed, especially in the dry season. Goats, cattle and, to a lesser extent, sheep may all obtain an appreciable proportion of their food from the leaves, flowers, pods, seeds, twigs, and even the bark, of a large number of species (Wilson, 1957a). In many places it is normal practice for native herdsmen to lop branches off certain trees and throw them on the ground or on a fence for their animals to eat in the dry season. (See Plate 39.)

In Africa probably about 75 per cent of the trees and shrubs are browsed to a greater or lesser extent by domestic animals. In Tanzania, Staples *et al.* (1942) found that all but four of 100 species occurring in an experimental plot were browsed by goats and most of these were eaten by cattle if grass was not available. Dougall and Bogdan (1958b), in an incomplete list for Kenya, name fifty-seven species most of which are eaten by cattle, although some only by goats, and give chemical analyses of the parts eaten. Much of this material is high in protein and minerals; twenty-nine of these species have crude protein contents exceeding 15 per cent. The trees and shrubs come into leaf early, before the rains, and the edible parts retain a fairly high nutritive value well into the dry season. Their fibre content may be rather high but it is no higher than the dry-season grass, or hay, and the few digestibility trials that have been made suggest that the feeding value of fodder from trees and shrubs is greater than that of grass grazed in the dry season (French, 1950). Referring to trees which occur in the veld of farms in Rhodesia, West (1950) states that four to eight large *Acacia albida* trees per acre produced an average annual yield of 1,000–2,000 lb of pods without serious reduction in the yield of grass, while ten to twenty smaller trees, such as *Acacia subulata*, may yield 250–500 lb of pods per acre. Ten pounds of pods and 10 lb of veld hay provide a maintenance ration for a 1,000 lb animal. Fodder from trees and shrubs is also valuable in many other parts of the tropics: for example in India, Pakistan and Ceylon (Commonwealth Agric. Bureaux, 1947), and in many of the Central and South American countries, for which Roseveare (1948) has listed 385 species of trees, shrubs and other browse plants.

Bush thinning

While emphasizing the part played by trees and shrubs in providing fodder it is not, of course, suggested that they should be encouraged at the expense of grass. Goats thrive on a mixture of grass and browse but cattle undoubtedly do better on grasses and herbs and the cattle-carrying capacity of pastures is normally increased by reducing the number of trees to the minimum required for shade. Thus in Rhodesia the carrying capacity of woodland-grassland, with widely spaced trees of species of *Brachystegia* and *Julbernardia*, is only one steer to 20 to 30 acres, but when the trees are reduced to the minimum needed for shade the cleared veld will carry a steer to 8 or 10 acres. In many drier areas the presence of numerous trees and shrubs not only reduces the carrying capacity but also leads to the development of bare land and erosion, owing to the fact that the rainfall is insufficient to maintain the grass in competition with woody vegetation better adapted to the semi-arid climate. For example, Staples (1945) found in Tanganyika that pasture in which trees and shrubs were reduced carried a beast to $1\frac{1}{2}$ to 3 acres, whereas adjacent 'bushland' pasture, although only carrying a beast to 14 acres on a deferred grazing system, showed obvious signs of sheet erosion. An initial reduction in the stand of trees must be obtained by felling, or poisoning; burning will be ineffective, since mature trees of most species are relatively fire-resistant.

Control of bush regeneration and encroachment by burning

After the initial thinning it will be necessary in many types of grasslands, especially those in the thorn woodland and thicket areas, to prevent the regeneration, spread or encroachment of bush. Very commonly the only practicable and economic means of doing this is by burning, but it is important that burning should be minimized, done at the proper time, and combined with good management.

As a rule the best time to burn, for the purpose of bush control, is right at the end of the dry season, when the grass is still dormant but the trees and shrubs have begun to produce a flush of young leaves and are therefore in a vulnerable condition. In many places a few showers, known as the 'grass rains', occur almost at the end of the dry season and are followed by a short, dry spell before the main rains begin. A common practice is to burn the grass for bush control immediately after the first showers. This avoids the risk of a long dry spell after the burn owing to the late arrival of the rains. Burning earlier in the dry season results in the grass making transient growth at the expense of its food reserves and is often done by native herdsmen to produce a little green grass for hungry cattle. It is a bad practice because it actually encourages bush encroachment, and it weakens the grass so that it cannot make vigorous growth when the rains come. The result is that soil exposure is increased and erosion encouraged.

A fierce fire is needed to control bush; consequently, pastures must be rested from grazing for some time before burning in order that there may be a good accumulation of dry grass. In the drier areas the grass cover is usually sparse and it is commonly necessary to rest the pasture for nearly a year, through most of the growing season and the whole of the dry season. Previous grazing intensity will affect the quantity of grass available; where there has been heavy over-grazing, resting for longer than a year may be needed to accumulate enough grass for a good burn.

After burning, the pasture must not be grazed at the beginning of the rainy season but allowed to grow unchecked for a time in order to allow for some replenishment of the food reserves expended in initial growth. The requisite period of rest at this time varies with local conditions, but is usually one or two months.

For effective bush control, burning must be done regularly at suitable intervals, so that young tree seedlings which may survive one fire will not have grown up sufficiently to escape a second. The required frequency of burning will vary from place to place and with the condition of the pasture resulting from management. For example, West (1947) suggests that in Matabeleland the veld may need to be burned once in three years until a fairly stable open parkland has been established but thereafter, if it is stocked correctly, it should not be necessary to burn more often than once in five or six years. For most grasslands that need it, burning is usually done about once in four years.

Certain shrubs occurring in grassland cannot be controlled by fire alone. Examples are *Tarconanthus camphoratus* (locally known as 'leleshwa'), which has heavily and extensively invaded pastures in the Kenya Rift Valley, and *Acacia pennata*, which is common in semi-arid pastures in several parts of Kenya.

Burning may be useful for purposes other than bush control. In 'sour' or mixed veld, where it is not possible to graze uniformly every season without deterioration in the pastures resulting from overgrazing, there is always a good deal of grass left uneaten. If mowing is impracticable, it is desirable to rid the land of this coarse, unpalatable herbage by burning periodically, perhaps once in two years. Fire may also be the best way of getting rid of undesirable species of grass which have invaded a pasture owing to inappropriate management. For example, a single, good burn at the end of the dry season proved the most effective way of ridding a Kenya pasture normally dominated by *Cynodon dactylon* and *C. plectostachyum* of a massive invasion of a poor annual grass, *Aristida kewensis* (Bogdan and Storrar, 1954). Fire was also found by Edwards (1942) to be the best means of preventing the replacement of *Themeda triandra* by the useless coarse grass *Pennisetum schimperi* in the 'highland grassland and forest' zone of Kenya, or by *Digitaria abysinnica* in marginal *Acacia-Themeda* country.

Where fire is used for purposes other than bush control, it may often be unnecessary and undesirable to burn at the end of the dry season. Fires at this season are hot and fierce, and there is always danger of them getting out of hand and causing damage, unless careful precautions are taken to control them. In some places they may also kill out some of the desirable grass species in the swards and tend to increase the proportion of coarse, unpalatable, fire-resistant species. Consequently, where bush control is not the main problem, a safe, early, light burn to remove unpalatable material may be all that is required.

Other means of controlling bush

Digging out or uprooting shrubs and trees by hoeing or other means is effective in controlling bush, but is expensive and laborious because of the extensive roots of many species and their ability to produce shoots therefrom. Merely cutting down trees and slashing shrubs is usually ineffective, because many plants regenerate from the base of the stem or the roots. Repeated slashing at frequent intervals may weaken and kill out some plants, but with many species it is a protracted and expensive procedure. Cutting combined with burning will deal with some species, especially if the cut material is stacked on the root stocks, allowed to dry and then burned *in situ*.

Where it is practicable, and while it is done regularly, mowing will keep down regeneration of many species but it does not usually eliminate them and they come away strongly as soon as regular mowing ceases. Thus, on the higher rainfall sandveld in Rhodesia, annual mowing for ten years kept down coppicing of bush but as soon as mowing was stopped regeneration occurred (West, 1956). Mowing is very commonly impracticable owing to the large areas involved, the uneven nature of the ground and the presence of rocks, anthills or trees.

Many trials have been made with arboricides for initial thinning of bush, or for controlling regeneration. Sodium arsenite is effective but normally cannot be used on account of its high mammalian toxicity. Sodium chlorate has not usually given satisfactory results. Power paraffin, or diesolene, generally only kills thorn trees when liberally painted on to a 6–9-inch band at the point where the stem joins the root and, as this involves removing several inches of soil, it is too expensive (West, 1947).

The most useful chemicals at present are 2,4-D and 2,4,5-T and the latter is usually rather more effective than the former. (Ivens, 1958, 1959, 1960). Both give the best results when applied to a 6–9 inch band at the base of the stem (after frilling or ring-barking) as ester formulations in oil, emulsions in water being less effective, presumably because they penetrate less readily. Ivens found that most species of *Acacia* and *Commiphora* were killed by concentrations of 0·5–2·0 per cent in oil, 1·0 per cent or less being adequate for the majority, but *Combretum* spp. were less susceptible and required rather higher concentrations of 1·5–2·0 per cent. In the *Brachystegia-Pseudoberlinia* woodlands the above-ground parts of more than 90 per cent of the species were killed by concentrations of 1·0–2·0 per cent, but many trees subsequently regenerated from their roots, as was also observed by Cleghorn (1958) in Rhodesia and Jackson (1955) in Malawi. Some species of the main genera occurring in other types of grassland and woodland are also capable of regeneration after the above-ground parts have been killed by these chemicals, although few of the *Acacia* spp. do so. Certain common species, for example *Tarconanthus camphoratus*, *Euclea divinorum*, *Acacia pennata* and *Dichrostachys glomerata*, are very resistant to the chemicals and even repeated spraying may not prevent regrowth.

It seems that, while these chemicals are useful to effect an initial thinning of trees, the present formulations and methods of application are not very satisfactory for maintaining a pasture free of bush thereafter. Regeneration usually requires several treatments, and since the cost is at present high, spraying is only worth while if a considerable increase in the carrying capacity of the land results, which is unlikely to be obtained in many of the drier areas.

GRAZING MANAGEMENT

It will now be apparent that, in order to achieve the aims stated on page 256 in the management of pastures of the seasonal rainfall areas, the following requirements will have to be met as far as possible:

1. Adjustment of stocking rate to carrying capacity, which will usually mean relatively light stocking well within the average productivity of the pasture.
2. Periodic rests from grazing during the critical growth periods.
3. Bush control, usually by fire and involving a need for rest before and after burning.
4. Some special provision for dry-season feed.

These requirements are usually met by adopting a system of deferred rotational, or seasonal, grazing by which the use of part of the pasture is delayed until the herbage has reached a relatively mature stage. The pasture is fenced into paddocks, each normally provided with a water supply, and these are grazed in rotation in a manner which provides for periodic rests and burns, and also ensures that sufficient grass or standing hay is available for stock all the year round. Details of the system to be adopted will depend on local climatic conditions, the type of grassland and whether or not burning is needed for bush control, and on economic considerations. There are many such systems, some simple, some rather complex; a number of examples are given by Scott (1947), Semple (1952), Kennan *et al.* (1955) and West (1948, 1958). Only two will be mentioned here by way of illustration.

A very simple system is that known as the 'three herd: four paddock' system. Each paddock is grazed continuously for three years and in the fourth year grazing is stopped just after the beginning of the rains, the accumulated grass is burnt off at the end of the dry season, and the paddock is then rested for a further one or two months at the beginning of the next rainy season. Owing to the long period of continuous grazing involved, and the need to keep some grass standing in the dry season, the stocking rate has to be rather conservative.

A 'one herd to four paddocks' system is described by West (1958) as being suitable when a higher carrying capacity justifies more expenditure on fencing and water supplies. It has advantages in being flexible, and permitting heavier stocking rates, because it provides ample reserves for droughts and bad seasons. The rotational details are shown in Table 27. Each paddock is used for growing season grazing in one year. In the next it is rested for burning. After burning at the beginning of the third year it has a growing season rest and is then grazed in the dry season. In the fourth year it is again only grazed in the dry season. It is advisable to wait after the onset of the rains until the grass is growing strongly before putting the herd into whichever paddock is to be grazed in the growing season that year. At the end of the rains the animals are moved from this paddock into the two dry-season paddocks.

Table 27. *One herd – four paddock system*

Paddock	1st year	2nd year	3rd year	4th year
A	Graze growing season	Rest for burning	Burn, graze dry season	Graze dry season
B	Rest for burning	Burn, graze dry season	Graze dry season	Graze growing season
C	Burn, graze dry season	Graze dry season	Graze growing season	Rest for burn
D	Graze dry season	Graze growing season	Rest for burn	Burn, graze dry season

The adoption of a suitable system of deferred rotational grazing usually brings about an improvement in the pasture and its carrying capacity as compared with continuous grazing. However, in the poor grasslands of the drier areas there may be no appreciable increase in production from the use of such a system instead of continuous grazing. Indeed, there are large areas which are inherently of such low productivity that they do not justify expenditure on bush thinning and control, fencing and water supplies.

The use of mixed grazing systems employing cattle, with sheep or goats
Owing to differences in the grazing habits of cattle, sheep and goats, mixed grazing may be preferable to stocking with cattle alone. Sheep are very selective in their grazing; they prefer the finer grasses and nibble the growing leafage, neglecting stems and coarser grasses, so that a pasture grazed by sheep alone tends to become very rough. Cattle graze more uniformly, but disregard certain coarse and unpalatable grasses. In temperate countries mixed grazing of permanent pastures with cattle and sheep has generally been found advantageous in its effect on the composition of the sward and on productivity. Probably the same applies to natural grasslands in the tropics, but there is little evidence on this point.

In grasslands containing trees and shrubs mixed grazing with cattle and goats may often be advantageous. In contrast to cattle, goats will obtain the greater part of their food from the trees and shrubs, thus aiding in bush control. Goats are not very selective as regards plant species but have a strong preference for succulent shoots and leaves that are slightly above their normal head level. Given the opportunity, they will browse mainly on trees and shrubs. They do not normally graze low down and, though they may eat the shoots and inflorescences of the taller grasses, they neglect the growing points near the ground and the shorter grasses (Wilson, 1957a). In experiments conducted by Staples *et al.* (1942) in Tanganyika, goats kept down coppice and seedling growth of most bush species, with the result that a good, low ground cover was maintained. By contrast, grazing by cattle alone concentrated mainly on the grasses, thus relieving the bushes of competition and encouraging the formation of thicket with little ground cover. Similar results were obtained in experiments in Kenya (Bogdan, 1955).

THE USE OF FERTILIZERS

The response to fertilizers depends very much on the rainfall and the type of grassland. For example, fertilizer experiments on natural grasslands in the Kenya Highlands showed that, although responses were generally obtained to both nitrogen and phosphate, the yield increases were directly related to rainfall and that, whereas fertilizers were profitable on pastures dominated by good grasses, such as Kikuyu or star grasses, they gave only small and unprofitable yield increments where poorer species were predominant (Dougall, 1954).

Most of the natural grasslands are in drier areas where responses to fertilizers are marginal, or too small to be profitable. However, profitable responses may be obtained in some places where the total rainfall is rather low but is concentrated in one rainy season. For example, at Frankenwald in South Africa, where an average annual rainfall of 31 inches falls in a five-month rainy season, the response of the natural veld to nitrogen was linear up to a fairly high level of application and there was a substantial profit from the use of this fertilizer (Hall *et al.*, 1950; Rose, 1952).

In the more limited areas of better rainfall, fertilizing natural grassland can often substantially increase the yield of herbage and improve its content of protein, phosphate, potash and lime. Nitrogen is the element that is most commonly needed and which produces the biggest yield increases, but response to phosphate is also common provided that sufficient nitrogen is available. Potash responses are less common.

Fertilizer application will usually increase yields in the good rainfall areas, but it is far easier to increase grass growth by this means during the wet season than during the dry season. The use of fertilizers thus tends to accentuate the already marked difference between wet- and dry-season productivity, rather than to even out the production curve. However, late fertilizer application can, if successfully and skilfully

Table 28. *Effect of fertilizers on veld and of replacing veld by fertilized, sown pastures of selected species in Rhodesia*

West (1956)

	Live weight production in lb per acre per annum	
	1953/4	1954/5
Veld, no fertilizer	90	93
Veld, 300 lb nitro lime, 150 lb supers, 50 lb muriate of potash p.a.	121	147
Veld, 600 lb nitro lime, 150 lb supers, 50 lb muriate of potash p.a.	167	152
Veld, 900 lb nitro lime, 150 lb supers, 50 lb muriate of potash p.a.	177	162
Cultivated pasture, 300–524 lb nitro lime, 400 lb supers, p.a.	459	–
Cultivated pasture, 200 lb nitro lime, 250 lb sulphate of ammonia, and 250 lb supers, p.a.	–	425

carried out, produce its main effect at the beginning of the dry season and thus extend the growing period somewhat at this time. But under good rainfall it may well prove that better responses can be obtained to fertilizers, and more profitable use made of the land, if the natural grassland is replaced by cultivated pastures of selected species. This was found to be the case in some of the higher rainfall areas in Rhodesia, as can be seen from the figures of Table 28 (West, 1956).

AIDS TO THE PROVISION OF DRY-SEASON FEED

It has already been mentioned that rotational grazing systems normally provide for some paddocks to be rested in the growing season in order that 'standing hay' may be available to feed stock in the dry season. This standing hay is frequently of such poor feeding value that it will not even provide the protein needed for maintenance. Consequently it is desirable to provide additional food for the dry season by any other means that may be practicable.

One possibility is to make hay from the surplus, high protein grass available in the growing season, but in many places this is impracticable for the following reasons:

1. Much of the land is unsuitable for mowing owing to its topography, uneven surface or the presence of rocks, anthills, trees and bushes.
2. The low yields per acre make hay-making uneconomic.
3. The grass reaches the appropriate stage for hay at the height of the rains. At this time the weather either makes hay-making impossible, or difficult and liable to heavy losses.

Nevertheless, hay-making is practicable in some places especially as the difficulty of the weather can to some extent be overcome by grazing an area early in the growing season and then closing it, so that the grass reaches the hay stage in drier weather towards the end of the rains. Although the veld hay may not be of very high quality, it will usually be of better feeding value than dry over-mature grass left standing in the field into the dry season. This is illustrated by a comparison of the figures in Table 29 for 'whole seasons growth' of mixed grasses with those for 'first growth' and 'aftermath' of the same material (French, 1943a). Provided that veld hay is not cut later than at the flowering stage, that loss of the more nutritious leaves does not occur through rough handling, and that it is not exposed to bad weather, it will generally show lower crude protein and higher crude fibre values than medium quality British hay, but will have about the same energy value (French, 1956c). As can be seen from Table 29, some species, such as *Cynodon plectostachyum*, *Cenchrus ciliaris* and *Panicum maximum*, are superior to others, but a considerable proportion of veld hay is likely to be of species with relatively low crude protein and energy values and rather high crude fibre.

Table 29. *Dry matter composition and digestible nutrients of hays* (1)–(10) French (1943a, 1956c); (11), (12) Dougall (1960)

Species	Maturity stage	Crude protein	Crude fibre	S_1O_2 free ash	Dig. crude protein	S.E.
1. Mixed grasses	First growth	8·8	34·6	3·6	5·5	37·6
2. Mixed grasses	Aftermath growth	8·5	34·2	4·3	4·8	30·9
3. Mixed grasses	Whole season's growth	6·1	33·5	4·1	1·6	22·5
4. *Cynodon plectostachyum*	1st cut, 2' high, early flowering	9·2	34·9	5·2	5·7	38·6
5. *C. plectostachyum*	Aftermath, 2' high, early flowering	7·3	37·1	4·5	3·8	31·7
6. *C. plectostachyum*	Cut mid dry season	5·2	39·7	—	1·2	18·0
7. *Chloris gayana*	1st cut, 3' high, flowering	3·7	43·5	3·8	1·1	21·2
8. *Chloris gayana*	Aftermath 3' high, flowering	3·7	42·0	3·4	1·1	20·2
9. *Cenchrus ciliaris*	2½–3' high, flowering	7·4	35·2	3·8	4·0	37·9
10. *Panicum maximum*	2¾–3' high, flowering	8·9	37·0	3·8	4·6	39·1
11. *Chloris gayana*		7·0	33·1	—	3·1	—
12. *Setaria sphacelata*		6·2	38·1	—	2·5	—

Silage is made to a limited extent with surplus grass from the more productive natural grasslands of the better rainfall areas. In the drier areas it is sometimes possible to make silage or hay from limited areas of high-yielding, cultivated grasses and legumes grown on dry land during the rains, on seasonally swampy or flooded areas, or under irrigation. These fodder crops are dealt with in Chapter 13.

Table 30. *Composition and feeding value of cereal stovers*
French (1943b)

	Maize	Sorghum	Bulrush millet
Composition, % on dry matter			
Crude protein	4·99	4·30	4·29
N-free extract	52·50	46·99	41·39
Crude fibre	34·69	37·16	43·60
Total ash	6·71	10·30	9·32
Digestible materials and S.E. per 100 parts dry matter			
Crude protein	2·46	1·31	0·55
N-free extract	34·43	26·44	19·98
Crude fibre	24·56	23·03	24·96
Starch equivalent	29·51	24·82	21·71

Crop residues, which become available after harvest when grazing is scarce and poor, deserve maximum use as supplementary feed during the dry season. These are available in limited amounts in both ranching and native pastoral areas and consist principally of cereal stovers. French (1943b) has examined maize, sorghum and bulrush millet stovers and his results are given in Table 30, from which it will be seen that all are at least equal in starch equivalent to mature grass left standing in the dry season (cf. figures for 'Mixed grasses – whole season growth' in Table 29). He also considers that maize and sorghum stovers contain rather more digestible protein than dried grass, although bulrush millet stover has less. Finally, there is, of course, the possibility of feeding a supplementary ration of a high protein concentrate, the principal kinds likely to be locally available in the tropics being cotton seed, cotton seed cake, coconut cake and groundnut meal.

CHAPTER 12

The Use of Natural Grasslands by Native Pastoralists

The land used by pastoralists for the extensive grazing of their herds on the unimproved, natural herbage is mainly in regions of low and unreliable rainfall, experiencing a severe dry season, where it is impracticable to grow crops. The grasslands of these areas are those derived from, or associated with, the drier types of broad leaved woodland, thorn woodland and subdesert scrub (see Chapter 11). This chapter discusses the use of these kinds of grasslands by native herdsmen. Pastoralists are also found, to a limited extent, in the wetter grasslands derived from the moister broadleaved woodlands and montane evergreen forests but, as crop production is practicable in these areas, relatively little land is nowadays used solely for grazing. Brief reference to pastoralism in these wetter regions is made at the end of this chapter.

As already mentioned in Chapter 11, most of the drier types of grassland consist of a rather sparse cover of grass with scattered trees and shrubs, many of the latter making an important contribution to stockfeed, especially in the more arid regions or where the grass cover is further reduced by poor management. The year-round carrying capacity is low; under native pastoralism it is commonly of the order of one beast to 20 to 25 acres but may be very much lower where deterioration has resulted from prolonged mismanagement. There is marked seasonal variation in carrying capacity owing to the annual cessation of growth of the grass in the dry season, during which period pastures used by native herdsmen are almost invariably overstocked. As the rainfall is unreliable, years of severe drought are not uncommon, and at such times overstocking is still more pronounced and many animals may die of starvation. Water supplies are often inadequate and ill-distributed. Apart from causing hardship and loss of stock during the dry season, scarcity of water may result in poor pasture utilization because concentration of stock near watering points during the dry season results in overgrazing and erosion in these localities, while other pastures, remote from water supplies, may not be fully utilized. Over much of the African pastoral regions the vegetation is of a kind that forms a suitable habitat for tsetse flies and considerable areas are thus rendered unsuitable for grazing for a part, or the whole, of the year.

It is thus evident that a variety of environmental factors restrict grazing either seasonally or geographically and are thus conducive to overstocking of those areas that are available for the use of the herds. As will be seen below, some of the customs and habits of the pastoralists themselves also make for overstocking.

Features of Pastoralism

Most pastoralists are nomadic, or seminomadic, not merely because of inherited custom or tradition, but as a direct consequence of the limitations of their environment. The movements of the people and their stock are mainly dictated by the need to go where fodder and water are available and to avoid areas with disease hazards, such as those infested with tsetse, *Stomoxys*, or other biting flies (see Chapter 15, p. 368 and Chapter 16, pp. 373–5).

Generally the land held by pastoral tribes is owned and grazed communally. On the other hand, individual or family ownership of livestock is normal and it is often usual for each owner to aim at keeping as many animals as possible, irrespective of the quality of the beasts or the availability of pasture. This is partly because livestock are regarded as wealth, and a man's social position and prestige depend on the number of stock he has rather than on money, or other possessions. It is also because cattle are needed to fulfil certain obligations under tribal customs, such as the payment of bride price, which is a normal feature of the social life of many pastoral tribes. The head of a family may not be free to sell his cattle, since he holds them in trust for the family, and may be under an obligation to retain them in order to provide bride price for his sons or younger brothers. Large numbers of stock are also commonly kept as an insurance against years of drought and famine, on the mistaken assumption that the more cattle a man has the more are likely to survive a bad year. Little or no attempt is made to cooperate by restricting the numbers of stock held by individuals to the number which might reasonably be expected to survive on the available fodder and water resources. This is either due to failure to appreciate the advantages of such a procedure, or to reluctance to adopt it on the part of chiefs, who usually own the largest herds.

As a result of their traditional attitude towards stock and, in particular, their desire to possess as many cattle as possible, pastoralists are usually very reluctant to sell their animals. The material needs of primitive pastoral tribes were small: they lived on milk, blood and meat from their own stock, by hunting, by the collection of berries and other wild forms of food and by raiding cultivating tribes, from whom they stole grain and cattle. The raiding has largely ceased, but in other respects many pastoral peoples have changed but little. Their needs are still small; they barter milk and skins for grain and they will only sell stock when they are short of cash to pay taxes, or to purchase limited essentials, or when there is an immediate prospect of losing many animals through drought and famine. There are, naturally, exceptions to this generalization. For example, the Fulani of Northern Nigeria have long been accustomed to selling considerable numbers of stock to supply meat to people to the south. Other African tribes have also recently begun to develop in the same way but in general a commercial attitude is lacking.

The combination of communal ownership and grazing of the land with unrestricted individual or family ownership of stock leads to an almost complete absence of pasture management. It is an extraordinary fact that neither the individual herdsmen, nor the chiefs, nor the community, appear capable of taking any effective interest in this vital matter. Stock numbers are not adjusted to carrying capacity and available water supplies, little or no effort is made to organize rotational grazing and resting of pastures, and there is no attempt to control bush encroachment. Matters are made worse by the indiscriminate setting of fires to facilitate hunting, or to promote a short-lived growth of green herbage during the dry season, a procedure which provides a negligible amount of feed and only serves further to exhaust grass plants already weakened by overgrazing. These fires are usually uncontrolled and, sweeping across large areas, they destroy useful standing hay, leaving the ground bare and exposed to erosion at the onset of the following rains. The result of this gross overstocking, lack of pasture management and uncontrolled burning is destruction of the grass cover, invasion of bush, and erosion on a large scale.

The pastoralists' level of animal husbandry is generally low when judged by modern standards, although certain qualities of stockmanship are sometimes developed to a surprising degree. The nature of the environment seldom makes it possible to avoid animals suffering an annual loss in weight during the dry season, but this and other checks to the development of the stock could be reduced by better husbandry. Seasonal mating, so that the majority of calves are dropped as soon as there is adequate grazing for the cows, would reduce calf mortality and aid the early growth of the calf. This is not often practised, with the result that calf mortality is generally high and calves are often small and underfed; the cow often does not give enough milk to support a healthy calf and part of what it does give may go for sale or human consumption. Pasture management and careful herding would, in many places, make it possible to keep the stock on the better grazing at weaning and fattening, but this is hardly ever done. Correction of mineral deficiencies is needed in many areas but mineral licks, apart from rock salt, are not commonly used. Herdsmen often take little interest in the measures recommended, or facilities provided, by veterinary departments for the control of disease. For example, dipping schemes can only be effective if progressive reduction in the tick population of an area is brought about by regular dipping of *all* cattle, but it has often been found difficult or impossible to persuade herdsmen fully to cooperate in such schemes.

The cattle kept by pastoralists, though hardy, capable of surviving and reproducing under harsh conditions, and showing some resistance to certain diseases, are of very poor quality, undersized, of poor conformation, inherently slow maturing and inefficient producers of meat and milk. This is in part due to the custom of retaining all cattle irrespective of quality; as a rule no culling of heifers or castration of poor bulls is practised, and breeding is quite indiscriminate. Some pastoralists, however, for example the Fulani and Abahima, do effect some control and selection in breeding. (See Chapter 16, pp. 370–2.)

Failure to follow good husbandry practices and, above all, failure to adjust stocking rate to carrying capacity, results at best in slow growth, with annual dry-season checks. Cattle often take from four to seven years to produce a poor, light-weight carcass. In drought years, as a consequence of gross overstocking, large numbers of cattle die of starvation and thirst. Alternatively the herdsmen, threatened with heavy losses, may offer many immature beasts for sale. The sale and slaughter of the latter before they have fattened is wasteful, and obviously represents a loss to the stockmen.

Results of Pastoralism

Primitive pastoralism, which continues to be widely practised with little departure from ancient, traditional methods, has been responsible for tremendous damage to huge areas of land in Africa and Asia. This is primarily due to the inherent deficiencies of the system, which have been described above. The resulting mismanagement of the unstable subclimax vegetation types that form the drier tropical grasslands rapidly destroys the grass cover and causes bush encroachment and erosion. The damage is primarily due to overstocking. This has always tended to occur in primitive pastoral systems that combine communal grazing with unrestricted individual ownership of stock, but in recent times it has become still more prevalent. Both the pastoralists and their beasts have become more numerous as a result of the abolition of tribal warfare and slavery, reduction in the incidence of famines, and better control of disease. At the same time the areas available to pastoralists have tended to decrease owing to the greater land requirements of increasing numbers of cultivators, who have taken to growing cash crops in addition to subsistence food crops. Consequently, damage to land in the pastoral areas has been more disastrous and extensive.

The sequence of events which occurs when pastoralism is combined with increasing population pressure on the land is well known. Overgrazing, which often begins at the watering places (where animals congregate in the dry season) and spreads out from them, destroys the grass cover. The removal or reduction of competition from the grass hastens the invasion and spread of trees and shrubs that are better adapted to arid conditions, and this encroachment, together with continued overgrazing, prevents the re-establishment of the grass cover. Rain falling on the bare ground between trees runs off, erosion begins and continues at an ever-accelerating rate. Once the orderly absorption and percolation of rainfall through the soil no longer operates, the streams cease to flow throughout the greater part of the year and only come down in flash floods for short periods, the watercourses remaining dry and filled with silt for most of the year. Springs dry up, water for stock and vegetation becomes progressively scarcer and the end of the process is the creation of near-desert conditions. The rate and extent of denudation varies with local circumstances, being greater where human and stock population are dense, where steep topography favours erosion, or where the rate of regrowth of the vegetation is reduced by more arid climatic conditions.

The economic benefits from this disastrous method of land use have been negligible or non-existent. Quite apart from the reluctance of the people to sell their stock, the productivity of the system is extremely low. This might be expected, because relatively poor grasslands are used with little or no attempt at good management and because the livestock employed are often of poor quality and suffering from a variety of inadequately controlled diseases. On well-run ranches on grasslands similar to those occupied by native herdsmen in Tanzania the average calving percentage is 70–80 per cent, calf and herd mortality is low, average liveweight increases of $\frac{1}{2}$–1 lb per day are obtained and the average annual off-take of mature cattle is 18–20 per cent of the total stock population. Accurate data of a similar kind for the herds of African pastoralists are not available, but there is no doubt that their performance is much poorer. Estimates made in

Tanzania (International Bank, 1961) indicate that the average calving rate is only about 50 per cent, 20 per cent of the calves do not survive the first six months and a further 20 per cent fail to reach maturity and the average annual off-take is of the order of 7 per cent. In Northern Nigeria, Shaw and Colville (1950) estimated that the average calving rate was 40 per cent, calf mortality 15 per cent, herd mortality 5 per cent and the annual off-take 5·2 per cent.

Indigenous pastoralism in the tropics thus results in extensive destruction of natural resources and produces little economic return. The continuation of the pastoralists in their present mode of life, as stated by the East Africa Royal Commission (1955), '... portends both a danger and a deficiency. The danger is that they may turn their lands into desert; the deficiency that, without management of their herds, and in some cases better usage of their lands than mere pastoralism, they will contribute far less than their land's potential to the growing needs of the community.'

Possibilities for Improvement

It is obvious that what is needed in order to put land use on a sound basis is the application, as far as the environmental conditions allow, of the principles of ranching husbandry outlined in Chapter 11. This, however, is far from being an easy matter. Pastoralists are usually conservative and it is difficult to persuade them to change their ways unless the proposed changes are clearly beneficial and do not infringe religious beliefs and social customs. Furthermore, in Africa good husbandry can often only be satisfactorily employed if it is accompanied, or preceded, by clearing to control tsetse flies and by the provision of water supplies, facilities for disease control, roads and markets. In many places, also, denudation has already gone so far that special measures must be taken to rehabilitate the land before good husbandry can achieve satisfactory results. All these things involve Governments in capital expenditure, and a major difficulty is that the potential of much of the pastoral lands is so low that it cannot pay a reasonable return on the capital investment required. Certainly no adequate return will be obtained unless the investment is followed by good animal and grazing management.

ADJUSTMENT OF STOCKING RATES

Usually, the first essential in improvement is to reduce stock numbers in order to adjust them to carrying capacity. In some places where the existing stocking is clearly in excess of the potential carrying capacity of the land on any system of management, a permanent reduction in numbers is needed. More commonly, the problem is not so much overstocking in relation to the real potential of the land as overstocking under present lack of pasture management. Hence an initial reduction in stock numbers, if followed by good pasture management, will in due course enable the land to carry more stock than it does at present.

The first essential step of adjusting stock numbers to carrying capacity has generally proved to be extremely difficult to achieve. It demands a radical change in the pastoralists' way of life in that they must be persuaded to abandon many of their ancient customs, especially to cease keeping large numbers of animals for prestige, or for the discharge of tribal obligations. Instead, they must be persuaded to make an initial reduction in the size of their herds and thereafter to raise animals for profit, selling a suitable proportion of their stock annually. Efforts to get pastoralists to sell more of their animals have been made by administrative and technical officers for many years, but have made slow progress, and in many places education and persuasion will probably have to continue for some years before a solution to the problem is in sight. However, definite progress has been made in some areas. For example, the Fulani in Northern Nigeria have long been accustomed to selling animals for meat; in Uganda cattle auctions have operated for some years, and recently the Masai in Kenya have been selling increasing quantities of milk and meat. In other parts of Africa, pastoralists are also showing signs of becoming more interested in building houses, in cash for household requirements, clothes and bicycles and, consequently, in selling more stock.

MARKETING AND MOVEMENT OF STOCK

During the process of adjusting stock numbers and of the development of a commercial attitude by the graziers, it is clearly necessary that government, or other agencies, should find an outlet for the increasing number of animals offered for sale, and should provide means for facilitating the movement and marketing of livestock.

A discussion of the likely future demand for animal products from tropical pastoral areas cannot be undertaken here but it may be noted that in Africa an increase in the present very low average consumption of meat would be desirable on nutritional grounds. As the people are always extremely anxious to obtain meat, a big increase in the domestic demand may be expected in view of the increasing purchasing-power of African communities. Secondly, there is probably a fair prospect for some expansion in the production of canned meat and meat extracts, both for export and for local consumption in the urban areas.

To aid in increasing the off-take of animals it will be necessary to establish more markets in the pastoral areas, with access roads for buyers, and to provide more stock routes, furnished with watering points and fenced night-stops, so that animals can be trekked without suffering from shortage of food and water or from the depredations of wild game. Where large numbers of immature cattle are sold in dry years, it may be possible to reduce the waste resulting from their slaughter before maturity (and at the same time to even out the supply of meat to the market) by the provision of holding grounds, possibly operated in conjunction with canneries, where the cattle could be fattened. A problem which is liable to arise, at any rate during the early stages of an improvement programme, is that of the disposal of a proportion of very poor quality cattle that are below the standard required for canning, let alone for fresh meat. Herdsmen are unwilling to bring in such beasts over long distances to markets when the chances are that they will be rejected. Yet these are the very animals which it is particularly necessary to eliminate. The problem has been met in Kenya by the provision of mobile abattoirs and meat processing plants, which travel into the more remote overstocked or drought-afflicted areas to purchase low-grade cattle and process them into biltong, meat meal, blood meal and bone meal. Even if these plants require an element of subsidy, which seems to be the case (Enlow, 1961), this seems worth while in order to remove a segment of the cattle population which is a distinct disadvantage to the industry, and an impediment to improvement in land use.

PASTURE MANAGEMENT

Pasture management will essentially involve the division of the land into defined grazing areas or 'ranges' (usually of relatively large size, and necessarily related to water supplies) and the introduction of some form of rotational grazing coupled with the controlled use of fire to prevent bush regeneration or encroachment. The rotational grazing system adopted will vary with local conditions (see Chapter 11) but it should be kept as simple as possible, minimizing the number of ranges and the frequency of stock movements. In a simple system used in Kenya the land is divided into four blocks; in each year three blocks are grazed in rotation for four months each, leaving one block for resting and burning.

Fencing is most desirable but is usually too expensive, and the grazing areas are usually defined by natural features, the stock being controlled by herding. Alternatively, the boundaries may be demarcated by lines of felled thorn trees, by cutting wide traces through the bush, or with live fences. The latter are often not very satisfactory as they take time to establish and may be unreliable due to breaks in continuity, or to developing an open base as they mature. In most areas a compromise is possible if the demarcation plan is carefully thought out. An area of land may be delimited by a river on one side and a mountain range on the other. Cheap transverse divisions may be possible by running barbed wire on live trees and bushes, selecting a transverse where the bush is thickest. When bush clearing is to be done, it is most important to have in mind a plan for the siting of range boundaries, as it may be possible to leave thin strips of bush which may be made impenetrable by a patchy and crude system of barbed-wire fencing, alternating with rows of heaped up thorn scrub. Provided the stocking rate is adjusted to the carrying capacity of the land, and animals are moved on before the herbage is in short supply, the standard of fencing required in extensive grazing systems can be much lower than that needed for intensive cattle management.

In view of the great disadvantages of communal grazing, it is often suggested that a change to individually owned and operated holdings is desirable in order to obtain good pasture management. However, over a great deal of the pastoral country individual holdings are impracticable for several reasons. First, seasonal migration of stock is often essential because large areas of pasture can only be used for part of the year. Secondly, provision of water to individual holdings is often impracticable or uneconomic in areas where water is scarce. Thirdly, individual holdings would often be unacceptable because few pastoral regions are uniform. Under individual tenure some people would get only poor grazing, whereas under the communal system all have equal access to good and poor areas. Fourthly, the fencing of pastures and water supplies on individual holdings would be expensive, and in many places the absence of fencing materials would make it impracticable. In some circumstances, it may be practicable and beneficial to divide large communal grazing areas into smaller units that can be effectively used and managed communally by small groups of people.

It has been mentioned earlier (p. 267) that mixed grazing with cattle and goats is often advantageous in grasslands containing shrubs and trees, but this involves certain practical difficulties. In order to achieve the best effect on the composition and productivity of the pasture it is necessary to adopt a suitable ratio of cattle to goats, and this ratio will not necessarily remain constant. For example, a ratio of one steer to twenty goats may initially be suitable on grassland containing much thorn bush, but after ten years the goats may have virtually eliminated the thorn bush, and a ratio of one steer to five goats would then be appropriate. Goats are rather difficult to confine within a range, since they can leap over 5-foot fences and scramble through hedges adequate to deter cattle. Goats are also more liable to suffer from the depredations of wild animals, and are more readily stolen, than cattle. Consequently it may be necessary to provide shelter for goats in places where this is not essential for cattle.

REHABILITATION OF DENUDED AREAS

In many places denudation has gone beyond the stage where improvement can readily be effected by introducing controlled stocking and pasture management, and these need to be preceded by special measures for the rehabilitation of the grazing areas. Thinning of rather dense bush may be needed, either for tsetse control or to improve the growth of grass. Control of erosion, particularly on hillsides where gulleying has occurred, may be needed on a considerable scale. This is normally done mechanically, but should form part of a coordinated programme of improvement. For example, there is little point in putting in contour terraces on badly eroded land unless simultaneous steps are taken to ensure establishment of a good grass cover, since in the absence of the latter the terraces will quickly silt up. Where natural grasslands have been severely overgrazed it may be necessary temporarily to close them to grazing in order to allow regeneration of the grass cover. This can usually be done without difficulty and with the consent of the people if the areas concerned are small, as may often happen in the case of small hills or steep slopes which need to be regrassed to check erosion affecting surrounding lower land. If the denuded areas are more extensive, the problem of stock feed may render closure impracticable, at any rate for more than a short period, and it may be necessary to expedite revegetation by planting grass.

Planting of grass is, in any case, likely to be needed over limited bare eroded areas, or where the good perennial grasses have either been destroyed, or so reduced that there are insufficient plants left for natural reseeding to provide a cover within a reasonable period of time. Reseeding will normally necessitate some preparatory cultivation, partly to enable the seeds to be covered, but also because a common difficulty in denuded areas is the presence of a hard, eroded soil surface from which the rain runs off. As a result of this hard crust little water percolates into the soil, and seed or seedlings are liable to be washed away by heavy storms. Cultivation should break this crust, permitting water percolation and providing tilth for seeds, without unduly increasing the risk of soil wash. In view of the large areas of low potential which have to be covered, such cultivation must also be cheap. Methods adopted include scratch ploughing on the contours at intervals of from 3 to 12 feet, disc or tine-harrowing strips on the contours, or chopping small holes at intervals with hoes. Mulching with cut brushwood after scattering seeds on the surface is sometimes done. The brushwood affords some protection to the soil surface from heavy rain, checks run-off, increases percolation and provides better anchorage for young seedlings. When patches of land require to be reseeded within a larger area which is not closed for grazing, mulching with brushwood will also give localized protection against grazing and trampling. Another method which may be used to prevent washing away of seed and seedlings is to allow the natural development of annual and ephemeral grasses of little grazing value to occur before broadcasting the seed of perennials. This procedure, however, it not always possible as sometimes undesirable weed species develop and become dominant.

The seed required will usually be collected near the area to be reseeded, from locally adapted indigenous perennial species. Since viability of the seed may be markedly affected by the time of collection and the period of storage before use, its germination capacity should be tested. It is also often desirable to treat it with an insecticide, such as DDT or gamaxane, as a precaution against the depredations of insects, expecially of harvester ants. Planting grass vegetatively may be more successful than seeding in some places and is inevitable with some species, but it is always more laborious and costly.

Sometimes cultivation to increase water percolation will in itself be sufficient to effect the recovery of the vegetation without the necessity of reseeding or replanting. Oates (1956) has described the effect of ripping on Rhodesian thorn-veld where, as a result of overgrazing, the grass cover had become exceedingly poor and a hard surface soil crust had developed. Ripping was done on the contour with two tines 6 feet apart penetrating to a depth of 12 inches, and with 15 foot intervals between runs. Ripping increased the cover by 9 to 18 times after one season and by 18 to 27 times after four seasons, while in the whole period the cover on unripped plots increased only up to maximum of 3·8 times.

Reseeding, or replanting, is liable to be expensive in relation to the returns which can be expected from many of the natural grasslands but this, and other measures for rehabilitation, become economic if the pastoralists themselves can be persuaded voluntarily to cooperate in the work. This has been achieved in the Kitui District of Kenya where large areas of denuded and eroded grazing land have been rehabilitated by clearing or thinning bush regeneration, scratch-ploughing and seeding. In 1956 alone, 61,000 acres were cleared or thinned of bush and 29,000 acres scratch-ploughed and seeded (Jordan, 1957).

WATER SUPPLIES

Very commonly an essential prerequisite for improvement will be the provision of additional and well-distributed water supplies. Williamson and Payne (1965) recommend that watering points should be so distributed that each serves grazing land within a radius not exceeding five miles. Although this is desirable, in practice it is rarely realized in most of the South American and African pastoral regions. The natural tendency is for animals to graze intensively within a radius of about two miles from water points. Beyond that the grass is not evenly grazed unless grazing is controlled, or feed is very scarce. In many parts of Africa the ground is grazed so tightly near water that the watering points are surrounded by virtually bare ground for a radius of a quarter of a mile or so.

It is imperative that new watering points should form part of a coordinated plan for improved land use. They must be sited in relation to the situation and carrying capacity of available pastures, and they must be accompanied by control of stock numbers and the introduction of a proper system of rotational grazing. If this is not done, then more water supplies will merely make matters worse by allowing overstocking and localized overgrazing to continue with increased severity. Tanks, hafirs and small earth dams, collecting run-off from a catchment area during the rains and storing it for dry-season use, are the cheapest forms of water supply. Such reservoirs should be fenced, and the water piped from them to troughs; otherwise large numbers of cattle approaching the water's edge are liable to break down the banks and create swampy patches. It is important to avoid overgrazing of catchment areas, with consequent erosion and silting up of the reservoirs, but this is difficult to achieve because herdsmen are liable to keep cattle near the reservoir during the dry-season as long as any water remains. Boreholes, with pumps to bring underground water to the surface, although more expensive, are more satisfactory since the water can be turned off when it becomes desirable to force the movement of stock elsewhere in order to prevent overgrazing. The same applies to the supply of water by pipelines from perennial streams and springs. These can often be used to bring water from mountains, where the rainfall is good, to neighbouring lower and drier lands.

The cost of providing water supplies is an important factor, particularly as there are large areas where the pasture is so poor that it will not pay for even a moderate outlay on water, or where the water is so inaccessible that its supply becomes very expensive. Cost will obviously vary greatly with local circumstances. It will generally be desirable that watering points should not be further than five to seven miles apart, both to restrict the distance over which animals have to move between grazing and watering and to avoid erosion near a watering point by limiting the number of cattle using it to 1,250–1,500 head. On this basis it was estimated that in Tanzania the average cost of providing water is about £2 per beast (International Bank, 1961). An investment of this size will not be justified unless the cooperation of the pastoralists in other measures for better land use can be relied upon. (See also Chapter 16, p. 379.)

ANIMAL HUSBANDRY AND DISEASE CONTROL

The most important improvements in husbandry that are needed are probably culling and castration (to permit of selective breeding), control of calving times, provision of dry-season feed and the correction of mineral deficiencies. Feeding mineral licks is of widespread importance as the soils of many tropical grasslands are deficient in phosphorus and calcium. Possibilities for the provision of dry-season fodder are strictly limited in these low rainfall areas. But where seasonally flooded or swampy areas are available, or where small areas of irrigated fodder crops can be grown, every advantage should be taken of such possibilities. The alternative of supplementary feeding with a high protein concentrate is hardly a practicable proposition at present as such feed is not available in quantity or at a price which the pastoralists could afford. However, it has been shown (Smith, 1961) that where adequate low quality standing hay is available the dominant feature of dry-season malnutrition is deficiency of protein, which limits the forage intake and depresses its digestibility. Feeding protein concentrate (e.g. 2 lb per head per day of groundnut cake) during the dry season greatly increases the forage intake, converts an energy deficient diet into a productive one and enables weight gains to be made. Consequently, if and when improvement of pastures results in more standing hay being available in the dry season, the use of protein supplements might become economic.

BETTER QUALITY STOCK

In view of the inherent low quality of the pastoralist's animals, it is evident that the development of a profitable livestock industry will demand the provision of better quality, earlier maturing and more productive stock. Breeding programmes on a modest scale are in progress in most countries and small numbers of improved bulls and heifers are already available to pastoralists. But the provision of better quality stock can only be of value if it goes hand in hand with improved pasture and animal husbandry, adequate water supplies and better disease control. It is these latter matters which form the priorities in any improvement programme, but a genetic improvement programme should proceed at the same time in order that full advantage may be taken of the environmental improvements once they have been effected. (See Chapter 17, pp. 424–48.)

PROVISION OF ADDITIONAL LAND

Improvement in the pastoral regions may necessitate the provision of new, unoccupied land, either to absorb people and stock in excess of the real potential of an overcrowded area, or to provide 'elbow room' to get the process of rehabilitation started. The unoccupied land now available is usually bush-covered, lacking water, often with poor soil, rather remote and, in Africa, frequently tsetse infested. Rendering it fit for occupation involves considerable initial expenditure on bush clearing, tsetse control, water supplies and roads. At best, costs are high in relation to the potential of the land, and very often the latter is too low to pay a reasonable return on the capital investment needed. Furthermore, it is obvious that unless the pastoralists are prepared to change their ways and practise good husbandry, spending money on providing them with new land only enables them to extend their disastrous destruction of natural resources and serves to create a bigger problem. Unfortunately, the latter is just what has usually happened hitherto when Governments have taken steps to provide additional land. Consequently it would seem wise, in future, if Governments were to limit expenditure on providing additional land to those places where the pastoralists can be relied upon to practise good husbandry, to areas of relatively high potential, or to places where only limited further clearing is needed to safeguard, or extend, tsetse free areas.

IMPROVED LAND USE IN THE WETTER PASTORAL AREAS

At the beginning of this Chapter it was mentioned that although the majority of the pastoralists live in drier zones, pastoral tribes are found to a more limited extent occupying grasslands derived from the wetter broad-leaved woodlands and montane evergreen forests. In general, although not invariably, the degree of denudation which has occurred under these latter conditions is much less than in the drier regions. Improvement in these regions of better rainfall, which usually have fairly fertile soils and are quite suitable for crop production, should clearly be in the direction of more productive land use than can be achieved by pastoralism or ranching, and might be along the lines of dairy farming, mixed farming or the production of perennial crops. A good example of a change from pastoralism to quite intensive mixed farming can be seen in the Kipsigis country in Kenya. These tribesmen, who formerly merely grazed their herds extensively and communally over their lands, have been persuaded entirely to forsake their old customs. They have divided the land into enclosed, individual holdings, with fenced or hedged fields on which they practise rotation of grass and arable crops, using their cattle for ploughing and provision of manure as well as for the production of meat and milk.

CHAPTER 13

Cultivated Fodder Crops and Pastures

Fodder Crops

Certain fodder crops, capable of giving high yields when properly managed under fertile and moist conditions, are grown on a moderate scale in the tropics for stall feeding or ensilage. Some of them can also be grazed, but the majority do not persist well under long-continued grazing and give lower yields than when they are cut for fodder. The plants most commonly used are tall, stool-forming grasses, such as *Pennisetum purpureum* (Elephant, Napier or Merker grass), *Tripsacum laxum* (Guatemala grass), *Panicum maximum* (Guinea grass), *Setaria sphacelata* and *S. splendida*, but to a lesser extent more prostrate, stoloniferous grasses, such as *Brachiaria mutica* (Para grass), are also used. Legumes and grass-legume mixtures are also sometimes cultivated for soiling and ensilage, but pure stands of legumes are more usually grown for hay, common species for this purpose being lucerne (*Medicago sativa*), velvet bean (*Stizolobium deeringianum* or *Mucuna utilis*), and certain varieties of soya bean (*Glycine max*).

These crops can be very valuable for the production of dry-season feed in regions of seasonal rainfall. In conjunction with the utilization of natural grasslands in the drier regions, it is often possible to grow limited areas of fodder crops, either on dry land during the rains, or in favourable places that are seasonally swampy or flooded, or under irrigation. They are also useful in the moister parts of the seasonal rainfall regions where mixed farming, involving the use of natural or sown pastures, is practised. Here, rain-grown fodder crops can be cut for silage or hay during the wet season and, if manured after cutting in the latter part of the rains, will go on growing into the dry season to provide a certain amount of grazing or a late cut.

In the humid tropics, where a dry season is absent or slight, fodder crops can also be a valuable aid in the development of more intensive mixed farming, because the natural grasslands are limited and poor and, in most places, suitable species for sown pastures have yet to be found. For example, in many rice-growing areas cattle subsist on the grazing or cut fodder, provided by limited areas of poor, coarse natural grasslands during the rice-growing season, and during the off season they depend mainly on the poor grazing afforded by the fallow rice fields. Butterworth (1962) has indicated the poor feed value of the rice stubbles, and it is well known in many places that, after their period of fallow grazing, the animals are in poor condition for the work of preparing the fields for the next rice crop. In these humid zones, and also under irrigation, fodder grasses can give high yields; their cultivation on limited areas would greatly improve the supply of cattle food.

FODDER GRASSES

Cultivation

Most species have to be propagated vegetatively from stem cuttings or rooted stool pieces (e.g. Elephant and Guatemala grasses) or from stolons (e.g. Para grass) but some, such as *Setaria sphacelata* and certain strains of *S. splendida*, can be established from seed. It is usually best to plant in rows in order to facilitate weeding during the establishment stage. The best spacing will obviously vary with the species, the soil and rainfall, whether the crop is to be irrigated or not, and whether cultivation is to be mechanized. In Trinidad, Paterson (1939) recommended 2½ feet by 1 foot to 3 feet by 2 feet for Guatemala grass, 3 feet by 1½ feet for Elephant grass and the planting of Para grass stolons in continuous rows 2 to 3 feet apart. In a humid climate or under irrigation, fairly close spacing in rows about 3 feet apart may be expected to give the best yields, provided that sufficient nitrogen is supplied, but spacing can be varied within the limits of 2 to 4 feet between rows without greatly altering yield. In a drier climate, wider spacings may be advantageous where cultivation can be mechanized, since they involve little or no loss of yield after the first year. For example, in an experiment with Elephant grass in Rhodesia row spacings of 2 and 4 feet gave higher yields than 6 and 8 foot rows in the first season but not thereafter (Rowland, 1955). However, in some circumstances it may be desirable to have close spacing in order to reduce the labour, or cost, of weeding by obtaining a maximum smothering effect from the grasses.

After planting, cultivation for weed control will be needed for a few months, or perhaps throughout the first year, but thereafter it should become unnecessary since, if the crop is well manured and managed, it should suppress weeds. On normal soils cultivation is not needed except for weed control, and deep tillage is unlikely to be beneficial. For example, in a Rhodesian experiment with Elephant grass on a rather stiff soil, annual cultivation 9 inches deep close to the stools had no effect, while subsoiling between the rows, either once a year or every third year, appeared to depress yields (Rowland, 1955).

The effects of cutting intervals and season on yields and composition

When fodder grasses are cut regularly, increasing the intervals between cuts (thus allowing the grasses to reach a more advanced stage before utilization) results in increased yields of green or dry fodder, increased percentage contents of crude fibre, carbohydrates and dry matter, but decreased contents of crude protein and ash. This is illustrated by the figures of Table 31 for the composition of three species in Tanzania at various stages of growth (French, 1943a; van Rensburg, 1956), by those of Table 32 showing the effect of four cutting intervals on the dry matter and crude protein contents of four species in Trinidad (Paterson, 1936), and by those of Table 33 showing the effect of four cutting intervals on Elephant grass in Nigeria (Oyenuga, 1959). With Elephant grass more than about three weeks old the stems mature more quickly than the leaves, resulting in a rapid deterioration in the nutrient value of the plant as a whole but, since animals do not normally eat the mature stems, it is the nutrient content and yield of the leaves which is important.

Table 31. *Variation in composition of fodder grasses with stage of growth, in Tanganyika*

1 and 2. van Rensburg (1956): ash figures = total ash,
3. French (1943a): ash figures = soluble ash

Species	Stage of growth (Height in feet)	Percentage of dry matter			
		Crude protein	Crude fibre	N-free extract	Ash
1. *Pennisetum purpureum*	1	20·2	27·4	28·7	20·1
	4 (silage stage)	11·2	33·4	38·2	14·3
	8	5·9	35·4	42·3	14·3
	11 (with bare stem)	2·5	37·7	39·3	18·3
2. *Setaria sphacelata*	1	21·3	26·4	32·3	16·2
	2	15·3	35·2	32·0	14·6
	4 (flowering)	8·7	42·0	36·1	10·4
	6 (after seeding)	4·1	36·3	50·5	7·8
3. *Panicum maximum*	1–1½	9·2	31·2	—	7·4
	2½–3	9·3	34·5	—	8·4
	6	5·6	41·8	—	5·6

A cutting interval must be selected which gives a satisfactory compromise between nutrient value and palatability on the one hand and yield on the other. This will vary with the species and also with the

Table 32. *Effect of cutting interval on dry matter and crude protein content of four fodder grasses in Trinidad*

Paterson (1936)

Species	Percentage dry matter (Cutting interval, days)				Percentage crude protein in dry matter (Cutting interval, days)			
	45	90	120	180	45	90	120	180
Elephant grass	13·38	21.53	24·35	23·25	9·61	5·93	6·06	4·87
Guatemala grass	16·40	16·45	16·95	22·82	8·12	5·74	3·88	3·31
Guinea grass	16·16	19·37	24·33	31·85	8·37	5·43	5·87	3·75
Para grass	18·05	17·42	22·08	17·15	11·08	6·87	6·65	4·00

Table 33. *Effect of cutting interval on Elephant grass (unfertilized) in Nigeria*

Oyenuga (1959)

Cutting interval, weeks	3	6	8	12
Whole plant:				
Dry matter, per cent	16·5	19·0	21·7	25·9
Yield, green fodder, tons/acre in 11 months	29·05	38·64	39·04	52·91
Yield, dry matter, tons/acre in 11 months	4·79	7·34	8·46	13·72
Leaves only:				
Crude protein, per cent of dry matter	14·74	11·39	10·51	9·58
Silica-free ash, per cent of dry matter	10·19	8·88	7·41	5·98
Crude fibre, per cent of dry matter	25·66	28·08	29·50	29·56
N-free extract, per cent of dry matter	40·9	43·1	44·8	47·6
Yield, crude protein, tons/acre in 11 months	0·72	0·45	0·52	0·64
Yield, ash, tons/acre in 11 months	0·51	0·36	0·35	0·39
Yield, total carbohydrates, tons/acre in 11 months	3·40	3·10	3·67	5·29
Leaf: stem ratio, by weight	12	1·3	1·2	0·9

weather, especially the rainfall, which will affect the rate of regrowth of the grass, so that intervals normally need to be longer in the dry season than during the rains. It should also be noted that the dry matter content may increase by as much as 50 per cent in the dry season (see Table 35; Paterson, 1939), and the protein content of the fresh material also rises, with the result that a maintenance ration at this time is provided by about two-thirds of the weight of fresh grass required in the wet season. The effect of cutting frequency on the rate of regeneration and persistence of the grass must also be borne in mind. Too frequent cutting will weaken the grass, reduce the rate of

25 Mist spray propagation of rubber cuttings

26 Terminal stem cutting of rubber beginning to root

27 Rooted rubber cutting in veneer tube ready for planting

28 Budding-rubber; removing the bud patch from the budwood

regrowth and result in lower yields, and perhaps in death of a proportion of the stools. Suitable cutting intervals are usually regarded as varying between six and ten weeks, depending on species and seasons. In Trinidad, Paterson (1939) recommended intervals of six to eight weeks for Para grass, seven to eight weeks for Elephant grass and eight to ten weeks for Guatemala grass, while in Malaya six-weekly cutting seemed to be best for Guinea grass (Keeping, 1951). From his results for the nutrient yields of the leaves only of Elephant grass (Table 33), Oyenuga (1959) concluded that cutting every three weeks gave the highest yields of crude protein and ash, twelve-weekly cutting gave the maximum yields of carbohydrates and total nutrients, and six- or eight-weekly cutting was not profitable for either purpose.

Height of cutting

The height of cutting may also affect yields, rate of regrowth and persistence. In the experiment mentioned above, Paterson (1936) found that for all four species cutting at ground level reduced the rate of regeneration and gave lower yields than cutting at 4 to 6 inches or 9 to 12 inches, and tended to cause death of stools, especially if combined with frequent cutting at intervals of forty-five days. In a subsequent experiment (Paterson, 1938) he compared cutting at 3 inches and 12 inches above ground with Elephant, Guatemala and Para grasses, and found that the higher cut gave significantly better yields with Guatemala grass but significantly lower yield with Para grass, while with Elephant grass there was no significant difference between the two heights (Table 34). For most species except Guatemala grass, cutting at 3 inches above ground will be satisfactory, but it may be desirable to cut at ground level occasionally to remove trash and old stumps of stems.

Table 34. *Effect of cutting height on yields of three grasses in Trinidad*
Paterson (1938)

	Yields of fresh grass, tons per acre per annum	
	Cut at 12 inches	Cut at 3 inches
Guatemala grass	42.40	32·52
Elephant grass	62·35	63·78
Para grass	35·48	39·93

Manuring

Liberal manuring is likely to be necessary in order to maintain the vigour and yield of the grass over a period of several years under a system of regular and fairly frequent cutting. The main need is usually for nitrogen, and with high-yielding grasses responses to this nutrient are usually linear up to fairly high levels: for example, Elephant grass in Rhodesia responded linearly up to at least 1,000 lb of sulphate of ammonia, and in Puerto Rico up to 800 lb of N per acre per annum. The lower yielding Guinea grass gave good responses up to 6 cwt of sulphate of ammonia per acre per annum in Malaya (Henderson, 1955). Nitrogen will also increase the protein content; with Elephant grass in Rhodesia an increase of 0·83 lb of protein was obtained for every pound of N applied over a fair range of applications. Phosphate is likely to be required on many soils; it is best to give a good application at planting time as subsequent top dressings are less effective. Potash may also be needed on some soils especially for Elephant grass, which is a gross potash feeder, but other grasses may also need it, e.g. in Malaya 3 cwt of sulphate of potash per acre more than doubled the yield of Guinea grass. Interactions between the three major nutrient elements may be important. Phosphate and potash may fail to produce significant effects for want of nitrogen, but may lead to further increase in yield in the presence of liberal supplies of nitrogen.

Fairly frequent fertilizer applications are needed. Paterson (1939) recommended a dressing of 1 cwt of an NPK mixture every two months in Trinidad. Even if frequent small applications of fertilizers do not give higher yields per annum than the same amount applied less frequently, they will usually give more uniform yields from cut to cut throughout the year provided that there is no severe dry season (Henderson, 1955).

Yields

Yields vary greatly with species, soil, fertilizer practice and moisture supply. Average yields obtained by Paterson (1939) with Guatemala, Elephant and Para grasses in Trinidad, on fairly good soil with 6 cwt of fertilizer per annum, and under an average rainfall of 70 inches with a moderate dry season of four months, are given in Table 35. At the other extreme, Kennan (1950) under low rainfall in Rhodesia obtained average fresh weight yields over two years of only 6·93 tons per acre per annum for Elephant grass, 5·59 for *Setaria sphacelata*, and 4·57 for Guinea grass. High yields can be got with irrigation; thus van Rensburg (1956) obtained average yields per acre per annum over four years of 103 tons of Elephant grass, 97 tons for *Setaria splendida*, and 63 tons for *Setaria sphacelata*.

Ensilage of these grasses in either pit or tower silos presents no real difficulties, and is commonly the best way to preserve fodder for use in the dry season, but it involves some losses. Thus, McWilliam and

Fodder Crops

Table 35. *Yields, composition and feeding value of fodder grasses in Trinidad*

Paterson (1939)

	Guatemala grass	Para grass	Elephant grass
Yield (tons per acre)			
Fresh, weight	42·4	40·0	63·8
Dry matter	8·09	9·74	9·31
Crude protein	0·569	0·600	0·624
Percentage of fresh herbage			
Dry matter, wet season	18·64	21·60	13·65
Dry matter, dry season	27·69	32·65	19·48
Crude protein, wet season	1·31	1·29	0·90
Crude protein, dry season	1·88	1·69	1·33
Percentage of dry matter (average for whole year)			
Crude protein	7·03	6·07	6·68
Crude fibre	34·14	34·78	33·36
Carbohydrate and fat	48·03	48·82	44·99
Total ash	10·80	10·33	14·97
CaO	0·410	0·664	0·902
P_2O_5	0·764	0·880	1·222
Na_2O	0·034	0·259	0·036
Starch Equivalent, wet season	10·6	12·9	7·8
Nutritive ratio	11	13	11

Duckworth (1949), who made silage from Elephant grass in tower silos in Trinidad, found that wastage averaged 21·2 per cent of the dry matter and that the silage was rather less palatable and had a slightly lower feeding value, than the fresh grass.

Species

Pennisetum purpureum (Elephant grass) is the most widely used fodder grass in Africa as it is easily propagated from stem cuttings and usually gives the highest yields of herbage and nutrients. It is very succulent and palatable but has a high moisture content; about 80 lb of fresh herbage is required to provide maintenance for a 1,000 lb cow in the wet season (Paterson, 1939). It needs cutting every 6 to 8 weeks in the wet season, or even more frequently at the height of the rains, when it quickly reaches the stage of forming 'canes' which cattle will not eat. It can be grazed, if managed carefully, and it is often convenient to do this at the beginning of the rains, followed by several silage cuts, and then to graze again early in the dry season. It is suitable for cultivation up to 6,000 feet. There are many varieties, some of which are undesirable because they are too hairy.

Tripsacum laxum (Guatemala grass) is propagated by rooted stool pieces. It is not as high yielding as Elephant grass but is more leafy and palatable, persists well under appropriate cutting and maintains its yield well during the dry season compared with most other species. It should be cut at 9 to 12 inches above ground, usually at intervals of eight to ten weeks. Paterson (1939) gives 60 lb as a wet-season maintenance ration for a 1,000 lb cow. It does not persist well under grazing as cattle tend to uproot it. *Panicum maximum* (Guinea grass) exists in a very large number of strains, varying greatly in vigour and habit, some of which are more suitable for pastures than as a fodder crop. Types used for cutting are usually propagated as rooted stool pieces, although some produce viable seed rather sparsely. It is not as quick to establish itself and smother weeds, nor as high yielding, as Elephant and Guatemala grasses. It can be cultivated up to 6,000 feet and should be cut at 3 inches above ground level about every six to eight weeks. Paterson (1939) gives information on *Brachiaria mutica* (Para grass), but this grass is usually grazed and seldom used for cut fodder. Both *Setaria splendida* and *S. sphacelata* are leafy and succulent grasses with higher protein contents than the previous species. They can be grown up to 7,000 feet, but require fertile soil to do well. The former has normally to be propagated vegetatively (although one variety selected in Kenya can be established from seed), the latter produces seed abundantly.

FODDER LEGUMES AND GRASS-LEGUME MIXTURES

Legumes are normally of better nutritive value than grasses, having higher contents of crude protein, calcium and phosphorus, and often slightly lower crude fibre values. They are not grown as fodder crops in pure stand as extensively as grasses, mainly because they give much lower yields but also partly because they are generally unsuitable for ensilage unless mixed with other crops. Nevertheless, in stock-keeping areas experiencing seasonal rainfall, a most valuable use for limited areas of irrigable land is for the production of protein-rich leguminous hay or green fodder. For example, in Tanzania irrigated lucerne gave 35 to 45 tons of green fodder per acre per annum containing 20 per cent of crude protein on the dry matter. In some irrigated areas occupied by smallholders legumes are quite extensively grown to provide green fodder for stock feeding. Thus in Egypt about 20 per cent of irrigable cropped land grows berseem (*Trifolium alexandrinum*); in the Punjab about 14 per cent of irrigable land is under berseem or Bokhara sweet clover (*Melilotus alba*), and in the Gezira area of the Sudan *Dolichos lablab* is extensively grown.

To a limited extent legumes are also grown in pure stand for the production of hay without irrigation. In South Africa, Rhodesia and many parts of South America, lucerne is probably the most widely used species for this purpose, but in Africa soya beans (*Glycine max*) cowpeas (*Vigna sinensis*), velvet beans (*Stizolobium deeringianum*) and *Dolichos lablab* are also grown. Figures for the composition of hay of all these species are given in Table 36. In Rhodesia certain varieties of soya bean (e.g. Hernon, Biltan) have been specially selected for hay and these are capable of giving higher crude protein contents than those indicated in the table, probably up to 19 per cent.

Table 36. *Leguminous hays: percentage composition of dry matter*

1. French (1943a); 2. Dougall (1960); 3, 5, 7, 8. Chemistry Branch, S. Rhodesia (1947); 4, 6. Foster and Mundy (1961).

Species	Crude protein	Digestible crude protein	Crude fibre
1. Lucerne, Tanzania	16·5	10·7	31·4
2. Lucerne, Kenya	18·9	13·6	28·9
3. *Stizolobium deeringianum*, Velvet bean	13·3		27·6
4. *Stizolobium deeringianum*, N. Nigeria	11·2	7·3	34·2
5. *Vigna sinensis*, Cowpea, Rhodesia	11·1		28·7
6. *Vigna sinensis*, Cowpea, N. Nigeria	10·9	7·4	27·4
7. *Glycine max*, soya bean, Rhodesia	13·8		32·2
8. *Dolichos lablab*, Rhodesia	12·1		21·3

Although the fodder grasses previously mentioned can give high yields, they have the disadvantage of a relatively low protein content; it would be a great help if this could be counteracted by growing them in mixtures with suitable legumes. Unfortunately it seems to be very difficult to find mixtures in which a satisfactory balance can be maintained between the two species and which are also capable of giving yields as high as pure grass stands. Success depends not only on finding compatible perennial legume and grass species but also on determining sowing times, spacings and cutting procedures that will enable a balanced mixture to be maintained. A satisfactory combination of these factors does not seem to have been worked out in most places. A good deal of work has been done in Kenya, but although *Dolichos falcatus*, *Dolichos* sp near *lablab* and *Dolichos* sp ex Kilima Kiu all show promise for this purpose, the problem of combining them satisfactorily with the tall fodder grasses has not yet been fully solved (Strange, 1955). In Rhodesia a mixture of *Glycine javanica* and Elephant grass has proved satisfactory for silage and some grazing in the higher rainfall Melsetter area, but elsewhere mixtures have not been satisfactory. At altitudes of 5,000 – 8,000 feet in Bolivia lucerne has been successfully combined with Guinea grass and also with cultivated oats; the mixtures are both grazed *in situ* and cut for hay. In Puerto Rico mixtures of Kudzu (*Pueraria phaseoloides*) with Elephant, Para or Guatemala grasses have been used and it is stated that an acre of unfertilized mixture yielded as much as an acre each of the two components in pure stand (Warmke *et al.*, 1952). But it is also reported that a pure stand of Elephant grass, fertilized with 300 lb of N per acre, gave much higher yields of dry matter and crude protein than an unfertilized Kudzu – Elephant grass mixture (Caro-Costas and Vincente-Chandler, 1956). As fodder grasses alone are much easier to establish and manage than mixtures, it may be that the use of the former, with applications of nitrogen to increase yield and protein content, will generally prove to be the best answer.

Cultivated Pastures

From what has already been said in Chapter 11, it will be evident that the majority of the natural grasslands of the tropics are low-yielding and of poor quality. Where it is practicable to do so, it would be advantageous to replace many natural grasslands with permanent pastures or temporary leys of improved species. These would be higher yielding, of better nutrient quality, and might also be more responsive to fertilizers, recover more rapidly after grazing and provide a longer grazing season.

Cultivated pastures, however, are only practicable in regions of favourable rainfall, which are principally those where the climax vegetation is lowland or highland forest, or the wetter types of the broad-leaved woodland. Over much of the vast areas of drier natural grasslands the establishment of sown pastures is very difficult and their chances of surviving the dry season are slender. In further marginal areas their use may be uneconomic because they yield little or no more than the natural pastures. For example, on rather poor and eroded soil at Makavete in Kenya, where annual rainfall averaged 26 inches but varied from 10 to 53 inches, it was found possible, by cultivation methods designed to conserve moisture, by the use of fertilizers and by good grazing management, to establish and maintain pastures of three grass species for seven years (Pereira and Beckley, 1953; Pereira *et al.*, 1961). But over this period the average carrying capacity of the best of these pastures, *Cenchrus ciliaris*, was not significantly better than that obtained from an adjacent area of natural grassland that was fenced, cleared of bush, and put under controlled grazing. The relatively low and very unreliable rainfall was the chief factor limiting productivity.

For several reasons, peasant farmers have so far made little use of sown pastures, even where the climate is suitable. In many places population density is so high that almost all the land is required for the direct production of crops for human food. Land tenure systems under which the land is communally owned, or held by individuals on short-term, insecure leases, may also make farmers disinterested in establishing permanent pastures or leys. Fragmentation of holdings, resulting in small, scattered and unmanageable plots of land, has a like effect. Above all, the poor productivity of the stock and the unorganized nature of the livestock industry commonly mean that the farmer does not receive a sufficient return to justify any appreciable expenditure on pasture establishment or improvement. Selection and breeding to produce better animals, and the provision of better communications, abattoirs, organized markets and inducements to produce good quality meat, are needed before a satisfactory return can be obtained from investment in better pastures.

Most of the above circumstances do not normally apply to large farming enterprises, such as those operated by Europeans in some parts of the tropics. There is, however, a further limiting factor of overall application, namely that there is insufficient knowledge of suitable species for improved pastures and of satisfactory methods for their establishment and management.

Herbage species for cultivated pastures need to be selected for the following desirable features:

1. They should be capable of easy and rapid establishment. Seeding is usually the easiest way to establish a pasture but selection need not be confined to species which are good seed producers since some grasses are readily, though more expensively, propagated by stolons, rhizomes or stem cuttings, and this procedure can often be partly mechanized.
2. They must be high-yielding, palatable and nutritious, which means that they must be vigorous, leafy, with good leaf quality and preferably late flowering within the growing season.
3. They must persist under intensive grazing – for at least three years for leys and appreciably longer for 'permanent' pastures.

4. In most places they need to be capable of surviving a dry season, and preferably of providing grazing well into the dry season.
5. If they are to be used for temporary leys, the herbage species must be capable of fairly easy eradication.

In Europe grasses and clovers have been selected for desirable features for a long time, and selection has been for much less variable climatic conditions than those occurring in the tropics. In the tropics such work started only recently, and it is necessary to find species and strains to suit a wide range of environmental conditions. The latter is true even of single territories: for example, in Kenya or Tanzania conditions vary from humid tropical in some coastal areas to cool temperate in the highlands.

In view of the circumstances outlined above it is not surprising that the use of cultivated pastures in the tropics is so far quite limited. In the wet tropics there is practically no use of leys, and only a limited development of permanent pastures in certain areas, such as Ceylon, Puerto Rico, the West Indies, several South and Central American countries and Queensland. A little more progress has been made in regions of seasonal rainfall but, everywhere, cultivated pastures are mostly confined to the larger farms at higher altitudes, such as those operated by Europeans in East Africa and Rhodesia. With certain local exceptions, no satisfactory legumes have been found for inclusion in pastures. Tropical cultivated pastures are therefore generally of grass alone, usually of only one species of grass, and the remarks in the following sections refer primarily to such pure grass swards, further reference to pasture legumes being made later in this chapter.

ESTABLISHMENT OF PASTURES

Methods of land preparation will naturally depend to a considerable extent on the soil, climatic conditions, species to be used and the method of establishment. As in temperate regions, a weed-free seed bed, firm and consolidated to plough depth, is needed. When the climate and soil permit, it will usually be desirable to plough the land fairly early in the dry season so that weeds can be destroyed by subsequent cultivations during the dry weather, and the soil has time to settle and consolidate after ploughing. Usually a final cultivation will be needed after the rains break to deal with germinating weeds and to prepare the final tilth. If seed is to be sown, a moderately fine tilth is needed, but the seed bed should be left rougher than in temperate regions as high intensity rainfall is likely to effect further breakdown of small clods and to lead to packing of the soil surface. If the grass is to be propagated vegetatively, a fine tilth is unnecessary and undesirable. Rolling to consolidate the land before sowing seeds has been found beneficial in some places. As mentioned below, under 'Fertilizers', mineral deficiencies must be corrected when planting and, if the soil is very poor in organic matter, it may be advantageous to plough in a green manure crop before attempting grass establishment.

Where the rainfall is seasonal the best time to sow or plant pasture will usually be as soon as possible after the rains break, but in some places with a long rainy season it may be desirable to establish a pasture after an arable crop has been taken in the early part of the season. Seeded pastures may be established by undersowing in a companion crop grown for grain, silage or grazing, or by sowing the grass alone. The former practice, which is normally confined to leys, has the advantage that it reduces the cost and labour of extra cultivations and minimizes the period during which the land is unproductive and during which the bare soil is exposed to the elements. Also, undersowing may give earlier grazing. However, there is inevitably competition for nutrients and moisture between the two crops, and reduction in light intensity for the grass, which will often result in some reduction in the yield of the arable crop and a less satisfactory take of the grass. Sowing the grass alone, though more costly, will usually give better establishment and is certainly preferable in drier areas, or where the soil fertility is poor. The choice between the two methods depends on local conditions and economic considerations.

The best results will usually be obtained by drilling the seed in rows, which will not normally be more than 18 inches apart, rather than by broadcasting. Row planting allows weeding during the early stages of establishment and permits placement of fertilizer with the seed, where it will be more readily available to the young seedlings. Depth of sowing is important; the majority of tropical grass seeds are small and require sowing to a depth of only a quarter to half an inch, although some larger ones can be sown up to one inch deep. Shallow sowing can be accomplished with seed barrows, modern grass seed drills or by sowing through the fertilizer box of a combine drill, but grain drills will not usually sow less than an inch deep. If drilling is not possible, broadcasting by hand, by a fertilizer distributor, or through grain drills with the seeding hoses hanging free, can often be satisfactory.

Undersowing in a cereal crop grown for grain has been successful in Kenya (Grassland Research Station, Kitale, 1959). For this purpose wheat and barley have both proved satisfactory (with some reduction in the normal seeding rates) and the grass seed can be sown at the same time as the cereal. Maize is fairly satisfactory as a companion crop but should not be undersown until it is 3 or 4 feet high, otherwise its yield will be reduced by competition from the grass. There is some advantage in spacing the maize rows widely, at 6 to 7 feet, in order that mechanical cultivation may continue until the grass seeds are sown, but if this is done the maize must be planted densely in the rows, to give a minimum population of about 12,000 plants per acre, as otherwise its yield will be reduced. Oats as a companion crop was found to be too smothering if left to mature, but can be usefully sown with grass to provide early grazing and to help in weed suppression. Sorghum for silage has been successfully used as a companion crop when sowing grass in Uganda (Harker, 1954) and there is no reason why it should not be equally successful if grown for grain.

Seed rates will vary with species, germination capacity, method of sowing, soil and rainfall. The germination capacity of many species is relatively low and varies with the conditions of harvesting and sowing: consequently it is important to know the germination capacity and to sow sufficient pure germinating seed per acre. Seed rates will usually be higher for broadcasting than drilling and less in dry areas than in wet. After sowing, the seed must be only lightly covered, using a brush harrow, light tine harrow or disc harrow; if the small seeds are buried too deeply many seedlings will fail to reach the surface. Rolling after sowing has sometimes been found beneficial.

Planting grasses that are vegetatively propagated may involve a considerable amount of labour in the collection, preparation, transport and planting of the material. This is not as serious a disadvantage as it sounds, because on peasant holdings a rather high demand for labour does not rule out the use of such grasses, and on larger holdings the procedure can usually be partly mechanized. Most species are propagated by stolons or rhizomes and, if the rainfall is good and reliable, they can be planted by broadcasting on the soil surface and discing in. Under less favourable conditions it may be necessary to plant by hand in holes made with a hoe, or in furrows made by a plough or cultivator. Mechanization of the collection of planting material may also be possible. With some stoloniferous grasses planting material can be obtained simply by mowing a pasture very low at an appropriate stage of growth. In the south-eastern United States rhizomes of Bermuda grass are collected by dragging them from sandy soils with a chisel-type cultivator (after burning the top growth), and windrowing them with a side delivery rake or a drag harrow. As with seeded grasses, some vegetatively propagated grasses require shallow planting but others, particularly rhizomatous species, prefer deeper planting. Where deep planting is desired, planting material can be placed in furrows opened with a mould-board plough.

MANAGEMENT

The management of the new sward must aim at obtaining as quickly as possible a full, vigorous and weed-free stand of the sown or planted grass, and at achieving consolidation of the soil. Although grass weeds should have been largely eliminated by dry-season cultivation, or herbicide application prior to sowing, a strong growth of other weeds will often compete with the sown grass during the establishment stage. During the early period a combination of several methods of weed control is likely to be needed. Interrow cultivation will be possible if the grass has been sown or planted in rows. Weed-killers, such as 2,4-D or 2,4,5-T, can be used to control many susceptible weeds, especially as tropical pastures do not usually contain broad-leaved legumes, which are susceptible to most herbicides. Mowing or brush cutting will remove the competition and shade of aggressive dicotyledons and encourage the grass. Light grazing should be started as soon as possible as this will consolidate the soil and encourage the tillering of some grasses, or the development of others that are spread by runners, which are trampled into the soil and root rapidly to form a turf. The stage at which grazing can begin and its intensity will naturally depend on the species sown and the environment. Vigorous creeping grasses which spread by stolons or rhizomes may be grazed as early as four to eight weeks after planting, but many tussock-forming species are easily uprooted when young and should not be grazed until they have developed a good root system. Heavy grazing in these early stages is usually undesirable, as the stock will select the more palatable sown grass, and reject the more obnoxious weeds, thus favouring the development of undesirable species. Usually a combination of several treatments, properly timed, will be needed to control weeds and promote the development of the grass. For example, six to eight weeks after planting pangola grass in Trinidad it is grazed lightly, followed by the use of a brush cutter, and finally sprayed with 2,4-D (Forster et al., 1961).

After the establishment phase the aims of management will be those already stated on p. 256, and the main problem will be to relate the stocking rate to the seasonal variation in carrying capacity and nutrient value of the pasture, while at the same time maintaining the sward in vigorous and pure condition. The seasonal changes in composition will be similar to those described for natural grassland on p. 258, the nutrient value declining, and the yield of dry matter increasing, as the grass grows towards maturity. It is therefore necessary to graze at a stage which achieves a satisfactory compromise between yield and nutrient quality, bearing in mind also that too heavy or too frequent grazing will reduce the yield and vigour of the grass, causing deterioration of the sward and invasion of weeds or the occurrence of bare patches. Undergrazing which results in much grass becoming over-mature, and in selective grazing, must also be avoided. However, in sown pastures, usually of one species, selective grazing will be less serious than in natural grasslands.

There will be much more grass available in the rainy season than in the dry season, and it will not usually be possible to vary the number of stock kept to the same extent. There is no organized trade in store and fattening cattle in the tropics, as there is in temperate countries. Therefore there is generally no ready means of adding to, or subtracting from, the livestock population on the farm, which consequently must remain fairly fixed throughout the year. Hence it will be necessary to try to keep the number of stock fairly constant at a level between the wet- and dry-season carrying capacities, to conserve some of the surplus grass of the wet season as hay or silage and to extend grazing as far as possible into the dry season, either by careful management of general purpose pastures, by irrigation, or by having special purpose pastures specifically to provide dry season grazing. These procedures are much more practicable with sown pastures than with natural grasslands, since the former are established in areas of more favourable rainfall and it is profitable to manage them more intensively than the latter.

Continuous grazing of good quality pastures will clearly not be satisfactory. The stocking rate would have to be relatively low with the result that during the wet season some grass would be left ungrazed to grow coarse and tall and, although it might eventually be eaten, it would then be of low feeding value. Consequently, rotational or strip grazing is desirable.

The essential feature of rotational grazing is that, when the pasture has reached its optimum stage of growth, it is intensively grazed down for a short period of up to a week and is then rested for a longer period to allow the herbage to grow back again to the same stage before the animals return. The land is subdivided by permanent fences or hedges into a number of fields of manageable size, depending on their carrying capacity and the size of the herds, and each of these is grazed and rested in turn. During the main growing season all the fields may not be needed for grazing and some may be used to produce silage or hay. The appropriate lengths of the grazing and resting period, especially the latter, will vary with the climate, soil, season of the year and type of pasture. It is particularly important to allow the herbage sufficient time to make good regrowth between grazings as this is essential for maintaining the vigour of the pasture. It is usually desirable to remove highly productive animals from a paddock before they start any regrazing; a common practice is to allow dairy cows to graze for a short period and then to follow them with dry cows or heifers to clean up what is left.

Strip grazing is an intensification of rotational grazing. By the use of an electric fence the animals are confined to an area sufficient to provide feed for one or two days only, and the fence is moved forward every day or so to provide a fresh strip of grazing. This results in more uniform grazing than a less intensive rotational system and reduces wastage and spoilage by fouling. The latter, however, is by no means eliminated, and in some places it has been found advantageous to shut the fields up for silage after two or three grazings in order further to reduce the effects of fouling. Strip grazing requires rather more labour, and more watering points, than ordinary rotational grazing; on the other hand, the use of the electric fence reduces the need for permanent fencing into small fields. Experiments in Europe have shown increases of the order of 15 to 20 per cent in the efficiency of pasture utilization by dairy cows from strip grazing as compared with less intensive rotational grazing. It is also considered to be the most efficient way of grazing sown pastures with dairy cows in the higher rainfall areas of Rhodesia (West, 1956) and has been adopted elsewhere in the tropics; for example, on some permanent pastures of pangola grass in Trinidad and on leys of various species in Kenya. It is, however, only likely to be satisfactory with productive pastures grazed at the optimum stage. On poor quality or over-mature pastures restricting the stock and forcing them to consume all the herbage, irrespective of quality, will be less productive than allowing more extensive and selective grazing by high-quality stock and then grazing off the residual herbage with followers, or with sheep. Strip grazing is unlikely to find application on peasant farms, where fields are small and few animals are kept, since, apart from the cost of the equipment, stock can be adequately controlled by tethering or herding.

Deferred grazing may be practised in order to produce dry-season feed. This may merely consist in closing a field after grazing early in the rains and allowing the grass to mature for grazing off as poor quality standing hay in the dry season. More commonly, grazing may be stopped a few weeks before the end of the rains, a dressing of nitrogen given, and the pasture rested to produce relatively young grass for consumption early in the dry season.

In some places where rainfall is fairly good, for example in Kenya and Rhodesia, special purpose leys to provide fresh grazing or standing fodder during the dry season are established with species possessing some drought resistance. Elephant grass (*Pennisetum purpureum*),

Sudan grass (*Sorghum sudanense*), Guinea grass (*Panicum maximum*) and Star grass (*Cynodon plectostachyum*) have been used for this purpose (Grassland Research Station, Kitale, 1959). These leys are generally managed by taking one or two grazings, or a silage cut, early in the rains, resting through the main growing season, then cutting or grazing, applying nitrogen, and leaving the regrowth for grazing early in the dry season.

Whatever grazing system is followed, mowing or brush cutting is likely to be needed periodically to tidy up the pasture by removing patches of coarse, uneaten grass and controlling weeds. Most tropical grasses, especially tufted species, tend to lack persistence under frequent close grazing or mowing, and therefore the use of the mower should be minimized and the grass should not be cut too closely. However, a close cut at the end of the dry season, to clean up the pasture thoroughly before the new season's growth, will have no adverse effect.

With crossbred zebu × temperate breed cattle it has been found that the proportion of their grazing which they do by day or by night can be modified by providing day and night pastures of different quality. Providing better quality pasture by day results in increased daytime grazing and *vice versa*. With crossbred zebu × Holstein cattle on pangola grass in Trinidad, providing better pasture by day resulted in 67·2 per cent of the grazing hours being in the daytime while better pastures by night increased night grazing to 54·2 per cent, there being no significant difference in total grazing times (Wilson, 1961b). Since daytime grazing increases the radiation heat load on the animal, night grazing is to be preferred, and the provision of high-quality night pasture can be advocated under conditions where the heat load is important. (See also Chapter 15, p. 361.)

THE USE OF FERTILIZERS

1. *For establishment*

The most widespread requirement at the time of establishing a pasture is for phosphate, since so many tropical soils are deficient in this element in an available form. Numerous experiments in various parts of the tropics have shown that application of phosphate at sowing time commonly has a marked effect in improving the establishment and early growth of grass, and an even greater effect on that of legumes in the limited areas where these are included in leys. Usually applications of the order of 2 or 3 cwt per acre of single superphosphate will be adequate, but in some areas the soils are so markedly deficient in phosphate that much higher dressings are needed in order to ensure satisfactory establishment. For example, in parts of the Kenya

Highlands, Birch (1959) found that successful ley establishment required an application of at least 3 cwt per acre of double superphosphate, which makes it uncertain whether leys could be economic. It is usually best to use a soluble phosphatic fertilizer and to drill it with the seed, or place it just below the seed, as phosphate is relatively immobile once applied. The residual effect of such dressings varies: in some cases there is a prolonged effect, in others the effect on growth and yield soon disappears, although that on protein and phosphate content of the herbage may last longer.

Responses to potash at sowing time are confined to limited areas where soils are markedly deficient in this element. Although the majority of tropical soils are acid, lime is not generally used in establishing grass but it has been found beneficial in certain localities. For example, at Entebbe in Uganda, Ledger (1950) found that, on a sandy soil of pH $4 \cdot 5 - 4 \cdot 8$, ploughing in lime before sowing Rhodes grass improved establishment and yields, but on a soil of pH $5 \cdot 6$ at Serere there was no response to lime. Similarly in Kenya lime has only been found necessary on the more highly acid soils. It is nearly always beneficial to apply a small dressing of nitrogen at sowing time but large dressings at this time are not usually worthwhile as they encourage weed growth and, as grass seedlings are unable to absorb large quantities of nitrogen until they have become established and developed a more extensive root system, much of the fertilizer may be lost by leaching.

2. *For established pastures*

When pastures are grazed, part of the nitrogen and mineral elements removed from the soil will be permanently lost in the form of animal products sold off the farm, but the greater proportion will be returned to the land in the form of dung and urine. If the pasture is cut for hay or silage, the depletion of fertility will be much greater, since all the nutrients taken up by the herbage will be lost to the soil. With grazed pastures on fertile soil, the loss of available mineral nutrients, due to leaching and removal in animal products, may largely be compensated for by the slow but continuous release of nutrients from the reserves held in the soil, in which case only nitrogen will need to be applied as fertilizer. The majority of tropical soils, however, are not highly fertile, and intensively grazed pastures will often require application of phosphate and, less frequently, of potash.

Nitrogen is almost invariably the main nutrient which limits production. Although most of the nitrogen that is eaten in the grass is excreted in the dung and urine of stock, it is probable that much of it does not become available to the herbage. Under temperate conditions it has been found that 70 per cent of the excreted nitrogen is in the urine, but from 12 to 20 per cent of this is lost by volatilization as ammonia (Walker *et al.*, 1954). Under the higher temperatures of the tropics losses are probably much greater. It must also be remembered that the nitrogen in urine is applied at high rates over small patches of the sward only. The nitrogen in the dung, representing about 30 per cent of the nitrogen in the excreta, is not immediately available and part of it may be retained as organic matter in the soil. Again, part is lost by volatilization as ammonia. Bearing in mind that some nitrogen is leached, and that tropical pastures do not usually contain any considerable proportion of effective nitrogen-fixing legumes, it is only to be expected that fairly liberal application of nitrogenous fertilizer is needed to maintain productivity at a reasonable level.

Many experiments, with a number of different grass species in various parts of the tropics, have amply demonstrated that profitable responses in yield and increased crude protein content may be obtained from the application of nitrogen up to a high level. For example, Dougall (1954) and Birch (1959) have reported marked responses to nitrogen applied to several species at intervals during the growing season in the highlands of Kenya; Ripperton and Takahashi (1948) found that application of up to 400 lb of sulphate of ammonia after each cut or grazing of *Panicum maximum* or *Paspalum dilatatum* gave an almost linear response in yields of dry matter and crude protein; Romney (1961) obtained responses with *Digitaria decumbens* and *Hyparrhenia rufa* in British Honduras; and Whyte *et al.* (1959) give figures showing increases in yield of dry matter for various levels of nitrogen application to *Chloris-Paspalum* pastures in South Africa and to *Cynodon plectostachyum* pastures in Rhodesia. Yields of dry matter and crude protein may be increased by 200–300 per cent and the crude protein content of the dry matter increased by 3–4 per cent. The residual effect of nitrogen is shortlived, and it is usual to apply moderate dressings at least two or three times during the growing season, or to give a dressing after each grazing. A dressing late in the rains will often assist in maintaining growth into the early part of the dry season.

After nitrogen, phosphate is the fertilizer most generally required in grazed tropical pastures, especially if a high level of nitrogenous fertilizing is practised. The efficacy of phosphate applied as a top dressing is somewhat variable and in some places it has been found advisable to disc it in rather than merely to distribute it on the surface. It is not normally necessary to apply phosphate more frequently than once a year. On the whole the soluble phosphate fertilizers tend to give the best results, but in many places there is little difference between superphosphate, basic slag and the various forms of rock phosphate;

29 Budding rubber; removing wood from the bud patch

30 Budding rubber; inserting the bud patch on the prepared stock

31 Budding rubber; t[he] bud patch tied and shad[ed] after insertion on the sto[ck]

32 Budding rubber; sci[on] growing from bud patch

consequently, the choice of the type to use must depend on local experience and on costs. Apart from increasing yields on phosphate-deficient soils, these fertilizers also raise the phosphate content of the herbage, which is of considerable importance in some areas where it may fall to dangerously low levels, especially during the dry season. Heavy dressings of nitrogen tend to depress the phosphate content of herbage but if phosphate is also applied a normal content should be maintained.

SPECIES AND STRAINS OF PASTURE GRASSES

The main features that grasses need to possess to render them satisfactory for use in cultivated pastures have already been stated. Many of the tropical grass species exist in a number of ecotypes, or strains, which differ in characteristics of agronomic importance, such as adaptability to different environments, vigour, persistence, growth habit, leafiness, leaf quality, seed production, or productivity at different seasons of the year under similar environmental conditions. Consequently, when seeking a good pasture grass for a given locality, it is often not sufficient merely to try a single sample or introduction of a species; it is particularly important to obtain and test a number of different strains. In some places an introduced species has proved outstandingly successful, but more commonly work aimed at the development of improved pasture grasses has achieved the best results by the selection of good strains, adapted to local environments, from within indigenous species. Good progress has been made by selection within the rich, indigenous grass flora in Kenya and Rhodesia, especially as strain selection and multiplication is facilitated by the fact that the majority of the important species are self-fertile and apomictic (Bogdan, 1959).

It is impracticable to list here the very large number of species and strains which have been used for cultivated pastures in various parts of the tropics, but for the majority of species information is given by Whyte *et al.* (1959). By way of illustration, brief mention is here made of nine species, all used extensively, and including examples of grasses used for both permanent and temporary pastures, at high and low altitudes and under low, medium and high rainfall. The crude protein and fibre figures for these species, given in Table 37, give some indication of the range found in cultivated grasses.

Pangola (*Digitaria decumbens*) and Kikuyu (*Pennisetum clandestinum*) are both low-growing, leafy grasses of high nutrient value, which are vegetatively propagated and form highly productive permanent pastures under good rainfall. The former, a native of South

Table 37. *Composition of some cultivated grasses at the grazing stage*

1, 2 and 3 from Edwards and Bogdan (1951); 4 and 6 from French (1943a); 5, 7, 8, 11–14 from Dougall (1960); 9, 10 from Butterworth (1961).

Species	Percentage on dry matter basis		
	Crude protein	Digestible crude protein	Crude fibre
1. *Brachiaria brizantha*, Kenya, 3 weeks' growth	10·7	—	26·1
2. *Bothriochloa insculpta*, Kenya, 3 weeks' growth	9·1	—	31·8
3. *Cenchrus ciliaris*, Kenya, 3 weeks' growth	12·5	—	31·4
4. *Cenchrus ciliaris*, Tanzania, 1–1½ feet, early flowering	11·0	8·2	31·9
5. *Chloris gayana*, Kenya, 9 inches high	12·6	7·7	29·2
6. *Chloris gayana*, Tanzania, 1–1½ feet	9·5	6·6	32·5
7. *Cynodon dactylon*, Kenya, 6 inches high	18·9	14·0	24·6
8. *Cynodon dactylon*, Kenya, 10 inches high	16·5	11·5	25·3
9. *Digitaria decumbens*, Trinidad, 10 days after grazing	14·9	10·6	31·0
10. *Digitaria decumbens*, Trinidad, 15 days after grazing	13·8	9·0	29·6
11. *Melinis minutiflora*, Kenya, 6 inches high	8·8	4·7	24·3
12. *Pennisetum clandestinum*, Kenya, July	20·8	15·9	22·8
13. *Pennisetum clandestinum*, Kenya, December	10·9	6·5	22·3
14. *Setaria sphacelata*, Kenya, 9 inches high	15·1	10·3	22·9

Africa, is suitable for low altitudes, and has been extensively planted in Florida, part of Central America and some of the Caribbean Islands (Oates *et al.*, 1959; Forster *et al.*, 1961; Romney, 1961). The latter, which is indigenous to limited upland areas of East Africa (Edwards and Bogdan, 1951) is only suitable for medium to high altitudes; it has been introduced into many parts of the tropics. *Brachiaria brizantha* (Signal grass), also an African species, propagated by division of rootstocks, has been widely used for permanent pastures, under good rainfall, from sea level up to 7,000 feet, in Africa, Queensland and the West Indies, and is regarded as the most successful pasture grass in Ceylon (Anker–Ladefoged, 1955). It is high-yielding, palatable and tolerant of shade, but only of moderate nutrient value. There are many varieties of the pan-tropical *Cynodon dactylon* (Common Star grass, Dhoob or Bermuda grass), most of which are vegetatively propagated, and this species has been very widely planted, both for permanent pastures (rhizomatous strains) and leys (stoloniferous strains), at altitudes of up to 8,000 feet and under rainfall from 20 inches upwards. It is highly persistent under grazing, responsive to fertilizers, and gives high yields of herbage of above average nutrient value. *Chloris gayana* (Rhodes), *Melinis minutiflora* (Molasses) and *Setaria sphacelata* are all good seed producers and are the species most widely used for leys (Edwards and Bogdan, 1951; Bogdan, 1959). Molasses grass gives good yields but is below average nutrient value and will not stand hard grazing. Rhodes, of which there are a number of varieties in commercial use, is very easily established and gives high yields in areas of 30 to 50 inches of rain per annum with a moderate dry season, but most varieties tend to be rather fibrous and lack persistence, lasting only two to four years. The better varieties of *Setaria* are rapidly increasing in importance as ley grasses on fertile soils in East and South Africa and Rhodesia, because of their ease of establishment, persistence, palatability and high nutrient content, although they have a tendency to go to seed too early. *Bothriochloa insculpta* (Sweet pitted grass) and *Cenchrus ciliaris* (African Foxtail or Buffel grass) are both good seed producers, palatable, persistent under heavy grazing and notable for their drought resistance. They are particularly useful for either leys or permanent pastures in areas of low and erratic rainfall up to altitudes of 6,000 feet. The former is of average nutrient value but the latter, which has been widely used in Northern Australia, is of above average protein content.

PASTURE LEGUMES

A good deal of effort has been expended in the search for suitable legumes for inclusion with grasses in tropical pastures, but so far this has met with little success and legumes are used for this purpose only in quite limited areas. Some progress has been made at higher altitudes, where conditions are not really tropical, in countries where European-style farming has been practised. For example, at high altitudes in Kenya, introduced species such as lucerne (*Medicago sativa*), Ladino or Louisiana white clover (*Trifolium repens*) and sainfoin (*Onobrychis viciifolia*) are grown to some extent (Grassland Research Station, Kitale, 1959). At slightly lower altitudes, the perennial Kenya white clover (*Trifolium semipilosum*) is the best and most widely used pasture legume, but several other indigenous clovers are also of some promise (Bogdan, 1956). Other legumes finding limited use in pastures include

Clitoria ternata, Glycine javanica, Stylosanthes gracilis and *Melilotus alba*. At high altitudes and under favourable rainfall in Rhodesia, New Zealand wild white clover (*Trifolium repens*) has proved successful in combination with grasses such as Kikuyu, Star and *Paspalum dilatatum* (Williams, 1953), but elsewhere the long dry season precludes the use of clovers. In the latter areas a search is being made for either an annual legume capable of perpetuating itself by seeding during the dry season, or a perennial which will survive the dry season. For the former purpose Korean Lespedeza is so far the most promising and for the latter purpose pigeon pea, sown in rows 30 feet apart in a grass pasture, gives appreciably more keep in the dry season than grass alone (West, 1956).

In the humid tropics there are a few species that have been extensively tried as pasture legumes in mixture with grasses and are now finding limited use in certain countries. For example, *Pueraria javanica* is used in Puerto Rico, Ceylon, Queensland and the West Indies; *Centrosema pubescens* in Queensland and Nigeria; *Stylosanthes gracilis* in Ceylon, Queensland and Nigeria; *Indigofera* spp in Ceylon and Puerto Rico. Some use has also been made of species of *Alysicarpus, Desmodium, Glycine,* and *Dolichos*. An account of these and other species that have been tried in pastures will be found in Whyte *et al*. (1953).

As legumes have so far only been used to a limited extent, there is little information on the value of including them in cultivated pastures in the tropics. It is therefore worth while briefly to consider the advantages obtained by including clovers in temperate pastures and assessing the likelihood of legumes conferring similar benefits to tropical pastures.

Firstly, under favourable conditions in the temperate zone, there is no doubt that clovers fix nitrogen, a considerable amount of which is returned to the soil in the excreta of grazing animals or liberated by the decay of rootlets and nodules of the legumes. In a well-balanced grass-clover sward this nitrogen increases the growth, yield and protein content of the grass, and reduces the need for applying nitrogenous fertilizers. The amount of nitrogen contributed by clovers in the temperate zone varies with climatic conditions. In parts of New Zealand, with year-round grazing and a long growing season under favourable temperatures and moisture conditions, large amounts of nitrogen are fixed and the need for fertilizer nitrogen is relatively small. Most estimates of nitrogen fixed by clovers in mixed swards exceed 100 lb N per acre per annum, and as much as 539 lb per acre per annum has been recorded, which would probably be as much nitrogen as the grasses could use (Sears, 1953). In Britain, where the growing season is shorter, estimates of the amount of nitrogen fixed are usually in the range of 50 to 200 lb N per acre per annum, and there is a greater use of nitrogenous fertilizers, particularly as much of the animal excreta is not returned to the pastures during the winter period of indoor feeding.

There is insufficient information on the extent to which nitrogen is fixed by pasture legumes in the tropics. It has been suggested that many tropical legumes may not fix appreciable amounts of nitrogen, either because effective symbiosis is prevented by soil acidity, or deficiencies of major or minor mineral nutrients, or because of the absence of strains of *Rhizobium* bacteria specific to the legumes concerned, or on account of the presence in the soil of antagonistic bacteria, actinomycetes or fungi.[1] It is likely that the performance of legumes may be unsatisfactory on some more acid soils, but this probably does not usually occur unless the pH is below 5·0. There is evidence that some tropical legumes are able to grow more satisfactorily on acid soils, and are less susceptible to calcium deficiency, than temperate legumes (Andrews, 1962). On some soils legume performance may also be impaired by deficiencies of mineral nutrients, but these can be diagnosed and corrected. It is certainly true that the efficiency of some introduced legumes, such as lucerne, lupins and clovers, may be reduced by lack of the appropriate strain of nodule bacteria, since the growth of these species has been greatly improved in Kenya and Rhodesia by inoculation with suitable strains of *Rhizobium*. This probably also applies to some species of *Trifolium* indigenous to tropical highlands. On the other hand, Norris (1956) has pointed out that the 'cowpea type' of *Rhizobium* is effective with a wide range of tropical legumes and has suggested that, since this type is capable of flourishing in acid soils low in calcium and phosphate, the majority of legumes are not restricted in growth and efficiency by lack of effective strains of nodule bacteria. Although some species may be strain-specific for nodule bacteria, there seems to be no reason to suppose that the majority of tropical legumes fail to fix nitrogen, unless the soil is highly acid or markedly deficient in essential nutrients.

There is evidence from a number of experiments that the inclusion of a legume in a mixed sward increases the yield and protein content of the grass component, probably indicating a beneficial effect of nitrogen from the legume on the growth of the grass, although the benefit to the grass might also result from the legume making little or no demand on soil nitrates. For example, in an experiment in Kenya in

[1] For a concise, general account of symbiotic nitrogen fixation and the importance of bacterial strains in relation to groups of leguminous genera, see Whyte *et al.* (1953), Chapter 9.

which each of four grasses (Rhodes, Star, Molasses and Nandi Setaria) were grown either alone or in association with Kenya white clover (*Trifolium semipilosum*), the average crude protein content and dry matter yield of the grasses was significantly greater when they were grown with clover than when they were grown alone. (Table 38; Strange, 1961). Similar results were obtained in Ceylon when the grass *Brachiaria brizantha* was grown alone or in association with either *Centrosema pubescens* or *Pueraria phaseoloides* (Fernando, 1961). In Nigeria, Moore (1962) compared pastures of pure Star grass (*Cynodon plectostachyum*) and of mixed Star grass and *Centrosema pubescens*, and found that the nitrogen content of the grass in the mixed sward (2·4 per cent) was significantly higher than that of the grass in pure stand (1·8 per cent). By analysis of soil samples taken to a depth of 30 cm beneath the pastures he also obtained evidence that the *Centrosema* had fixed nitrogen at the rate of 250 lb per acre per annum.

Table 38. *Yield of grasses grown alone or in association with clover*
Strange (1961)

	Grass grown alone	Grass grown in association with clover
Dry matter yield, cwt per acre, 1957	16·3	22·5
Dry matter yield, cwt per acre, 1958	15·5	24·5
Crude protein content, per cent, 1957	7·7	9·6

There appear to be virtually no other reliable quantitative estimates of the amount of nitrogen fixed by pasture legumes in the tropics, and it clearly cannot be assumed that Moore's figure for *Centrosema* is generally applicable. In many places the restriction of the growing season by the occurrence of a dry season may reduce the nitrogen contribution made by legumes in tropical pastures compared with that made by clovers in the more favoured parts of the temperate zone. Furthermore, there is some evidence that fairly high soil moisture favours nodulation and that legumes may nodulate less in some parts of the tropics than in the temperate zone because of the drier conditions (Masefield, 1952).

A second benefit resulting from the inclusion of clovers in temperate pastures is that, as they are richer than grasses in crude protein, phosphate and calcium, their presence improves the overall feeding value of the herbage and gives higher yields per acre of these nutrients. In some places legumes have certainly proved beneficial in this way in tropical pastures. For example, mixtures of the four grasses mentioned above with either Kenya white clover, or clover and lucerne, gave higher yields and contents of digestible crude protein than the grasses alone (Table 39; Strange, 1961). Similarly in Ceylon, mixtures of *Brachiaria brizantha* and *Centrosema pubescens* or *Pueraria phaseoloides* gave higher yields of crude protein, calcium and phosphorus than the grass alone (Fernando, 1961). However, the inclusion of legumes will only be advantageous in this respect if they are palatable and if they grow vigorously to form a considerable proportion of a well-balanced sward. Many tropical legumes are not highly palatable; for example Moore (1962) noted that the *Centrosema* in the mixed sward mentioned above was little eaten by the stock. It has also often been found difficult to maintain a good proportion of legumes in a mixed sward, as the grasses tend to become dominant under grazing. For these reasons it may well be that higher yields of dry matter and crude protein, or greater liveweight production, can be obtained by fertilizing a pure grass sward than from a mixed grass–legume pasture, especially in view of the very high potential yield of some tropical grasses.

Table 39. *Mean yield and protein content (1957 and 1958) of grass and grass-legume mixtures*

Strange (1961)

	Grass alone	Grass and clover	Grass, clover and lucerne
Dry matter yield, cwt per acre	15·9	32·2	35·5
Digestible crude protein, lb per acre	54	270	280
Digestible crude protein, percentage	3·3	7·1	6·9
Crude protein, percentage	7·3	11·8	11·6

Thirdly, clovers usually aid in maintaining a more uniform level of production in temperate pastures because they grow strongly in summer, when the grass is less vigorous than it is in the earlier part of the growing season. This benefit is not likely to occur in the tropics unless pasture legumes are found which are more productive in the dry season than the grasses, and so far it has usually been found that the grasses are more drought resistant than the legumes.

Finally, clovers help in providing a 'bottom' in temperate pastures, that is, their low-growing, creeping habit aids in providing a close cover for much of the soil between tufts of grasses. The trampling of the ground by cattle when tropical pastures, which are mainly of tufted grasses, are heavily stocked is liable to result in deterioration in soil structure. Provision of a 'bottom' by legumes would therefore be an advantage but, except in favoured areas where introduced or indigenous clovers can be used, this is unlikely to be possible, since the majority of the promising pasture legumes do not possess the necessary low, close-growing, creeping habit.

On the whole it would seem that, except in limited highland areas with good rainfall, the inclusion of legumes in tropical pastures is unlikely to prove as beneficial as it is in the temperate zone. At present, it is generally more practicable to increase yields of herbage and of crude protein by fertilizing pure grass swards with nitrogen. However, bearing in mind the relatively high cost of nitrogenous fertilizers in the tropics and the low returns generally obtainable from the existing poor quality livestock, it will be desirable to continue the search for efficient legumes that can be maintained in vigorous condition in a mixed sward.

THE LEY AND SOIL FERTILITY

In the temperate zone the use of grass–clover leys in a system of alternate husbandry is capable of building up soil fertility and maintaining a high level of crop and stock production. It appears to be reasonably well established that the most important single factor in the beneficial effect of the ley itself is the nitrogen supplied by the clovers, although the improved physical condition of the soil after ploughing in the ley is also usually of value. The ley itself is by no means the only factor contributing to the maintenance of fertility in the alternate husbandry system, which also involves the use of considerable quantities of farmyard manure and fertilizer and the feeding of imported foodstuffs to livestock.

In the tropics the use of leys in alternate husbandry is at present almost entirely restricted to very limited highland areas where European-style farming is practised and where conditions are markedly different from those obtaining in the vastly greater areas at low and medium altitudes. Information on the value of the ley for fertility maintenance under the latter conditions is confined to the results of a comparatively small number of well-designed experiments, of rather limited duration, conducted in areas of monsoon climate. Almost all these experiments have been with leys of grass alone (since suitable pasture legumes were not available) and they have not included the application of fertilizers to the grass, nor supplementary feeding of livestock.

An example of one of the best of these experiments is that conducted at Serere, in Uganda (Martin and Biggs, 1937; Jameson and Kerkham, 1960). This trial was of phasic design and compared five resting covers, each of which was included as a one-, two- and three-year rest in a five-year rotation, giving fifteen rotations in all. A comparison of

0, 2 and 5 tons of farmyard manure, applied once in five years, was superimposed. The resting treatments were as follows:

A. Natural regeneration, mainly of grass, herbs and leguminous shrubs, ungrazed.
B. Planted grass, ungrazed (Star grass in the first two cycles, Rhodes grass in the third).
C. Velvet bean, lightly topped in August, dug in the following March.
D. As B, but grass cut in the first cycle, subsequently grazed.
E. Velvet bean dug in and resown in August, and dug in again in the following March.

Consideration of the rotation yields over three cycles (fifteen years) for all crop: rest sequences and all levels of manure (Table 40) reveals that there were no significant differences between types of resting cover. Yields from rotations including rests under natural regeneration or grazed grass appeared slightly better than the others, but not significantly so, and whether the grass was grazed or not made no significant difference. In the absence of farmyard manure, yields declined with a crop: rest ratio of 4:1, when about three tons per acre of produce were taken from the land in five years, but yields appeared to be maintained by a 3:2 ratio, when about two tons per acre were removed in five years. Yields of individual crops were greater after the longer rest periods but these increases did not compensate for the loss of crop while the land was resting. There was a big response to farmyard manure, the residual effects of which could be traced through to the fifth year after application. With an application of 5 tons of manure every fifth year it appeared possible to crop four years in five without apparent decline in yield, and this system gave the highest total yields of produce.

Table 40. *Rotation yields in lb per acre for all crop: rest sequences and all levels of manure at Serere*

Jameson and Kerkham (1960)

Type of resting cover	1st cycle	2nd cycle	3rd cycle	3rd cycle minus 1st cycle
A. Natural regeneration, ungrazed	4,522	5,119	5,033	511
B. Planted grass, ungrazed	4,738	4,863	4,921	183
C. Velvet bean, topped	4,589	4,822	4,825	236
D. Planted grass, grazed	4,604	5,098	5,011	407
E. Velvet bean, dug in	4,417	4,844	4,698	281

The results of a number of other experiments have been similar and, in particular, have indicated that the effects of pure grass leys (grazed or ungrazed) on soil fertility are no better than those of natural regeneration. For example, Dennison (1959) found in Northern Nigeria that there were no significant differences between yields of crops following three-year rests under planted grass (*Andropogon gayanus*) or natural regeneration (Table 12, p. 180); Clarke (1962) found three years rest under grazed Star grass no more beneficial than three years cassava-weed fallow in Kenya (Table 13, p. 180); and experiments in Malawi comparing bush fallows and Rhodes grass leys showed no significant differences in following yields (Nyasaland 1957–8). A number of experiments at Ukiriguru in Tanzania (Peat and Brown, 1962) have shown that crop yields following three- or five-year rests under a grazed grass ley were not significantly better than those following a grazed tumbledown fallow of similar duration (Tables 15, 41 and 42). Furthermore, in these experiments the yields following three- or five-year periods of continuous cropping, during which moderate dressings of farmyard manure or phosphatic fertilizers were applied, were not significantly inferior to those following the resting treatments (Tables 41, 42). The effects of grass or tumbledown fallows were thus about equivalent to those of reasonable dressings of manure or phosphate but, since continuous cropping with manure or fertilizer does not involve any loss of crop during a resting period, it must clearly have given greater total crop production from the land than rotations including leys or fallows. Planting grass is not considered worth while at Ukiriguru, where it is difficult to establish and forms a sparse sward, and where the effect of resting covers on the structure of the coarse, sandy soil must be negligible. It should be mentioned that the grazed plots in the Serere and Ukiriguru experiments were only grazed during the day time, the cattle being kraaled at night and not normally given any supplementary food, except a little roughage. This would necessarily be normal farming practice in many parts of Africa but it means that a high proportion of the animal's food was taken from the experimental plots and much of their excreta was dropped in the kraal at night, so that there was some transference of fertility away from the experimental area.

The beneficial effects of grass fallows on fertility are not primarily due to their improving soil structure. This has been demonstrated by Pereira *et al.* (1954) who showed that, although grass leys of three years duration do effect significant improvements in water-stable aggregation, freely drained pore space and rates of infiltration of rainfall, these improvements rapidly disappear during the arable cultivations that follow and are largely lost before the end of the first

Table 41. *Effect of three-year rests at Ukiriguru: yield of cotton test crops as percentage control*

Peat and Brown (1962)

	1st Cycle			2nd Cycle		
Type of 'rest'	1948–49	1949–50	1950–51	1954–55	1955–56	1956–57
1. Control, continuous cropping	100	100	100	100	100	100
2. As 1 with 3 tons compost, 3 yearly	164*	130*	143*	154*	165*	127
3. As 1 with phosphate fertilizer	171*	133*	158*	151*	146*	117
4. Planted grass, grazed	130*	137*	130*	155*	166*	146*
5. Tumbledown fallow, grazed	85	157*	164*	150*	170*	152*

* Denotes significant difference from control.

Table 42. *Effect of five-year rests at Ukiriguru: yields of cotton and millet test crops as percentage control*

Peat and Brown (1962)

	1952–53		1953–54	
Type of 'rest'	Cotton	Millet	Cotton	Millet
1. Control, continuous cropping	100	100	100	100
2. As 1, plus cattle manure	144*	147*	135	137*
3. As 1, plus phosphate fertilizer	146*	121	120	138*
4. Tumbledown fallow, grazed	149*	125*	109	134*
5. Planted grass, grazed	127*	113	118	149*

* Denotes significant difference from control.

cropping year, whereas the beneficial effects on crop yields are maintained longer.

The very low nitrate levels in the soil under grass, coupled with the incorporation of a large bulk of grass tops and roots of low nitrogen content when the sward is ploughed in, often result in a deficiency of available nitrogen for the first crop following a grass ley, especially if the crop is a cereal. This tendency can be counteracted to some extent by grazing management that maximizes the return of dung and urine by the animals, and especially by heavy grazing shortly before ploughing, so that there is a final, relatively heavy, deposit of excreta and only a comparatively small quantity of grass tops is ploughed in along with the mass of root material. Nevertheless, cereals often suffer from lack of available nitrogen even when they follow a well-grazed grass ley. This was stressed by Ellis (1953), who found in Rhodesia that average maize yields following a grazed, four-year Rhodes grass ley were little better than those from continuous maize cultivation. The difficulty can be overcome by following the ley with a legume (either as the first crop or as a green manure) or by applying nitrogenous fertilizer to a cereal crop. However, it is usually impracticable to solve the problem by applying nitrogen to the grass during the ley period, since residual effects are only apparent when very high dressings are given.

The existing evidence indicates that in regions of relatively good seasonal rainfall, and on soils that are not inherently infertile nor exhausted by intensive cropping, a rotation of about three years pure grass ley and three years cropping can maintain fertility at a moderate level for a prolonged period. The grass ley probably has a modest direct beneficial effect on the nutrient and physical status of the surface soil and, compared with continuous cultivation, a rotation including a grass break is likely to be advantageous in reducing erosion and decreasing the incidence of weeds, pests and diseases in the arable crops. However, unfertilized grass leys are no more effective in maintaining fertility than are fallows of natural regeneration. Hence it is unlikely to be worth while expending labour and money on planting grass unless an appreciably better return can be obtained from animal production on the ley than from natural regeneration. Furthermore, although increased crop yields are normally obtained after a grass ley, these increases do not usually compensate for the loss of cropping during the ley period, so that the maintenance of fertility is, in part, achieved at the expense of a reduction in the total amount of crop products removed from the land. Clearly, alternate husbandry will only be satisfactory if crop yields during the arable break and animal production from the ley can be raised to levels that together more than compensate for the loss of cropping during the grass break. It seems unlikely that this can be achieved without a use of fertilizers which most tropical farmers cannot at present afford, or without a much greater ability in husbandry than the majority at present possess.

Where it is possible to establish an efficient, nitrogen-fixing legume as a palatable component of a vigorous, well-balanced, mixed sward, it may be expected that the beneficial effects of the ley on soil fertility and crop yields will be considerably enhanced, and the need for nitrogenous fertilizers reduced. Moore (1962) has shown in Nigeria that the top 30 cm of soil under a Star grass–*Centrosema pubescens* pasture two and a quarter years old contained 1,600 lb per acre more organic matter and 560 lb per acre more nitrogen than that under a pure Star grass ley, indicating that the *Centrosema* had fixed nitrogen at the rate of about 250 lb per acre per annum. Such figures suggest that pasture legumes can play an important part in maintaining fertility but, so far, legumes have only been used in cultivated pastures to a very limited extent and there is virtually no reliable information on the effect of mixed swards on the tropical soil.

CHAPTER 14

Classes of Tropical Livestock

This chapter is confined to a brief consideration of those species of farm animals which are relatively unimportant in temperate agriculture, but which are of great importance in all, or several, parts of the tropics. Sheep, horsekind, pigs and poultry will not be discussed, since all these domesticated livestock are common to both temperate and tropical agriculture and they are adequately dealt with in many standard reference works. They will also be referred to in the appropriate sections of subsequent chapters, when some aspects of the management of tropical livestock are considered in more detail.

The distribution of the different species of farm animals differs in the tropics from that in temperate regions. The Indian zebu (*Bos indicus*) cattle are more abundant than the non-humped, European cattle (*Bos taurus*). The dual-purpose goat is generally more important than the single-purpose, hair sheep. Draught animals, bovines as well as horsekind, are still of great importance in many areas and there is a large number of species of lesser importance employed for this purpose, ranging from the Indian elephant to the South American llama, and including the important beast of burden of North Africa and the Middle East – the Arabian camel or dromedary. In Asia the water buffalo ranks as a most important triple-purpose animal, and in the subtropical parts of North America the wild American bison has been successfully crossed with European-type cattle to form a hybrid beef-type animal known as the cattalo, which has since spread in a limited manner into certain parts of Central America.

Swine and poultry are becoming increasingly important in many tropical countries, but as management systems change over from crude forms of 'backyard enterprises' by subsistence level peasant farmers to more intensive husbandry practices managed by progressive agriculturists, so we find a swing away from the so-called 'indigenous' breeds to widespread use of imported temperate-type breeds and strains. This chapter will deal primarily with those classes of livestock which differ, in some important manner, from their temperate counterparts. Readers seeking a detailed knowledge of pigs, poultry and horsekind are advised to consult standard temperate literature, such as Jull (1951) and Davidson and Coey (1966).

Cattle

The F.A.O. estimate that the present world cattle population is of the order of 850 million. Of this total figure, approximately 500 million cattle are distributed in tropical countries and about 350 million in temperate zones. A very rough breakdown of this distribution by continents is shown below.

Table 43. *Approximate distribution of world cattle population* (millions)

Temperate Regions		Tropical Regions	
Europe	108	Central America	48
U.S.S.R.	67	South America	72
North America	88	Asia	275
South America	60	Africa	96
South Africa	10	Oceania	8
New Zealand	15		
Total	348	Total	499

No accurate figures of the numbers of European-type and zebu-type cattle are available, nor is the size of the crossbred population (*Bos taurus* × *B. indicus*) known. Rough estimates place the zebu-type cattle in a distinct majority with a population of the order of 460 million; European-type cattle probably number about 360 million, leaving a minority group of approximately 27 million crossbreds. As we shall see later, it is almost certain that the numbers of crossbred cattle will increase markedly during the next few decades, at the expense of a diminishing zebu-type cattle population, since crossbred breeding programmes are now being vigorously pursued in most tropical countries.

In spite of their numerical superiority, the zebu-type cattle found in tropical countries contribute much less than one half of the cattle production consumed by man. Two factors are responsible for the relative low productivity of zebu cattle. First, the genetic capacity of tropical cattle, whether for milk or meat, is less than that of European-type cattle. Secondly, a relatively small percentage of tropical cattle owners tend their stock in order to exploit them for commercial reasons. Each cattle-keeping community raises stock for a variety of purposes, and the importance given to each purpose varies widely, sometimes within the confines of quite a small geographical area. It is true that, in the tropics as elsewhere, cattle are essential as sources of food, but even when we deal with this basic reason for cattle-keeping we encounter religious and social customs which, for instance, prohibit the eating of cattle-meat by Hindus and which encourage the drinking of cattle-blood by African nomadic tribes such as the Karamajong and Masai of East Africa.

It is most important to understand and appreciate the significance which cattle, more than any other class of farm animal, have to the majority of the inhabitants of Africa and Asia. Cattle are part of their way of life, an integral part of their religion and a dominating factor in their social organization. In the Ankole District of Uganda, for example, the whole basis of the social structure of the tribe revolves around the unit of the cattle kraal. A workable number of cattle is between 100 and 200 head, and if the cattle numbers increase then a son will be encouraged to marry, leave home and start a new kraal, taking with him for this purpose about 100 head of cattle. If a herd owner dies without issue, then near relatives of the deceased will be ordered to take over responsibility for the kraal. Human social behaviour, who and when to marry and where to live, is largely dictated by the husbandry requirements of the cattle herd. It follows that any attempt to change the system of cattle husbandry will almost certainly interfere, to a lesser or greater degree, with the social customs and religious beliefs of the people. There is therefore a very real need for every animal husbandry adviser to have at least a fair working knowledge of the social customs of the cattle-owning peoples if he is to be successful in bringing about technical change without encountering opposition. This knowledge is more important in tropical countries than in temperate regions, since cattle play a much greater part in the social life and religious beliefs of the tropical peoples.

In addition to the important part which cattle play in the social and religious life of many tropical peoples, cattle are generally prized more for their quantity than for their quality. A great deal of social prestige is attached to the ownership of large herds, irrespective of the breeding value and productive efficiency of the animals comprising the herd. This is not surprising when it is understood that cattle are regarded as a 'bank account on the hoof'. Ten poor quality, old cows are, in the eyes of many tropical communities, worth precisely double five healthy and productive heifers. The reason for this state of affairs is that, until comparatively recently, the cattle population fluctuated widely as drought, disease and natural predators decimated numbers, and the more cattle one possessed before calamity struck the more survivors one was likely to retrieve as conditions improved. Large cattle numbers were, therefore, a form of insurance, especially so as large herds were often split up and distributed over a wide geographical area so that if disease broke out in one location, animals in an adjoining location were likely to escape unscathed.

EVOLUTION OF CATTLE

All domesticated cattle belong to the genus *Bos* of the Bovidae family. The exact relationship of the various species of the genus is not known with any certainty, and the following account is merely a survey of the literature on the subject. There is an urgent need to employ the new genetic methods of 'blood grouping' to the diversified cattle populations of the world, in order to trace their origins and migratory movements with a greater degree of accuracy.

The genus *Bos* as represented today may be conveniently subdivided into four subgroups, known as Taurine, Bibovine, Bisontine and Bubaline.

1. *Taurine subgroup*.

This includes the two most important cattle species which exist today, *Bos taurus*, or European-type cattle and *Bos indicus*, or zebu-type cattle. The former are thought to be descendants from urus (*Bos primigenius*), one of the first wild animals to be domesticated by man and exploited as a triple-purpose animal. The evolution of *B. taurus* cattle from *B. primigenius* is thought to have taken place in mid- and north Europe and south-west Russia. Urus originated on the west coasts of the Pacific ocean and spread through Asia and Europe to the eastern coasts of the Atlantic. Urus cattle can be traced through about one million years of geological time, from the Pleistocene period to 1627, when the last known living representative died in captivity in Poland. *Bos indicus* cattle are said to be descendants of *Bos nomadicus*, a native of Asia which is thought to have been domesticated somewhat later than *B. primigenius*. *Bos indicus* probably evolved from *B. nomadicus* in south-central Asia. Another member of this taurine subgroup is *Bos opisthonomus*, a non-domesticated species once native to North Africa. It is the only taurine animal indigenous to the African continent but is now extinct, having been displaced by the various migrations of other cattle species from the east, as we shall see later. No satisfactory explanation has yet been given of the ancestry of zebu-type, non-humped shorthorn cattle which are found today in parts of the north Mediterranean coastline, China and Mongolia. These cattle form a quite distinct species, classified by most authorities as *B. brachyceros* (Jeffreys, 1953; Mills, 1953).

2. *Bibovine subgroup*

This group includes several semi-domesticated species found in the Far East, in particular the gayal (*B. frontalis*) of Assam and parts of Burma and the banteng (*B. sondaicus*) of Borneo and Indonesia. Another representative of this subgroup is the gaur (*B. gaurus*) which has not been domesticated and which is found wild in parts of India, Burma and Malaysia.

3. *Bisontine subgroup*

The only important member of this subgroup is the North American bison (*B. bison*) which has been successfully crossed with *B. taurus* cattle to form the new hybrid, the cattalo. Records of cattalo productive performance are not easily come by, but it appears that cattalo calves are smaller at birth than purebred *B. taurus* calves, but make better gains from birth to weaning. Cattalo slaughter stock apparently obtain lower grades than European stock, but they generally have higher dressing percentages due to the fact that cattalos have lighter hides and lower weights of non-edible viscera. Other, non-tropical members of this subgroup are the yak (*B. grunniens*) which is an important domesticated draught animal in Tibet, and the European bison (*B. bonasus*).

4. *Bubaline subgroup*

This subgroup is of great importance in India and other parts of the Near East and South-east Asia. The nomenclature of the different species is still in dispute by systematic zoologists, but Williamson and Payne (1965) recognize three distinct species, of which the most important is classified as *B. bubalis*, commonly known as the water buffalo because of its habit of wallowing in the shallow water of rivers and swamps. The other two species of lesser importance are *B. mindorensis*, the tamarao or carabao indigenous to the Philippines, and the Celebes buffalo, *B. depressicornis*, also known as the anoa or the dwarf buffalo of Borneo. Lydekken (1912) includes the African buffalo (*B. caffer*) in the Bubaline subgroup. It should be noted in passing that various workers have stated a case for a more systematic approach to the utilization of the large reservoir of wild game animals, present in Africa and elsewhere, which belong to the Bovidae family. It is suggested that these animals could be 'ranched' in the wild state, and regularly 'cropped' as young stock reach slaughter-weights. If this proposal finds acceptance with those responsible for the policy affecting wild game, it may well be that *B. caffer* (African buffalo) and possibly also *B. bonasus* (European bison) will have to be regarded in the future as semi-domesticated livestock (Ovington, 1963).

It is assumed that most readers of this book will be familiar with the characteristics of European cattle (*B. taurus*), and no specific descriptions of this species will be given here. There is, in any case, a prolific literature on the subject. A brief description of zebu (*B. indicus*) cattle is appropriate at this stage, and those interested in a more detailed treatise should consult the works by Mason (1951), Faulkner and Brown (1953), Joshi and Phillips (1953), Faulkner and Epstein (1957), Mason and Maule (1960). An excellent bibliography of the subject is provided in Mason and Maule's book, which lists 251 references.

Just as there are numerous well-defined breeds of European cattle, there are also many different breeds and types of zebu cattle recognized. However, the classification of the zebu cattle population is arbitrary, the same type of animal often being given a variety of different names in different parts of the tropics. No attempt will be made to classify or define the various breeds of zebu in this brief survey, but it is useful to divide the species into two distinct subgroups on the basis of the anatomy of the hump (Thorpe, 1953; Milne, 1955).

a. Neck (or cervico-thoracic) humped zebu. This subgroup has the hump placed in position anterior to the forelegs. The hump is composed of extensions of the *rhomboideus* muscle, with little intermuscular or intramuscular fat. The hump appears to function as an aid to the traction of the animal, giving a greater leverage to the action of the forelegs. This subgroup almost invariably possesses long horns, and for this reason is sometimes referred to as 'longhorned zebu'. The best known example of this type is the Africander of South Africa (Bonsma and Joubert, 1952).

b. Chest (or thoracic) humped zebu. This subgroup has the hump placed in position just over, or posterior to, the forelegs. The hump is composed of both muscular and adipose tissue, extensions of the *rhomboideus* and *trapezius* muscles ramifying into a large quantity of subcutaneous fat. The structure serves little or no tractive purpose, and is regarded by most authorities as a food-storage organ. This subgroup generally possesses short horns, and is sometimes referred to as 'shorthorned zebu'. There are numerous Indian and African breed names given to this type, some of the more important being the Sahiwal, Hallikar, Gir and Sindhi cattle of India and the Nkedi and Nandi cattle of East Africa. (See Plates 45–48.)

There is some evidence that when either of these two subgroups of zebu cattle is crossed with the humpless, hamitic longhorn cattle found in parts of the Middle East, North Africa and Eritrea, a stabilized crossbred is produced, known as the sanga. When neck-humped zebus are crossed with humpless longhorns, the resultant crossbred may be described as neck-humped sanga, the hump being much reduced in size compared to the zebu type. The best known example of the neck-humped sanga is the Ankole breed of cattle found in Uganda. Similarly when the cross is made between the chest-humped zebu and humpless longhorns, the resultant crossbred is known as chest-humped sanga, several examples of which may be found in Ethiopia.

The above classification may well be an over-simplification of the true position, and it should be noted that no two authorities agree on either the nomenclature of the sub-grouping of the *B. indicus* species or on the definition of the term 'sanga' (see Payne, 1964). Once again, the need for further investigation of the affinity of the various cattle types by the use of blood-grouping techniques is apparent. Similar reservations must also apply to a later section of this chapter, which seeks to trace the major cattle migrations which have taken place from Asia and the Middle East into the African continent. For more detailed descriptions of African and Indian *Bos indicus* breeds the writer is referred to Mason (1951), Williamson and Payne (1965), Joshi and Phillips (1953), Mason and Maule (1960), and Olver (1938).

The following points of comparison bring out the chief differences between *B. taurus* and *B. indicus* cattle:

European (*B. taurus*) cattle	Zebu (*B. indicus*) cattle
1. No hump.	Hump present in thoracic or cervico-thoracic region.
2. Rounded ears, held at right-angles to the head.	Long drooping ears, pointed rather than rounded.
3. Head short and wide.	Head long and comparatively narrow.
4. Skin held tightly to body, dewlap, umbilical fold and brisket small.	Skin very loose, often falling away from body in folds. Dewlap, umbilical fold and brisket extensively developed.
5. Skin relatively thick, average thickness 7–8 mm.	Skin relatively thin, average thickness 5–6 mm.
6. Large amounts of subcutaneous fat, especially in mature stock.	Relatively small amounts of subcutaneous fat, especially in mature stock.
7. Backline straight or relatively straight.	Backline high at shoulders, low behind hump, high over pin bones, sloping down markedly over tail-bud.
8. Hip bones wide and outstanding.	Hip bones narrow and angular.
9. Thoracic ribs well sprung away from body.	Thoracic ribs poorly sprung, forming an angle with the vertebral column.
10. Udder long, with a flat sole, well suspended between and behind the hind legs.	Udder more rounded with a curved sole, poorly suspended and carried in front of, rather than between, the hind legs.

11. Hair fibres non-medullated, so that they are held limply on body surface.	Hair fibres usually medullated so that they tend to stand more erect and away from body surface.
12. Hair relatively long, rough and double-coated. Seasonal difference in hair length. Average population density of 800 follicles per sq cm.	Hair relatively short and smooth-coated. Little or no seasonal difference in mean hair length. Average population density 1,700 follicles per sq cm.
13. Hair and skin generally both pigmented or both non-pigmented.	Skin usually pigmented irrespective of colour of overlying hair.
14. Legs short. Slow moving.	Legs longer. Faster moving.
15. Skin and hair attractive to most cattle ticks.	Skin and hair less attractive to cattle ticks.
16. Fast maturing. Full mouth in 4 years.	Slow maturing. Full mouth in $5\frac{1}{2}$ years.
17. Milk yield, lactose content and nitrogen content drop when ambient temperatures reach or exceed 24C°(75° F).	Milk yield, lactose content and nitrogen content do not drop until ambient temperatures reach or exceed 35°C (95°F).
18. 'Comfort Zone' of species 4°–15°C (40°–60°F).	'Comfort Zone' of species 15°–27°C (60°–80°F).
19. Adult animals relatively large; fully grown bulls commonly reaching 2,000 lb liveweight.	Adult animals relatively small; fully grown bulls of most breeds rarely exceeding 1,500 lb.

CATTLE MIGRATIONS FROM THE MIDDLE EAST AND ASIA INTO AFRICA

It has already been noted that the only Taurine species indigenous to Africa, *B. opisthonomus*, has now become extinct. The various breeds and types of cattle now found on the African continent, numbering approximately 100 million head, have been introduced from Asia, the Middle East and, in more recent years, from Europe. It is instructive to trace the approximate timings of, and routes taken by, these various migrations, since such studies throw a great deal of light on the origins of the African cattle population and the reasons for the present distribution of related breeds. Once again, it must be stressed that the following account cannot be substantiated by really critical scientific data, since parts are based on anthropological evidence, such as the crude paintings of cattle found on the walls of caves, but the main outlines of the story are probably accurate enough for our present purposes.

First migration (*about* 5000 B.C.)

Introduction of *humpless, hamitic longhorn* cattle from the East, across the Nile Delta, thence westwards along the north coast of Africa or southwards down the Nile into Ethiopia. This introduction took place in the Badarian period, about 5000 B.C. Prior to this migration there were no domesticated cattle on the African continent. This same migration, crossing the Straits of Gibraltar northwards into Spain, gave rise to the 'Spanish longhorn' cattle of Spain, Portugal, and, via the Spanish conquest of Central and South America, the American continent (see Esperandieu, 1952).

Second migration (2500–2000 B.C.)

Introduction of *humpless, shorthorn* cattle (*B. brachyceros*) from Asia during the period 2500–2000 B.C. These cattle displaced the *humpless hamitic longhorn* in certain areas, presumably because of their greater adaptation to African conditions. These cattle are still the dominant type found in North Africa and Egypt, and also in certain islands off the E. African coast, such as Pemba and Mafia.

Third migration (*about* 1500 B.C.)

Introduction of *cervico-thoracic longhorned zebu* from Asia via the Red Sea. These cattle were well represented in the Persian Gulf from 3000 B.C. and reached the Sudan by about 1500 B.C. This group of cattle was taken southwards, westwards along the Congo River basin, then again southwards down the west coast of Africa by the Hottentots in their extensive migration from the great Lakes to South Africa. In 1652 the Dutch East India Company acquired these stock from the Hottentots, and they have since been improved in South Africa into the important breed now known as the Africander.

Fourth migration (1500 B.C.–A.D. 500)

The evolution on the African continent, of the *cervico-thoracic-humped sanga* by the crossing of the *humpless, hamitic longhorn* with the *cervico-thoracic-humped zebu*. This cross is thought to have taken place in Ethiopia, from whence it has spread westwards across the Great African Plains to West Africa (where it has given rise to the Fulani cattle of Northern Nigeria) and southwards down the Great Lakes leaving behind 'islands' of isolated cattle breeds, such as the Dinka cattle of the Sudan, the Ankole cattle of Uganda and the Bechuana cattle of Bechuanaland. The cross is thought to have been made about 1500–1000 B.C., but the migrations of these cattle to the west and south are thought to have taken place about A.D. 500.

Fifth migration (*about* A.D. 700)

Introduction of *thoracic-humped shorthorned zebu* from Asia by the Arabs during their invasion of the eastern seaboard of Africa from about A.D. 669. These cattle were transported by sea, and spread inland from the ports along the main Arab slave and trade routes. The southern migration stopped abruptly at the Zambesi, since this river was never crossed by the Arabs during their travels in East and Central Africa.

Sixth migration (about A.D. 800–1000)

The evolution, on the African continent, of the *thoracic-humped sanga* by the crossing of the *humpless, hamitic longhorn* with the *thoracic-humped shorthorned zebu*. This cross is also thought to have taken place in Ethiopia, but it has not spread from this point so extensively as has the *cervico-thoracic-humped sanga* (see Fourth migration, above).

Seventh migration (about A.D. 1700–*present*)

Introduction of *B. taurus* animals from Europe during the period of the European invasion of West, South and East Africa. The earlier introductions were notably unsuccessful, and mortality was high due to the susceptibility of *B. taurus* cattle to tropical diseases and tropical parasites. More recent introductions, carried out under improved conditions of hygiene and disease control, have been far more successful and many herds of purebred European stock are now performing well in most African countries.

Eighth migration (about 1920–*present*)

Introduction of improved breeds of *thoracic-humped shorthorned zebu* from India and Pakistan into East, Central and South Africa. These introductions, mainly of Sahiwal and Sindhi cattle, have been made from government livestock farms in India and Pakistan in an attempt to utilise the successful breeding programmes carried out in Asia during the last hundred years. The productive performance of the stock thus introduced has often been about double that of the local breeds of zebu cattle, which have been relatively little improved since their first introduction from Asia. (See Plates 46 and 48.)

The future picture is likely to become far more complicated than that outlined in the above chronological account of the eight major cattle migrations from Asia and Europe onto the African continent. Crossbreeding programmes (*B. taurus* × *B. indicus*) are being carried out in most countries, and the increased use of deep-frozen semen as a cattle-breeding technique will enable these programmes to be extended into areas where purebred *B. taurus* cattle can only be kept alive with great difficulty. This point will be dealt with in greater detail in Chapter 17, 'Livestock Improvement'.

WATER BUFFALO (*B. bubalis*)

The domestic water buffalo is the third most important species of the *Bovine* family, with a total world population of just under 100 million. The present breeds are thought to be descended from the wild buffaloes which still inhabit the swamps and forests of India, Pakistan and Malaysia. The domesticated species are now widely distributed throughout the northern tropical and subtropical regions of Asia, the countries bordering the eastern Mediterranean, and in isolated islands such as the Philippines and Trinidad. The largest concentration of water buffalo is found in India (about 45 million) and the second largest in China (about 22 million). It is not generally realized that the species is also found in Europe, some 600,000 buffaloes being distributed in southern European countries, particularly Italy and Greece (Mammericks, 1960).

The water buffalo occupies an important place amongst the domestic animals of the tropics, and may be regarded as a truly triple-purpose animal. The meat, from young slaughter stock, is of good quality and is palatable even to those who are mainly used to high quality beef from European beef cattle breeds. The milk yields are comparatively high by tropical standards and the butterfat content is about double that of European milk breeds. The animal is strongly built and achieves large mature liveweights and is capable of pulling very heavy loads through conditions which would prove impassable to horsekind or taurine cattle.

The water buffalo species may be conveniently divided into two major types, the river buffalo and the swamp buffalo. These two types have marked differences in anatomical characteristics and habit and they are usually non-miscible, crossbreeding between the two types being uncommon.

The swamp buffalo is the smaller animal, and it possesses a small hump situated over the shoulder. It is chiefly found in Malaysia where it is employed both for draught purposes and as a milch animal. According to MacGregor (1941) the swamp buffalo is a semi-aquatic, nocturnal animal which spends the hotter period of the day, from 10 a.m. to 4 p.m., semi-submerged in natural swamps or self-made wallows. Attempts to impose a different diurnal behaviour-pattern on the swamp buffalo, or to deny them access to swamps of wallows, are usually unsuccessful and lead to a number of ill-effects, such as cessation of breeding, increased calf mortality, incidence of 'joint ill' and a reduction in growth rate. The usual management of swamp buffalo is therefore designed to conform to their normal behaviour-pattern as closely as possible, and they are commonly used for such tasks as puddling rice paddies and hauling timber during the early and late parts of the day. They are used for lighter work at any hour of the day.

The river buffalo is a larger and more versatile animal and most of the important domesticated breeds belong to this type. The hump is not so well defined as in the case of the swamp buffalo. The behaviour-pattern is similar to that of taurine cattle and swamps and wallows, although greatly appreciated, do not appear to be necessary for the physical well-being of these stock. The major breeds of riverine water buffalo may be distinguished by their horn characteristics and facial profile, and to a lesser extent by their size and the colour of the hair. The five most important breeds are as follows:

1. *Murrah (or Delhi)*. The most important Indian breed and a most efficient milking breed. The home of the breed is mainly in Punjab and Delhi, but pure-bred herds are also found in the United Provinces. This breed has a deep, massive frame with a short, broad back and a comparatively light neck and head. It has short, characteristic tightly-curled horns. The udder is well-developed and the average milk yield in recorded herds is about 300–400 gallons per lactation (I.C.A.R., 1950). (See Plate 34.)

2. *Nili and Ravi*. These types are found in the valley of the rivers Sutlej and Ravi in the West Punjab. There is no essential difference between the two types and they are now officially treated as a single breed. The breed possesses a medium-sized, deep frame with an elongated, coarse head. Horns are small with a high coil and the neck is long and fine. The common hair colour is black but brown animals are also found. Average lactation yields are quoted as being 375–475 gallons (Houghton, 1960). (See Plate 33.)

3. *Surti*. This breed is mainly found in Bombay State between the Mahi and Sabarmati rivers. The animals are well proportioned and of medium size, with short legs. The horns are of medium length and sickle-shaped. This breed has an exceptionally straight backline, and the udder is well formed with squarely placed teats of medium size. It is very early maturing, but it is not regarded as a good breed for draught purposes.

4. *Jaffrabadi*. This is the largest breed in body size, selected bulls weighing as much as 3,000 lb. The home of the breed is in the Gir forest of Kathiawar in Bombay State, where large numbers are bred primarily for ghee production. The most noticeable feature of the breed is the very prominent forehead and heavy horns which droop on each side of the head and turn up at the extremities. The dominant colour is black. The breed is relatively late-maturing, maximum weights being reached about one year later than in the Surti breed.

5. *Nagpuri.* This breed is found mainly in central and southern India. The most noticeable feature is the characteristic formation of the long sweeping angular horns. The animals are of light build and possess comparatively fine limb bones and smaller feet. Some authorities regard this breed as being poor for milk production and mainly used for draught, but Kothavala (1935) considers that the Nagpuri shows the best combination of milk and draught qualities.

Kartha (1959) draws attention to the fact that the water buffalo has a remarkable capacity to adapt itself to extreme climatic conditions. He states that the better class water buffaloes are mostly confined to the Punjab where summer temperatures reach 46°C (115°F), and winter temperatures drop to 4°C (40°F), or even lower. In India the poorer types of water buffaloes are found in the wetter areas where temperature variation is not so great. In such places, the water buffaloes are preferred to taurine cattle because of their better milking capacity and their remarkable performance for draught in areas where bullocks would find it difficult or impossible to work efficiently. Kartha points out, however, that the buffalo cannot stand abrupt changes of temperature and that the species requires time to acclimatize itself to varying climatic conditions. It is of interest to note that there have been no major importations of water buffalo into Africa from Asia, although most African breeds of cattle originated in Asia.

The beef produced from water buffaloes has a distinct bluish tinge and the fat is white in colour. Young buffalo steers slaughtered for beef at about two years of age yield very acceptable carcasses. A series of 'palatability dinners' reported by Wilson (1961c) revealed that fresh young water buffalo meat was preferred to the meat of deep-frozen, high quality beef steers on grounds of better flavour and more attractive colour of the fat. The muscular development of the water buffalo is particularly good and the hind quarter is better developed than that of *B. indicus* cattle, and compares favourably with that of *B. taurus* beef stock of similar weight. The water buffalo meat is not well marbled, and fat is laid down but sparingly in both the intramuscular and intermuscular regions. This presumably renders the meat more difficult to cook evenly and well without desiccation (MacGregor, 1941). The mature buffalo fattens very rapidly and large quantities of subcutaneous and abdominal fat are laid down (Kothavala, 1935). The skin of the water buffalo is thicker than that of cattle and can be tanned into a very tough and durable leather, known in the trade as 'hog hide' or 'hog leather'.

No critical studies of the basal metabolism of the water buffalo are reported, but temperature, pulse and respiration rate are all lower than the equivalent figures for taurine cattle. (Temperature 37°C:

(98·8°F), compared to 38·6°C: (101·5°F) for cattle; pulse rate 40 per minute compared to 50–55 for cattle; respiration rate 16 inhalations per minute compared to 20–25 for cattle; MacGregor, 1941.)

Buffalo bulls are sexually mature at two years of age though even earlier reproductive activity is reported by Hafez (1952). Cows usually breed at 2–2½ years and continue to calve to twenty years of age or more. The gestation period is about 10½–11 months, that for river-type cows being about two weeks longer than that for swamp-type cows. The oestrus cycle is twenty-one days in length and the oestrus period lasts for between three and four days. Bulls frequently become infertile by the seventh year even though sexual desire remains and muscular strength increases well past that age. Water buffaloes are relatively slow maturing and growth frequently continues up to the tenth year of age although it is comparatively slow, and often only seasonal, after the fifth year.

Mature weights for swamp buffaloes are around 1,600 lb for males and 1,100 lb for females. River buffalo bulls have an average adult weight of between 1,200 lb and 2,500 lb varying according to breed, and river buffalo cows average between 1,000 lb and 1,600 lb (Kothavala, 1935).

It has already been mentioned that the milk yields of water buffaloes compare favourably with the records of zebu cattle maintained under similar conditions. Dastur (1956) quotes a range for the average lactation yield as being 150–330 gal. Maule (1953–4) presents data obtained from a survey in seven cattle and buffalo breeding areas in India, and his findings are summarized below.

Table 44. *Data for mean lactation yield of various breeds of water buffalo and zebu cattle in India*

Maule (1953–4)

	Water buffalo (*Bos bubalis*)	Zebu cattle (*Bos indicus*)
No. of records	6160	4310
Mean daily yield (lb)	7·82	3·74
Estimated lactation yield (lb)	2160·4	943·0
Average length of lactation (days)	300	264
Calving interval (months)	18	18·2

Maule (1953–4) provides further data showing the results obtained in well-managed government-owned herds for some of the major breeds of river buffalo. The Nili and Ravi breed topped the list with average yields of 458 gallons per lactation, followed by murrah (358 gallons) and Surti (305 gallons). Exceptional yields as high as 800–1,000 gallons/lactation have been recorded in some instances, and lactation yields of 600 gallons are not uncommon.

Dastur (1956) has reviewed the information on buffalo milk quality. The most important feature is the comparatively high butterfat percentage which averages 6·7 per cent but may exceptionally reach 15 per cent. The solids-not-fat percentage averages between 9 and 10 per cent, whilst the mineral content is similar in most respects to that of the milk of zebu cattle.

Maule (1953–4) refers to the widespread use of water buffaloes in India in 'town dairies' such as those supplying milk to Bombay. In the Bombay scheme 15,000 buffaloes are maintained in units averaging 300 cows each. Each unit has about 50 acres of grass for exercise and the provision of some forage, but the buffaloes are mainly intensively fed. Their milk is processed at a central milk depot where it is 'diluted' with skimmed cows' milk to give a standard product averaging between 3·5 and 4·0 per cent butterfat and 9 per cent solids-not-fat.

DROMEDARIES, LLAMAS AND ALPACAS

These three types of domesticated livestock belong to the family *Camelidae*. The camels belong to the genus *Camelus* whilst the smaller, South American llamas and alpacas belong to the genus *Auchenia*. There are two species of camel, the two-humped, temperate Bactrian camel (*C. bactrianus*) and the single-humped, tropical camel, more correctly known as the dromedary (*C. dromedarius*).

The llamas and alpacas of South America are mainly confined on the 'altiplano' of the Andean mountain range, at altitudes averaging over 10,000 feet. They are small-bodied animals, the trunk and legs roughly resembling large, long-legged goats, but they possess a long neck so that the head is borne about two feet away from the shoulders. Llamas are used as pack animals (they are rarely used to pull wheeled implements or carts) and the wool is spun and woven into blankets and wearing apparel. The animals are capable of withstanding very great extremes of diurnal variation in temperature (ambient temperatures ranging from below freezing to above 27°C (80°F) in the short space of 24 hours) and they can maintain themselves on sparse, xerophytic vegetation high in fibre and low in digestible nutrients. The total world population of llamas and alpacas is not known with any certainty but probably does not exceed one million.

The dromedary is an important domesticated animal of the tropics and its wider distribution and larger total population (about 8 million) demands that it should receive reasonable attention in any textbook on tropical agriculture. Unfortunately this demand is seldom met, possibly because of the dearth of scientific literature on the subject, even though its temperate relative, the Bactrian camel, has received a great deal of attention from Russian authors.

Two distinct types of dromedary may be recognized, although there are no true 'breeds' and they are often named after the tribes which breed them. On the one hand there is the heavy, thick-boned, slow-moving 'baggage camel' which is used as a pack animal. (See Plate 35.) On the other there is the sleeker, longer-legged, fine-boned 'riding camel'. Both types are less heavily built and possess longer legs than the Bactrian camel, and both have a much softer and thinner hairy coat. Dromedaries are well adapted for hot, even arid, climates but they rapidly lose condition in the humid tropics and are particularly susceptible to attacks by biting flies which are common in the wetter regions of the tropics.

Most of the world's population of dromedaries is to be found in North Africa, the chief concentration (1·8 million) being in Somalia. Asia has about 2·8 million dromedaries, and a few isolated herds exist in places as far removed as Australia (3,000), the Canary Islands (about 5,000) and, up to the end of the last century, in Cuba, Jamaica and Texas, U.S.A. In North-east Africa the southern limit of the breeding area of the camel is said by Mason and Maule (1960) to be the 15°N parallel, although herds are common in the Northern Frontier Province of Kenya, adjacent to Somalia. The distribution in the Horn is limited to the Somalia areas, approximately east of the 41st meridian. In most of this area the mean annual rainfall is less than 350 mm. (12.8 in.).

The dromedary is thought to have evolved in South-west Asia, possibly in Persia or Arabia. There are abundant accounts of the species in very early Egyptian, Jewish and Greek writings. Its spread along the North African coast is due to the Arab migrations across the deserts to the west coast of Africa, with incursions across the Straits of Gibraltar and southwards into what is now the Northern Province of the Federation of Nigeria. Muhammad was a camel lover, and there are many references in the Koran and other Islamic writings to the virtues of the camel, for instance, 'God has created no better animal than the camel'; and again, 'At the third blast of the trumpet on the Day of Judgment the truly faithful will be borne to Heaven on winged camels as white as milk, with saddles of fine gold'.

The dromedary is a fatty humped, hornless ruminant. The upper lip is divided, and is a very sensitive and motile organ, capable of performing amazing feats of labial dexterity. The dentition is most abnormal, the dental formula being:

$$\frac{|\ 1:1:3:3}{|\ 3:1:2:3}$$

The young camel calf has three temporary incisors on the upper jaw, but only the third of these is replaced by a permanent tooth when the milk teeth are lost. The canine teeth are long and pointed, and are used with great effect as an offensive weapon.

The dromedary, even as a young calf, grows horny callosities instead of skin under the sternum, and at the elbows, knees and stifles. This fact once gave rise to much debate between biologists, as it was quoted as partial evidence for the hypothesis of the inheritance of acquired characters. The feet consist of two digits united by a single, horny 'sole' common to both. The feet are very large in area, a fact which enables the dromedary to travel across open deserts without the feet sinking into the loose sand. The gall bladder is absent in the camel – a curious anatomical deficiency which this species shares in common with the horse.

Male dromedaries possess a pair of 'poll-glands' which omit a strong smelling liquid in the rutting season. The opening of the sheath is very small and points backwards, so that the male animal urinates backwards between the hind legs. It is not wise to stand directly behind any large domestic animal, but this is another good reason for not standing close behind a male dromedary.

The dromedary cow has an udder divided into four roughly equal quarters, each bearing a teat, similar to taurine cattle. The gestation period varies between 370 and 375 days, but females only breed every second year, so that calving is strictly seasonal and biennial. The lactation period varies according to the nutrition of the animal, but commonly extends for as long as eighteen months. Dromedary cows may be mated at four years of age, calve when they are five, and can continue reproducing in alternate years until they are twenty or so years of age. A good cow, therefore, is capable of bearing eight calves in a lifetime, but the average is much less.

The dromedary is essentially a bush-browser, but stock will graze on grasses if no shrubs or trees are available. The usual behaviour-pattern for the dromedary is to feed and rest during the day, and to travel at night. Desert caravans often move during the coolest part of the night, from 2 a.m. to 8 a.m. The ability of the dromedary to go for long periods without water is well known. In Somalia, dromedaries moving over the desert routes are generally provided with water once every 3–7 days; in the Sudan stock are watered once every 3–6 days;

in Arabia once every 3–4 days, whilst Algerian dromedaries are generally provided with water once every 2–3 days under similar conditions. A well-documented and classic endurance record for the dromedary was set up in Australia during the year 1891–92, when a troop of soldiers, under Tietkins, travelled with a caravan of dromedaries across 537 miles of desert in thirty-four days without water. A quarter of the caravan survived!

During the northern winter, or rainy season, the moisture content of the trees and bushes browsed by the dromedary is quite sufficient to supply the total water requirement of the animal, and water may not be drunk at all during this period. In the dry season the vegetation is desiccated and extra water, over and above winter requirements, is required for heat regulation. Adult working stock require on average between 3 and 10 gallons of water per day under dry season conditions.

The misconception about the hump of the camel being a water-storage organ is so commonly quoted that it is necessary to restate the fact that the hump of the dromedary, like the hump of thoracic-humped zebu cattle, is composed of fatty-tissue and is a form of energy reserve. Fat provides more energy per unit weight than any other foodstuff or reserve tissue, and hence fat is the most economic form of energy reserve. The amount of fat contained in the hump of a dromedary averages about 40 lb, and on complete oxidation this could produce about the same weight of oxidation water. The total fat contained in the hump would therefore only be sufficient to supply a dromedary's water requirements for about one normal working day.

The thermal regulation of the dromedary, and the physiological characteristics which result in such a high level of water economy, have not yet received sufficient critical scientific study. There is a great need for more research in this field. However, the following points summarize the most important factors which make the dromedary one of the most efficient desert animals yet studied:

1. *Insulation.* The thick, wool-type hair, the thick hide and the deep layer of subcutaneous fat effectively insulate the vital body tissues from the radiant heat-load received from the sun.

2. *Body temperature range.* The dromedary has an extremely wide diurnal variation in temperature. Normal early morning temperatures average about $35 \cdot 6°C$ ($96°F$) and normal midday temperatures average about $41°C$ ($106°F$). The dromedary thus has a daily temperature range of about $5 \cdot 4°C$ ($10°F$), compared to about $0 \cdot 6°$–$1 \cdot 2°C$ (1–$2°F$) in most other mammals.

3. *Dehydration of body tissues.* The dromedary can lose up to about 40 per cent of its body water before pronounced physiological disturbance, resulting in 'explosive' increases in body temperature,

occurs. Most mammals, including man, die after the loss of body water has reached a critical value of about half this amount.

4. *Water drinking capacity.* When most mammals have experienced severe thirst and are given access to unlimited quantities of water, they can only drink small quantities comparatively slowly or they suffer a physiological disturbance known as 'water intoxication'. The dromedary, under similar conditions, is able to drink large quantities of water very rapidly without experiencing ill-effects. There is a record of a dromedary, 650 lb in liveweight, drinking 103 litres of water in less than ten minutes. The water thus drunk is evenly distributed in the vital body tissues in less than two days, and there is no excessive loss due to excretion in the urine. Most mammals take several weeks to re-hydrate their vital body tissues after periods of severe desiccation.

5. *Maintenance of appetite during conditions of thirst.* Most mammals lose their appetites completely when experiencing severe thirst. The dromedary does not do so, and is capable of maintaining its food intake at reasonable levels, providing food is available, in spite of pronounced thirst and consequent dehydration of its body tissues.

6. *Low respiration rate.* The normal respiration rate of a resting dromedary is 5–8 respirations per minute, one of the lowest respiration rates for any mammal; consequently the evaporative water losses from the buccal cavity and upper respiratory tracts are reduced to a minimum.

The dromedary calf is exceptionally delicate in comparison to the hardiness of the adult animal. Large numbers of calves die before attaining three weeks of age. Many stock owners consider that the rich colostral milk is responsible for early digestive disorders, and it is a common practice to 'ration' the calf to only one quarter of the udder, the other three teats being milked by hand and the milk used for human consumption. No data for the composition of the colostrum of the dromedary are available, but the average percentage composition of bulked milk is: water 86·9 per cent; lactose 5·8 per cent; casein 3·7 per cent; butterfat 2·9 per cent; minerals 0·7 per cent.

The dromedary is chiefly used as a pack and riding animal, but the meat is eaten by Arab peoples, that of young castrated males being very fine in texture and flavour. The milk is frequently drunk by man, most dromedary cows producing yields well in excess of the requirements of their calf providing nutrition is adequate. The hair possesses qualities of durability, is exceptionally fine and light and is used for weaving blankets and cloth. The dromedary hide forms a rather poor-quality leather, which is usually used for making saddlery. The long leg bones are used for making tent-pegs,

Cattle

and the dung of the dromedary, which is exceptionally dry, is used as a fuel. Finally, the ticks which commonly infest the skin of the dromedary are hand-picked and used as a food for falcons – falconry being a major sport of many Arab communities.

GOATS

The F.A.O. estimate that the world goat population is of the order of 325 million, of which approximately two-thirds are distributed in the tropics. A very rough breakdown of the distribution in the tropics is shown below:

Table 45. *Approximate distribution of tropical goat population*

(millions)

Africa	91
India, Pakistan, Ceylon	67
South America	16
North America	7
Asia (east of India)	8
Asia (west of India)	3
Total	192

All domestic goats belong to the genus *Capra*, but the systematists are undecided as to whether all goats belong to the same species or whether different species and subspecies should be recognized. The origins of most domesticated goats were in eastern Europe, the Middle East and western Asia. Probably all the breeds known today are descended from *C. aegagrus*, although certain of the less common wool-bearing types of goat may be related to the wild species *C. falconeri*. Williamson and Payne (1965) classify goats according to their function, and they recognize meat, milk and wool-type goats. This classification breaks down when goats are used for two or more purposes, and it is thought that a more helpful classification is one based on the recent origin of the breeds. Such a system of classification recognizes four major goat-types, European, Oriental, Asiatic and African.

European type

Most of the European-type goats originated in central Europe, particularly in the mountainous regions of Switzerland and Austria. Four of the more important individual breeds, organized into breed societies in many different countries both in and outside Europe, are:

1. *Toggenburg*. The home of this breed is in North-east Switzerland.

It is a large animal, distinguished by its fawn or chocolate colour with white or cream stripes and markings. The ears are generally dark in colour but are marked with characteristic white edges. Both sexes are polled. In common with other European breeds, the ears are borne erect and many individuals possess a pair of tassels – small pendant outgrowths of skin – below the lower jaw at its junction with the neck.

2. *Saanen.* This breed originated in west Switzerland. The animals are white, light grey or cream in colour and the breed is often named by its colour in countries other than Switzerland, such as 'Netherland White' in Holland and 'White German' in Germany. This breed is very popular in tropical and sub-tropical countries, and has been widely distributed.

3. *Alpine.* This is a very handsome animal, primarily black in colour but bearing white, cream or fawn markings which enable it to be very readily recognized. It is widely distributed throughout Europe although its home is in the Swiss and Austrian Alps. Breed societies have been formed in many countries, and it is variously described as the French Alpine, Italian Alpine, British Alpine, etc.

4. *Old English.* This breed is rough-coated and horned. It is said to be indigenous to Britain, possibly related to the wild Welsh goat. Old English goats provide mascots for several British army regiments, and it is said that the breed has been distributed, albeit in a limited fashion, in those territories, including tropical countries, where British regiments have been stationed. The breed is variously coloured, and milk yields are said to be much lower than for the other breeds mentioned above.

Oriental type

This type of goat originated in the eastern Mediterranean and the Middle East. It is more suited to drier, arid conditions and for this reason it has now become widely distributed throughout the drier tropics. It has been extensively crossed with European breeds, the heat tolerance of the Oriental combining with the productive capacity of the European to form a well adapted, tropical, dual-purpose goat much sought after in many continents. The two most important Oriental breeds are:

1. *Nubian.* The home of this breed is the eastern Mediterranean and North-east Africa, but especially in the Sudan where Chalmers (1954) estimates there may be a population of about $2\frac{1}{2}$ million Nubian goats. It is a polled breed, with a convex facial profile. The ears are long and pendant or 'drop-eared'. The colours are various, black being perhaps the most common, but dark and light markings are frequently found on the same animal. It is a very hardy breed, capable of standing up to

very harsh conditions and of living on fibrous, xerophytic vegetation for long periods.

2. *Angora.* This breed originated in Turkey but has now been widely spread into other dry, tropical areas but especially the Republic of South Africa (where is is known as the 'sybokke') and into Texas (where it is called the 'mohair goat'). The breed is white in colour, and bears long lateral horns in both sexes. The hair or wool is very valuable as it is strong and takes dyes very readily; it commands a high price as a high quality yarn. The fleece, averaging 30·5 cm (12 inches) in length, hangs in separate and well defined ringlets. It may be regularly combed out (in which case the annual yield of mohair is between 3 lb and 12 lb per annum) or it can be shorn off, in which case the fleece varies more in length but the yield can be increased up to a maximum of about 16 lb per annum. The average 'mohair' yield in Texas, where very large herds of angora goats are kept under systems of extensive management, is about 8 lb per annum.

Asiatic type

About half the total world population of goats is found in Asia, divided almost equally between the tropical part of the continent (especially India and Pakistan) and the temperate part to the north or at higher altitudes. The most important breed is the Kashmir or Kashmiri, but numerous other Asiatic breeds are recognized, many taking district or provincial names which they share in common with breeds of cattle and water buffaloes. Three Asiatic breeds will be briefly referred to below:

1. *Kashmir or Kashmiri.* This breed originated in Central Asia. It is white in colour, occasionally black and white, and the breed is much prized for its long, fine hair. The goats are double coated, the outer coat containing hairs up to 12–13 cm (4 or 5 inches) in length, and an inner coat of high quality with very short hairs averaging only 2·5 cm (1 inch) in length. The inner coat yields 'cashmere hair' (usually spoken of as cashmere wool) average crops being as low as $\frac{1}{4}$ lb to $\frac{1}{2}$ lb per goat per annum.

2. *Jumna Pari.* This is a large animal with a characteristic convex face, a rather foreshortened muzzle and long, lop ears. It is primarily a milch goat, although it is also used for meat in certain areas. The average liveweight of the males is 175 lb and females average 120 lb. Kaura (1943) provides data for the average lactational performance of this breed, but the figures are not very impressive compared to those of European type goats, being of the order of 54 gallons per lactation with a maximum recorded yield of only 124 gallons.

3. *Cutchi.* This is a long-legged, meat type goat found in West India

and particularly in Bombay. The average weight of full grown slaughter-stock is about 80–90 lb, substantially less than that of the Jumna Pari breed, and the killing-out percentage averages around 45 per cent, although percentages in excess of this figure have been achieved in well grown and well proportioned animals.

African type

Mason and Maule (1960) suggest that a useful subdivision of the African type goats may be based on the relative length of the ear. They classify the goats of East, Central and South Africa as being either long-eared or short-eared. Other authors, such as Epstein (1953) place more importance on the size of the animal, and recognize the African dwarf goats as forming a separate group from the normal-sized types. Williamson and Payne (1965) as has been noted already, classify goats solely as milk or meat types. This is a difficult distinction to make in Africa, where the same breed of goat may be milked by one tribe but never milked by another because of social custom and religious taboo. In this chapter we will merely note in passing a few of the more distinctive African type breeds.

1. *Benedir*. This breed is found in the southern half of Somalia, but north of the River Juba. They are large in size and diverse in colour, but red or black spotted goats are common and white goats are rare. The coat is usually short and smooth but sometimes long and coarse. The ears are long and pendant with turned up tips. The horns are so placed that they point backwards, and run parallel along the back of the animal. The legs are long and the skin is thick. Rosetti and Congiu (1955) subdivide the breed into Bimal, Garre and Tuni varieties.

2. *Galla or Somal*. This breed is found in the Northern Frontier Province of Kenya and adjacent areas. It is white or off-white in colour, and is a meat type. It does not thrive in wetter areas and is rarely found in the more temperate areas adjacent to Mount Kenya.

3. *Nigerian*. This is a conglomeration of large goat breeds found in West Africa, and subdivided by some authorities into a large number of sub-breeds known by their colour and district, e.g. Bornu white; Damagaran dapple grey; Kano brown and Red Sokoto. The breed is primarily a meat breed but certain types are also milked.

4. *African dwarfs*. One group of these goats is said by Epstein (1953) to be a genetic strain possessing a gene for recessive pituitary hypoplasia which results in the adult goat weighing about 35–65 lb at maturity, instead of the normal liveweight for an African goat of 100 lb or more (see Plate 38). Mason and Maule (1960) query the use of the term 'dwarf' for the small goats common to East Africa (see Plate VII) and described in detail by Wilson (1958a, 1958b and 1960) whilst

presumably allowing the term for the very small 'true dwarfs' found in West Africa (see Plate 39). Another type of African dwarf goat is the 'disproportionate dwarf' of West Africa, possessing the genes for recessive achondroplasia which cause stunting of the leg-bones (as in the Dachshund breed of dog).

5. *Boer*. This South African breed was evolved by the Dutch settlers from local Bantu goats, possibly with the addition of some European and Asiatic blood. Van Rensburg (1938) describes this breed as compact, well proportioned and short-haired. The facial profile is straight, the ears vary greatly in size but are usually long and the horns project backwards from the head. The meat is coarse-grained but soft and palatable, and in some areas the Boer is also an important milk producer. The skin is regarded as a valuable by-product.

Many types of goat now found in the tropics are crossbreds, formed by breeding highly productive European-type goats with heat tolerant breeds, such as the Nubian. The best known example of a stabilized crossbreed is the Anglo-Nubian, thought to contain Nubian, British and also possibly Jumna Pari blood. This cross was first made in the nineteenth century, and Anglo-Nubians are now very widely distributed in the tropics and subtropics. In the U.S.A. they are highly favoured and the term employed for the breed is 'Nubian', which must not be confused with the true Nubian breed of Oriental type. Anglo-Nubians may be of almost any colour. They possess short, fine coats, a convex facial outline and the long, drop-ears of their Nubian ancestor. They are extremely good milkers, yielding a milk very high in butterfat.

The composition of goats' milk compares very favourably with that of cattle and water buffaloes. The average butterfat content is about 6·0 per cent, higher figures being obtained towards the end of the lactation. The casein (milk protein) percentage averages 3·3 (compared to about 3·0 per cent for the cow) and the albumen content is about 0·7 per cent (compared to 0·4 per cent for the cow). The milk of the goat is in some respects closer in composition to human milk than the milk of the cow, and for this reason goats' milk is an excellent baby food. It is slightly alkaline in reaction whilst cows' milk is slightly acidic. The fat globules are small in size, and are digested in the human stomach about four times as fast as are the larger fat globules of cows' milk. Finally, the milk from the milch goat has a much higher degree of initial cleanliness than milk from dairy cows. This is due to the fact that the faeces of the goat are pelleted and hence there is little or no contamination of the udder with bacteria of the *Bacillus coli* types and also because tropical goats are not so commonly infected with tuberculosis as are tropical dairy cattle.

Goats lactate well in the tropics when compared to cattle or water buffaloes weight for weight, but there is a very great difference in milk yield between breeds and between strains within a breed. The best lactation yields are obtained by European-type goats in the cooler tropics and by crossbred goats, such as the Anglo-Nubian, in the hotter tropics. These breeds can average 100 gallons per lactation with a range extending up to about 250 gallons, compared to zebu cattle and water buffaloes which average about 150 gallons with a range extending much higher (sometimes approaching 1,000 gallons). The following table compares the average performance of goats with cattle and water buffaloes on a weight basis.

Table 46. *Comparison of milk yield of tropical milch animals on the basis of liveweight of animals*

Type of stock	Average lactation yields (gallons)	Average live-weight (lb)	Yield (gallons) per 100 lb weight
Goat			
African and Asiatic	50	100	50
European crossbred	100	110	91
Purebred European	150	120	125
Cow			
Unimproved zebu	100	800	13
Improved milch zebu (Africa)	250	800	31
Improved milch zebu (Asia)	500	800	63
European crossbred	500	1,000	50
Water buffalo			
Unimproved breeds	250	1,100	23
Improved breeds	400	1,100	36

As food intake is closely related to liveweight it will be seen that the goat is a comparatively efficient converter of stock feed into milk in the tropics. In areas where food is in short supply and animals must fend for themselves, the goat is often able to live and lactate in areas where there is an insufficiency of food for larger types of farm livestock.

Naturally the highest lactational records for milch goats are held by European-type goats maintained in temperate climates. Such records frequently exceed 650 gallons/lactation and yields of over 700 gallons are occasionally recorded. The lactation yield usually increases from the first up to the third lactation, and thereafter declines slightly at each succeeding lactation. The maximum daily yield is not reached until between the eighth and twelfth weeks of the lactation, which results in the lactation curve of the goat having a much flatter appearance when compared to the curve for dairy cows which has a sharper peak, with the maximum daily yield obtained between the fourth and eighth weeks.

The gestation period of the goat is about five months in duration. That for European type goats is generally given as 150 days, but slightly shorter gestation periods (147 days) have been recorded for smaller, African type goats. Mating takes place early, sexual maturity being reached soon after the sixth month. The oestrus cycle is three weeks in duration, and the oestrus period lasts for about forty-eight hours. The female goat will accept service from the male whilst pregnant, and this action occasionally leads to abortion. It is therefore advisable to separate pregnant females from the male whenever possible (Wilson, 1957a).

In spite of the obvious importance of the goat in many parts of the tropics, there is a great deal of controversy as to whether the goat is mainly beneficial or mainly detrimental from the standpoint of its effect on tropical ecology. On the one hand, writers such as Maher (1945) condemn the goat as being the major cause of deforestation and hence of erosion. Examples of drastic ecological change are quoted in support of this viewpoint from East Africa, Cyprus, southern Italy, St Helena, Ethiopia and Palestine. On the other hand, workers such as Staples *et al.* (1942), Hornby and Van Rensburg (1948) and Wilson (1957a) have shown conclusively that the goat, because of its browsing habits, can be extremely useful in certain tropical regions by preventing or restricting bush encroachment. The encroachment of bush into grassland areas is another highly undesirable ecological change common to many regions of the tropics, and it is primarily brought about by continuous overgrazing by cattle. Hornby and Van Rensburg (1948) have stated: 'Whenever in bushland country there is some sort of ground cover of grasses and herbs which can be expected to extend if given a chance, such an area may be improved even by heavy goat browsing, whereas the same area would be damaged by anything more than the lightest grazing by cattle.' Staples *et al.* (1942) reported the results of an experiment in which goat browsing, cattle grazing, and the effects of mixed stocking by both cattle and goats, were compared over a four-year period. At the end of the experiment, the cattle plots had developed a flora consisting of trees and open thicket, 1·2–3·7 m high (4–12 ft). There was only a very poor grass cover, and there was much bare ground. The goat plots, on the other hand, developed a flora consisting principally of grasses, with occasional grazed bushes, between 0·6–1·5 m (2–5 ft) in height. The percentage of bare ground in the goat paddock was very low. It was concluded from these experiments that goats were not so responsible for soil erosion as cattle, since the primary cause of soil erosion is removal of the ground cover of grasses and herbs rather than a reduction in the population of trees and bushes. After cattle have removed the ground cover by extensive overgrazing, the vegetation can no longer support anything except the very lightest stocking by cattle, whilst it can continue to support the browsing goat. The result is that the goat is the animal found on land which is eroding for lack of a ground cover, and the uncritical observer is misled into thinking that the goat has been primarily responsible for the whole ecological change which has resulted in erosion. The tropical goat is thus made the scapegoat for the consequences of overgrazing by cattle and, possibly, by sheep.

The arguments frequently advanced against the practice of mixed stocking by cattle and goats are that goats are difficult to control in numbers and to confine by means of cattle fences and hedges. The former is a criticism of the stock-owner rather than the goat *per se*, and the latter point, though it has much substance, can be partly met by the technique of placing a 'collar' round the necks of the goats, formed by a triangle of sticks, which effectively prevents the wearer from scrambling through, or jumping over, normal cattle fences.

Milch goats are best kept under more intensive conditions, such as by running them in yards with simple protective shelters. In cases where it is necessary to allow the goats access to pasturage or browse as the major item in their diet, it should be noted that the total daily intake of nutrients can be obtained in approximately two to three hours of browsing. This compares very markedly indeed with the equivalent requirement for cattle, since cattle need to graze for a minimum period of about eight hours each day. Allowing for walking and idling time, the total period at pasture for cattle cannot be reduced much below ten to twelve hours without severely restricting herbage intake and hence productive performance.

CHAPTER 15

Adaptation of Livestock to Tropical Environments

The early attempts to export temperate breeds of livestock to the tropics met with dismal failure. After a relatively short time in the tropics, the productivity of many breeds of exotic stock decreased, their condition deteriorated and they were more prone to attack by tropical diseases. It is only during the last two decades that the agricultural scientist has devoted attention to the subject of the adaptation of livestock to hot climates. Many of these studies have been indirectly aided by the military and medical professions, since they are likely to be relevant to the problems besetting the acclimatization of man to hot conditions encountered in tropical warfare. This chapter will seek to review some of the more important agricultural findings which have emerged from this work, with particular reference to the adaptation of bovines to hot climates. The basic principles of thermal adaptation are common to all vertebrates, although the different types of external covering (hair, wool, feathers and bristles) and the different types of underlying skin structure, give rise to variations in the mechanism of heat-loss and the maintenance of homeothermy. The reason for focusing attention on the bovine is that much of the early critical work has been carried out on this genus and because the successful thermal adaptation of cattle is probably of greater importance than the adaptation of any other class of livestock.

A temperate animal taken to a hot climate is affected in two distinct ways: directly by the influence of heat radiation and, possibly, humidity on the animal itself, and indirectly by the effect of heat on the animal's environment.

Direct Effects of Heat on Tropical Livestock

Brody (1945) stated that, 'an animal's main task is to maintain its internal environment constant'. In the case of vertebrate farm animals this 'task' may be reduced to maintaining the internal temperature of the animal constant, or nearly constant. The 'normal' diurnal variation in the body temperature of most farm animals is of the order of 0·6°–1·2°C (1°–2°F). Rises in temperature of more than 1·2°C

(2°F) are indicative of ill health or of poor adaptation to hot conditions. One notable exception to this general rule is the Arabian camel, or dromedary, which has a 'normal' diurnal variation in body temperature of about 5·4°C (10°F). (See Chapter 14 for a more detailed description of the physiology of the dromedary.) This maintenance of an almost constant internal body temperature is known as homeothermy, and it is necessary for the efficient and normal functioning of the brain tissues. When these delicate tissues are overheated, many cells are destroyed beyond repair, and the animal eventually dies.

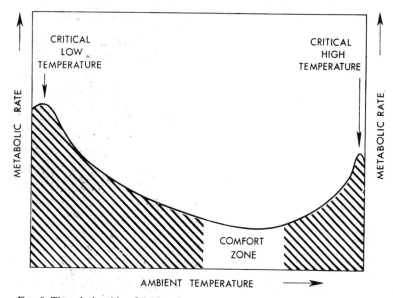

FIG. 9. The relationship of the basal metabolic rate to the ambient temperature in homeotherms, showing the position of the 'Comfort Zone'

Every vertebrate animal has a particular range of environmental temperature to which it is adapted and in which it is able to live most efficiently at a minimal basic metabolic rate. This temperature range is spoken of as the 'comfort zone' and it varies widely from species to species and between different breeds within a species. *Bos indicus* cattle for example, have a comfort zone of 15°–27°C (60°–80°F) whilst *Bos taurus* cattle perform best at lower temperatures, with a 'comfort zone' of 4°–15°C (40°–60°F). When animals are kept at temperatures below or above their comfort zones their metabolic rate is increased, either to keep the animal warm (shivering) or to assist in heat dissipation (panting). This point is illustrated in Fig. 9.

In order to follow the different steps in the maintenance of homeothermy, let us consider the reactions of a cow (whose normal body temperature is approximately 38·6°C (101½°F) exposed to direct tropical sunlight, and to an air temperature of about 43°C (110°F). We will assume that the cow is unable to alleviate the heat stress by seeking shade.

1. The skin temperature will rise. This is a direct result of the external surface of the animal receiving direct sunlight, and, therefore, radiant heat-energy. Because the cow has a layer of subcutaneous fat underneath its hide, there will be a heat-gradient between the outer surface of the animal and the underlying vital tissues. Skin and blood temperatures will differ markedly at this initial stage.

2. The rise in temperature of the skin will have been detected by the nerves which penetrate into the hide of the animal, and messages to this effect will be communicated to the brain. This will set up a chain reaction, involuntarily controlled by the brain, each step of which is an attempt to reduce the heat-stress of the animal. The first involuntary response will be that the animal reduces her activity, thereby lowering her metabolic rate. Walking will slow down or stop; grazing will be reduced or cease.

3. As a reduction in the activity in the animal can only reduce the metabolic heat load and will not enable it to dissipate the extra heat received from the sun, a more complicated series of responses will now take place. The most important of these is that the animal will transpire. Transpiration can take place in two distinct ways: First, sweat can be secreted in the apocrine sweat glands of the animal, and pass up the sweat ducts to the skin surface, from whence the aqueous fraction of the sweat can be evaporated. Secondly, water can be diffused through the epidermal layers of the animal on to the surfaces, from whence it can be evaporated. These layers can be outside the animal (skin), or inside the animal (buccal cavity and respiratory tracts). In practice, it is difficult to determine the relative importance of evaporative losses of 'diffusion water' and 'sweat'. Transpiration is the primary and most important heat-regulating mechanism in all farm animals other than poultry. It can be clearly recognized on animals with ecrine sweat glands secreting large droplets of sweat, such as equines, but it is not so readily apparent in animals having small apocrine sweat glands, such as the cow or goat. (In the case of cattle which are reasonably adapted to a hot climate, the chain reaction will stop at this stage 3.)

4. Assuming that the evaporation of water and sweat from the surface of the body has not restored homeothermy, the animal will evaporate large quantities of water from the upper part of the

respiratory tracts and the buccal cavity. This is assisted by a rise in the respiration rate, which is achieved by increasing the number of respirations per minute with a corresponding decrease in the tidal volume. At this stage, therefore, the amount of air entering and leaving the lungs per unit time is not significantly altered.

5. After a period of high respiration rate, the animal will be tired and depressed. She will probably lie down, but as there is no shade this will serve merely to increase the heat-stress on the animal by exposing a greater body surface to the direct sunlight. From now on each progressive reaction of the animal, instead of alleviating heat-stress, will have the exact opposite effect.

6. The heart rate will rise. This rise in the heart rate is necessitated by the increased respiration rate. The muscles concerned with respiration will require an enhanced oxygen supply because of their increased activity. This will be achieved by increasing the blood supply through increased activity of the heart. The higher pulse will increase the muscular heat produced by the animal itself and this will lead to a further warming of the blood. The animal is now overheated both by external causes and by internal causes.

7. The true body temperature will commence to rise. The true body temperature is generally taken to be the temperature of the blood leaving the heart.

8. A little later, due to the lagging effects of the contents of the alimentary canal, the rectal temperature of the body will also rise. Generally, the rectal temperature will be between $0.1°$ and $0.3°C$ ($0.2°$ and $0.5°F$) less than that of the true heart temperature measured in the dorsal aorta (Bligh, 1955).

9. The cow is now in a condition of marked distress. The true body temperature and the rectal temperature continue to rise. Eventually the 'tidal volume' will increase. In other words, the amount of air entering and leaving the lungs per unit time will become greater than normal. This is sometimes spoken of as 'deep panting' or 'second phase breathing'.

10. This second phase breathing will result in a marked increase in body temperature due to the increased muscular activity of the respiratory system. This in turn will result in yet warmer blood being pumped to the brain. Eventually damage of a permanent nature will be done to the sensitive brain tissues and when this has occurred the animal will go into a coma and death will result.

We must regard the evaporative losses of water from the body surfaces as being the most fundamental of all the processes concerned with heat regulation. For every kilogram of water evaporated from the animal, either from the external skin surface or from the mouth and upper respiratory system, 580 kilocalories of excess heat will be lost.

At low environmental temperatures, non-evaporative cooling (i.e. loss of heat due to conduction, convection and radiation) is responsible for more heat-loss than evaporative cooling. At freezing point, about 75 per cent of the heat lost is lost through non-evaporative cooling. As

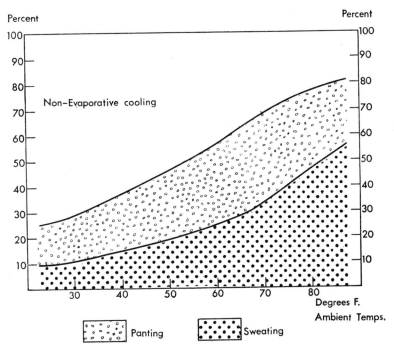

FIG. 10. The proportion of non-evaporative to evaporative cooling in cattle, at different ambient temperatures

the environmental temperature rises, the proportion of evaporative cooling rises, and the proportion of non-evaporative cooling falls. At environmental temperatures around 32°C (90°F) about 80 per cent of the heat-loss is due to evaporative cooling, and only 20 per cent to all other factors. These changes in proportions of non-evaporative and evaporative cooling are illustrated in Fig. 10.

Early research workers were sceptical about the importance of sweating in the heat tolerance mechanism of the bovine. Worstell &

Brody (1953) claimed that cattle, like rabbits, do not sweat and could only evaporate water from the mouth, tongue and upper respiratory tracts. Dowling (1958) indicated the importance of sweating by simple experiment. He covered some Australian shorthorn cattle almost completely with polythene bags, which effectively prevented any vaporization of moisture from the body surface of the animal apart from the legs and head. He found that the rectal temperature of animals treated in this fashion rose significantly above the temperature of unbagged, control animals, showing that if the animal is unable to lose heat by surface evaporation, the body temperature will rise significantly. Another demonstration of the importance of sweat glands has been given by Brook and Short (1960a and b) working with merino sheep. Four ewes were found congenitally lacking sweat glands. The temperature response of these abnormal ewes was compared to that of normal sheep in a psychometric chamber at an ambient temperature of 40°C (104°F). It was found that the temperature of the ewes without sweat glands was about 1°C (2°F) above the corresponding temperature of the control group.

SWEAT GLANDS

The sweating mechanisms of most farm animals are relatively inefficient when compared to that of man. This is borne out by Table 47, in which statistics relating to the efficiency of the sweat glands of sheep, European cattle and man are compared.

Table 47. *Approximate values for efficiency of sweat glands of sheep, cattle and man*
Brook and Short (1960)

Animal	Sweat glands		Rate of sweating		Ratio of sweat output per gland/hour to gland volume ($mg/hr/mm^3$)
	No. per cm^2	Volume of secretory part (mm^3)	Sweat per gland (mg/hour)	Total sweat ($g/m^2/hr$)	
Sheep	290	0·004	0·01	32	2·5:1
Cattle (*B. taurus*)	1,000	0·010	0·06	588	6·0:1
Man	150	0·003	1·3	2,000	433:1

A diagrammatic sketch of an apocrine sweat gland of a cow is shown in Fig. 11. Sheep possess similar apocrine glands. It will be seen that each sweat gland is relatively deep seated, and is closely associated with a hair fibre. The histology of the apocrine sweat gland has been studied and reviewed by Nay (1959). Nay grouped the sweat glands found in Australian cattle into three distinct types:

1. *Tubular coiled glands*, of variable length, but with small diameters (found usually in imported European cattle).
2. *Baggy glands*, of variable length but with large diameters (found usually in imported zebu cattle.)
3. *Club-shaped glands*, of more constant length with a wide, lower end and a narrow, semi-coiled, upper end (found usually in crossbred *B. taurus* × *B. indicus* cattle).

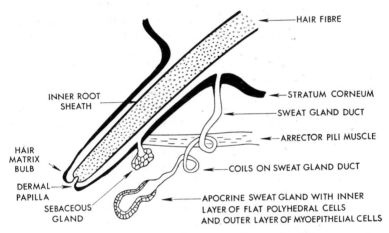

FIG. 11. Diagram of a section through the skin of a crossbred *Bos taurus* × *Bos indicus* cow, showing the relationship of a hair follicle with its associated apocrine sweat gland

Nay indicated that the ranking of different breeds of Australian cattle for heat tolerance agreed closely with their ranking for mean sweat gland volume. Baggy glands had the greatest volume, followed by club-shaped glands, with the tubular coiled glands having lowest volumes. He inferred that the volume of sweat gland of the bovine could be taken as an indication of potential heat tolerance. However, the study of 14 different breeds of *B. indicus* cattle by Walker (1960) suggests that the rate of sweat secretion is a better index of heat tolerance than the volume of the sweat gland. This point may be illustrated by comparing the similar sweat gland volumes, but vastly different sweating rates, of sheep and man in Table 47.

The duct of the apocrine sweat gland is relatively long and narrow, and in general the structure cannot be regarded as a very efficient means of secreting a liquid capable of evaporation, and conveying it to the surface of the animal. Moreover, as each sweat gland is associated with a hair follicle, any increase in the sweat gland population is accompanied by a corresponding increase in the number of hairs. The nature and the length of the hair fibre will also influence the sweating efficiency of the animal. This point is dealt with in greater detail below.

The number of sweat glands present on an animal is probably fixed at birth, though some workers claim that there is an increase up to the third year of age. However, it is probable that the efficiency of each apocrine gland may increase throughout the life of the animal. If the number of glands is fixed at birth, it follows that the stretching of the skin during postnatal growth will tend to reduce the sweat gland population per unit area as the animal develops.

The number of sweat glands per unit area at any given stage of growth is not constant over the whole surface of the animal. Some parts of the body are better provided with sweat glands than others.

Data for the different quantities of water evaporated from the surface of purebred and crossbred animals reveal most startling differences. At 42·1°C (108°F) at a relative humidity of 28 per cent, the water evaporative loss of crossbred zebu × Jersey cattle was 320 g/cm²/hour; the rectal temperatures were normal. By comparison, purebred Ayrshire cows, kept under the same environmental conditions, exhibited a water evaporation loss of only 116 g/cm²/hour, and the rectal temperature of these European cattle rose to 40·2°C (104·6°F).

When cattle are exposed to high temperatures the chief cause of heat-loss from the animal is by the evaporation of water from the internal and external body surfaces. Water evaporation is not however the only means by which animals may lose heat. The following equation, often spoken of as the 'homeothermic equation', indicates all the various ways by which an animal may lose or gain heat from its environment:

$$M - E \pm F \pm Cd \pm Cv \pm R = 0$$

where M = the metabolic heat produced by the living animal;

E = the heat lost through evaporation of water from the body surfaces;

F = the heat lost or gained by the consumption of food which is either hotter or colder than the body temperature;

Cd = the heat lost or gained by conduction (*i.e.* the transfer of heat energy from particle to particle by increased molecular activity);

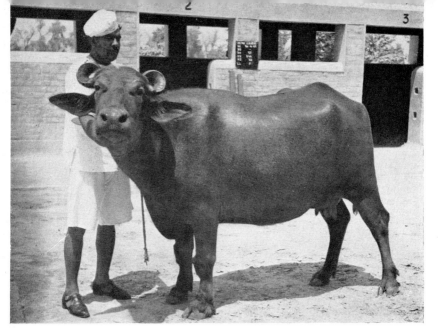

33 Ravi type Water Buffalo cow, at the Okara Military Farm, West Punjab, Pakistan. This cow gave over 600 gallons of milk in her first lactation. Note medium size, deep frame and small coiled horns. (see Chapter 14)

34 Crossbred Murrah type Water Buffalo steers emerging from a wallow at the University Field Station, St. Augustine, Trinidad. These steers were bred at Caroni Sugar Estates, Trinidad. Note large frame, relatively short back and curled horns. These steers averaged 850 lb. weight at 18 months; and were gaining weight at $1\frac{3}{4}$ lb./day when the picture was taken. (see Chapter 14)

35 Baggage Camel, or Dromedary, used for farm transport in the Canary Islands. Note the height of the animal compared to that of the farm worker. The wooden box fits over the single hump, and very heavy loads may be carried provided due attention is paid to correct balance. (see Chapter 14)

36 Mule, imported from the U.S.A. into the West Indies, being used for light farm transport at the University Field Station, St. Augustine, Trinidad. Mules are extensively used throughout the Caribbean region, especially for the transport of sugar cane. (see Chapter 14)

Cv = the heat lost or gained by convection (*i.e.* the transfer of heat energy by a circulation of heated material, usually air);

R = the heat lost or gained through radiation (*i.e.* the transfer of energy across space without heating the space through which it passes).

Where the environmental temperature is above the normal body temperature, a cow can clearly not lose heat to its environment by convection, conduction or by radiation. In fact, the animal will be gaining heat by the radiant energy received from the sun, by the convection currents of warm circulating air and by conduction of heat to the animal from its hot surroundings.

The animal may possibly lose heat if, during the period of heat-stress, it ingests large quantities of water at lower temperature than the body of the animal, such as river water. Normally, however, water which has been standing in an exposed trough or shallow pond will be at or near body temperature in the hot tropics.

It is pertinent to add a comment at this stage concerning the various ways by which the animal is affected by the radiation received from the sun. Radiation may be divided into infra-red or heat radiation (long wave lengths); visible light (medium waves); and ultra-violet radiation (short waves). The proportion of long and medium waves in the total radiation will increase as one moves from the poles towards the thermal equator. The proportion of short waves present in the radiation will increase as one moves from sea level to high altitudes. It follows that the greatest radiation intensity is found on farm animals kept at a high altitude on the thermal equator.

Most of the infra-red radiation received by the animal will be absorbed by the skin. The characteristics of skin colour, hair length and hair colour will influence this absorption, since light, sleek hairs will enable a proportion of the infra-red radiation to be reflected. The visible light falling on the animal can be more readily reflected, the amount also depending upon the physical characteristics of the coat surface.

DEFINITION OF HEAT TOLERANCE

Now that we have considered the various ways by which an animal receives a heat-load from its environment, and also something of the chain of the events which takes place when this heat-load is recorded by the peripheral nervous system of the animal, we can attempt to define the term 'heat tolerance'.

An old definition of a heat tolerant animal is 'one which eliminates large amounts of excess heat and allows productive processes to continue at a high level at high air temperatures'. In other words, emphasis used to be placed upon the elimination of excess heat by animals kept in hot environments.

This definition is now considered to be too narrow, and modern research workers would define a heat tolerant animal as 'one which has a high efficiency of energy utilization and allows productive processes to continue at a high level without the production of excessive amounts of heat'. In other words, emphasis is now placed more upon the first term (M) in the homeothermic equation. The greater the amount of metabolic heat produced by the animal itself, the greater the need to eliminate this heat from the animal by some means or other. Conversely, at lower basic metabolic rates less heat is produced. The result is a greater adaptation of the animal to its environment.

It should be noted that both definitions place emphasis upon the continuance of productive processes. This is most important. We cannot describe the animal as being heat tolerant if it lives in a tropical environment but fails to produce calves, milk or liveweight increase. Unfortunately, most attempts to measure heat tolerance by some means or other usually have failed to take into account the parallel measurement of the animal's production. There is a very real need to devise a new index, easily measured, which incorporates an objective assessment of productivity.

HEAT TOLERANCE INDICES

The first heat tolerance index was evolved by Rhoad (1944) at the Iberia Research Station in the U.S.A. This is a simplified field test, only the rectal temperature being measured, with no account taken of the animal's productivity. The animals on test are kept in direct sunlight, with environmental temperatures between 29·4° and 35°C (85° and 95°F) between the hours of ten in the morning and three in the afternoon. The heat tolerance index is then defined as follows:

$$\text{Heat tolerance index} = 100 - 10\,(t_{3\,\text{p.m.}} - t_{10\,\text{a.m.}})$$

where $t_{3\,\text{p.m.}}$ is the rectal temperature of the animal at 3 p.m. and $t_{10\,\text{a.m.}}$ is the rectal temperature of the animal at 10 a.m. measured in degrees Fahrenheit.

Thus if an animal whose morning body temperature is 38·2°C (101°F) is exposed to a high air temperature for five hours so that its temperature at 3 p.m. has risen to 40°C (104°F), then the heat tolerance index would be:

$$\text{Heat tolerance index} = 100 - 30 = 70$$

An animal which shows no increase in rectal temperature during the five-hour exposure period would be regarded as well adapted and would have a heat tolerance index of 100. Rhoad calculated the heat tolerance of different cross-breeds and obtained the following results:

Purebred zebu = 89; $\frac{1}{2}$ zebu $\frac{1}{2}$ Angus = 84; Santa Gertrudis = 82; $\frac{1}{2}$ Africander $\frac{1}{2}$ Angus = 80; Jersey = 79; $\frac{1}{4}$ zebu $\frac{3}{4}$ Angus = 77; Grade Hereford = 75; Purebred Aberdeen Angus = 59.

Payne (1952) working in Fiji has shown that the range of heat tolerance index found within a breed is extremely large, and that the

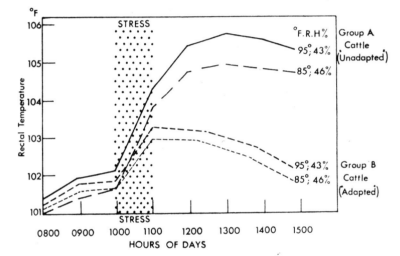

Fig. 12. Diagram illustrating the concept of the 'Rainsby Heat Tolerance Test'. Cattle are subjected to thermal stress for a short period (1000–1100 hours), and the patterns of their rectal temperature response are compared.

differences between breeds may in fact be less than the differences of individual animals within breeds. Payne showed that the range within purebred Nellore (zebu) was 70–84; within grade Friesians 58–78; and within a group of Grade Jerseys 74–86. As with other aspects of animal husbandry, it is the characteristics of the individual animal that are important rather than the mean characteristics of the breed from which it is drawn.

Other workers using the Rhoad heat tolerance test have shown that the level of the index increases up to the second year of life. It has also been shown that the mean heat tolerance index of a herd of cattle does not show appreciable variation from year to year, and that it is unaffected by either the stage of lactation or the stage of gestation.

Further work by Dowling (1956) has indicated that the rate of cooling after exercise may be a better indication of ability of an animal to dissipate heat than the measurements of the temperature of an animal standing in the sun which is merely absorbing heat from its environment. Dowling has suggested that the Rhoad heat tolerance index may merely measure the ability or inability of the animal to reflect solar radiation efficiently. He has suggested the use of another index, which measures directly the ability of the animal to recover from heat-stress after the stress has been removed. In Dowling's test, known as the 'Rainsby Test' (after the station at which it was devised), a group of animals are vigorously exercised for a period of at least one hour, after which the rectal temperatures of the animals are measured throughout the subsequent cooling period. The results are then presented in graphical form (see Fig. 12). An animal which is heat tolerant will begin to revert to normal body temperature as soon as the heat-stress has been removed. Conversely, animals which are ill-adapted to a hot climate will continue to increase in temperature for some time after the heat-stress has been removed.

It must be repeated that both these tests fail to take into account the all-important measurement of the productivity of the animal, without which no test can be said to be really meaningful in strictly agricultural terms.

COAT CHARACTERISTICS

Early work conducted in South Africa by Bonsma and in America by Rhoad, reviewed in detail by Findlay (1950), stressed the importance of coat colour in reflecting or absorbing solar radiation. Hair fibres which are light in colour reflect more solar radiation than hairs which are dark. The amount of sunlight reflected can be as much as 20 per cent in the case of a white-haired animal, but as little as 3 per cent with a dark-coloured beast. The importance of coat colour is greater during the summer period of high light intensity than during the winter period of lower light intensity.

Bonsma (1940) and Bonsma *et al.* (1940, 1943) also focused attention on the texture of the coat. Rectal temperatures and respiration counts of woolly-coated animals were invariably higher than those of animals with fine, glossy coats. Bonsma showed that the coat texture, as well as the coat colour, affected the 'mean effective absorptivity', i.e. the heat produced at the surface of the body by conversion of radiant solar energy into heat-energy, expressed as a percentage of the theoretical maximum. White-haired zebus had a mean effective absorptivity of 49 per cent, red Africander cattle 78 per cent and black Aberdeen Angus 89 per cent.

Dowling (1959) in Australia has shown that the presence or absence of a central medulla in a hair fibre may be of paramount importance. The hair fibres of cattle exhibit marked seasonal variation in the length, diameter and weight of hair per unit area. Large and significant changes are also found in the incidence and degree of medullation of the hair at different seasons of the year. The presence of a central medulla in the centre of a hair fibre enables the hair to stand more rigidly and therefore it assists in the evaporation of surface water, since a semi-erect coat allows freer air movement on the skin surface than a tangled, limp coat. Dowling argued that the hair medulla must therefore be regarded as a most critical characteristic of the coat in the regulation of heat-dissipation. There is a significant correlation between the presence of medullated hair fibres and the maintenance of near-normal rectal temperatures. This point is borne out in the following table:

Table 48. *Percentage medullation and rectal temperature of different types of cattle in Australia*

Dowling (1959)

Type of cattle	Percentage medullation	Rectal temperature	
		°C	°F
Brahman × shorthorn	100.0	38.5	101.3
Illawara shorthorn	79.9	38.6	101.5
Australian shorthorn, group A	62.3	38.8	101.7
Australian shorthorn, group B	46.4	39.2	102.6
Long-coated shorthorn	29.5	40.9	105.6

More recently Turner and Schleger (1960) have shown that the coat characteristics of cattle are directly related to productive performance. A large number of cattle in Australia were typed for 'coat score' by visual observation and manual handling. 'Coat score' was an attempt to define the length, thickness and texture of hair in quantitative terms without actually measuring the hairs in the laboratory. When these 'coat scores' were plotted against subsequent liveweight gain, it was found that they had a low, but statistically significant, negative correlation with liveweight gain during the period 16 months–28 months of age. ($r = -0.395$). They also established a statistically significant negative correlation between coat score and skin temperature, and showed that the correlation of coat score to succeeding liveweight gain was greater than the correlation of liveweight gain in the second year of life to liveweight gain in the first year. They therefore suggested that an evaluation of coat score would give a better indication of meat production, as measured by liveweight gain, than would an examination of liveweight gain data during the first year of life.

If these conclusions of Turner and Schleger can be verified with other breeds of cattle in other parts of the tropics, then we are much nearer to the goal of defining certain factors or characteristics which are directly related to productive performance at high temperatures. Coat scores are not only of greater repeatability than measurements of body temperature, but they can be assessed at any time without the need to conduct field experiments to enable measurements to be taken during periods of actual exposure to heat-stress.

This work has shown that the coat type affects the heat-regulating mechanism of the animal. It does not make clear whether the thriftiness of the animal is the effect of the temperature-regulating system of the animal or, conversely, whether the coat type of the animal is the result of inherent thriftiness. Three possible paths of cause-and-effect relationships can be postulated:

1. Coat type ----→ temperature regulation ----→ thriftiness.
2. Thriftiness ----→ coat type ----→ temperature regulation.
3.
$$\text{coat type} \longrightarrow \begin{array}{c} \text{Thriftiness} \\ \nearrow \quad \searrow \\ \end{array} \text{temperature regulation.}$$

It is well known that an animal which is unthrifty, for instance one suffering from a heavy worm burden, has a rough 'staring' coat. It would therefore appear that thrift has an effect on coat type, (2) and (3). The experiments described in the above paragraphs show that coat type is related to temperature regulation, (1) and (2) and (3). It is also known that an animal which can readily maintain homeothermy in a hot climate will be more thrifty in the tropics than one which is often exhibiting high body temperatures, (1) and (3). It would therefore appear feasible that the last path of cause-and-effect relationship (3) is the closest approximation to reality of the three.

PIGMENTATION OF THE SKIN

The skin of farm animals can be either pigmented or non-pigmented. Usually, the breeds of domesticated livestock which have been evolved in the tropics, such as zebu cattle, possess pigmented skins, irrespective of hair colour, whilst those evolved in temperate regions have skins which are non-pigmented. In some breeds, such as the Friesian, both pigmented and non-pigmented skins are found on the same animal. Black hairs overlie the pigmented areas, whilst white hairs overlie the non-pigmented areas.

A pigmented skin is most desirable in the tropics, since it is less susceptible to sun-burn and photosensitivity disorders. A good example of this is provided by white-faced Hereford cattle, which are prone to an eye disorder known as epithelioma due to the high photosensitivity of their unpigmented eyelids. Similarly, non-pigmented breeds of pigs, such as the large and middle-white breeds, are particularly susceptible to sun-burn. Many instances are on record of such white-skinned pigs dying after relatively short periods of exposure to direct tropical sunlight.

EFFECT OF TROPICAL CLIMATE ON ANIMAL BEHAVIOUR

The behaviour patterns of farm animals, both in temperate areas and in the tropics, have received a great deal of study in recent years. The behaviour pattern of farm animals is usually determined by watching a group of animals continuously for a seventy-two hour observation period, and recording the behaviour of each animal at regular time intervals.

It has been found that the grazing behaviour of temperate cattle in a temperate area is very similar to that of tropically adapted livestock in a tropical area. However, when unacclimatized temperate livestock are observed in the tropics, large differences from normal behaviour can be detected.

Cattle normally spend between seven and nine hours grazing in every twenty-four hour period. This grazing activity is broken up into three distinct and separate periods, one between dawn and mid-morning, a long period between midday and dusk, and a further, much shorter, period around midnight.

Unadapted cattle generally exhibit a behaviour pattern with an abnormally short grazing period and an abnormally long resting period. Moreover, unadapted livestock exhibit less inclination to graze during the main afternoon period when ambient temperatures and solar radiation intensity are greatest. This point is illustrated by the following figures from Trinidad for mean grazing, ruminating and idling times of pure bred zebu heifers and high grade Holstein heifers observed simultaneously in the same pasture.

These figures show that the less-adapted animals spend approximately two hours less in grazing and one hour less in ruminating than the well-adapted purebred zebus.

Breed	Grazing hours	Ruminating hours	Idling hours	Hours spent drinking
Pure-bred zebus	8·8	9·3	5·8	0·1
High grade Holsteins	6·8	8·4	8·7	0·1

Experimental work conducted in Louisiana (Seath and Miller, 1947), Fiji (Payne et al., 1951) and Trinidad (Wilson, 1961b) has stressed the importance of providing tropical livestock with good night grazing. This is contrary to common practice in many tropical countries, where herds of livestock are often placed into a night corral, completely devoid of any vegetation (Wilson, 1961a).

An experiment conducted in Trinidad (Wilson, 1961b) has shown that it is possible to alter the ratio of day to night grazing by varying the quality of the pastures on offer to the cattle by day and by night. The system in which the best pastures are provided by day and worst pastures by night resulted in 4·5 hours day-grazing and 2·9 hours night-grazing. Conversely, a system of management in which the best pastures were offered by night and poorer quality swards offered during the day resulted in 3·1 hours grazing by day and 4·2 hours grazing by night. Since the heat-stress on an animal in the tropics is much greater by day than by night, it follows that it is preferable to induce an animal to adopt a system of behaviour in which the maximum night-time grazing is practiced. Such systems should be far more widely advocated and practiced in the tropics than they are at present.

EFFECT OF HOT CLIMATES ON GROWTH, MILK PRODUCTION AND REPRODUCTION

Livestock growth rates are high when the animals are kept within their particular comfort zone. When livestock are maintained at higher ambient air temperatures, the appetite is depressed, the feed intake is reduced and consequently the animals grow more slowly (F.A.O., 1955).

In spite of the large amount of work conducted in the last twenty years with identical twin calves, so far only one major experiment has been conducted in which monozygotic twins have been separated immediately after birth, one twin being reared in a tropical environment and the other in a temperate climate under conditions of nutrition and management which were either identical or made as similar as possible. In this experiment, reported by Hancock and Payne (1955) and Payne and Hancock (1957), half the twins were reared in

New Zealand and the remainder in Fiji, and their growth and performance were studied from 7½ months of age until the end of their first lactations. At point of calving, the heifers reared in the temperate zone were approximately 10 per cent heavier than the heifers reared in the tropics. This difference was substantially reduced later in life. It was noted that the only appreciable differences in growth rate occurred when the ambient temperatures in Fiji were high (above 29·4°C, 85°F). The reduction in growth of the tropically raised heifers was reasonably uniform, affecting equally all body measurements other than that of belly girth. The greater belly girth of the tropically reared heifers was attributed to their significantly greater water intakes.

The effect of high ambient temperature is to decrease both the quantity and quality of the milk produced by *Bos taurus* cattle. No studies have been made up to 1964 on the effect of temperature on the milk production of *Bos indicus* cattle. Williamson and Payne (1965) state that the optimum temperature for milk production in temperate breeds appears to be 10°C (50°F), and that the milk production will decline steeply after different critical temperatures have been reached in different breeds.

The experiment carried out by Hancock and Payne (1955) and Payne and Hancock (1957) in which monozygotic twins were divided between New Zealand and Fiji, indicated that the effect of climate on milk quality is more marked on the butterfat constituents of milk than on the solids-not-fat. These two workers showed that the average milk production of the New Zealand twins was 44 per cent higher than that of the co-twins in Fiji, and that the total quantity of butterfat produced was 50 per cent greater in the temperate climate.

Cobble and Herman (1951) have shown that high ambient temperatures also result in a rise in the chloride content of milk and a fall in the milk sugar and total nitrogen content. This and other work on the effect of increased environmental temperature on the milk production of cattle has been reviewed by Findlay (1954) and Hancock (1954).

High ambient temperatures appear to affect the reproductive efficiency of the male rather than of the female. At low ambient temperatures, the testes are held close to the abdominal wall so that their temperatures can approximate to the true body temperature of the animal. As temperature increases, the testes descend lower into the scrotal sac. When the body temperature of the animal is slightly above normal whilst the ambient temperature is less than body temperature, the temperaure of the testes will be reduced. However, when ambient temperatures are such that the temperature of the surfaces of the animal which are exposed to solar radiation are higher than 38·6°C (101½°F), the testes will be above the true body temperature of the animal. As spermatogenesis is adversely affected by high temperatures, the effect of high ambient temperatures on non-adapted *Bos taurus* bulls is to decrease their reproductive efficiency. Waites (1961) working with sheep has shown that the scrotal cutaneous receptors are capable of influencing the general body temperature, thus giving a much broader thermo-regulatory role to the scrotum than the purely local one previously considered.

Several workers have suggested that the reproductive efficiency of cows may decrease at an earlier stage of life in a tropical environment compared to a temperate environment. An examination of the lifetime records of many European cattle maintained in the tropics supports this hypothesis. The increase in calving index of such cattle at each subsequent lactation is a very common feature of the records of dairy herds kept in a tropical environment.

EFFECT OF SEASONAL VARIATION IN DAY LENGTH

In temperate regions certain classes of livestock, especially sheep and goats, exhibit seasonal reproductive behaviour. A decrease in day length and a corresponding increase in night length results in the onset of oestrus. Sheep and goats can be made to breed at abnormal times of the year by artificially reducing the day length, although hormone treatment is also necessary in order to ensure high levels of fertility out of season.

In the tropics, especially near to the equator, seasonal variation in day length is a matter of minutes rather than of hours. Many workers have therefore assumed that such small differences would be insufficient to cause a seasonal reproductive pattern, and it is well known that most breeds of livestock can reproduce at any month of the year in tropical latitudes. However, careful analysis of the parturition data for different species of livestock in the tropics usually reveals a seasonal incidence of conception and, therefore, of parturition (Barrett and Bailey, 1955; Wilson, 1957). It has yet to be shown whether these seasonal rhythms of reproductive behaviour are dependent upon the small differences in day length, or whether they are imposed by other factors, such as seasonal differences in nutritional quality of food. It is, however, important for the tropical animal husbandryman to realise that these seasonal variations in reproductive behaviour do exist, as it may well be advisable to plan seasonal breeding programmes which are coincident with the normal rhythm.

It is well known that the artificial provision of extra lighting for intensively managed livestock, such as poultry, results in an increase in production in temperate areas. Limited work carried out in the tropics. indicates that the provision of artificial lighting may prove beneficial even in those areas where the length of day does not vary much from an average figure of twelve hours. Poultry houses constructed with low hanging eaves, designed to keep out driving tropical rain, effectively cut out the early morning and late evening sun. The effective day length inside such houses may be as little as eight hours. The provision of artificial light in such instances may well prove economic and lead to a significant increase in animal productivity.

EFFECT OF TROPICAL CLIMATE ON WATER CONSUMPTION

The effect of high temperature is to increase the water requirement of all forms of livestock in the tropics compared to their normal requirement in temperate areas. Part of the increased water intake is required to replace water lost from the body by evaporation. The increased water requirement can be as high as 200 per cent of the requirement of the same type of animal in a temperate climate. For instance, the intake of drinking water of temperate dairy cows is approximately half a gallon per 100 lb liveweight per day, whilst the same type of animal maintained in the tropics will have a free-water intake approximating to one gallon per 100 lb. liveweight.

In the case of intensively managed livestock with simple stomachs, fed on dry meals, the extra water consumption is rarely in excess of 50 per cent. In the case of animals whose diet consists partly of grass or forage crops, the amount of water drunk is directly dependent upon the water percentage of the grass and forage consumed, as well as on the ambient temperature.

Significant differences in the free-water intake of cattle at different seasons of the year are often found in the tropics where there are pronounced wet and dry seasons. This is illustrated by the data presented in Table 49, which refer to the seasonal differences in water intake of zebu cattle in Uganda and crossbred Zebu-Holstein cattle in Trinidad.

It will be noted that the free-water intake during the wet season in Trinidad, which is situated in the humid tropics, is very low. In fact, the values are lower than those usually found in a temperate climate. This is because the dry-matter percentage of pasture grasses in the humid tropics during the wet season is extremely low, sometimes 10 per cent or less of the total fresh weight. The result is that a very large proportion of the total water intake of cattle under these conditions is obtained from the grass itself and not from the free-water

Table 49. *Seasonal differences in free-water intake*

Wilson (1916a) and Wilson et al. (1962)

Type of Stock	Wet Season Intake		Dry Season Intake	
	Water drunk per cow/day	Water drunk /100 lb liveweight	Water drunk per cow/day	Water drunk /100 lb liveweight
Zebu heifers at Serere (Uganda)	22·3	5·1	44·8	8·8
Crossbred Holstein × zebu cows at Centeno, (Trinidad)	18·5	1·9	81·5	8·1

drunk. When the figures are expressed as total water intake (both free-water drunk and feed-water consumed) the seasonal difference is greatly reduced. In Trinidad, for instance, in the experiments referred to in Table 49 the total water intake per 100 lb liveweight was 12·1 lb per cow per day in the wet season and 14·1 lb per cow per day in the dry season. In places such as Trinidad with a climate mainly dependent upon moist easterly air currents blowing in from maritime high pressure cells, the chief climatic differences are found in the rainfall rather than in ambient temperature. The heat-stress on the animal is therefore reasonably constant throughout the year, with day temperatures almost invariably in the upper twenties (C) or in the eighties (F). The total water requirement of livestock is therefore relatively constant from one season to another. Under these conditions, the differences in the amount of drinking water which the farmer must provide for his grazing animals is dependent mainly upon the water content of the herbage, and this is itself mainly dependent upon the seasonal variation in rain precipitation.

ADAPTATION IN BIRDS

The 'normal' body temperatures of birds are in general much higher than those of mammals, thus the normally reported figure for geese is 40·8°C (105°F), for ducks and turkeys 41·1°C (106°F) and for domestic fowls 41·1°–41·8°C (106°–107°F) (Hutchinson, 1954). It follows that birds have relatively large differentials between their body temperatures and ambient temperatures when they are maintained in temperate climates, and much lower differentials when kept in the tropics.

The actual body temperature of birds varies markedly according to the state of activity of the animal. Thus the rectal temperature of certain wrens is usually 42·2°C (108°F) on capture, but after a period of rest and fasting the temperature is reduced to between 40° and 40·6°C (104° and 105°F) (Baldwin and Kendeigh, 1932). Conversely, the body temperature rises quite rapidly after ingestion of food, and is higher in poultry maintained on a high plane of nutrition (Robinson and Lee, 1947).

As with mammals, avians experience a diurnal variation in body temperate, and this variation may be independent of the diurnal pattern of environmental temperature. Thus in nocturnal birds, such as the owl, the body temperature is highest at night, lowest by day.

Birds have no sweat glands, but they are able to lose heat through evaporative cooling by panting, and in addition a little water diffuses through the skin from whence it can be evaporated provided the skin is not covered by so many feathers that air movement is minimal. Panting starts in fowls in still air at ambient temperatures within the range of 27°–33°C (80°F to 90°F). Well adapted breeds of poultry are capable of maintaining extremely high respiration rates, up to 400 respirations a minute, without, apparently, suffering from thermal stress. Unadapted breeds of poultry are incapable of achieving, or maintaining, high respiration rates of this order and it is probable that this essential difference in respiration rate is the chief factor determining tropical adaptation with poultry.

As birds are being intensively managed in the tropics on an increasing scale, it follows that a great deal of attention must be devoted to the construction of suitably designed poultry houses. The essential requirements are shelter from sun and rain, and the maximum air flow through the poultry house. The enclosed 'sweat box' type of deep-litter house designed to rear and maintain poultry through a temperate winter is totally unsuited for the tropics. Instead, open-sided houses, with due regard to adequate roofing or walling along the side facing the prevailing wind, are desirable. Preferably, houses should either be built with relatively tall roofs or else they should be constructed with roofing materials which reflect, and do not absorb, solar radiation. Corrugated aluminium is expensive but ideal, and similar effects can be obtained by painting less expensive corrugated iron roofs with several coats of glossy white paint. The provision of adequate drinking water, and water-trough space sufficient for the needs of each bird housed, are essential points, since birds can go for several days without food, but usually die in hot climates if denied access to water for much longer than twenty-four hours.

Indirect Effects of Climate on Tropical Livestock

The most important indirect effect of climate on tropical livestock has already been mentioned during the above discussion on water requirements. The quantity and quality of the feed on offer to tropical livestock is primarily dependent upon the climatic factors influencing, and possibly limiting, plant growth. A much fuller treatment of these climatic effects on tropical vegetation will be found in Chapters 1 and 3.

The second most important indirect effect of climate on farm animals is its influence on the distribution of the major pests and diseases, and the arthropod vectors which are responsible for their spread. A good example of this is given by the disease Trypanosomiasis, which is mainly spread by the tsetse fly (*Glossina* spp). The distribution of the tsetse fly is directly related to the presence or absence of suitable breeding sites, and these are themselves influenced by the climate of the region. If the climate or the vegetation pattern can be modified so that the tsetse fly is denied suitable breeding grounds, then Trypanosomiasis can be rapidly eliminated from tropical areas where it is endemic. This subject will be discussed further in Chapter 17.

CHAPTER 16

Cattle Management in the Tropics

Usually, in a textbook of this nature, the chapter on cattle management is a detailed description of the systems which, in the opinion of the author, are best suited to the production of milk and meat on farms of different sizes. The systems described are determined primarily by the two factors – the commodities for sale and the scale of farming operation in which they will be produced. In the tropics this premise is generally invalid. Cattle are not kept solely for the production of saleable commodities; they are maintained, to a very large degree, because of their intrinsic social value or for work. This point has already been referred to earlier in this book, in Chapters 4 and 12.

For this reason, a distinction between 'beef production systems' and 'dairy production systems' is not altogether valid. Before tropical cattle can be efficiently and effectively managed for the production of saleable commodities such as milk and beef it is important to have a clear grasp of the characteristics of indigenous systems of livestock management in the tropics, many of which need to be modified before they can be regarded as economically efficient.

Although great strides have been made in the last two decades to assist cattle-owning peoples to change over from a subsistence economy (based on livestock) to a cash economy (in which stock are used as a means of production) it will take many more years before this agricultural revolution has been completed. Customs, such as the payment of the 'bride price' with so many livestock, die slowly outside the townships and the tropical agriculturist ignores such customs at his peril. It is important to realize that herein lies an important distinction between crop and animal husbandry in the tropics. Crops are grown for two main purposes, to eat or to sell. Livestock are kept for a variety of reasons, only a few of which have anything to do with eating and selling. Just as a rich landowner in temperate regions may maintain a beautiful garden of no commercial value whatever, so do many tropical peoples maintain vast herds of cattle for similar, nonproductive purposes.

A full realization of the importance of this point can only come from experience of working with tropical cattle owners. An attempt will be made to illustrate it by providing in this chapter a brief and somewhat sketchy description of the system of cattle management once practised by the Abahima of Ankole District, Uganda. This example has been chosen since the system of management of the Abahima has been well documented (Mackintosh, 1938), and it is a good illustration of the interdependence of livestock husbandry and human sociology. With such peoples as the Abahima it is a moot point as to whether the cattle management system was made to fit in with human requirements, or whether human behaviour was adapted to suit the needs of the cattle herds.

The Abahima managed their herds of sanga-type, Ankole cattle on the inland plains bordering the north-west shores of Lake Victoria. The altitude varies between 4,000 and 6,000 feet, and the mean annual rainfall is between 40 and 50 inches. The vegetation is open grassland, with light thorn scrub, and the dominant species is red oat grass, *Themeda trianda*.

The Abahima considered four factors to be of the greatest importance when judging the value, or 'desirability', of an animal. First, the colour should be a uniform dark red, and calves with other colour-patterns are generally slaughtered at birth. Secondly, the horns should be long and white, with their tips bent forward. Breeding for this character has resulted in stock with excessively massive horns, which often weigh the head of the animal down in such a manner that movement is slow and the head is rarely held erect. Pairs of horns weighing 150 lb have been recorded. Thirdly, the hump should be large. Selection for this factor has resulted in many stock possessing humps which are so large that they topple over, giving the animal an asymmetrical appearance. Lastly, the animals were selected for fatness, obese bulls in particular being highly prized. It will be noted at once that none of the factors mentioned above has any direct economic significance. On the contrary, it can be seen that continuous artificial selection for large horns, humps and excess fat could well result in a lowering of the efficiency of the stock for production and reproduction.

The herdsmen were always present at the birth of a calf. The tribal magician (*Omufumu*) was called in in cases of difficult parturition, and charms were fixed to both the unborn calf and its dam. In cases of extreme difficulty the tribal doctor (*Omuzazi*) could be called upon to dissect the calf *in utero* so as to save the life of the dam. Female calves were especially prized, and were kept in the owner's own hut for the first week, before being placed in the communal calf-house. Calves did not run with their dams, but were allowed to suckle before the morning and evening milkings – a system of calf management which is common in many parts of the tropics. The young calves were pastured near to the kraal and their herding was the responsibility of young children. Calves were invariably brought in during the six

37 Ox cart in Northern India. Note the crude construction of the heavy cart with solid wooden wheels. This type of cart has been little changed in design over the last 2000 years. Note the method of yoking the oxen—the yoke is pushed forward by the animals' hump and is held in position by wooden-peg spacers and a light rope harness

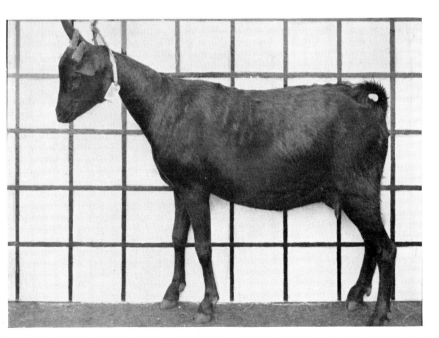

38 East African Goat, at Serere Experiment Station, Uganda. This female weighed 70 lb. at 2 years of age. (The background wall is marked with 6 in. squares.) These small goats are seldom milked, but their flesh is highly prized and live goats form part of the 'bride price'. Note the small horns and short hair. (see Chapter 14)

39 West African Dwarf Goats, at Kumasi, Ghana. These adult goats average between one-half and one-third the weight of normal mature goats of other breeds, and are common in West Africa. Note the typical goat stance, standing on the hind legs with the fore legs raising the body to allow high level browsing. The goats in the picture are eating cut forage suspended over a split-bamboo fence. (see Chapter 14)

40 Intensive grassland farming at the University Field Station, St. Augustine, Trinidad. In the foreground may be seen a flock of Black-head Persian sheep, originally imported from South Africa. In the middle distance is a herd of crossbred Holstein Zebu milking cows, averaging 500 gallons per lactation. Both classes of stock are rotationally grazing Pangola grass pastures (*Digitaria decumbens*, Stent) which carry an average of two adult cattle per acre throughout the year. Note teak post and barbed wire fencing, cambered beds to afford drainage, and Saman shade trees. (see Chapter 16)

hottest hours of the day to protect them from the sun and from attack by biting flies such as *Stomoxys*. Calves did not join the main herd until after they have been weaned at about eight months of age. Heifers were not ' named ' with their final adult name until after the birth of their first calf, and many different nouns were used to describe the various stages of growth through calfhood to first parturition.

The organization of the Abahima tribe was determined in large measure by the need to group a certain number of people at each cattle kraal, sufficient to carry out the various duties connected with tending cattle. There was an optimum ratio of able-bodied men to number of cattle tended, and the division of a herd into two parts, when it had grown too large to be conveniently managed at one kraal, was attended by the division of the family group into two parts, one to tend each of the herds. If a stock owner died, leaving a small herd of 100 cattle and four sons to inherit his property, it is likely that the four sons would remain together, living in a communal kraal and tending a communal herd, until natural increase allowed the herd to be sub-divided between them. However, if a herd reached a size of 800 cattle, it is likely that the owner would have expected his sons to marry, leave home and each form a new kraal with a section of the herd, as 800 cattle based on one site would lead to overstocking and a greater disease risk. The social organization of the Abahima therefore revolved around four factors: (1) the number of head of cattle at each kraal; (2) the carrying capacity of the land and water supplies; (3) the need to spread disease risk by cattle dispersal, and (4) the size of the cattle owner's family.

The ceremony of marriage could take place without the physical transfer of cattle from the groom's family to that of the bride. Cattle formed the ' bride price '. This was more properly regarded as a sort of insurance against the risk of marriage failure, since the bride price was usually returnable, with interest, should the marriage fail through some fault on the part of the bride. Before the wedding took place, the bride was fattened on a diet rich in milk, since obesity was regarded as a sign that milk is plentiful, which itself was a good indication that the bride was from a rich family – one owning many cattle. The chief rite in the marriage ceremony was an act of ' milk spitting ' between bride and groom. Milk was regarded as the symbol of union – akin to the ring still used, though of pagan origin, in Christian marriage.

The dung of cattle was regarded as semi-sacred – nothing associated with cattle was regarded as being unclean – and it is for this reason that the Abahima buried their dead in the dung-hill which was a central feature in the kraal. It may be readily judged from this fact that attempts by zealous agricultural officers to persuade the Abahima to use farmyard manure for agricultural purposes have met with conspicuous lack of success. The corpse was prepared for burial by smearing it with butter on the face and body and with dung on the arms and legs. Special milk from the ' lead cow ' was poured into the mouth of the corpse, so that the gods might observe that the deceased was well fed up to the last – a symbol that he owned many cattle and therefore came of a good family.

This brief glimpse at some of the customs of the Abahima is sufficient to indicate that, to cattle-owning tribes of this type, livestock are not merely regarded as a means of production but are an integral part of social and religious life. Their religious beliefs, superstitions, taboos and customs are all closely bound up with cattle. It follows that any alteration in the system of cattle management is likely to have side-effects on human social organization, and it is imperative to have a full appreciation of the possible consequences of attempting to bring about changes in livestock husbandry. This is not to say that changes cannot be made. Cattle are so important to many tropical peoples that any new ideas which are obviously beneficial may be welcomed and implemented, provided they do not unduly disturb the social organization of the tribe or infringe religious beliefs and social customs. By and large, tropical livestockmen are neither more nor less conservative than members of agricultural communities in most other parts of the world. Customs die hard, but the more progressive members of any agricultural community are not as slow in adopting improvements as is often made out.

It must be realized that the social organization of many tropical communities is undergoing drastic change. The impact of mass education will modify social customs and will increase man's capacity to absorb new ideas. However, some of the cattle-owning peoples of India and Africa will probably be the last to be affected by modern civilization; nomadic peoples last of all. It will be many decades before all tropical peoples move over to a cash economy, and before all tropical livestock are regarded as a means to production, and not as an end in itself.

We will now consider four different systems of cattle management in the tropics:

  A. Extensive systems. 1. Nomadic pastoralism.
                             2. Ranching.
  B. Intensive systems. 1. Small-scale mixed farming (peasant agriculture).
                             2. Large and medium scale cattle farming (yeoman and estate agriculture).

Nomadic Pastoralism

The classic nomadic pastoralists are to be found in East and West Africa, the best known examples being the Fulbe (or Fulani) in West Africa and the Masai, Suk, Turkana and Karamajong in East Africa. Further north, nomadism is practised by the Borana and Somali, these tribes keeping mixed herds of cattle and dromedaries. The Turkana, previously a cattle-owning nomadic tribe, are slowly changing over from cattle to dromedaries. Most nomadic peoples also tend flocks of sheep and goats, the small stock usually being the responsibility of the women-folk.

Often nomads depend on a diet in which milk and blood figure predominantly, the blood being obtained by puncturing the jugular vein with a sharp instrument, such as an arrow. The classical nomadic tribes of Africa are non-Bantu. Many of them share a love of cattle with a tendency towards war, and tribal rivalries are strong.

The chief reason for nomadism is the seasonal grazing requirement of the cattle herds. Some argue that many nomadic tribes were once sedentary, but that an increase in cattle numbers, coupled with a decrease in grazing quality and a reduction in the availability of drinking water, forced the tribes to migrate once each year to 'dry season grazing lands' in which grass and water were in more abundant supply. In recent times, the need to escape from foci of animal disease, from biting flies such as the *tsetse* and from the levies imposed by the tax collector, have also served to keep nomadic peoples continually on the move. This movement naturally results in more attention being devoted to the stock than to the husbandry of the grazing resources. The effect on the quality of the natural grasslands of the area may be quite disastrous, a point already made in Chapter 12.

Most authorities (Gulliver, 1955; St Croix, 1945; Mares, 1954) regard nomadic pastoralism as being on the decrease, and it is generally thought that this primitive system of cattle management will eventually die out. Many tribes, previously completely nomadic, are now either sedentary or are passing through the various transitory stages between nomadism and permanent settlement. These stages may be summarized as follows:

1. Seasonal grazing system in which wet- and dry-season grazing lands are completely separate (usually by a distance of about 100 miles or more) and change in location each year. In each grazing ground, the animals are rotated around the various watering points in such a manner that it is common for the animals to be driven ten to twenty miles away from water for one day (overnight), and then graze back towards water on the following day. Occasionally when stocking rates

are heavy this system may be extended so that the animals go for three whole days without water. With dromedaries the period without water can be further extended, up to about a week (see Chapter 12). Under this system the wet-and dry-season grazings are continually being shifted, both within and between seasons. There is therefore no permanent settlement, no village life and the nomads live in tents or temporary portable hutments. Trading is reduced to a minimum, the chief commodities purchased are clothing and salt for both human and cattle consumption. *Examples:* the Jie, Turkana and Somali of east and north-east Africa.

2. Seasonal grazing system in which the wet- and dry-season grazing lands are separate, usually by a distance not exceeding 100 miles, but with the wet-season grazing area permanently located in one place. This system allows permanent houses to be built in the wet-season grazing area, with the result that some form of village life usually develops. Trading takes place throughout the wet-season, when the pastoralists may sell appreciable quantities of milk and buy a large range of consumer goods. During the dry-season these semi-nomadic pastoralists often migrate to agricultural areas in order to graze the stubbles of cereal crops such as millet and sorghum. Provided the herds of cattle are kept under control, the farmers concerned often encourage this practice, since they value the manure which the cattle leave behind them, and the removal of the stubble makes subsequent cultivation easier. Semi-nomadic pastoralists are usually partially integrated into the social and political life of their country. They pay taxes, and some may even take part in elections and have their representatives in the government. *Example:* the Fulbe (Fulani) of West Africa.

3. More or less permanent settlement in the wet-season grazing area, the problem of the dry-season being overcome by conserving herbage or growing fodder crops wherever possible for utilization when grass is unobtainable. There may still be a certain degree of movement in and around the wet-season grazing area, and each kraal or camp may have an effective life of between five and ten years, after which the centre of operations may be shifted a few miles to an area where grass is more abundant and overstocking less acute. *Examples:* The Abahima and Karamajong of Uganda.

It is likely that some nomadic pastoralists will pass through these transient stages and that eventually they will become permanently settled and combine the growing of crops with the tending of cattle where soil and climate permit. A good example of a tribe which has now reached this last stage is the Iteso tribe of eastern Uganda, which is now permanently settled and which combines the growing of cotton and food crops with the management of large cattle herds. When this final stage has been reached it is quite common to find the tribe employing herdsmen from other, less advanced communities to take over the responsibility for cattle herding (Wilson and Watson, 1956; Wilson, 1958a).

As might be expected, the knowledge of livestock and the standard of husbandry practised by nomadic pastoralists is sometimes of a high order. Many nomadic tribes carry out crude but effective vaccination of their animals against diseases such as bovine pleuro-pneumonia and rinderpest. This is done by inoculating virulent material from infected animals into the faces of healthy stock, and then restricting the severity of the local reaction by cauterization. Unfortunately the technique of cauterization with hot irons is carried out too freely, and many animals are maimed or killed in this primitive attempt at practising veterinary medicine. The common 'cure' for East Coast Fever is to cauterize the inflamed lymphatic glands, and many animals are permanently crippled by this procedure. All these procedures are carried out with the best intentions, and nomadic peoples may be willing to accept new ideas providing they do not run counter to accepted beliefs and providing the efficacy of the new innovations can be fully proven. Great harm can be done by introducing, for instance, modern vaccination programmes (which always result in a few deaths of old and unhealthy stock) without warning the cattle owners that vaccination cannot work wonders overnight, and that some stock are bound to die before the beneficial results are manifest. Once the confidence of nomadic peoples has been lost, a very great deal of time and effort must be expended before the lost ground can be regained.

Ranching

Ranching is carried out on a large scale in South America, Central America and the tropical and subtropical regions of the southern States of the U.S.A., in parts of East Africa, particularly Kenya and Tanzania, in Central Africa, particularly Rhodesia, the Republic of South Africa, the northern regions of Australia and certain regions of Asia.

Ranching has not been evolved as an 'indigenous' system of cattle management in the tropics. It is a system imported from Europe, and modified to suit tropical conditions. Perhaps the greatest influence in the spread of ranching techniques has been Spanish, since ranching was carried out on a large scale in Spain for many centuries, and has spread into those areas colonized by the Spaniards from the fifteenth and sixteenth centuries.

Ranching is usually considered to be a system of extensive beef production, but in many areas of Africa, South America and in the island of Puerto Rico herds of dairy stock are ranched, the milking usually being carried out on a portable bail system which is moved around the ranges with the cattle herds. Much of the fresh milk supply for Caracas city, Venezuela, is produced in low rainfall ranching areas adjacent to the Maracaibo sea, from whence it is conveyed some 300 miles by refrigerated milk tanker (see Plate 44). Milk ranches usually consist of at least five or six self-contained herds of dairy cattle, each milked separately on the open range. The milk is then brought to a central depot, cooled, and conveyed by tanker to the fresh milk market. The dairy cattle found on South American milk ranches are mainly of Criollo type (improved Spanish longhorn) but crossbred Friesians and Brown Swiss cattle are popular in some areas.

Most of the ranches in tropical Africa, Asia and Australasia are beef ranches, the cattle generally being zebu (as in the case of the Africander cattle ranches in South Africa and Boran ranches in Kenya), but increasing numbers of crossbred cattle ($B.taurus \times B.indicus$) are being raised on progressive ranches. The beef ranches in the Americas used to be stocked almost exclusively with Spanish longhorn cattle, but during the present century the impact of European stock has been very great and most American ranches are now stocked with crossbred animals (such as the Santa Gertrudis) or with purebred European stock, as in the case of shorthorns and Herefords in Argentina. (For fuller descriptions of the crossbred cattle mentioned in this chapter, reference should be made to Chapter 17.)

Unfortunately, as with nomadic pastoralism, most of the land at present utilized for ranching purposes is either land of low fertility (as in the case of the Rupununi savannahs in British Guiana) or land of very low rainfall (as in the case of the cattle ranches in parts of Kenya and South Africa). The result is that the stocking rates are very low and often more conveniently quoted in terms of beasts to the square mile rather than beasts per acre or acres per beast. In the Rupununi savannahs in British Guiana, for instance, the overall stocking density is about ten to twenty beasts per square mile (32–64 acres per beast). It follows that the area required for an efficient cattle ranch is very large indeed, and a great deal of capital needs to be expended in the erection of fences, stockades, slaughter-houses, cattle dips and crushes, and in the provision of water and the improvement of communications. In many parts of South America ranching is in the hands of large companies, able to supply capital and technical knowledge as and when required, and also able and equipped to export the beef stock economically and efficiently from the producing areas to the consuming countries.

In ranching, the profit per beast is extremely low, and a very large annual throughput of slaughter stock is required. This makes it difficult, if not impossible, for the native tropical nomad to change over from nomadic pastoralism to a modern ranching system. Various authorities, for example East Africa Royal Commission (1955), have suggested that cooperative ranching might be possible, with several cattle owners combining their resources, including their herds of cattle, in order to make possible the provision of the essential capital development. Up to the present time, however, there are few, if any, examples of successful cooperative ranching, although the picture may change in the future. Again, because so much capital is necessarily tied up in breeding stock, it is difficult for ranchers to obtain agricultural credit, since cattle are not usually regarded as adequate security for the granting of loans.

Some of the major limiting factors of cattle ranching in the tropics are as follows:

1. *Lack of stratification.* In advanced ranching systems, the different phases in the management of cattle (such as breeding, rearing and the production of meat or milk) are carried out on separate, specialized farms. In the tropics such a stratified system of cattle management is unusual, and most ranches carry out all these operations on the same holding. Furthermore, there is little or no stratification within the ranch itself, and instead of the best ranges being set aside for breeding herds and the fattening mobs, whilst the poorer ranges are used for running store cattle, the whole ranch tends to be utilized by all classes of stock more or less indiscriminately. This point has been mentioned in Chapter 12.

2. *Poor communications.* Many ranching areas in the tropics are geographically remote, and can be reached only by air or river. Examples of such areas are the Rupununi savannahs in south-west British Guiana (reached only by air since the closing of the long cattle-trail in the mid 1950s) and the El Beni Department of Bolivia (reached mainly by air, but also connected with the Atlantic Ocean by some 1,500 miles of river via the Amazon). The consequence of such poor communications is that the freightage rates are extremely high, and this results either in a very low net return to the producer or else in the necessity to reduce bulky livestock products by processing. Thus meat may be marketed as canned meat or meat extract, and milk may be sold as butter, ghee or cheese.

3. *Lack of good local markets.* Many tropical ranching areas are situated in countries with a small human population (e.g. North Australia and Bolivia) with the result that the local market is limited and production must be aimed at an export market. Production for export is usually more complicated than production for local consumption. For example, meat must usually be deep-frozen and the disease control requirements of most meat-importing countries are extremely rigid. Furthermore, the export trade in meat tends to be managed by large shipping companies which own cold storage facilities, and hence the control of the market tends to pass from the producer to the merchant.

4. *Marked seasonality in growth and productivity.* Most of the ranching areas in the tropics are situated where either climate, soil or both are limiting for crop production, and where grass and animal growth is often markedly seasonal. (The implications of seasonality of grass growth have been fully discussed in Chapter 11.) It is quite common for tropical ranch animals to gain in weight for about eight or nine months of the year, and to lose weight for the remaining dry period. After a prolonged drought which has caused the stock to lose weight, the growth and productivity of the animals in the first part of the subsequent rainy season is often better than at any other time of the year. This enhanced growth is spoken of as 'compensatory growth' and has been discussed in detail by several workers such as Wilson and Osbourn (1960) and Hodnett and Smith (1962).

5. *Low availability of crop by-products, forage crops and concentrates.* As has been pointed out in Chapter 11, tropical ranching areas are often situated on lands unsuited for arable cropping, and as such regions often suffer from poor communications, it follows that it is often difficult, if not impossible, to supplement the natural range vegetation with other livestock feeds. The making of hay and silage is not usually practicable, and the freightage on imported livestock feeds from other regions is generally prohibitive. For this reason, fodder conservation in most tropical ranching areas merely consists of closing up one or more ranges for a season. Such crude conservation measures enable a supply of energy foods to be maintained throughout a dry season in the form of 'foggage', but these methods invariably result in a deficiency of protein. Under such conditions, it may well be advisable to introduce seasonal breeding, so that cows calve during those months of the year when the best quality herbage is available.

6. *Difficulty in giving individual attention to ranch stock.* The point has already been made that the size of a tropical ranch is large, due to the low profit per beast and the low carrying capacity of the land. The result is that stock management is rendered very difficult indeed,

and most animals can only be given the minimum of personal attention. A stockman can generally only get to know his animals individually when they are run in groups of 100 head or less. On many ranches the units (or mobs) of cattle number 500 head or more, and only a few outstandingly good or bad individuals will therefore be known individually. On many ranches stock are only seen at close quarters once or twice a year, when they are rounded up for dipping, culling, vaccinating and the branding and castration of calves. Animals on open range tend to be very dispersed and it is rare for a mob to keep close together. It is consequently difficult, if not impossible, to locate stock which become injured, which have calving difficulties or which are in some way or other in need of individual attention.

7. *Wild temperament of ranch cattle.* Stock which are infrequently handled tend to be wild compared to stock which are in regular contact with man. It follows that great skill is required in managing large mobs of ranch stock if damage to the stock and injury to the stockmen is to be avoided. Good stockmen handle cattle in such a way that the natural fear which most animals have for man is minimized. This can be achieved by curtailing the use of the stockwhip and stick to the minimum, and by encouraging the animals to approach the stockmen by dispensing 'range cubes' (a pelleted compound ration), molasses, or salt and minerals during each visit to the various mobs of cattle on the range. Even with these elementary measures of good stockmanship, the construction of efficient facilities for confining cattle is all-important. Kraals, camps and stockades must be constructed with strong materials, and proper races and crushes are essential for restraining animals whilst they are being vaccinated and branded, spayed or castrated. There are many instances of heavy losses of stock due to inadequate security measures taken in the kraal or stockade. Ranch stock are excitable and panic easily; a weak gate or defective perimeter fence may result in a cattle charge and consequent mortality due to the stock piling up and trampling over one another whilst trying to escape. The details of the construction of suitable cattle enclosures and crushes have been adequately described by Williamson and Payne (1965) to which work the reader is referred for a fuller treatment of this subject.

8. *Shortage of water.* In most ranching areas supplies of drinking water for the stock are deficient, and steps have to be taken to provide additional water by conserving rain water in dams during the wet season, or by drilling bore holes. Each range on a ranch should have its own supply of drinking water and large ranges require several drinking places if they are to be efficiently utilized. The amount of

water required will depend upon the mean size of the stock, the water content of the herbage, the mean maximum ambient temperature and relative humidity and the evaporation rate of water from the surface of the dam or trough. In very general terms, a rough calculation can be made by assuming that ranch stock will require about one gallon of water per day for every 100 lb liveweight in the dry season, and about half this amount in the wet season (Wilson, 1961a and Wilson, Barratt and Butterworth, 1962). Therefore a mob of 500 store cattle averaging 800 lb liveweight would require roughly $500 \times 8 = 4,000$ gallons of water per day in the dry season. If stock have to be watered from troughs instead of from river banks or dams, there is a very real problem in providing quantities of water of this order in a relatively short time. Troughs should be so constructed that they are long and narrow, so that many stock can drink simultaneously. If water can only be provided slowly from its source, such as by windpump from a borehole, then it is important to construct storage troughs large enough to provide at least half of the required daily amount of water at any one time. In the example cited above, the capacity of cattle troughs should be at least 2,000 gallons. (See also Chapter 11.)

It can be seen from the above summary of the chief limiting factors to ranching systems in the tropics, that the principal food of range cattle is the natural vegetation of the range. The efficient husbandry and utilization of range grasses is absolutely vital to efficient ranching. As ploughing and reseeding of the range is usually impracticable and uneconomic, the rancher must develop grazing systems which lead to a favourable ecological succession. As the application of fertilizer on a vast scale is usually uneconomic, the rancher must use the resting period rather than the manure distributor to maintain fertility and to increase grassland productivity. (See Chapter 11 for further information on grassland management.)

The most important single factor under the direct control of the rancher is the number of livestock carried on his land. Overstocking is the most common mistake made by ranchers. However, it is often difficult to determine the carrying capacity of a new area in advance, and hence the rancher often does not realize that he is overstocked until the detrimental effects of overstocking become apparent in terms of sward deterioration and reduced livestock productivity. For these reasons, the build up of cattle numbers on a new cattle ranch should proceed with caution, and immediate steps should be taken to destock as soon as it is realized that the optimum stocking rate has been exceeded. A useful technique which enables a reduction in stock numbers without the disposal of growing and breeding stock, is to spay (remove the ovaries of) all the elderly breeding cows and cull cows. Spaying, by preventing further pregnancies, will enable these animals to gain in weight in such a manner that they can be marketed for beef at some profit to the rancher. The resultant reduction in the calf crop will, in time, stabilize or reduce the livestock population to a safer level. This technique has been employed with great success in parts of British Guiana, and it is a technique which can be recommended to other ranching areas faced with problems of acute over-stocking. The actual operation of spaying can be quickly learnt, rapidly performed in the field, and once skill has been acquired by the operators the resultant mortality from the minor surgical operation should not exceed 5 per cent.

The foregoing account of some of the major limiting factors of largescale cattle ranching may appear to be pessimistic and somewhat discouraging. Whilst it would be unwise to minimize the very real difficulties which confront the tropical rancher, the future prospects and rewards of efficient cattle ranching are both promising. The world demand for, and price of, beef follows closely the increase in numbers and standard of living of the human population. The labour requirement in cattle ranching, per beast or per acre, is probably lower than in any other form of tropical agriculture. Land suitable for ranching but unsuited to other forms of land use is often available at very low cost (1 U.S. dollar per square mile in parts of South America). Once the original capital investment in terms of land and breeding stock has been amortized, the recurrent costs of cattle ranching are very low. Finally, it is likely that the productivity of ranch-type cattle will be markedly increased during the next few dacades by the application of better breeding techniques, and it is also likely that the efficiency of long distance transport, both of live cattle and processed meats, will be made more efficient and less expensive.

Small-scale Mixed Farming

The successful introduction of mixed farming techniques in temperate agriculture has led many authorities to recommend mixed farming in appropriate areas of the tropics. In certain areas, such as the Kipsigis district of Kenya, intensive mixed farming has been introduced in a relatively short period with marked success. In many other parts of the tropics, attempts to introduce mixed farming into peasant agricultural systems have been anything but successful. In the following paragraphs, some of the most important managerial difficulties of smallscale mixed farming with cattle will be examined.

SIZE OF FIELD

The size of most tropical fields or plots has been determined more by the amount of land a family can cultivate with a hand hoe than by the amount of land which can be cultivated with the aid of the draught animal or tractor. The smaller the area of land planted down to a grass ley or permanent pasture, the greater the proportionate cost of providing stock-proof fences or hedges. Apart from cost, intensive grazing techniques become difficult when field size and number are so small that herd size is severely limited and rotational grazing is ruled out. In general terms, fields of less than one acre in size are unlikely to repay the heavy capital and recurrent costs of confining cattle to them, and a series of at least five paddocks is probably required to enable adequate grass regrowth and to reduce reinfestation with parasites to manageable proportions. It follows that an area of about 5 acres of grass is probably the minimum required for small scale intensive cattle management, but there are many areas of the tropics where the mean farm size is considerably less than 5 acres, and the ratio of grass and resting land to arable is generally weighted in favour of arable. It may well be highly undesirable for governments to encourage by subsidization the planting of pasture grasses on peasant farms where the area available for this form of land use falls below a certain minimum figure.

HIGH COSTS OF FENCING

Live hedges are uncommon in the tropics, and where suitable species are available they are difficult to maintain in a stock-proof manner for large periods. Fencing materials are relatively expensive, since only a small number of timbers are capable of being termite-proofed for reasonable periods and these timbers usually command high market values. For example, the current costs of teak posts in the West Indies is over five shillings per post. Wire, whether barbed or plain, is usually imported and carries high transport costs. The result is that a traditional-type post and wire fence costs about £250 per mile, plus erection costs. In some areas the use of electric fencing has been successfully introduced at great saving in cost, but permanent fencing is usually highly desirable round the perimeter of the grass acreage, particularly where pastures are adjacent to neighbouring arable farms. In certain areas of the tropics where improved but disease-susceptible breeds of cattle are used intensively, double perimeter-fencing is necessary. This further increases the capital and recurrent costs, and also leads to a slight reduction in the area available for grazing.

The high costs of fencing have led many workers to consider the prospects of zero grazing systems, in which the grass is cut and carried to the cattle which are kept in yards or cattle sheds, pastures being used merely for exercising purposes. Indeed, in many tropical countries such systems, known as 'cut and carry' systems or 'soilage grass' systems, were introduced before the advent of suitable species of improved pasture grasses. (See Chapter 13.) Such systems tend to break down as labour rates rise and the costs of cutting and carting become prohibitive, but there are still many areas of the tropics where they are still practised. The advent of zero grazing has led to a revival of interest in the cut and carry system, but the machinery involved is expensive and requires a minimum acreage, usually in double figures, for economic operation. The capital costs, and minimum acreages, are usually prohibitive for small scale peasant operations.

PROVISION OF PRODUCTIVE CATTLE

Intensive livestock production is not possible with ranch-type cattle suitable for more extensive systems. Productive types of stock are an absolute essential, and in most cases this means that tropically adapted breeds of *Bos taurus* cattle, or crossbred *Bos taurus* × *B. indicus* cattle, must be employed. Unfortunately many tropical countries, for reasons of disease prevention and control, denied such cattle to indigenous small-scale farmers, and based their national planning schemes on the use of native *B. indicus* cattle which, in many instances, were completely uneconomic.

The breeding, selection and multiplication of productive cattle of the required standard is usually, if not invariably, outside the scope of the small-scale producer. The most that the small farmer can be expected to do is to cull his own stock on preformance as well as on appearance, and even this requires a minimum standard of herd recording, such as for milk yield or calving interval, which is more than can be reasonably expected from an illiterate peasant producer. The government, or some efficient livestock producers' cooperative, must organize the breeding and supply of livestock on a national basis. Useful adjuncts to this work are regulations to limit the indiscriminate breeding by scrub bulls, and the development of a nationwide artificial insemination service which can spread the use of the nation's best proven bulls even to the smallest producer owning only one or two breeding cows. In this respect tropical countries can profit from the experience gained in the early days of A.I. in temperate countries, and refrain from placing responsibility on the selection of stud bulls on the shoulders of those having a passable 'eye for stock' but no knowledge whatsoever of the true meaning of the term 'animal breeding'. This point will be referred to in greater detail in Chapter 17.

The definition of the production levels required in tropical cattle will vary from place to place and from time to time according to differing costs of production and market values of the end products. A common mistake in the past has been to think in terms of existing production levels, and of ways by which these might be slowly increased, instead of dealing in economic terms from the start, and to consider the minimum acceptable production levels for economic forms of land use with cattle. Where the problem has been tackled from the former standpoint, figures of 100–200 gallons of milk (per cow per annum) or $\frac{1}{4}$–$\frac{1}{2}$ lb liveweight gain (per head per day) have been quoted for milk and beef production. Such figures are realizable from *Bos indicus* cattle. However, where the problem has been tackled economically, it is readily apparent that such low production targets are generally unacceptable, and in many instances could only result in the farm running at a loss, particularly if a fair value is placed upon land and interest charges levied on the theoretical capital investment. Maule (1962) has suggested the following objectives for intensive cattle production levels in the tropics. The implications of accepting Maule's suggested targets are very far reaching.

Beef production: Animals should be able to subsist without supplementary feeding and make reasonably good growth without excessive fluctuations, and be fit to kill at about four to five years with a dressing percentage of at least 50. Good breeding cows should rear three calves in four years.

Milk production: Cows should be able, with good management, to withstand difficult climatic conditions and give a yield of at least 400–600 gallons of milk a year and to breed a calf every twelve to fourteen months from the age of three years.

As Maule stresses, these are generalized recommendations, and there are many instances where they would be completely inapplicable. His standards could not be met where disease ruled out the possibility of using European or crossbred cattle for milk, and his target for beef would be inappropriate in the many tropical areas which experience a marked dry-season and consequent wide variations in the quantity and quality of herbage.

SUPPLEMENTARY AND DRY SEASON FEEDING

In most tropical areas the seasonality of grass growth is such that some degree of supplementary feeding is required, particularly for dairy cattle, during the latter part of the dry-season and possibly also during the excessively wet conditions which accompany the onset of the rains. This need for extra feeding becomes more acute as breeds of cattle with higher productive potential are used by the small-scale farmer. Unfortunately, many of the traditional dry-season feeds are coarse roughages, very low in digestible protein. The taller soilage-type grasses, whose deep rooting systems allow a degree of dry-season growth, also come into this category, with digestible crude protein levels generally below 5 per cent. In many parts of the tropics the dry season is not so much a period of total shortage of edible herbage as acute shortage of digestible protein. It is generally relatively easy to conserve 'foggage' or standing hay, but such feeds do little or nothing to rectify this basic nutrient inbalance. This point is clearly illustrated in the data detailed in Table 50 below, which compare the nutrititional quality of a common pasture grass suitable for intensive small-scale cattle production, in the wet- and dry-seasons of the year under West Indian conditions.

Table 50. *Seasonal variation in nutritional quality of Pangola grass pastures*

Butterworth *et al.* (1961)

Season	Dry matter percentage	Composition expressed as a percentage of dry matter*						
		Crude protein	Crude Fibre	Ether extract	Ash	NFE	DCP	TDN
Wet	23·4	11·1	30·5	3·0	8·4	47·0	6·8	59
Dry	39·3	6·8	29·5	2·1	7·8	53·8	3·2	62

* For explanation of the nutritional terms used in this table, the reader is referred to Chapter 17, p. 395, section dealing with Improved Nutrition.

Table 51. *Intake of nutrients from Pangola grass pastures compared with requirements of a 1,000 lb heifer yielding 1½ gallons milk per day*

(lb per day)

Butterworth *et al.* (1961)

	DCP	TDN
Wet season	2·0	17·7
Dry season	1·2	22·9
Requirements for maintenance + 1½ gals.	1·6	11·8

It will be seen from the figures above that, whereas a high quality Pangola grass pasture can provide more than enough digestible crude protein and total digestible nutrients for the needs of an average dairy cow in the wet season, the intake of protein in the dry season is about 25 per cent below the theoretical requirement. This picture is true of many other grasses in many other parts of the tropics, and it is essential that the small scale peasant farmer is educated in the need to supply some form of protein supplement at this time. In some areas this can be provided by judicious use of local by-products, such as copra meal, soya bean meal, spent grains etc., which are relatively high in crude protein. For further information on this subject, the reader is referred to the standard works on the feeding value of foodstuffs, including tropical foodstuffs, such as Morrison (1956).

HOUSING

Although the need for housing is much less in warm tropical climates than in the colder, temperate regions of the world, nevertheless some cattle buildings are necessary for efficient small-scale production. It is often traditional to house, or kraal, all stock at night and although this is not essential for the well-being of the cattle it may well be necessary to reduce losses by theft or by predatory animals. In mixed farming areas, the danger of stock breaking out from night pasture and damaging neighbouring arable crops is often sufficiently great to induce the owner to think in terms of night housing. Wherever possible, however, the aim should be to encourage night grazing systems in order to maximize the intake of herbage. Various workers (Harker *et al.*, 1954; Wilson 1961a, b) have shown that up to 40 per cent of the total grazing time may occur during the 12 'night' hours, and that denial of night grazing is never fully compensated by more intensive day-time grazing.

There is much less need to house calves in the tropics than is generally recognized by small-scale farmers in mixed farming areas, where the 'calf house' is often regarded as an integral part of the homestead. If the various dangers of night grazing detailed above can be discounted or overcome for adult cattle, there is usually no good reason why calves should not also be kept out at night. On the other hand, the need of the calf for night grazing is less than that of the adult, as can be seen from the data provided in Table 52.

Perhaps the chief need for some form of cattle building in the tropics in the systems of small-scale intensive agriculture lies in suitable milking accommodation for small dairy herds. All too often the type of buildings recommended are miniature milking sheds or milking parlours suitable for family farms in temperate climates. In the tropics milking accommodation has to meet two main requirements. First, to provide minimum protection for the milker during the rains; secondly,

Small-scale Mixed Farming

Table 52. *Comparison of grazing behaviour of female cattle at different stages of growth (hours)*

Wilson (1961b)

Behaviour	Calves		Heifers		Cows	
	Day	Night	Day	Night	Day	Night
Grazing + Walking	6·0	0·6	6·1	1·0	4·9	4·4
Ruminating	0·7	3·5	1·4	3·9	2·5	4·0
Idling	5·3	7·9	4·0	7·1	4·6	3·6

to provide reasonable facilities for the storage of milk and milking utensils and for the cleaning and sterilization of these utensils. A simply constructed lean-to shed, sited on well-drained land with protection from the prevailing wind and driving rain, is quite sufficient. If cheap local materials are employed, then the structure can be rebuilt on a different site every few years so as to prevent excessive fouling and a build-up of parasites in the vicinity of the shed. During the dry season there is no reason why milking should not be carried out in the pasture itself, with the cows tied to temporary stakes driven into the ground. In this way not only are building costs reduced to the minimum but the cattle benefit by reduced disturbance to their normal grazing routine. The tropical dairy should have good facilities for keeping the milk clean and cool prior to disposal, and there should be a small room, preferably with a concrete floor and walls which can be painted or regularly cleaned, in which the utensils can be cleaned, sterilized and stored when not in use. The tropical sun is a most effective sterilizing agent, and after washing thoroughly it is a good practice to expose buckets, churns and other metal utensils so that they are dried and sterilized by the sun. The dairy should be supplied with running water if at all possible, and care must be taken to see that flies are effectively excluded.

It should be stressed that tropical dairy buildings do not have to be constructed in expensive materials in order to be efficient. A dark, damp concrete structure, with cracked walls and broken floor, situated in a valley subject to occasional flooding, is far less desirable than a cheap, open wooden shelter sited on well-drained open land.

MILKING PROCEDURES AND CALF REARING

The common system of milking practised in most parts of the tropics by small-scale producers is twice-a-day milking, with calf at foot. The calf is usually separated from the dam during the night, and is allowed to suckle for several minutes immediately prior to the morning milking, which generally takes place within an hour of dawn. The cows and calves usually graze separate pastures by day, and the calf is allowed its second daily feed with the dam immediately prior to the afternoon milking, which may take place at any time from about 3 p.m. to as late as 7 p.m.

This system of milking is often used with purebred zebu cows, which let down their milk more readily in the presence of the calves. However, it often results in too little milk being drunk by the calf, with the consequence that calf growth and development is restricted. The calf is thus weak when it is exposed for the first time to tick and flyborne disease and to internal parasites. Part of the high incidence of calf mortality in the tropics is probably due to the poor nutrition of the calf, which is related to the system of milking with calf at foot. (Further information on the nutritional requirements of calves in the tropics is given in Chapter 17, section on Nutrition.)

The system of milking with calf at foot is not essential, and there are many recorded examples of successful attempts at complete milking-out of zebu cattle coupled with the bucket feeding of calves (Williams and Bunge, 1952). However, in an unselected herd of zebu cattle, there are generally a small percentage of cows which are difficult if not impossible to milk out completely without the presence of the calf. In such cases, the normal reflex action by which milk is released from the milk secretory tissues into the milk sinus (as a result of a chain reaction triggered off by the sight, smell and touch of the calf) cannot be conditioned by man into a modified pattern in which the stimulus provided by the calf is replaced by a stimulus provided by the milker and the act of hand-milking. In order to achieve efficient results, these 'difficult milkers' must be culled from the herd. There is some evidence that this undesirable characteristic may be inherited, and if this is the case then it is essential to spot and eliminate any bulls which may be transmitting this genetic character to their daughters.

With crossbred, or *B. taurus*, cattle, the normal milking procedures adopted in temperate regions, and adequately described in textbooks on dairying in those regions, can be employed. In the case of zebu cattle it is highly desirable that calf-at-foot milking should be replaced by complete milking-out, but it must be expected that this change-over will necessitate higher than normal culling rates for the first one or two cow generations. More attention must be paid to the nutritional needs of the growing calf than is usually practised in the tropics, and a system should be aimed at in which about 50 gallons of whole, skimmed or reconstituted milk or milk substitute is fed to each calf during its first three to four months of life. It is usually helpful to adopt a system in which milk, or milk substitute, is gradually replaced by a balanced concentrate, which is itself gradually replaced by grass, the total period between birth and full grass grazing lasting not less than five and not more than eight months. *B. indicus* calves should be expected to grow at a rate not less than ½ lb/head/day for their first year of life, crossbred calves at a rate not less than ¾ lb/head/day and *B. taurus* calves at a rate not less than 1 lb/head/day. Rates such as these are rarely, if ever, realized in management systems in which cows are milked with calf at foot, with little if any attention given to the nutritional needs of the calf during its first, most vital, year of life.

An alternative milking and calf rearing system advocated in some parts of the tropics (Parsons, 1958) is, 'once-a-day milking'. In this system the calf is separated from its dam all night, and the latter is then milked out by hand as completely as possible in the morning. The calf may be 'at foot' at this milking, but is only allowed a few sucks, sufficient to stimulate let-down. After the morning milking the calf is allowed to run with its dam for the rest of the day, being separated again at dusk. It is calculated that, under such conditions, the milker obtains about 55 per cent of the total milk yield and the calf about 45 per cent. This ensures that the calf has an adequate milk supply, but with cows giving over 150 gallons it is wasteful in that the calf will obtain more milk than is required, and too small a proportion of the total yield will be available for sale off the farm.

The actual act of hand milking is often carried out under very unhygienic conditions in the tropics, and there is a very considerable proportion of milk produced by small-scale farmers which is commercially unacceptable because of adulteration, high bacterial count, or contaminated with solid matter. It is difficult to enforce a higher standard of personal hygiene in the dairy than is practised in the home, and the problem of clean milk production must be tackled by a concerted approach by all interested parties – milk buyer, agricultural officer and health inspector. One undesirable milking practice found in both African and Asian countries is 'wet milking', during which spittle is conveyed by the milker to the teats of the cow, and the teat is then milked by a stroking action between thumb and forefinger instead of by a 'squeeze and release' action with the teat held in the palm of the hand. The use of the strip-cup to detect mastitis, and the use of a strainer, such as butter-muslin, to separate solid-matter from the liquid milk before it is poured into churns, requires active encouragement. It may well be desirable for the milk purchaser to take the initiative by providing his suppliers with these materials and utensils free of charge, and to pay a slightly lower price for the milk.

MARKETING

The small scale producer, unless efficiently organized into well-run producer-cooperatives, is at a disadvantage when marketing his cattle and cattle produce. Highly perishable animal products naturally command lower prices at farm gate than at the condensary, processing plant or abattoir. However, the differentials which exist under the uncontrolled operation of the market by private enterprise are generally excessive, and this is accentuated by the high transport charges levied by middlemen.

For these reasons, the regulation of the market by some form of direct or indirect government control is generally desirable, at least in the early stages of development of an animal industry amongst small-scale producers. Unfortunately, although this point is readily accepted when it comes to the marketing of cash crops for export, it is often not taken in respect to the marketing of produce primarily intended for local consumption. On the contrary, it is often said to be in the national interest to market such staple items of diet as meat and milk at the lowest possible prices in order that they are within the limit of purchasing power of the working classes. At the same time, governments in the tropics are less willing and able to subsidize the farmer in similar fashion to the heavy subsidization of, say, the British milk and beef producer. Tropical governments are also unwilling, for similar reasons, to levy import restriction or tariff control upon cheap milk and meat products brought in from overseas if by so doing they cause a rise in the cost of living index.

Producers of meat and milk in the major cattle-raising countries balance their net returns by selling some produce fresh, at high prices, and some for processing, at much lower prices. Small-scale peasant farmers have great difficulty in improving their scale of operations and their efficiency if they must sell all their produce in direct competition with cheap processed world surpluses. The dairy producers of Great Britain and the beef producers of Australia would quickly react if they had to market all their produce on these harsh terms, yet this is what the tropical peasant farmer is forced to do in many of the development plans sponsored by their governments. It is not within the scope of this chapter, or this book, to deal with this subject in any detail, but the point cannot be overstressed that there is frequently a conflict between cheap food for the masses and good financial returns and inducements to the primary producer. It is difficult to assist the primary producer by merely giving technical advice, if government are committed to a 'cheap food' policy which precludes support for the local producer by some form of control over the market.

Large- and Medium-Scale Cattle Farming

The proportion of farms falling into this category is small, expecially if the tropical parts of North America and northern Australia are excluded. However, large areas of the tropics are suitable for this form of land use, and it is likely that the numbers of farms falling into this category will increase markedly in future where political pressures do not restrain the normal and natural increase in mean farm size due to technical improvement and enhanced labour efficiency.

Some regions of the tropics, such as the wetter, medium altitude areas of Africa, certain of the West Indian islands and the better drained and more fertile areas of tropical South America, are capable of intensive large-scale grassland production which can rival the best temperate grasslands of the world in terms of animal productivity. At present this potential is relatively little exploited, since the fertile soils of the tropics have tended to be used for cash crops rather than grassland production. For instance, many parts of the West Indies have been widely used for the production of sugar-cane rather than the grass crop, since the profitability of sugar production has exceeded the potential profitability of livestock production over the last three hundred years or so. There are, however, signs that in certain parts of the tropics this pattern may change in future. One factor which may bring about this change is the rising cost of labour which is operating against high labour-intensive crops, such as hand-harvested sugar-cane, which command relatively stable world prices. Another factor is the rising world demand for livestock products, particularly meat, which are comparatively less labour-intensive and which command continually rising prices on the world markets. For these and other reasons, the last decade has seen an increase in grass acreage at the expense of cane in Jamaica (Nestel and Creek, 1964a) and an increase in grass and a decrease in maize production in certain parts of South America and Africa.

The techniques of cattle management on farms of this category are very similar to those employed on farms of similar size in temperate agriculture, and for this reason a detailed description will not be given in this book. Attention will be drawn to a few important points which may be overlooked in considering this form of tropical land use.

First, the capital required for efficient and intensive grassland farming is comparatively large. The costs of cattle, fencing materials, buildings and equipment are often more expensive in the tropics than in temperate countries, and this high capital cost is not always compensated for by reduced costs of land. Figures on the actual investment capital required for intensive large-scale cattle farming are difficult to come by, although some useful detailed figures are now emerging from the work of the Livestock Research Division of the Sugar Manufacturers' Association of Jamaica. A recent survey (S.M.A., 1963) showed that the average size of the grass-beef farms operated by the sugar estates in Jamaica which cooperated with the research division of the S.M.A. was 2,203 acres, and the average capital investment was about £64 per acre, of which 53 per cent was in land, 14 per cent in fixtures and 33 per cent in stock. Further work reported in Trinidad by Squire (1964) confirms this point.

Secondly, the systems of cattle management at present employed on these large-scale, intensive farms are not sufficiently intensive for maximum efficiency and profitability. One of the chief criticisms of existing livestock estates is that they are often understocked, and their general technical efficiency in many other ways leaves much to be desired. Thus Nestel and Creek (1964b) have shown high and statistically significant positive correlations between liveweight gain per acre and stocking rate, and also between economic output per acre and stocking rate. The average farm examined in the S.M.A. survey (loc. cit) ran 1,049 cattle (903 'adult animal units') on 2,203 acres, which is less than one beast to two acres. The carrying capacity of many of these farms was thought to be better than a beast to the acre, and in many parts of the tropics it can exceed two beasts to the acre (Wilson, 1963).

Thirdly, the response of tropical pastures to heavy dressings of nitrogenous fertilizer is often underestimated. There is a tendency for temperate concepts of grassland husbandry and standards of grassland productivity to be transported into the tropics by estate managers. In most temperate countries, grass growth is limited by light, temperature and soil water. In the wetter tropics these limiting factors are less important, or operate for shorter periods, and nutrient supply is often one of the chief limitations to maximum grass growth. Economic responses to dressings exceeding one half ton of sulphate of ammonia per acre per annum have been recorded, but the average fertilizer applications on large-scale intensive farms are a small fraction of this figure. All tropical grassland farmers are aware that water is the limiting factor to growth in the dry season, yet few appear to realize that nitrogen is one of the chief limiting factors for the remainder of the year, especially when grass is grown in the absence of a grazing legume.

Lastly, the seasonality of grass growth due to the dry-season may well require a planned programme of seasonal calving, seasonal milk production and seasonal beef marketing, just as the seasonality of grass growth in New Zealand has forced the New Zealand dairy farmer into becoming a summer milk producer, with the whole herd dry for the two most severe winter months. At present, most tropical grassland farmers breed all the year round and attempt to even out the seasonal fluctuation in grass growth by irrigation or grass conservation. More attention should be devoted to the advantages of seasonal breeding and seasonal productivity, not only of dairy and beef cattle but of other tropical livestock such as sheep and goats. This subject will be dealt with more fully in the chapter on Livestock Improvement which follows.

CHAPTER 17

Livestock Improvement in the Tropics

The productivity of tropical livestock can be improved in two ways: by improving their management or by breeding better animals. Better management may effect improvement relatively quickly. For instance, more skilful milking techniques can be adopted overnight, and milk production thereby increased in a matter of days. Genetic improvement is of necessity a much slower process, since at least one complete animal generation must elapse before any results are forthcoming, unless improvement is brought about merely in a negative manner by culling stock of poor quality. Dramatic production-increases often result in the next generation after cross-breeding or outcrossing, but the annual rate of increase due to genetic improvement within a relatively stable population of farm animals, such as within a closed flock or herd, is usually painfully slow. For instance, in the case of a herd of milking cows from which the worst animals have been removed, genetic improvement of this sort may lead to an average increase per cow of about a gallon of milk a year. It follows that a very long period of time, measured in decades rather than in days, must elapse before any notable success has been achieved.

As some improvement is obtained much more quickly from better management than by breeding, many people have placed more weight on management than on genetic improvement. But in practice both methods need to be carried out simultaneously. If genetic improvement is neglected, then a stage will soon be reached where little further advance can result from better management. This is because the stock will then have approached their genetically controlled 'ceiling' of productivity, which must be raised before any further increase in production can be achieved. It may take, say, ten years for the major limiting factors of management to be overcome and for the livestock to approach their genetic ceiling. If no breeding work is carried out during this period, there may then be a long pause whilst attention is diverted to raising the genetic ceiling. This 'start and stop' system of livestock improvement is obviously undesirable, and it is far preferable for the genetic ceiling to be slowly but progressively raised whilst each of the major limiting factors of management is overcome.

Although improvement in environment and improvement in the animal's genotype are considered separately below, the reader should remember that in practice both methods of raising animal productivity should proceed simultaneously.

Improvement through Management: I. Nutrition.

The purpose of the farm animal is to convert food into some form of animal product. The manner in which it performs this function is a measure of the efficiency of the particular farm animal on the one hand and of the skill of the farmer on the other. The range in food-conversion efficiency of animals within a breed is relatively limited, although small differences of a few per cent can be of great economic importance in the case of farm animals which are intensively fed on purchased feeds, such as broilers or pigs. The range of foodstuffs which the tropical farmer can offer to his livestock is often less limited, but it is vital that the right feeds, in the right proportions, are fed to his animals. A deficiency of one item in the diet may cause ill-health and low productive output (Abrams, 1950). A surfeit of one ingredient may result in similar disadvantages, or in unnecessary expenditure on unwanted food. Indigenous livestock often feed upon a wider and more diverse collection of foodstuffs in the tropics than the more sophisticated breeds maintained in temperate countries. This fact has led some workers to suggest that tropical animals are more efficient converters of food, especially roughages, into animal products. This point has yet to be proven, and investigations into the nutrition of tropical animals are at present hampered both by lack of reliable data on the nutritional value of many tropical feeds and by insufficient knowledge of the basal metabolism of most kinds of tropical livestock (Rogerson, 1960).

The nutritional requirements of farm animals can be divided into two distinct types. These are the maintenance requirements to keep the animal alive and in good health, and the production requirements to enable the animal to put on weight or to produce milk, eggs or offspring. The amounts of nutrients required for both maintenance and production for the various classes of farm animal are reasonably well known under temperate conditions, although perhaps not with the precision which will satisfy the research scientist, but only crude estimates, based on very little critical work, are available for the tropics. It would appear that the maintenance needs of well-adapted tropical stock are somewhat less than those of temperate animals, because the former are more lethargic and spend less energy in maintaining body heat. It would also appear that the temperate standards for production rations can be taken as a reasonable working guide for the tropics.

Under systems of small-scale subsistence agriculture the cost of food is low, since little if any concentrate food is bought and little or no expenditure is incurred in pasture improvement or the growing of specific forage crops. Under systems of nomadic pastoralism, for instance, the food cost can be virtually equated to the cost of buying rock-salt as a mineral supplement. However, as large- and small-scale tropical agriculture becomes more efficient the cost of livestock feed must increase and will soon become one of the major items of expenditure. The cost of feed often becomes greater in proportion to the total costs of production in those areas where the costs of land, labour, livestock and buildings are low. It is therefore important that livestock feeding programmes should be efficiently planned, and that the greatest possible use should be made of local foodstuffs, correctly supplemented where necessary. The balanced rations thus devised must be fed in amounts appropriate to the liveweights and production levels of the animals in question. This exercise requires a knowledge of the basic principles of animal nutrition. This subject can only be dealt with in general outline in a book such as this, and the reader is referred to the standard textbooks on the subject, and especially to the work of Morrison (1956), which covers fairly comprehensively the majority of the common tropical foodstuffs.

A brief account will now be given of the main kinds of substance that are found to a greater or lesser extent in most foodstuffs. These substances may be listed as: Water; carbohydrates; proteins; non-protein nitrogenous substances; fats and oils; minerals; vitamins. We will deal briefly with each in turn.

WATER

All foodstuffs contain some water. The common bulky foods consumed by ruminant farm animals contain large amounts of water. Often 85 per cent of the fresh weight of young succulent grass is water, and so tropical cattle, sheep and goats can at times obtain all their water requirements merely by eating grass, without drinking any water at all. On the other hand the water content of most tropical cereals is between 10 and 25 per cent, and the water content of bought-in dry concentrates may be 10 per cent or even less. Clearly, livestock fed intensively on cereal diets have a large requirement for drinking water, and the supply of sufficient quantities of clean water is extremely critical. Tropical poultry raised on a deep litter system, for instance, will die if they are denied water for much more than twenty-four hours, whereas they can live for several days without food without suffering much more than a temporary setback to their growth or productivity.

The amount of water required will depend not only on the water content of the food, but also on prevailing climatic conditions, on the class, breed and type of animal, and on the animal's physiological state. Milking cows require more water than dry cows, and laying poultry more than poultry which are broody or moulting. Because of these facts it is impossible to lay down anything but a rough guide to the water requirements of tropical livestock. The data tabulated in Table 53 are taken from Leitch and Thompson (1944) and the values can be more than doubled when climatic conditions are hot and dry or where the diet of the animals is low in water content.

Table 53. *Approximate daily requirements of different forms of livestock for water*

Leitch and Thompson (1944)

Class of livestock	Daily requirements		Ratio of dry food to water
	lb	kg	
Working cattle	70	31·8	1:4
Milking cows	70 plus 3 lb for each lb milk produced	31·8 plus 3 kg for each kg milk produced	Varies widely
Idle cattle	50	22·7	1:4
Sheep and goats	2·5–9·0	1·1–4·1	1:2·5 to 4
Lactating sows	Up to 50	Up to 22·7	Varies widely
Bacon pigs	Up to 5	Up to 2·3	1:3

The figures given in Table 53 are nothing more than a general guide to minimal water requirements. Large differences in water consumption are found for adapted and unadapted stock maintained under the same tropical climatic conditions. For instance, French (1956b) recorded that indigenous African zebus, about $1\frac{1}{2}$ years of age, drank 130 lb of water per day when given access to water once each day, and an average of 89 lb of water per day when watered only every second day. On the other hand unadapted Grade-Ayrshire cattle, under similar conditions, drank 242 lb and 214 lb of water respectively. The unadapted cattle therefore drank approximately $2\frac{1}{2}$ times as much water as the adapted cattle. It follows that water consumption data calculated for a particular breed cannot be freely employed to estimate water requirements of a dissimilar breed, even though other variables are held constant.

The water content of certain grasses during the wet-season may be so high that not only will the grazing animal obtain all, or almost all, its water requirements through its feed, but the intake of nutrients other than water may be restricted. The rumen of the animal may be filled to capacity, and grazing halted, simply by the intake of unusually high quantities of water in the food. On the other hand, an acute shortage of water will also decrease food intake, since most livestock, except camels and other desert animals, exhibit decreased appetites for food when they are thirsty (Phillips, 1961). It will therefore be noted that both extreme conditions, excess moisture content of lush tropical herbage on the one hand and acute shortage of drinking water on the other, have the same effect of limiting dry-matter intake thereby decreasing animal productivity.

CARBOHYDRATES

Carbohydrates are the main source from which energy is derived and fats are elaborated, although proteins and fats can also be broken down to provide energy if necessary. Since larger amounts of energy-producing food are required than other classes of nutrients, the provision of sufficient carbohydrate in a diet is important. Carbohydrates are usually supplied as sugars, starches, cellulose or constituents of fibre. Animals with simple stomachs mainly utilize carbohydrate supplied in the form of sugars and starches, although they are able to digest a limited amount of fibre and even that which is not digested assists in stimulating digestive secretions and allows easier movement of digesta through the gut. The main carbohydrate supplies for pigs and poultry are cereal grains and by-products, which contain about 70–90 per cent of starch, and the numerous tropical root crops, such as yams, sweet potatoes, cassava, etc., which contain 80–95 per cent starch.

The main carbohydrate supplies for ruminant farm animals are derived from grass and other bulky, so-called 'fodder crops' such as clovers, edible leaves of trees, and the various products resulting from the drying and processing of these crops, such as hay, straw, silage and dried grass. These crops are relatively high in fibre, especially when they are mature, and variable proportions of this fibre are reduced to fatty acids and simple carbohydrates by the action of the micro-organisms present in the alimentary canals of all ruminants, especially in the first stomach or rumen. The flora of the rumen is under detailed study in many temperate research centres, and it is known that it varies according to the type of food being fed. The changeover from one diet to another may necessitate a major change in the composition of the rumen microflora, and as such a change may take time it is usually best to alter diets slowly, over a period of days, rather than abruptly. Although little is yet known about the details of the microflora of tropical herbivores, it would seem probable that similar principles apply.

The value of carbohydrate foods is related to the amount of energy they can supply. This value may be expressed in various ways. The least satisfactory is to deduct from a given foodstuff its water, mineral, protein and fat or oil percentage and to label the remainder the carbohydrate percentage, or 'nitrogen-free extract' (NFE).

Another approach is to measure the Gross Energy of the food in terms of the heat it will produce when ignited in the presence of oxygen in a bomb calorimeter. This method provides the calorific value of the food in terms of calories or therms. (A calorie is defined as the amount of heat required to raise 1 g of water 1°C. 1,000 calories = 1 kilocalorie. 1,000 kilocalories = 1 megacalorie or 1 therm.) Medical dietitians usually employ calorific values for assessing the energy value of different foods, but animal nutritionists use a greater variety of methods.

A third method, first developed by Armsby (1917), is to make appropriate deductions from the Gross Energy value of the food to allow for the inefficiency of the animals' digestive system compared to that of the bomb calorimeter. The animal never utilizes all the energy fed to it in its diet as some will always be voided in the faeces and urine or lost in gaseous form. The difference between the Gross Energy input and output is defined as the Metabolizable Energy (i.e. that part of the energy content of the diet actually available for animal metabolism). A further refinement is to correct the Metabolizable Energy for the energy which must be expended during the actual digestive process. The residual value thus obtained is known as the Net Energy of the food, and as Armsby was responsible for calculating some of the earliest Net Energy values this constant is sometimes spoken of as the 'Armsby Net Energy Value'.

The energy content of the diet is used by the farm animal in one of two ways. First, to keep the animal alive and in good health, secondly to enable it to grow or produce a livestock product such as milk, eggs, wool. The first part of the energy intake is called the maintenance requirement, and the latter requirement is designated the production requirement. It follows that the Net Energy estimates can also be broken down into the Net Energy required for maintenance (NEM) and into the Net Energy required for production at a certain level (NEP). The total Net Energy, for both maintenance and production, is often referred to as NEM + P. Further complications arise since a given food will have different values for NEM and NEP, and different values for NEP at different levels of productivity. This fact must be borne in mind when tables listing Net Energy values are being used. Net Energy is usually expressed in such tables as megacal per 100 lb of foodstuff, and Net Energy requirements are usually expressed as megacal or kilocal per day.

The fourth method, widely used by British workers up to the present time, is to calculate a value which relates the energy value of the food in question to the Net Energy value of pure starch, taken as a standard fattening food. This relationship is expressed as the Starch Equivalent (or SE) of the foodstuff, where Starch Equivalent is defined as the number of pounds weight of pure starch which will provide as much Net Energy for a fattening steer as 100 lb of the foodstuff.

A number of other methods are employed on the continent, or have been suggested by British workers (Blaxter, 1962) but it is not possible to detail these in this book.

Starch Equivalents are linked with the name of Kellner (1905) who was the first to use this method of comparing the energy values of different foods. Starch Equivalents are subject to criticism since they refer, by definition, to the energy values of foods fed to fattening steers, yet they are used for determining the feeding values of feeds fed to other classes of livestock for a variety of other purposes (Brody, 1945).

From the standpoint of tropical feeding standards, this same criticism can be levelled against all energy values apart from the basic 'calorific value' or 'Gross Energy value'. It does not follow that an Armsby Net Energy value calculated for temperate cattle is applicable to zebu cattle in tropical conditions, and it does not follow that a Kellner Starch Equivalent worked out for temperate fattening steers will apply, say, to foods fed to tropical meat goats. All these various energy values should be taken as rough guides, showing the relative effectiveness of one foodstuff compared to another as a source of energy for farm animals. No attempts should be made to apply feeding standards and schedules devised in temperate regions too rigidly in the tropics. Much more work requires to be carried out in this field before real precision can be used to calculate diets for optimal economic production with tropical breeds of livestock.

Since the tropical animal husbandman is often confronted with nutritional requirements for energy quoted in one unit, and a list of the energy values of different tropical foods employing another unit, it is frequently necessary to translate values from one unit to another. This exercise is a dangerous one, since energy values are not strictly convertible, but rough calculations can be made which enable approximate conversions from one unit to another. Thus Net Energy values are usually equivalent to approximately 60 per cent of Metabolizable Energy values when utilized as NEP values for fattening meat animals, and a rough and ready conversion of Net Energy (NEM + P) into Starch Equivalents may be obtained by use of the constant 1,071 kilocal = 1 lb of SE (see also Crampton *et al.*, 1957; Harris, 1962).

PROTEINS, AND NON-PROTEIN NITROGENOUS SUBSTANCES

Proteins are chemical compounds of great complexity, containing about 16 per cent of nitrogen in addition to the three basic elements common to carbohydrates: carbon, hydrogen and oxygen. On hydrolysis, proteins break down into their constituent amino acids, and the nature and value of a particular protein may be expressed by enumerating the amino acids thus released.

The vital living tissues of all animals are mainly composed of proteins, hence the supply of the raw materials which allow proteins to be synthesized in the animal body is of vital importance, especially with stock which are actively growing and increasing in weight. Proteins are not absorbed as such, hence it is not necessary to feed protein to farm animals in exactly the form it will subsequently take as part of that animal's body. However, the supply of the right raw ingredients, in approximately the right ratio, is vital if the conversion of nitrogenous compounds into animal tissue is to be efficient. Non-ruminant farm animals usually obtain most of their protein by eating plant or animal proteins, breaking them down by a variety of enzymes and absorbing the various amino acids thus formed. In the case of the ruminant animal, however, it is possible to provide a protein-source more indirectly by feeding the animal on simple nitrogenous compounds, such as urea or ammonium salts in certain forms, and allowing the bacteria present in the rumen to feed on these substances and build them up into animal proteins inside their cells. These bacteria die, still inside the gut of the farm animal, and the bacterial proteins break down to amino acids which can then be absorbed in the normal manner. It follows from this that the actual protein content of a foodstuff is an important parameter in the case of non-ruminants, but that the nitrogen content of a foodstuff has more validity in assessing its feeding value to a ruminant. In point of fact the determination of the actual protein content of a foodstuff is less rarely carried out than the assessment of the nitrogen content by the Kjeldhal process (Wright, 1938). The nitrogen content is then multiplied by $6\frac{1}{4}$, since proteins on average contain about 16 per cent of nitrogen. This parameter is referred to as the Crude Protein percentage, and it will invariably be somewhat higher than the True Protein percentage, as it will include an allowance for non-protein nitrogen. In the case of ruminants, which use most of their nitrogenous foodstuffs, this yardstick is quite a reasonable one, but in the case of non-ruminants, which can utilize relatively less of the non-protein nitrogen contained in their food, it is likely to give an over-estimate of the protein value. For this reason, a standard parameter known as a 'Protein Equivalent' has been devised. This is defined as the percentage of true protein, plus half the non-protein nitrogen multiplied by the factor $6\frac{1}{4}$. Tables based on the work of British scientists often give values in terms of Protein Equivalents (PE). Tables compiled by American workers frequently give the Crude Protein percentage (CP), often misleadingly abbreviated to Protein Percentage.

It has already been stated that proteins are made up of amino acids, the nature and proportion of the amino acids varying in different proteins. Certain foods, especially those of plant origin, have a range of amino acids which differs from the range required in the synthesis of protein by the farm animal. Livestock may, therefore, require a large amount of food to enable them to synthesize a given amount of protein. In extreme cases, the food offered may be completely deficient in a certain amino acid with the result that however much is offered to the animal there can be no prospect of the animal synthesizing its own specific protein from that food alone. Proteins in food accordingly have different 'biological values'. The value is 'high' when the quantity required to produce the same amount of animal protein is relatively large. As a general rule, animal protein in a food (such as fish meal or meat meal) has a higher biological value than vegetable protein (such as soya-bean meal or a leguminous hay).

CARBOHYDRATE AND PROTEIN DIGESTIBILITY

Although carbohydrates cannot replace proteins in a foodstuff, the proteins can be broken down and certain of their derivatives that contain carbon, hydrogen and oxygen can be used as a source of energy, or for the elaboration of fat. A protein food can be used for either or both of these purposes if it is not required by the body for the manufacture of animal protein. It follows that part of the protein percentage in a foodstuff should be rightly included in any estimate of the food's energy value.

On the other hand, a mere chemical analysis of a foodstuff in a laboratory into its protein content and its energy content can be very misleading, because certain foods are relatively indigestible to livestock since their energy and protein fractions may be contained inside

41 Charolais cattle at the Star Farm, Point-a-Pierre, Trinidad. *Left:* a 20-month bull, imported from Florida, weighing 1350 lb. *Right:* a 7/8 bred Charbray (Charolais Brahman) heifer weighing 1200 lb. at 20 months. Note the good beef conformation of the hindquarters of both animals. The French Charolais breed is now widely used for crossing with Zebu breeds in many parts of the tropics. (see Chapter 14)

42 White Fulani cattle at the Animal Health Livestock Farm, Accra, Ghana. The heifer in the foreground is three years of age and is in calf for the first time. Note the uniform colouring and the upright horns. (see Chapters 14, 16 and 17)

43 Improved N'dama cattle at the Main Farm, Pong-Tamale Veterinary Station, Ghana. Note the short hair, upright horns and absence of the typical Zebu hump. Unimproved N'dama cattle are often very small, averaging about 400 lb. when adult. The N'dama breed is relatively more resistant to trypanosomiasis than other African breeds. (see Chapters 14 and 16)

44 Criollo cow on a milk ranch near Maracaibo, Venezuela. Large numbers of these cattle, which are descended from the Spanish Longhorn cattle imported from Europe, are found on the low-lying ranches adjoining the Maracaibo Sea. The fresh milk is transported daily by road in refrigerated tankers to Caracas, a distance of over 400 miles. (see Chapter 16)

woody, fibrous tissues which are not broken down readily by mastication or by bacterial action. For this reason, it is important to correct the percentages of Gross Energy or total protein in a food by multiplying them by their percentage digestibility, more usually referred to as their 'Digestibility Coefficients' (Armsby, 1887). These digestibility coefficients vary from one class of stock to another, and they should be calculated by experiments in which the energy and protein values of the faeces (but not of the urine and waste gases) are deducted from the energy and protein values of the food itself. However, it has been found by several tropical workers (Glover and French, 1957; French *et al.*, 1960; Butterworth, 1963) that good estimates of the digestibility can be derived from equations involving the 'richness' of the food in terms of carbohydrate and protein, together with the food's fibre content. Thus a 'rich' food, low in fibre, may be expected to have a high digestibility coefficient, (e.g. meat meal). Conversely a 'poor' food, high in fibre, will usually be found to have a low digestibility (e.g. standing hay or foggage).

Since most of the protein content of the food, all the carbohydrate, together with portions of other ingredients, such as fibre and oil, can be used as an energy source, certain workers employ yet another standard parameter to arrive at the best estimate of a food's energy value. This is the 'Total Digestible Nutrients' (or TDN value) and it is defined as the percentage of digestible fat and oil (multiplied by 2·3), plus the percentages of digestible crude protein, digestible carbohydrate and digestible crude fibre. Total digestible nutrients are widely employed by American nutritionists (and hence by tropical scientists trained in the U.S.A.), and the feeding-values of many foodstuffs analysed by American laboratories are expressed in terms of digestible crude protein and total digestible nutrients (DCP and TDN). In order that the reader may be able to refer to nutritional data compiled by different workers using different systems, it is important to be able to equate, albeit roughly, SE and NE and TDN. As a rough guide, reasonably accurate for fattening cattle but of somewhat doubtful validity for other stock, 1 lb of TDN is equivalent to about 0·91 lb of SE or to 970 kilocal of Net Energy.

The amount of digestible carbohydrate and digestible protein, and the ratio of these two dietary constituents the one to the other (sometimes referred to as the 'Nutrient Ratio' or NR) are generally regarded as the most important parameters when working out balanced diets, (see Morrison, 1964). These standard parameters, in terms of pounds of DCP and TDN, will be tabulated for the different classes of livestock at different stages of growth and different levels of productivity later in this chapter.

FATS AND OILS

Fats and oils are similar to carbohydrates in that they are primarily composed of carbon, hydrogen and oxygen. They differ since they have relatively less oxygen in their molecules than carbohydrates, hence more oxygen is needed for combustion and consequently about two and a half times as much energy is liberated per unit weight on complete oxidation.

The fats and oils present in most stock foods are compounds of glycerine with certain fatty acids. Fats and oils are broken down during the digestive process into these major constituents, which are then absorbed and either built up in the body of the animal to form a specific animal fat, usually a triglyceride, or oxidized to gaseous end-products if the animal uses its food fat as an energy source. This oxidation process also occurs every dry-season in the case of those animals, such as tropical ranch cattle, which lose weight during periods of food shortage by depleting their reserves of body fat in order to provide for their energy requirements.

The quantity and quality of fat laid down in a farm animal can be partially regulated by altering the nutrition of the animal. (Within narrower limits, the amount of butterfat secreted in the milk of cows and goats can also be controlled in this fashion.) It therefore follows that diets have to be so designed that the right sort of fat is laid down in the animal, namely, hard light fat. A century or so ago the most highly prized meat animals were, according to current standards, excessively fat. Certain tropical peoples still place great store on a high fat content of slaughter stock, but the usual current market tendency is to give premium prices for animals which have a reasonable, but not excessive, covering of subcutaneous fat, the minimum of abdominal fat and a comparatively generous amount of fat both between and within the muscle bundles. This latter fat, known as intramuscular fat or 'marbling fat', is partly responsible for determining the eating quality of meat, since its presence probably enables the meat to be roasted or baked without unduly drying out and thus becoming tough and unpalatable.

There is usually little or no problem in preventing excess fat deposition in ranch cattle. On the contrary, it is often difficult to obtain a good 'finish', in the form of an even covering of subcutaneous fat. However, there are definite problems of excess fat deposition with all intensively fattened tropical livestock. These are due to the relatively low maintenance requirements of such animals in hot climates which enable them to utilize a much higher proportion of their energy intake for the deposition of fat. If the energy requirements of such animals are fully provided for by carbohydrate foods, then there is a possibility that the oil and fat content of the diet may be primarily employed for fat deposition and that carcass grade will deteriorate as a consequence. This problem is especially acute in the case of pigs since the improved breeds of pigs, when imported into the tropics, invariably lay down more fat when fed standard temperate zone pork or bacon diets than they would do in a cooler climate (Houghton *et al.*, 1964).

The quality of fat deposited in an animal is usually judged by its degree of hardness and its colour. Hard fats are generally preferable to soft fats, and white or light coloured fats are usually preferable to yellow or dark coloured fats.

It used to be thought that feeding stock on diets high in oil or fat would lead to the formation of soft fat, whereas feeding animals on non-fatty diets, such that body fat must be synthesized from raw materials derived from carbohydrate sources, would result in the deposition of hard fat. It is now known that this is an over-simplification. Relatively high levels of fat and oil can be safely incorporated in the diet providing they are composed of the 'right' fatty acids, such as stearic and palmitic, which are associated with hard fat. Indeed, it is now known that a proportion of the carbohydrate ingredients in the diet can be replaced by certain fats and oils without serious effects on carcass quality (Edwards *et al.*, 1961). As a general rule, however, it is wise to avoid diets high in fat or oil in the case of tropical livestock, such as pigs, in which the problem of excess fattiness has not been resolved. Foodstuffs containing fats or oils with a high proportion of oleic and other fatty acids associated with soft fat should be avoided, particularly during the finishing stages of the fattening process. The hardness of fat can also be effected by the rate of growth. Quick-growing stock tend to have harder fat than slow-maturing stock, therefore high planes of nutrition, with consequent high rates of liveweight gain, should be adopted whenever they are economically practical.

Fat colour depends upon the presence or absence of pigments, or pigment precursors, in the diet. Foodstuffs high in carotene content, such as yellow maize, tend to produce yellow-coloured fat. This yellow pigment is not depleted at the same rate as the fat tissue itself during periods of substandard nutrition. Animals fattened over a long period, along a zig-zag growth curve on foodstuffs containing a lot of carotene, therefore tend to yield carcasses with an excessively coloured fat which will not find favour with the meat trade. For this reason the majority of ranch cattle, which take anything up to six or seven years to reach slaughter weight, often contain a highly coloured fat.

The colour of the fat is also determined by species and, to a lesser extent, by breed. Thus cattle fed primarily on grass will invariably contain fats with some degree of yellow colour. Water buffaloes, on the other hand, reared and fed under similar conditions, will contain fat which is almost pure white (Houghton, 1960).

MINERALS

The total amount of mineral matter, or 'ash', present in the body of animals is a small proportion of the dry-matter, but adequate quantities are essential for growth and production. The proportion of minerals in the diet is therefore important for all forms of livestock, but the mineral levels may be quite critical for stock, such as milking cows or laying hens, which are excreting large quantities of minerals in their milk or eggs. As the loss of certain minerals, such as sodium in the form of common salt, is much greater for tropical animals that are sweating or salivating profusely than it is for temperate animals, the standard requirements for certain minerals as calculated in the temperate zone are inadequate in the tropics. Unfortunately the requirements of stock under different tropical environments have not been precisely determined, and it is a sound and common practice to provide mineral licks so that they may regulate their own intake.

Minerals may be divided into two categories according to the amount required by the animal. Those minerals required in relatively large quantities are referred to as the 'major elements', and these include sodium, calcium, phosphorus, iron, chlorine, magnesium, potassium and sulphur. Minerals required in much smaller quantities but which are still essential for normal growth and development are known as the 'trace elements'. It is likely that the list of such trace elements will lengthen as more is learnt about the nutritional requirements of farm animals, but at present the list as as follows: iodine, copper, cobalt, zinc, manganese and selenium.

Although certain minerals, such as calcium and phosphorus, are 'stored' on the body (e.g. in the form of skeletal deposits), it is usually unwise to consider these deposits as 'reserves' on which the animal can draw for long periods. Depletion of the calcium and phosphorus levels of bones, for instance, can lead to a weakening of this tissue, giving rise to major deformation or to fragile bones which are easily fractured. Other elements, such as sodium and chlorine, are not stored in any appreciable quantity at all. Consequently, in places where minerals are in short supply, mineral supplementation must be regularly and methodically provided by some means or other. Unfortunately for the farmer, mineral excess can at times be as detrimental as mineral deficiency, and hence the correct balance of the mineral constituents in a ration is very important. A notable example of mineral-excess is the occurrence of fluorosis in Tanzania, due to an excess of fluorine in the water supplies of certain areas. Mineral bricks and blocks are usually made up in such a way that a correct balance is maintained, but there may be local mineral excesses or deficiencies due to differences in the composition of the soil and water supplies of the area. In these cases, it may be possible to compensate for these localized peculiarities by manufacturing specific mineral supplements in which the mineral imbalance is corrected. It is important that the tropical livestock producer should ascertain whether such a situation exists in his region, and try to ensure that the suppliers of mineral supplements have taken any major local peculiarities into account. In the majority of cases, unfortunately, the mineral status of the region will be unknown and the producer will have to use a standard mineral supplement.

In the case of intensively managed livestock, which are fed mainly or exclusively on bought-in concentrates, the mineral supplementation of the ration may have been taken care of by the manufacturer. The farmer should check that this has been done, especially if he is purchasing his livestock feed from a small local mixing plant which may not have the knowledge or facilities for adding small quantities of trace elements to the mixed ration. Even where the food is correctly supplemented, there may still be a need for the farmer to take additional steps to ensure that mineral intake is never limiting production. The milk of the lactating pig, for instance, is invariably low in iron content however good the sow's ration, and it is wise to give young piglets supplementary iron either in the form of injections or by painting the teats of the sow with iron in some suitably available form.

Browsing animals, such as goats and camels, are usually better able to meet their mineral requirements than selective grazers, such as imported, European cattle. The leaves of certain weeds and the leaves and pods of many tropical trees, are particularly high in mineral content, and livestock consuming them will obtain quite high levels of mineral intake from such sources. Animals which mainly subsist on grasses and fibrous fodder crops are likely to have reasonably high levels of calcium intake, but such diets are likely to be low in phosphorus. Conversely, stock which are maintained primarily on cereal diets, such as feed-lot beef cattle, will tend to have a good supply of phosphorus but calcium levels may be limiting. Animals which are fed on a grazing system with cereal supplementation are least likely to suffer from calcium–phosphorus imbalance, but this critical ratio should always be considered in the case of stock on grass-only or cereal-only diets.

In temperate agriculture, gross deficiencies or excesses in the mineral content of the herbage are generally corrected by appropriate fertilizer treatment to the soil. In the tropics such corrections are only possible under intensive farming conditions, where the productivity of the soil is potentially high. Under extensive ranching conditions little or no modification can be made to the mineral balance of the land itself, and all attention has to be devoted to remedying the situation by supplementing the diet of the animals.

Although very little work has been carried out on the trace-element requirements of tropical livestock compared to temperate livestock, it is probable that temperate standards of trace-element requirements will serve as useful guides. The following data have been calculated by 'averaging' the different standards suggested by various workers in different countries:

Copper (Cu) 3–5 ppm for pigs; 5–10 ppm for cattle and sheep.
Cobalt (Co) 0·04 ppm for pigs; 0·1 ppm for cattle, sheep and goats.
Iron (Fe) 30–60 ppm for pigs; 30 ppm for cattle, sheep and goats.
Iodine (I) 0·1 ppm for pigs; 0·1 ppm for beef cattle and sheep; 0·8 ppm for lactating cattle and goats.
Manganese (Mn) 10–30 ppm for pigs, cattle and sheep.
Selenium (Se) Possibly 0·1 ppm for all classes of stock (no accurate data available).
Zinc (Zn) 40–80 ppm for pigs; 30 ppm for cattle and sheep.

VITAMINS

The vitamin requirements of extensively managed tropical livestock are usually met fairly readily, since they are either present in the herbage or else they are synthesized in the alimentary canal of the animals themselves. However, as more intensive methods of animal husbandry are adopted in the tropics, especially intensive pig and poultry production, attention to the amount and potency of the vitamin content of concentrate rations will become very necessary. One factor that should always be taken into account is that many vitamins are less stable under hot, tropical conditions than they are in temperate climates. Many vitamins have their potency reduced, or destroyed, by heat or by chemical action such as oxidation. Feed stores in the tropics are usually very hot, often being constructed of corrugated iron which is a very effective conductor of heat. Again, supplies of food often have to be kept for much longer periods in the tropics, and hence the risk of vitamin degradation is more pronounced. The vitamins of importance to animal nutrition are A, members of the related B-complex which extends to vit. B_{12}, C, D, E and K. Each will now be briefly considered.

Vitamin A. This is a fat-soluble vitamin formed in livestock from carotene obtained by eating green foodstuffs. It is destroyed by oxidation, so that hay, straw and silage, which undergo some degree of oxidation during the curing process, supply very little vitamin A precursor compared to fresh grass or other herbage. Vitamin A is stored in the liver of animals, and for this reason the supply of vitamin A need not be continuous except in the case of animals such as milking cows and laying hens which are excreting vitamin A in their products. As vitamin A is stored in this manner, the livers of fishes, birds and mammals are a rich source and cod-liver oil and halibut-liver oil are commonly used as vitamin A supplements. Because of the loss of potency of this vitamin through oxidation, it is necessary to 'stabilize' the source of vitamin A in animal concentrates by adding anti-oxidants. Without this precaution, which is particularly necessary with concentrates for feeding to animals denied access to green vegetation, the declared percentage of fish oil (or other source) may be very misleading. Vitamin A deficiency has been reported in poultry fed on rations containing 3 per cent of non-stabilized cod-liver oil. Wherever practicable, the practice of hanging up bunches of green leaves inside intensive poultry houses is commendable. This precaution not only ensures that poultry have access to a potent vitamin A precursor, but it also lessens troubles due to feather-pecking and cannibalism.

Vitamin B complex. This omnibus title covers a group of water-soluble vitamins, including riboflavin, pantothenic acid, nicotinic acid, pyridoxine, biotin, folic acid, choline and the two vitamins usually still described as B_1 (Thiamine) and B_{12}. A complication in nomenclature arises from the fact that vitamin B_1 is known as vitamin F, and B_{12} as vitamin G, in the U.S.A.

Vitamin B_{12} has received a great deal of attention in recent years and this research has been reviewed by Brown (1960). It was previously known as the Animal Protein Factor (APF) and it is thought to be primarily responsible for the biosynthesis of protein from amino acids (Wayle *et al.*, 1958). It is a very complex organic compound, and its action is closely associated with the copper metabolism of the animal. It is a common practice to add vitamin B_{12} to a large range of livestock foodstuffs, but the other members of this B-group are fairly readily synthesized in the alimentary canal of ruminants and are present in varying amounts in many of the common ingredients of pig and poultry rations, so deficiencies should not often be encountered. Meat and bone meal, whey and dried yeast are rich sources of the B-vitamins, and incorporation of small percentages of one or more of these ingredients in concentrate rations will usually take care of normal requirements.

Vitamin C (Ascorbic acid). This vitamin, once spoken of as the anti-scorbutic vitamin, is synthesized by farm livestock and is therefore comparatively unimportant in animal nutrition.

Vitamin D complex. This group of vitamins is associated with the calcium and phosphorus metabolism, and deficiency of D indirectly gives rise to rickets. The D-vitamins are synthesized in the skin of farm animals through the action of solar radiation, either direct or reflected. For this reason vitamin D deficiency is extremely rare in animals which are exposed to sunlight in the tropics. Vitamin D deficiency is most likely to occur in the case of intensively kept poultry, housed in buildings with low-pitched roofs which effectively exclude all ultra-violet radiation. In such cases, supplementation of the diet with sources of vitamin D_3, such as animal fat, meat and bone meal or fish liver oils may be necessary. Vitamin D_2 is less readily utilized by poultry than vitamin D_3, consequently vitamin D supplementation of poultry diets should always be in the form of D_3. Vitamin D is fat-soluble, and it will be noted that many substances of animal origin which are rich in vitamin D are also good sources of vitamin A.

Vitamin E. This vitamin, alpha-tocopherol, is known as the anti-sterility vitamin, since a deficiency gives rise to degeneration of the gonads and eventually to sterility. It is present in green forage and in many cereal grains, and hence the occurrence of avitaminosis E amongst tropical livestock is rare. It is stored in the animal's body, so stock can withstand quite long periods on a vitamin E-deficient diet. By the same token, meat meal and animal fat can be used in concentrate diets to ensure adequate vitamin E intakes.

Vitamin K. The absence of this vitamin causes weakness of the walls of the blood capillaries and delays the time taken for blood to clot. It is synthesized by the micro-organisms of the alimentary canal of ruminant animals. Livestock with simple stomachs must obtain their requirements from one of the many common sources, the most important being cereal grains. Vitamin K deficiency has been found in tropical poultry, and the condition can be rectified by adding manadione and sodium bisulphite if good sources of naturally occurring vitamin K are unavailable.

FEEDING STANDARDS

While the foregoing discussion has provided a qualitative indication of the value of different food constituents for different classes of stock, it is clearly desirable that the farmer and feeding stuff manufacturer

Improved Management: I. Nutrition

should have quantitative information on how much of each nutrient to feed each type of livestock at each stage of growth.

Unfortunately, precise information of this nature applicable to tropical livestock is not available. Temperate requirements are set forth in various publications, but even here suggested requirements vary greatly from author to author. It is clear that many of these standards are misleading if applied to tropical livestock. Careful analyses of the diet available to many extensively-managed tropical stock reveal nutrient intakes which are so low that, were temperate data wholly applicable, the stock in question would all be dead. Clearly, therefore, any attempt to lay down tropical feeding standards at this stage of our knowledge must be made with many reservations, and the only claim which can be made with confidence about the figures which follow is that they will eventually have to be superseded in the light of further knowledge. The reader should bear these points constantly in mind when using the tentative figures in the schedules which follow.

Many feeding standards express the energy, protein and mineral requirements as percentages of the diet. The amount of dietary ingredient consumed is therefore dependent on the quantity of total food eaten. As feed intakes vary widely, it is preferable to express feeding standards in terms of actual quantities (by weight) of each dietary component. The schedules which follow have been drawn up on this basis.

SCHEDULE OF TENTATIVE NUTRIENT REQUIREMENTS FOR LIVESTOCK

1. *Cattle*

(a) *Calves.* Calves should receive colostrum, preferably from their own dam, for the first 3–4 days of life. Thereafter it is advisable to rear calves on whole milk for at least the first month, and usually for the first two months, then changing over to a milk substitute or skim milk with at least a three-week changeover period, subsequently weaning from milk substitutes to solid food and herbage over a period extending from about the third month to the fifth month. Early weaning systems are not often practised in the tropics, and then only with *Bos taurus* or crossbred stock, or *B. indicus* stock kept under efficient management. Whole milk should be fed at a rate not exceeding about one gallon per calf per day, and skim milk up to a maximum of about 1–1½ gallons per calf per day by the end of the third month. As milk substitutes decrease after the end of the third month, some form of concentrate ration can be fed at rates increasing to a maximum of about 4 lb per calf per day at weaning. A very rough guide to the total food requirement of calves is that they will normally consume about 10 per cent of their liveweight in weight of food. Where calves are reared on whole milk followed by skim milk or milk substitute, the composition of the concentrate supplement is not critical, but it should be relatively high in DCP (about 16 per cent) and low in fibrous material of low digestibility.

(b) *Growing cattle.* The nutrient requirements of growing-stock are closely related to their liveweight. The following brief schedule will allow the reader to interpolate intermediate values if required. TDN values have been tabulated at lower rates than those normally advocated for temperate stock. No attempt has been made to differentiate between the maintenance and production requirements of growing cattle. Table 54 is based upon liveweight gains between 1 and 1½ lb per head per day. Calcium and phosphorus contents are important to the growing animal, and approximate daily requirements have been included in Table 54.

Table 54. *Approximate nutrient requirements of growing cattle*

Liveweight (lb)	Daily intake of dry matter (lb)	Daily requirements of nutrients			
		DCP (lb)	TDN (lb)	Calcium (g)	Phosphorus (g)
250	6–7	0·6–0·7	3·5–4·5	13	11
500	10–12	0·9–1·0	6·5–7·5	16	14
750	14–17	1·0–1·2	8·5–9·5	21	19

(c) *Dairy cows.* The nutrient requirements for dairy cows will be dealt with in two parts. First, their requirements for maintenance are as set out in Table 55, which is based on the liveweight of the cow; secondly, the production ration for cows in milk, which should be added to the relevant maintenance ration (Table 56).

For each gallon of milk produced the dairy cow will require about

Table 55. *Approximate maintenance requirements for dairy cows*

Liveweight (lb)	Daily intake of dry matter (lb)	Daily requirements of nutrients	
		DCP (lb)	TDN (lb)
700	13 –16	0·4–0·6	5·0–5·5
800	14½–17½	0·5–0·7	5·5–6·0
900	16 –19	0·6–0·8	6·0–6·5
1000	17½–20½	0·7–0·9	6·5–7·0

3·2 lb TDN and 0·6 lb of DCP, assuming the butterfat content to be about 4 per cent. For high butterfat cows, such as zebus, add 0·05 lb TDN for every 0·1 per cent increase in butterfat. Pre-calving cows can be treated as cows yielding one gallon of milk a day. The calcium and phosphorus levels in the feed of milking cows are extremely critical, as so much of these minerals is excreted in the milk. It may be necessary to add these minerals to the diet in some such form as bone meal or dicalcium phosphate.

Table 56. *Approximate production requirements for dairy cows*

Daily yield of milk (gal)	Daily requirements of nutrients			
	DCP (lb)	TDN (lb)	Calcium (g)	Phosphorus (g)
1 at 4% b.f.	0·6	3·2	10	8
1 at 5% b.f.	0·6	3·7	10	8
2 at 4% b.f.	1·2	6·4	20	16
2 at 5% b.f.	1·2	7·4	20	16
3 at 4% b.f.	1·8	9·6	30	24
3 at 5% b.f.	1·8	11·1	30	24
4 at 4% b.f.	2·4	12·8	40	32
4 at 5% b.f.	2·4	14·8	40	32

(*d*) *Fattening beef cattle.* The maintenance and production requirements for liveweight gains of about 1 lb per head per day are as set out in Table 57. About 2 lb of extra TDN are required for each additional pound of gain in excess of this standard.

Table 57. *Approximate maintenance and production requirements for beef cattle*

Liveweight (lb)	Daily intake of dry matter (lb)	Daily requirement of nutrients	
		DCP (lb)	TDN (lb)
800	20–22	1·2–1·4	10·0–12·0
900	21–23	1·3–1·5	11·0–13·0
1000	22–24	1·4–1·7	12·0–14·0
1100	23–25	1·5–1·8	13·0–15·0

(*e*) *Working oxen.* The maintenance and production requirements for mature working oxen are presented in Table 58. These two requirements are additive, in the same way that the maintenance and production rations for dairy cows are additive in section (*c*) above.

Table 58. *Approximate production and maintenance requirements for working oxen*

	Daily maintenance requirements			Production requirements per working hour	
Liveweight (*lb*)	Dry matter (*lb*)	DCP (*lb*)	TDN (*lb*)	DCP (*lb*)	TDN (*lb*)
600	11–15	0·3	4·8	0·03	0·27
800	13–17	0·4	6·0	0·03	0·30
1000	15–19	0·5	7·2	0·03	0·33
1200	17–21	0·6	8·4	0·03	0·36

2. Milking sheep and goats

The maintenance requirements are set out in Table 59, and the production requirements in Table 60 below. These requirements are additive, in similar fashion to the maintenance and production requirements of dairy cattle and working oxen already described.

Table 59. *Approximate maintenance requirements for sheep and goats*

Type	Liveweight (*lb*) when mature	Daily requirements of nutrients	
		DCP (*lb*)	TDN (*lb*)
Temperate breeds used in tropics	about 150	0·07	1·7
Crossbred stock	about 100	0·05	1·2
Tropical breeds	about 60	0·04	1·0

Table 60. *Approximate production requirements for milking sheep and goats.*

Daily yield of milk (gals)	Daily requirements of nutrients	
	DCP (*lb*)	TDN (*lb*)
0·5 at 4% b.f.	0·35	1·6
0·5 at 5% b.f.	0·40	1·8
1·0 at 4% b.f.	0·70	3·2
1·0 at 5% b.f.	0·80	3·6
1·5 at 4% b.f.	1·05	4·8
1.5 at 5% b.f.	1·20	5·4

Tropical sheep and goats are usually suckled, hence suitable nutritional requirements for hand-reared lambs and kids are not tabulated.

3. Pigs

In temperate agriculture it is common practice to express pig rations in terms of 'pig meal' required at different growth stages, and of the 'nutrient ratio' of the meal at each stage. Thus young piglets should be weaned on to high protein meals (nutrient ratio 1 part protein to 4 parts carbohydrate) and can be fattened for pork or bacon on meals of lower protein content (nutrient ratio 1 part protein to 6 or 7 parts carbohydrate). Such standards are applicable where 'whole meal' feeding is the rule and pigs are intensively kept. In many parts of the tropics, however, pigs are raised semi-extensively and the problem is usually one of balancing a bulky diet, consisting of vegetable wastes and by-products, with a high protein concentrate. Such methods were commonly practised in Europe during the war when pig foods were scarce, and are often referred to as the 'Lehmann system' after the German worker who first described it. In the tropics, therefore, it is useful to discuss requirements of the pig in terms of total digestible carbohydrate and protein, rather than in pounds of a hypothetical dry 'pig meal'.

It should be remembered that the pig possesses a simple stomach, and is unable to obtain protein from the breakdown of bacteria which have utilized simple proteins or non-protein nitrogen. It is therefore important to ensure that proteins of high biological value, containing the essential amino acids, are provided. This can be done fairly readily by including in the ration high-protein foods of animal origin, such as fish meal, whale meal or meat meal. The main nutrient requirements for pigs of different weight and type are detailed in Table 61. In compiling these tables, the recent work of Houghton *et al.* (1964) has been taken into consideration. This work indicated that restriction of TDN intakes to about two thirds of the accepted values for temperate pigs resulted in good food conversions, and lower carcass fat contents, in Large White pigs kept in Trinidad.

Table 61. *Approximate daily nutrient requirements for pigs of different weight and type.*

Class and liveweight of pig (lb)		Daily requirements of nutrients			
		DCP (lb)	TDN (lb)	Calcium (g)	Phosphorus (g)
Weaner piglet	25+	0·3	1·5	8	8
Grower	50	0·3	2·0	10	10
Fattener	100	0·5	2·8	12	12
Porker	150	0·6	3·8	15	15
Baconer	200+	0·7	4·3	20	20

4. Poultry

Even under the crudest systems of poultry husbandry, chickens are regarded as grain-feeders, and the basic ingredient of poultry diets the world over is some form of cereal. Carbohydrate is therefore rarely a limiting factor in poultry nutrition, and for this reason tables setting out the TDN requirements of poultry at different stages of growth are unusual. Nutritional requirements are more usually expressed in terms of the daily (or weekly) consumption of 'dry food' such as a cereal grain, and the protein, mineral and vitamin needs of birds at different stages of growth.

Until comparatively recently the protein requirements of poultry were often expressed in terms of the overall protein percentage of the ration. Such standards have limited significance since it is the amounts of different amino acids that are of primary nutritional importance. Overall protein levels can be kept relatively low provided a full range of amino acids is present in approximately correct proportions. (See A.R.C., 1963 and N.R.C., 1960.)

Where poultry are maintained for moderate or low levels of production, as in small-scale peasant agriculture, adequate intakes of cereal and soluble and insoluble 'grit' will enable the flock to perform satisfactorily. However, as more efficient methods of intensive poultry production become more widely practised, attention to amino acid balance, mineral and vitamin content of the ration will be as critical in the tropics as it is in temperate areas. For this reason, the nutritional requirements of poultry will be expressed in two forms; firstly a simple table showing weekly dry-food requirements for poultry at different liveweights, secondly a somewhat more complex table listing the amino acid, vitamin and mineral requirements of the young 'starter chick', 'grower', 'broiler starter', 'broiler finisher' and the laying pullet. It will be noted that many of the requirements are not yet known with any degree of precision. It should also be noted that the requirements for amino acids and, possibly, for vitamins and minerals are not absolute requirements, but are related to the energy intake of

Table 62. *Approximate weekly requirements of dry-food by poultry of different liveweights.*

Liveweight of chicken (lb)	Total dry-food lb	oz	Liveweight of chicken (lb)	Total dry-food lb	oz
$\frac{1}{2}$	0	7	3	1	11
1	0	13	4	1	13
$1\frac{1}{2}$	1	3	5	1	15
2	1	7	6	2	0

Table 63. *Provisional requirements for amino acids, vitamins and minerals of intensively managed poultry*

Item	Chicks 0–8 wk	Growers 9–18 wk	Layers 18 wk+	Broilers (Starter) 0–4 wk	Broilers (Finisher) 5 wk+
Total Protein (g/kg)	200	140	150	240	190
Amino Acids (g/kg)					
Arginine	12	?	?	11	10
Lysine	10	?	5	11	10
Histidine	3	?	?	3	3
Methionine	5	?	3	4	4
Methionine + Cystine	8	?	6	8	7
Tryptophan	2	?	1	2	2
Glycine	10	?	?	9	8
Phenylalanine + Tyrosine	12	?	7	15	13
Leucine	14	?	10	15	13
Isoleucine	6	?	5	7	6
Threonine	6	?	4	8	7
Valine	8	?	?	8	7
Vitamins (i.u. or mg/kg)					
A (stabilized) (i.u.)	2600	2500	4400	2600	2600
B_1 (Thiamine) (mg)	2	?	?	2	2
Riboflavine (mg)	3	3	2	2	2
Pantothenic acid (mg)	9	10	8	10	10
Nicotinic acid (mg)	26	13	20	26	26
Pyridoxine (mg)	3	?	3	3	3
Biotin (mg)	0.1	?	?	0.1	0.1
Folic acid (mg)	0.6	0.3	0.2	0.6	0.6
Choline (mg)	1320	?	600	1600	1600
B_{12} (mg)	0.01	0.01	0.01	0.01	0.01
D_3 (i.u.)	200	200	400	200	200
E	?	?	?	?	?
K_1 (mg)	0.5	?	?	0.5	0.5
Minerals (g or mg/kg)					
Calcium (g)	2.0	2.0	5.5	2.0	2.0
Phosphorus (g)	1.3	1.3	1.3	1.3	1.3
Potassium (g)	0.5	?	?	0.2	0.2
Sodium (g)	0.3	?	0.2	0.3	0.3
Manganese (mg)	60.0	60.0	60.0	60.0	60.0
Iron (mg)	20.0	20.0	20.0	20.0	20.0
Iodine (mg)	1.0	1.0	0.5	1.0	1.0
Magnesium (mg)	500.0	?	?	?	?
Copper (mg)	2.0	2.0	?	?	?
Zinc (mg)	50.0	50.0	?	?	?

the birds. The total food requirements, and the provisional requirements for the essential nutrients for poultry are presented in Tables 62 and 63.

A pullet weighing 5 lb and laying a modest number of 180 eggs in her first year will have produced $22\frac{1}{2}$ lb or so of eggs, or about $4\frac{1}{2}$ times her own weight. The nutritional requirements of highly productive laying poultry are therefore very critical, especially so if the eggs are to be used for hatching since the well-being of the unhatched chick is dependent on the food intake of its dam.

Wherever poultry are fed on diets containing a quantity of unground cereal or other grains and seeds, it is important that they be provided with insoluble grit *ad libitum* in order to allow the bird to grind down hard food particles in its gizzard. Birds given access to free range will pick up their own grit, but birds kept intensively must have grit provided for them.

Improvement through Management: II. Health and Hygiene

Livestock cannot perform well unless they are maintained in a good state of health. Maintaining stock in a healthy state should be mainly under the control of the farmer, since many diseases can now be prevented or cured on the farm with an elementary knowledge of animal hygiene, or with the assistance of qualified veterinary advice. There are, however, other disease conditions, especially contagious diseases and diseases which are spread by common vectors (such as birds and flies) which the farmer cannot hope to combat on his own. For these diseases a national policy must be enforced, such as a tsetse eradication campaign or a rinderpest vaccination programme. Even though the farmer is powerless to tackle these countrywide problems singlehanded, it is important that he should be familiar with the symptoms of all diseases which come under this category, in order that he may take prompt steps to notify the authorities of outbreaks on his farm and so assist the veterinary officers in coping with the problem.

The cost of animal diseases to the farmer and to the nation is difficult to assess because there are at present many tropical animal diseases which exert a depressing effect on livestock health and productivity which remain undiagnosed and unrecorded. Any estimates are likely to be conservative estimates because of this fact. Bearing this in mind, it is salutary to note that Wooldridge (1954) estimated that the trypanosome carried by the African tsetse fly was responsible for

45 Kenana bull from the Sudan, at the Veterinary Research Station, Entebbe, Uganda. A small number of Kenana cattle were imported into Uganda in 1955. Many breeds of thoracic humped Zebu cattle, improved in one country, have been imported into neighbouring countries to assist livestock improvement programmes in Africa. (see Chapter 17)

46 Sahiwal Cow at the Indian Agricultural Research Institute, New Delhi, India. This cow first calved at 22 months at a weight of 816 lb. Her total milk yield over 8 lactations was 8053 gallons, her highest lactation (the 6th) yielded 1300 gallons. Sahiwal cattle have been exported to many parts of the tropics to improve milk production within the *Bos indicus* species (as in Kenya) and also to cross with European breeds (as in Jamaica to form the Jamaica Hope). (see Chapters 16 and 17)

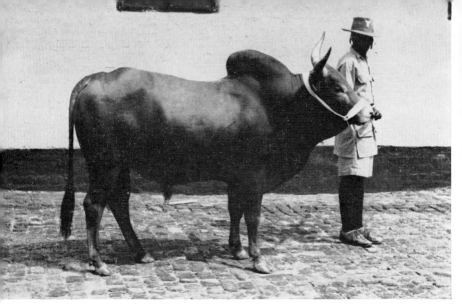

47 Angoni Zebu Bull at the Livestock Improvement Station, Malawi. This bull was 3 years 7 months and weighed 1022 lb. when the photograph was taken, and its height at withers was 49 inches. Note the short horns and large thoracic hump

48 Red Sindhi cows in the Mali herd, Karachi, Pakistan. Note the light conformation, and angular hindquarters, of this important dairy herd of Zebu cattle. Note also the method of tethering the cows by light chains to wooden stakes driven into the ground. This method of restraint is practised in many parts of the tropics, cows often being milked whilst tethered in the open field or yard. (see Chapter 17)

the loss of over £100 million in West Africa each year, and that South African livestock producers lost about £24m each year through disease.

The cost of entirely eliminating a major disease from a country, expecially a large continental country bordering other states, is often quite prohibitive. For this reason, it is often considered more realistic to 'live with the disease', and build up livestock populations which are immune or resistant to it, rather than to attempt the costly and difficult operation of complete eradication. Naturally enough, indigenous livestock are more resistant to the local diseases of the area than exotic stock imported from elsewhere, and for this reason many veterinary departments in the tropics have restricted the entry of 'susceptible exotic stock'. Unfortunately, though such a policy may be sound from the health standpoint, it may not be sound economically, since it may be thought more desirable to have, say, 1,000 head of stock producing 1 lb liveweight gain per day with a relatively high mortality rate averaging 20 per cent, than to have 1,000 head of local stock with a low mortality rate of 5 per cent but only giving a productive return of $\frac{1}{2}$ lb liveweight gain per day. The former policy, in spite of greater disease hazards, yields 292,000 lb of liveweight gain per annum, whilst the latter and safer policy only yields 173,375 lb liveweight gain per annum. Economic and political considerations being as important as they are today, it is found that many newly independent tropical countries will accept the greater disease risk in order to maximize the gross national product from which government revenue is derived. For this reason many tropical countries which previously legislated to prevent the importation of disease-susceptible exotic stock are now changing their policy and allowing importation, if not of exotic stock, at least of exotic semen.

This conflict between what may be scientifically preferable and what is economically desirable exists at the level of the farmer as well as at the level of national policy. For instance, it is generally conceded that it is scientifically preferable to feed animals on straight feeding stuffs which allow disease outbreaks to be relatively readily detected as and when they occur, rather than to feed animals on foods with added antibiotics which tend to mask the early manifestations of disease and hence delay its detection. However, the addition of antibiotics to pig, poultry and calf rations undoubtedly increases livestock performance and liveweight gain, and antibiotic supplementation is becoming as common in the tropics as it is in North and South America.

The ratio of qualified veterinary surgeons to the number of livestock present in the country is usually very low in tropical areas. In many places, the only veterinarians are those employed in government service and these officers are clearly unable to deal with cases of non-infectious disease in, or injury to, individual animals. This fact, coupled with the generally lower value of tropical livestock compared to temperate stock, means that the tropical farmer needs to become more self-reliant regarding animal health matters, since he will often be unable to obtain or afford professional veterinary advice. Unfortunately, due to a lower general standard of education, the tropical producer is often less informed and less capable of dealing with animal health problems than his temperate counterpart.

It is a common tenet of temperate animal husbandry that a successful farmer is one who devotes a great deal of personal attention to his livestock, treating each farm animal as an individual. The same principles apply in many tropical areas, and in certain cattle-owning tribes (such as the Abahima described briefly in Chapter 16) the standard of stockmanship is often high, even allowing for the many malpractices due to ignorance or tribal custom. This tenet has already broken down in the temperate poultry industry, where attention to the individual is impracticable and where the smallest unit dealt with is the group of birds occupying a single section of the poultry house. In this respect, certain specialized extensive tropical systems of animal production, such as ranching, are similar to large-scale poultry farming. It is uneconomic to devote too much attention to the individual and the sickly cow or steer on range must be despatched and disposed of hygienically, since it will rarely repay the expense of isolation, transport to a central point, treatment and possible cure. It is an unfortunate but inescapable fact that the relatively low profit margin per head of stock forces the tropical producer to consider his animals in groups rather than as individuals. There are naturally exceptions to this, the most important being in connection with elite breeding stock which merit regular individual attention.

Livestock buildings in the tropics have to provide shelter from driving rain, and possibly from direct sunlight, but having achieved this the chief attention should be devoted to making the buildings hygienic and easy to maintain in a clean condition. A concrete floor and an efficient system of drainage are of greater importance than high walls and expensive roofs. Direct sunlight is an extremely effective sterilizing agent, and for this reason open-sided structures which allow entry of the early morning and the evening sunlight are far preferable to the box-like farm buildings suitable for keeping livestock warm through temperate winters. Careful attention should be given to the efficient disposal of manure and effluents. Manure and compost heaps properly constructed on well-drained sites are not health hazards, but pools of stagnant urine and manure heaps constructed in secluded, low-lying

Improved Management: II. Health and Hygiene

places rapidly give rise to major fly problems and are a constant source of irritation and infection to man and beast alike.

Although there are a number of specific animal diseases which are peculiar to the tropics, such as East Coast fever which is confined to the coastal belt and hinterland of East Africa, by and large the types of livestock diseases are roughly similar in temperate and tropical regions. The basic textbooks of veterinary science and pathology, such as Lapage (1956), generally give adequate coverage of the major diseases of tropical livestock.

Many of the major contagious diseases in both temperate and tropical areas are caused by viruses which may be passed direct from animal to animal or indirectly through the agency of some vector, such as flies, birds, wild game or man himself. Viruses are not readily killed by drugs or chemicals, but they may be sensitive to high temperatures. The major virus diseases are often combated by taking steps to assist the farm animal to produce its own protection – in the form of antibodies – by the injection of vaccines and other related products. Thus rinderpest, an important virus disease in many tropical areas, is usually tackled by vaccination programmes, either employing a live but attenuated virus, or a virus killed by special treatment. The virus is thus reduced to a state of comparative impotence, so that when it is introduced into the bodies of farm animals it stimulates the tissues to react and produce antibodies, without causing disease.

Another important group of tropical diseases, such as anthrax, leptospirosis and tuberculosis, is caused by bacteria. Many of the tropical forms of bacterial disease are susceptible to treatment by the wide range of antibiotics now being marketed.

The livestock population in the tropics suffers from a much greater range and variety of protozoon parasites than is found in temperate regions. Many of these parasites are conveyed to the animal through the saliva of biting flies, such as the tsetse fly (*Glossina* spp), and *Stomoxys* fly, or by ticks. The control of biting flies is a difficult and expensive operation, usually dependent on a major modification of the vegetation so as to eliminate the bushes and trees which provide the fly with suitable breeding grounds. For this reason, more attention is now being given to the control of fly-transmitted diseases by the administration of drugs to the host animal, such as Antrycide or Dimidium bromide in the case of trypanosomiasis (sleeping sickness) (Robson and Wilde, 1954).

Ticks may be combated on the individual farm by dipping or spraying the livestock at frequent intervals. In many areas, efficient programmes of this sort have led to complete tick eradication, and therefore to the successful elimination of such diseases as tick paralysis, red water fever (Babesiosis), gall sickness (Anaplasmosis), spirochaetosis, East Coast fever (Theilariasis), biliary fever and heartwater. In other places programmes have been less successful due, it is thought, to a build-up of a population of ticks which are more resistant to the acaricide used in the dips and sprays. For this reason it is often prudent to adopt a programme in which the different basic chemical types of acaricide (arsenicals, chlorinated hydro-carbons, organo-phosphorus compounds) are used in rotation. Full details of the construction of suitable dips and spray-races for use in the tropics are given by Williamson and Payne (1965). It should be noted in passing that dipping and spraying will reduce the population of ecto-parasites other than ticks, and may also temporarily repel attacks from biting flies.

An interesting new development in biological control has recently been successful in eliminating an important ectoparasite of cattle, the screwworm (*Callitroga hominivorax*), from the semitropical, southeastern region of the U.S.A. This development consists of releasing large numbers of laboratory-reared, sterilized male flies during the peak mating season. These sterilized males mate with normal female screwworm flies, and provided the programme is so planned that the sterilized males released are in right ratio to the total screwworm population, complete eradication is possible within a few generations. The technique was first developed on one of the small Dutch islands in the Caribbean, and was successfully adopted on a large scale in Florida. It is possible that a similar approach will prove successful with other major ectoparasites of farm animals as a result of future research.

Another important group of animal parasites is the helminths. All three types of worm (roundworms or nematodes, tapeworms or cestodes, and flukes or trematodes) are well represented in the tropics. Few if any tropical livestock are completely free of some degree of infestation, and many million stock remain in an unthrifty, unproductive state because of chronic infestation by parasitic worms. The two main methods of dealing with helminth parasites are by breaking the worms' life-cycle and by directly attacking the worm by administering a helminthicide. Unfortunately insufficient is known about the duration of the different phases of the life-cycles of most tropical helminths, with the exception of the flukes, to be able to give definite advice on management systems which enable the former method of control to be completely effective. Under wet tropical conditions, grass growth is so rapid that the optimal speed of grazing rotation may be such that the herd is repeatedly reinfesting itself with worms which have been hatched from the eggs excreted in a previous grazing cycle. There may well be apparently conflicting needs from the point of view of good grass management and good animal hygiene.

With regard to the control of helminths by the regular administration of appropriate drugs, care must be taken to see that there are no untoward side-effects due to the action of solar radiation on the drug. Helminthicides are usually administered orally, but occasionally some of the drug finds its way to the skin and is broken down into a harmful toxic substance by 'photosensitization'. Because of this danger, it is important to keep stock out of direct daylight for one or two days after such a drug has been administered.

It should also be noted that there is a school of thought in certain tropical areas which favours minimal use of helminthicides on the basis that a healthy animal can cope with a 'normal' infestation of helminths and is capable of keeping the level of parasitism in check. The use of helminthicides, it is argued, makes the animal more susceptible and prone to subsequent reinfestation. This viewpoint does not appear to be backed by critical research, but it is true that a helminthicide programme, once embarked upon, must be maintained almost indefinitely if reinfestation is to be avoided.

One of the commonest anthelmintics in present day usage is phenothiazine, or one of the related chemical compounds. The use of this drug is a good illustration of the difficulties which surround the practice of veterinary medicine in the tropics. The phenothiazine is broken down in the body of the farm animal and eliminated as a red dye, and the sight of the blood-red urine, when unexpected, is exceedingly alarming. The drug has been introduced into certain tropical areas without sufficient warning being given to the native herdsman about these visible side effects, and the sight of red urine, pink milk and a discoloured fleece or hair has not unnaturally led the local inhabitants to believe their livestock were being poisoned on a large scale. Great harm can be done to the relationships between veterinarian, livestock officer and herdsmen if the full implications of such procedures as dosing with phenothiazine are not made very clear right from the start.

No attempt can be made in this chapter to go into the details of the biology of tropical pathogens, disease symptoms and control methods. A very brief summary of the major disease conditions is given by Williamson and Payne (1965), but the reader must consult standard veterinary texts, such as Dunne (1959), Lapage (1956), Morgan and Hawkins (1949), Monnig (1949) and Whitlock (1960) for fuller details of this specialized field of study.

Improvement Through Breeding

The term 'animal breeding' may be used in one of three different ways. It may be defined as 'sexual reproduction', and assessed by the efficiency of reproduction in terms of conception and birth rate or generation interval. It may be used to denote animal multiplication, such as the mass mating of inbred cocks of one breed with inbred pullets of another. It may be used to denote genetic improvement, and may be properly applied to systems in which every mating is designed either to assess the breeding value of an untested animal, or else to multiply the progeny of animals which have already been tested and shown to be genetically superior. All these aspects of animal breeding are pertinent to the theme of livestock improvement in the tropics, and each will be dealt with in turn.

SEXUAL REPRODUCTION

When male animals are run freely with breeding females, a detailed knowledge of the specific characteristics of the oestrus cycles of tropical livestock is unnecessary. However, as soon as seasonal breeding, controlled mating or artificial insemination is introduced, a knowledge of such matters is basic. The length of the oestrus cycle, the duration of oestrus and the time of ovulation in relation to the onset or termination of oestrus, vary greatly from animal to animal within a breed. It is therefore sometimes necessary to ascertain the characteristics of the individual animal, apart from those of the breed from which it is drawn, when strictly controlled mating is practised. It has been commonly stated, following an early report by Anderson (1943), that the duration of oestrus is much shorter in tropical species and breeds, such as *Bos indicus* cattle, than in temperate breeds. It is now thought that the external and visible signs of oestrus may be of shorter duration, but that the female is on heat, and will accept advances by the male, for periods of similar duration to those recorded for temperate breeds and species (Rollinson, 1955). With cattle, the onset and duration of oestrus can be traced experimentally in individual cases by taking small samples of the vaginal and cervical mucus and measuring its viscosity with an oestrometer. There is an abrupt change in viscosity as oestrus commences, preceded by a slow but steady rise in viscosity during the last half of anoestrus. The time of actual ovulation can also be detected by this means, thus enabling insemination to take place at the optimal time for conception (Crane et al., 1960).

Attempts to introduce A.I. with other classes of livestock have been relatively unsuccessful. It is probable that the technique will eventually be used in tropical pig improvement programmes, but to date the results with swine have been disappointing even in temperate countries, owing to the difficulty of detecting the different oestrus stages in the gilt and sow and to the problems associated with the dilution and storage of boars' semen (Aamdal, 1964; Bennett and O'Hagan, 1964).

The approximate parameters for the various phases of the reproductive cycles of tropical livestock are given in Table 64. It should be noted that the range is very wide, and that approximately three-quarters of the individuals within a species will conform to the 'norm' whilst approximately one quarter will have cycles which lie outside the limits of the normal values shown. Excellent reviews of this subject have recently been written by Hansel (1961) and Hancock (1962).

Table 64. *Average parameters of the reproductive cycles of tropical livestock*

Species	Length of oestrus cycle (days)	Duration of oestrus (hr)	Time of ovulation in relation to oestrus	Length of gestation (days)
Cow	20–24	12–18	9–14 hr from end	280–283
Buffalo	21–22	12–36	18–42 hr after onset	310
Ewe	14–19	24–36	24–48 hr after onset	147
Goat	18–21	24–36	30–40 hr after onset	151
Sow	21–22	48–72	35 hr after onset	114
Mare	21–26	120–168	2–3 days from end	336
Camel	21–28	72–96	Not known accurately	370–375

The chief variable in the reproductive cycle of each female is the length of the oestrus cycle. This parameter can be modified by subjecting the animal to some change in its environment, such as by altering the level of feeding or the management routine. The actual duration of oestrus, and the time of ovulation in relation to the beginning or the end of oestrus, is less variable.

It is often reported that, since the tropical breeds of livestock are generally slow maturing compared to temperate breeds, sexual maturity is also delayed. It is true that the average age at first calving of zebu cattle is often a year or more later than that of European stock, but there are nevertheless instances of zebu cattle coming into oestrus, and conceiving, at less than a year of age and of tropical goats reaching full sexual maturity at the age of a few weeks. It is therefore necessary to separate the sexes as soon as animals are weaned if the risk of promiscuous breeding is to be minimized.

The increased use of artificial insemination (A.I.) for cattle in the tropics will undoubtedly allow livestock improvement schemes to be accelerated, since A.I. enables an outstanding bull to sire several thousand calves in a year instead of the fifty or so possible under conditions of natural service. It should not be too readily assumed that A.I. only has a part to play in intensive farming areas, but no place under extensive ranching conditions. There are several reports, such as that by Horak (1960), of successful A.I. techniques employed in connection with tropical cattle ranching, and other workers have reported that advanced techniques, such as methodical pregnancy diagnosis, also have a part to play in the economic management of tropical beef cattle (Osborne, 1961).

Where A.I. is used to introduce *B. taurus* blood into tropical cattle populations, temperate experience on the techniques of A.I., such as semen collection, can be drawn upon with confidence. However, when *B. indicus* bulls are used, difficulties are sometimes encountered, as zebu bulls are often 'shy breeders'. For this reason, electro-ejaculation techniques may be desirable in place of the more normal use of live 'teaser' animals or dummy cows. Rollinson (1956) has described the details of a successful electro-ejaculation technique in regular use at Entebbe, Uganda.

Although A.I. is a very useful technique in breeding programmes, it will only effect improvement if the semen is derived from genetically superior bulls. The use of below-average bulls will exert a detrimental effect on the genetic value of herds on which they are employed. Unfortunately, the 'prestige value' of A.I. schemes has often led certain tropical countries to adopt large-scale programmes without ensuring that the bulls employed in them are genetically superior. Indeed, there is some evidence that certain schemes are definitely detrimental to livestock improvement, since the yields of daughters bred by A.I. show a tendency to be below those of the dams milking alongside them in the same milking shed (Wilson and Houghton, 1962). It is most important that A.I. schemes should be linked with performance testing or progeny testing programmes, in order that they may correctly be designated as parts of an overall livestock improvement policy.

It is argued in some quarters that A.I. schemes should precede the availability of genetically tested and superior stud animals, since it will take time to win the confidence of the tropical producer and to convince him that healthy, normal stock can be produced by A.I. It is said that there is not much point in purchasing or testing expensive bulls for an A.I. centre before the number of resultant inseminations justifies the heavy expenditure involved. It is often found that, in the early days of the A.I. programme, any available bull is used to provide semen. On the other side, it can be argued with greater force that the confidence of the producer will be lost if half or more of the calf crop is inferior to the average run of calves sired by natural service. It is only when A.I. programmes employ demonstrably better bulls than would otherwise be available to the stockman that the extra risks, costs and administrative complications are justifiable. The methods by which the bulls destined for possible use at stud in A.I. schemes can be genetically tested will be discussed later in this chapter.

The reproduction of certain temperate livestock, particularly sheep, is mainly governed by the season of the year. A decrease in daylight-length during the autumn causes the ewe to ovulate, and it is usually only by controlling the length of day by artificial means that sheep can be induced to breed at other times of the year (Eaton and Simmons, 1953). In most parts of the tropics there is much less variation in day-length, and it is not yet known with any certainty how much effect this reduced variation has on seasonal breeding. Within the latitude limits of 20°N. and S. sheep can breed all the year, but even so, there is still a tendency for seasonal reproductive activity. Also, an examination of the breeding records of tropical cattle often shows that the year can be divided into two halves, in one of which 60 per cent or more of the total calf crop is born (Wilson, 1957a; Alim, 1960). The mechanisms which operate this seasonal change in reproductive behaviour are not known with certainty. It may be linked to the rainfall distribution (which influences grass growth and hence nutritional factors) as suggested by Wilson (1957a) with reference to seasonal cattle breeding in Uganda, or it may be linked to seasonal changes in parasitism and animal health which affect reproductive efficiency. The point of interest is that the natural seasonal incidence in parturition can be used as a guide for artificial seasonal breeding practice. Whatever the causal factors which operate these patterns, it is clearly in the interest of the tropical livestockman to work with them rather than against them. When there is not a steady demand for animal products the year round, it may be preferable to adopt seasonal breeding programmes for both milk and meat animals, if by so doing reproductive efficiency is enhanced and productivity increased. There is a certain amount of evidence that such seasonal calving programmes may be very beneficial. Castillo *et al.* (1960) have shown that buffaloes calving in the Philippine dry-season yield an average of 620 kg of milk compared with only 436 kg for buffalo cows calving in the wet-season. Similarly Tomar and Mittal (1960) reported that, in India, June-calving cattle lactated for an average of 317 days compared to an average lactation length of 268 days for November calvers.

ANIMAL MULTIPLICATION

Due to the inherent low productivity of very many classes of tropical livestock, it is often desirable to replace them as rapidly as possible with animals of superior potential. The 'grading up' of flocks and herds on individual farms is an important aspect of this process, but it is often deemed desirable to hasten the replacement period by multiplying superior breeds and strains on stations under government control in order to release the stock in large numbers for use on private farms. The need for such multiplication centres is greatest in those tropical countries in which there are few, if any, alternative, commercial sources of superior stock. In many parts of Africa (other than the few European areas) there are no 'pedigree breeders' whose chief livelihood is obtained by raising and selling better breeding-stock to farmers with more limited facilities. There is some trade in breeding-cattle, but this is generally completely unorganized and there is little or no guarantee that the stock obtained from such sources are in any way superior to those which the farmer could raise himself. Indeed, once a tropical producer has established a local reputation for his breeding-stock, it is likely that demand will rapidly exceed supply and there is a danger that the unsuspecting buyer may be landed with 'culled' animals which are of lesser value than the mean of his own flock or herd.

Three important factors should be considered in connection with state-controlled animal multiplication schemes:

1. *Stock must be of consistent high standard.* Even with every possible attention to the choice of genetically superior sires and elite groups of breeding females, a certain proportion of below-standard progeny is bound to be born and care must be taken to see that these animals are culled and slaughtered and not passed on to the commercial producer. Rigid control of animal health is necessary, since it is a tragedy of the highest order when a government-sponsored scheme of this nature unwittingly becomes the means by which disease is spread to private farms. It is often useful to couple multiplication schemes with herd-book registration and the formation of breed societies so that the identity of all stock produced by such schemes can be retained. The use of suitable brand marks is preferable to ear-tagging, since it is easier to prevent unauthorized branding than unauthorized tagging. It is highly desirable to associate the producers themselves with these administrative aspects of multiplication schemes, so that farmers realize that it is in their own interests to maintain quality and to prevent unauthorised registration of stock.

2. *The future breeding programme must be agreed from the start.* It is

very easy to take one step forward followed by two steps backward in livestock breeding. The advantage gained by the supply of superior stock to farmers will rapidly be lost if these stock continue to be bred to average or below-average animals. For this reason, long-term planning is essential right from the very start. Two alternatives are possible. (a) The producer may be encouraged to refrain from breeding on his own, and to return repeatedly to the multiplication centre for future livestock needs. This policy is often adopted with poultry, since the birds supplied by the multiplication centre are often the product of mating several highly inbred lines, and further breeding from such birds will result in regression and a marked increase in genetic variability. Therefore poultry keepers, whether they are egg or meat producers, are often encouraged to return for fresh supplies of day-old chicks at regular intervals, and to refrain from breeding work on their own farms. However desirable, such a policy is less feasible with larger stock, since the multiplication centres would seldom be able to meet the heavy recurrent demands which such a policy of necessity entails. (b) The producer may be assisted with a breeding service, such as an A.I. scheme or a stud-loaning scheme, so that the improvement achieved through the supply of the original group of animals from the multiplication centre can be maintained or increased in future generations. Grave problems arise when crossbred stock, such as half *B. taurus*; half *B. indicus*, are released from the multiplication centre. These stock will exhibit hybrid vigour or heterosis, and as a consequence they will be unusually uniform. In the next generation, whatever animals they are bred to, genetic segregation will take place and the range of types produced in this generation will be relatively wide, with the result that fairly heavy culling will be necessary. For this reason, it is often better for the multiplication centre to retain halfbred animals for a further generation of breeding, so that the problem of segregation, with its attendant need for culling, can be tackled on the station and not on the private farm.

Where purebred stock, or crossbred stock which are no longer genetically segregating, are distributed amongst commercial farmers care must be taken to see that appropriate proven sires are available to prevent undesirable mating with scrub males or with males of the wrong genetic constitution. For instance, the multiplication centre may be producing seven-eighths bred stock (seven-eighths *B. taurus*; one-eighth *B. indicus*) for distribution to farmers. There may be good grounds for attempting to stabilize the population at approximately this level, since higher proportions of European blood may result in increased mortality and higher proportions of zebu blood may lower productivity. In this case, it is important that seven-eighths bulls are available (for purchase, for loan or via an A.I. service) so as to prevent the producer from being forced to grade up too high to the European-type or to grade too far towards zebu-type.

3. *Steps must be taken to ensure that stock distributed from multiplication centres are properly cared for and used for the purpose for which they were bred.* There are many instances of government multiplication schemes selling stock at subsidized prices with the result that, instead of finding their way to farms, they end up in the slaughterhouse or in the hands of unscrupulous dealers. It would seem justifiable to insist that the purchaser of improved stock, made available at government expense, should be required to enter into a written undertaking that he will keep the stock in good husbandry and in good health, and not dispose of them for any other purpose without special written permission. It may be unwise to subsidize the price at which breeding-stock are released from multiplication centres, since this may result in the producer assuming that he is receiving something inferior, instead of something superior. It is important to inculcate a proper sense of values into tropical livestock producers, and it is right that superior breeding-stock should command much higher prices than slaughter-stock. It is also in the best long-term interests of the livestock population of a tropical country to encourage responsible livestock breeders to supplement the resources which government can muster through its multiplication centres. If these private breeders are continually being undercut by the supply of subsidized animals from government centres, then they will inevitably have to lower their standards or be forced out of business. As in most agricultural matters of this nature, a sense of economic proportion must be preserved. Superior stock must not be priced out of reach of the progressive producer, neither should they be so heavily subsidized that they become artificially devalued and thereby encourage non-productive exploitation by the 'middleman'.

With regard to the need to insist that the producers receiving stock from government multiplication centres should be required to maintain these stock under good conditions of husbandry, the experience of the Kenya Government during the grade-Sahiwal multiplication programme is of interest. This programme sought to make available large numbers of three-quarter and seven-eighths bred Sahiwal × East African zebu stock to African peasant farmers, primarily for intensive milk production on small farms. The demand for these stock invariably exceeded supply, and producers could only obtain stock on the recommendation of the local agricultural officer. At time of purchase, the African producer had to sign a firm undertaking not to dispose of the animals without written permission, and to carry out the following five 'good husbandry practices':

1. To spray the stock regularly against ticks, for a minimum period of nine months.
2. To erect a perimeter fence around his farm and to maintain this fence in stock-proof condition.
3. To record all milk yields in a simple record book supplied by government.
4. To make adequate provision for a minimum quantity of dry-season feed (usually in the form of Elephant grass, *Pennisetum purpureum*).
5. To inoculate or vaccinate all cattle regularly against anthrax, blackquarter disease, brucellosis, and septicaemia on a full-payment basis.

If any of these conditions were not fulfilled, the farmer in question forfeited his right to any further help or assistance from government in connection with his livestock enterprise. The number of defaulters was very small, most producers being quite willing to comply with the regulations because they were anxious to benefit by receiving grade-Sahiwal stock of recognized superiority.

GENETIC IMPROVEMENT

Two different types of factor are responsible for the differences between individual animals. First, there are important environmental factors, due to climate, nutrition, health and management, which we may designate E, and which have already been considered in preceding chapters and in the first part of this chapter. Secondly, there are genetic factors, which may be designated G, which are factors due to the random sample of genes received from the two parental gametes – the ovum and the spermatazoon.

These types of factor act together, not separately, so that the total variation between two animals is equal to the sum of the effects of all the E and G factors, and the interaction between the two; thus:

Total variation $= S\,E + S\,G + S\,EG$ (where S means 'the sum of')

With a single individual, it is not possible to separate the effects of the various components of the above equation, and to estimate how much of its productive level is due to each. With groups of livestock, however, estimates of the relative importance of E, G and EG can be obtained (Lush, 1945; Falconer, 1961).

Clearly, there is little or no point in attempting to improve livestock by genetic means for a particular character if most of the variation of this character is environmentally determined. Genetic improvement of a trait can only take place if this trait is reasonably highly heritable. It is therefore of the utmost importance that the heritabilities of the various traits which figure in livestock improvement programmes should be known, in order that breeding programmes should concentrate attention on those of highest heritability. Unfortunately, heritability coefficients are difficult to calculate and in many cases an indirect assessment is made by measuring the 'repeatability' of a character which occurs more than once in the lifetime of an individual, such as milk yield (in successive lactations) or litter number (in successive litters of a pig). A further complication arises since heritability coefficients do not appear to be the same for different populations of related animals; thus the heritabilities of economic factors (such as milk yield or butterfat percentage) are different for *B. taurus* and *B. indicus* cattle, and different for various breeds within the species. A further difficulty arises since heritability estimates vary according to the environmental conditions under which the animals in question are kept. Harsh environments give rise to a greater degree of environmental variation (SE) and consequently to lower estimates of heritability (Wagnon and Rollins, 1959). However, although the heritability coefficients differ according to the population studied for purposes of calculation of the coefficient, the order or ranking of the heritabilities of different characters is generally fairly similar. Thus factors which are strongly inherited in *B. taurus* cattle will also tend to be strongly inherited in *B. indicus* cattle, even though the actual coefficients may differ widely in value. It is therefore instructive to glance through a representative table of heritability coefficients calculated for temperate livestock, using the methods described fully by Lush (1949). Some representative figures are tabulated in Table 65.

It will be seen from Table 65, based on estimates given by Rice *et al.* (1957); Duniec *et al.* (1961); Hale (1961); Smith *et al.* (1962) for temperate livestock maintained under temperate conditions, that carcass characteristics possess higher heritabilities than the weights at weaning or (with pigs) the size of the litter at weaning. Clearly, therefore, genetic improvement programmes can make faster progress in improving carcass conformation than in improving weaning weights of meat animals. It will also be noted that the efficiency of food utilization has a relatively high heritability. This factor, expressed as pounds of food required to provide a pound of liveweight gain, is of great practical importance and should be

Table 65. *Representative heritability coefficients calculated for European livestock*

(*Note:* Value of 1 = 100 per cent heritability. 0 = no evidence of any heritability)

	Item	Heritability coefficient
Beef cattle	Area of eye muscle (*L. dorsi*)	0·70
	Birth weight	0·45
	Carcass grade	0·39
	Efficiency of food utilization	0·39
	Slaughter grade	0·38
	Weaning weight	0·26
Sheep	Yearling body weight	0·40
	Clean fleece weight	0·38
	Birth weight	0·30
	Fleece quality	0·22
Pigs	Carcass back-fat thickness	0·68
	Carcass length	0·66
	Efficiency of food utilization	0·54
	Chemical fat in loin	0·50
	Dressing percentage	0·40
	Daily weight gain, weaning – 200 lb	0·41
	Market score at slaughter	0·33
	Yield of lean cuts in carcass	0·29
	Number of functional nipples	0·22
	Litter size at birth	0·19
	Weaning weight of litter	0·17
	Litter size at weaning	0·14
	Daily weight gain, birth to weaning	0·12
Poultry	Egg weight	0·71
	Pullet weight	0·61
	Chick weight	0·51
	Egg number	0·28

included in selection programmes for all meat animals whenever possible.

A certain amount of research on heritability in dairy cattle has been carried out in the tropics, and this has been well reviewed by Mahadevan (1958). Sikka (1933) estimated the correlation between the individual lactation yields of a cow and the highest lactation recorded. He found that the correlation increased from the heifer lactation ($r = 0.5$) to the fourth lactation ($r = 0.8$) and concluded that estimates of heritability would vary depending on how many lactation yields were used in the calculations, and also that there was little justification in selecting tropical dairy cattle on the basis of yield during the heifer lactation. This point was further emphasized by Robertson (1950) who reported that an analysis of records of the White Fulani cattle in Nigeria showed that the heritability of a single record of milk yield was only 0·32 whilst that of the mean of two records increased to 0·47. Wilson and Houghton (1962) showed that on occasions tropical milk cattle demonstrate a drop in yield during the second lactation compared to the first, instead of the steady increase from first, through second, to third lactation common among European cattle. This fact would further complicate the selection of dairy cattle on the basis of but a single lactation yield.

Heritability estimates obtained from tropical cattle are usually derived from daughter–dam comparisons and are frequently based on a small sample size. For this reason, the estimates so far recorded must be regarded as very tentative. Stonaker (1953) published a series of heritability estimates of Red Sindhi cattle at Allahabad, some of his results being as follows: Calving interval, 0·88; mature height, 0·16; age at first calving, 0·39; total butterfat yield, 0·25; birth height, 0·16. Stonaker's estimate of the heritability of butterfat percentage was only 0·09, which is suspiciously low in view of the known high degree of genetic variation in this character both in *B. taurus* and *B. indicus* cattle. However, heritability estimates from tropical breeds for the most important economic factor of all – total milk yield – are also very low. Thus Mahadevan (1955) and Alim (1960) estimated the heritability of milk yield to be of the order of 0·2–0·3 in zebu cattle in Pakistan, Ceylon and India. These results are somewhat surprising since they are lower than the general scale of values recorded for temperate cattle and it might be reasonably expected that the comparatively unselected and heterozygous *Bos indicus* cattle would provide higher heritability estimates than the comparatively highly selected and homogeneous *B. taurus* cattle. It would seem that the great genetic range in tropical livestock is masked by an even greater range in the operation of environmental factors. Thus where management practice is continually altering and evolving, as it does on many government stations in the tropics at the present time, the environment is not constant from one generation to another, and true assessments of the role played by genetics in livestock improvement are extremely difficult to obtain. As management becomes more uniform and environmental factors, such as grass quality and general nutrition, become more standardized from one generation to the next, it is possible that the hidden genetic differences between individuals and breeds will be revealed. This may be reflected in higher values being obtained for many heritability estimates.

In the meantine it is essential that the number of factors incorporated into livestock selection programmes should be kept as low as possible, and that the factors included should have reasonably high heritabilities. The rate of genetic improvement is inversely proportional to the number of factors incorporated into the selection programme, and very little, if any, progress can be made if more than two or three factors are selected for simultaneously. Priorities must be carefully allocated and with tropical cattle, meat quality, liveweight gain and food conversion efficiency are probably the factors most worthy of attention by the geneticist. Consideration of fancy breed points of little or no economic significance should be ruthlessly eliminated from any genetic selection programmes.

Unfortunately, because of the relatively low heritability of milk yield in tropical cattle already mentioned, selection for high milk yield within the *Bos indicus* species is likely to lead to disappointingly slow progress. Thus Robertson (1950) calculated that the average annual rate of improvement for milk yield with Fulani cattle in Nigeria by genetic selection would be of the order of 2 gallons/lactation, and a similar figure (1·85 gallons/lactation) was obtained by Alim (1960) for the Kenana cattle of the Sudan. For the present, it must be recognized that most of the variation which exists between the milk yields of different tropical cows is due to environmental differences, a significant part being under the direct influence of the stock owner and milker. Alim (1962), in an analysis of the factors influencing milk yields of Butara cattle in the Sudan, ascribed 74 per cent of the total variation to the effect of length of lactation, 10 per cent to age of cow, 7·5 per cent to the length of the calving interval and the remainder to the length of the dry period. Many wishful thinkers have endeavoured to ascribe the several notable instances of steadily rising average lactation yields within tropical cattle populations to the success of the breeding policy. The truth of the matter is that most of these successes reflect improvement in stockmanship, nutrition and grassland husbandry.

Unfortunately the meat animal, which lends itself more readily to genetic improvement because of the reasonable heritabilities of liveweight gain and meat quality (as judged by carcass characteristics) has received much less attention in the tropics than the dairy animal. There are important exceptions to this, as some of the work carried out in subtropical regions has also been of value in the tropics. For instance, a very fair measure of success has attended the efforts of American breeders to introduce *B. indicus* blood, of Brahman type, into the beef ranches of the southern U.S.A., particularly Florida and Texas. Greater attention should be devoted to the improvement of *B. indicus* cattle for beef purposes in the tropical parts of Africa.

SYSTEMS OF BREEDING

The two main questions that confront the tropical livestock breeder are, first, 'What breed should be improved?' and, secondly, 'What breeding system should be adopted to maximize the rate of improvement?' With regard to the first question, four different approaches are possible:

1. *Selection from within the indigenous tropical breeds.* As has already been mentioned in Chapter 16; 'Cattle Management', this approach is safe but slow. Disease risks are minimized, and there is no problem of acclimatization. On the other hand, the length of time taken to reach the desired goal is often excessive.

2. *Mass importation of temperate livestock.* This approach is likely to provide rapid results but it is expensive, often prohibitively so, and it suffers from major risks in respect to disease and acclimatization. When this approach is adopted, as with Friesian cattle in Puerto Rico, a large capital and recurrent sum must be earmarked for the control of ticks and other vectors of major tropical diseases. It may also be necessary to subsidize the importation of breeding stock over long periods. In Puerto Rico, for example, female replacements are still imported from the U.S.A. at regular intervals, and most stud bulls used for A.I. are still progeny tested in America and subsequently transported to the Caribbean.

3. *Mass importation of tropically adapted cattle from other tropical countries.* This approach has the merit of combining many of the advantages of (1) and (2) above, whilst minimizing most of the disadvantages other than cost. It enables certain tropical countries to benefit from the experience of, and advances made in, other tropical countries. (In this sense it parallels the pattern which formerly prevailed amongst temperate countries, in which the less developed countries relied heavily on a more advanced, exporting country for its supply of premier breeding-stock. Great Britain undertook the role of a major exporter of pedigree stock during the last few centuries, and the classic British breeds have been spread in this way to the Americas and the temperate parts of Australasia.) A good example of this approach is the importation of Sahiwal cattle from Pakistan to Kenya (Kenya Vet. Dept., 1961). Other examples are provided by the wide distribution of Santa Gertrudis cattle from Texas to many different tropical countries, and the spread of improved goat breeds, such as the Anglo-Nubian, from the eastern Mediterranean to tropical parts of the world. The major difficulty preventing any really widespread adoption of this policy is that most tropical countries are extremely short

Improvement through Breeding

of improved stock, and have little or no surplus left over for export.

4. *Crossbreeding of temperate × tropical livestock, and the evolution of stabilized or semi-stabilized crossbreeds.* This system is being adopted in the tropics on an ever-increasing scale. The results so far achieved in many different parts of the tropics are very promising, and it is likely that the proportion of the world's livestock population falling into this crossbred category will increase significantly during the next few decades.

The main decision which has to be made in order to implement a policy of cross-breeding followed by breed stabilization concerns the proportion of temperate and tropical genotype ('blood') which is deemed desirable in the end product. Wherever possible, this should be established by experiment, or by drawing on the experience of other tropical countries with similar environmental conditions, right from the very start. If this is not done, the breeding programme will follow the unfortunate pattern widely prevalent in the 1920s and 1930s, in which temperate males were top-crossed to indigenous tropical females until health and performance declined. This decline was rectified by changing the policy to one of back-crossing to tropical sires. In time this would result in a gradual regression to indigenous type, and a fall back to original levels of production, and so the process would be repeated *ad infinitum*. This aimless system of cross-breeding has been described as 'see-saw' breeding, and most tropical countries can furnish numerous examples of the failure of this method to meet their productive requirements. It is therefore important that the aim should be clearly defined from the start. Although it is dangerous to generalize too freely concerning the desirable aim for any given situation, enough is now known to give some useful pointers. Extremely hot and arid countries will probably require a stabilized crossbred with the greater contribution of the genotype coming from the tropical parental type. Areas in which the mean maximum temperature rarely exceeds $43.3°C$ ($110°F$) but often exceeds $32.2°C$ ($90°F$), and in which humidities tend to be high, will probably require a population stabilized at between five-eighths and three-eighths of temperate parental-type. As elevation, or a decrease in humidity, lessens the climatic stress, so the livestock population will perform best with increasing proportions of genotype derived from temperate sources. The disease situation and the standard of management at any one place will modify the picture, and there is no clearcut distinction between the requirements of one region and another. The summary in Table 66 must be taken as a rough guide to the situation likely to be found with cattle, but there will be many examples of successful exceptions to the sequence given. A great deal will depend on the choice and excellence of the breeds used to form the new cross-breed.

Table 66. *Suitable combinations of* B. taurus *and* B. indicus *for performance in the tropics*

Region	Proportion B. taurus (per cent)	Proportion B. indicus (per cent)
Very arid tropics	0– 40	100–60
Hot humid tropics	30– 60	70–40
Hot dry tropics	40– 70	60–30
Warm humid tropics	60– 90	40–10
Warm dry tropics	75– 95	25– 5
Subtropics and high altitude tropics	90–100	10– 0

A careful analysis of the production records of animals of varying proportions of tropical and temperate blood will enable a rough prediction of the desirable genetic proportions for that particular environment without conducting an expensive, long-term experiment. Thus the data given in Table 67, taken from a hot, dry region of India, would indicate that optimal performance in this particular environment will be obtained by stabilizing a crossbred population of cattle between the limits of one-half and five-eighths *B. taurus*.

Table 67. *Average production of milk of crossbred cattle in a hot, dry region of India*

Breeding of cow	Mean yield of milk (kg)
$\frac{1}{8}$ B. taurus	2,199
$\frac{1}{4}$ B. taurus	2,719
$\frac{1}{2}$ B. taurus	3,171
$\frac{5}{8}$ B. taurus	3,175
$\frac{3}{4}$ B. taurus	3,029
$\frac{7}{8}$ B. taurus	2,809

Care must be taken to see that the records are unbiased. Thus if it were shown that the sample of cattle used to provide the data in Table 67 were heavily culled for high milk yield in the one-half *B. taurus* and five-eighths *B. taurus* categories, but relatively unselected in the other categories, then it might well be that the optimal proportion of *B. taurus* blood was three-quarters *B. taurus* or higher. Care must also be taken to ensure that factors other than the production of saleable products are taken into account, and that reproductive efficiency and longevity are given adequate attention. This point is borne out by reference to the figures shown in Table 68. A glance at the first two columns might lead one to suppose that the pure-bred *B. taurus* type was the most suitable animal for this environment. A study of the last two columns would, however, confirm that it would be most unwise to proceed beyond the three-quarter *B. taurus* stage for purposes of forming a new stabilized cross-breed.

Table 68. *Data selected to illustrate the importance of taking reproductive efficiency and life-time performance into consideration when planning programmes for stabilizing cross-breeds*

Breeding of cow	Mean milk yield (gals)	Age at first calving (months)	Calving interval (days)	Total calvings
½ *B. taurus*	368	36	380	5
⅝ *B. taurus*	382	32	397	5
¾ *B. taurus*	435	31	456	3
⅞ *B. taurus*	567	30	534	2½
15/16 *B. taurus*	635	29	611	2
Purebred *B. taurus*	698	28	660	1½

Although it is clearly important to define the end-point of a cross-breeding programme, it is most unwise to place too much emphasis on 'fractional breeding' as an end in itself. There is a real distinction between aiming at a mean proportion of tropical and temperate genotype in a crossbred population, and aiming at a fixed proportion of tropical and temperate genotype so that stock which do not exactly conform are rigidly excluded. The designation 'seven-eighths *B. taurus* one-eighth *B. indicus*' can be given to a population to describe the average contribution to the genotype derived from each contributing species. The same designation is less meaningful when given to an individual, since in the course of the two back-cross generations following upon the original cross between a European bull and a zebu cow, the random assortment of chromosomes at each meiotic division will have given rise to individuals which possess either more or less genes derived from the *B. taurus* ancestor than the fraction 'seven-eighths' indicates. For instance, an animal which, on theoretical grounds, may be described as 'seven-eighths *B. taurus*' may in actual fact have considerably more than seven-eighths of its total genes contributed by *B. taurus* ancestors. This is very likely to be the case if heavy selection has been practised for productive factors which have been introduced into the genotype by the repeated use of European bulls. The only generation in a cross-breeding programme in which the proportional genetic contributions from *B. taurus* and *B. indicus* is known with precision is the F_1 hybrid, or halfbred, generation. In this generation precisely half the genes and chromosomes are derived from each species. In subsequent generations the proportions will be only approximate, and to exclude otherwise desirable and productive stock on the grounds that 'they possess the wrong proportion of European blood' is often unjustified. Because of the inherent variability of members of the animal kingdom, many individuals which appear to possess proportions of temperate and tropical genotype which are far removed from the desired average will be found to be well adapted to their environment, productive and obviously capable of making an important contribution to the breeding programme. This is illustrated by the data presented in Table 69 which refer to the composition of a crossbred herd being stabilized at the seven-eighths *B. taurus* level. The only types of animal not represented in this interbreeding population are purebred *B. taurus* and purebred *B. indicus* stock. It will be noted that, whereas the highest mean yield (624 gals) is obtained from seven-eighths *B. taurus* cows, certain animals within this seven-eighths group exhibit the lowest recorded yield (467 gals).

Table 69. *Types of crossbred represented in a tropical dairy herd being stabilized at the level of seven-eighths* B. taurus

Type of crossbred	Number	Mean yield (gals)	Range (gals)
$\frac{63}{64}$ *B. taurus*	2	576	556–595
$\frac{31}{32}$ *B. taurus*	7	589	541–703
$\frac{15}{16}$ *B. taurus*	19	610	510–748
$\frac{7}{8}$ *B. taurus*	73	624	467–927
$\frac{3}{4}$ *B. taurus*	21	582	499–882
$\frac{1}{2}$ *B. taurus*	3	499	481–533
$\frac{1}{4}$ *B. taurus*	1	504	504

The term 'stabilization' should not generally be interpreted to mean that the cross-breed in question has become a homozygous, true-breeding population with a series of standard breed characteristics.

Improvement through Breeding

There are few, if any, new tropical cross-breeds which would meet the requirements of such a definition, and it is not particularly important that they should do so. The term is used in this chapter to denote that inter-breeding is taking place in a crossbred population in which the majority of individuals are well adapted to their environment and capable of high levels of productivity, and in which the average proportions of temperate and tropical genotype are fixed. Where particular breed characteristics, such as colour, size, horn-type, can be incorporated and stabilized simultaneously with adaptability and productivity this is all to the good, but very often undue attention to superficial breed points will result in less attention being given to factors of economic importance. The new breed societies of the tropics should, wherever possible, try to avoid the early mistakes made by their counterparts in Europe and America. In many cases, undue stress on breed-type and too little emphasis on constitution and lifetime performance has led to the decline of hitherto well-known and important breeds. In temperate countries many people were, and still are, prepared to pay very high prices for show winners possessing all the essential breed characteristics. A similar situation is unlikely to occur in the tropics, where the differential between the price realized for elite breeding animals and for normal, commercial market stock is very much smaller.

Two examples of new tropical cross-breeds will now be considered, one developed for intensive milk production and the other for extensive beef production. These breeds are the Jamaica Hope and the Santa Gertrudis. (See Plate VIII).

Jamaica Hope. This is a new cross-breed combining *B. taurus* blood-derived from the Jersey (and, more recently, the Friesian) and *B. indicus* blood from the Sahiwal imported from India. It was formed in the island of Jamaica, and has spread through many of the West Indian islands. The breed was organized into a closed herd-book and stabilized at about the seven-eighths to five-sixteenths *B. taurus* level by Lecky in the early 1950s, but credit for the pioneering work must be given to Cousins, a former Director of Agriculture in Jamaica, who investigated crosses between Sahiwal, Jersey, Guernsey and Friesian breeds from 1910 onwards. The early days in the development of this breed were marked by the 'see-saw' breeding system, or lack of system, already discussed and the period of trial and error experimentation continued into the 1940s. By 1930 the Guernsey was eliminated, but the desirable proportions of temperate blood on the one hand and Sahiwal on the other were not known, and both purebred European and zebu bulls were in use as the pendulum swung from top-crossing with one species to top-crossing with the other. The grounds on which the Jersey was eventually chosen as the major contributor of European blood were (*a*) greater heat tolerance; (*b*) fat-corrected milk yield; (*c*) better reproductive performance and earlier breeding; (*d*) hardiness and more efficient utilization of rough grazing. The recent use of Friesian blood reflects the decrease in the importance of high butterfat and the increase in emphasis on total milk yield for the Jamaican market. Recent lactation yield data published by the Jamaican government (Jamaica, Ministry of Agriculture and Lands, Annual Reports (1958, 1959) show that, under good feeding and management, the Jamaica Hope can consistently maintain herd averages between 700 and 800 gallons, and yields of over 1,000 gallons are not uncommon.

Santa Gertrudis. This tropical beef breed was developed on the King Ranch in South Texas. The history of the ranch stretches back to 1853, when its first owner, Richard King, introduced Spanish longhorn cattle. Between 1880 and 1910 various cross-breeding experiments with improved *B. taurus* type bulls were conducted, but it was not until cross-breeding work with shorthorn and Brahman cattle commenced in 1910 that notable progress was made. This latter programme, carried out under the vigorous leadership of Kleberg, resulted in the production of an outstanding bull, Monkey, who turned the scales at 1,100 lb at one year of age. Monkey was the result of a mating between a Brahman bull and a high-grade shorthorn cow, and he was widely used for breeding with an elite herd of crossbred Brahman × Shorthorn cows. Monkey sired many outstanding bulls, the best known of which, Santa Gertrudis, eventually gave his name to the new beef breed.

The Santa Gertrudis is noted for its early maturity, hardiness and its propensity to fatten when provided with the most unpromising grazing. It is also capable of excellent performance when maintained under feedlot conditions, and has often topped the official American '140-day-gain performance test' run by agricultural experimental stations in the southern states of the U.S.A. In July 1957 an animal from the King Ranch on one such test gained at an average rate of 3·8 lb per day, which is an exceptionally high figure for ranch type cattle.

The Santa Gertrudis is now widely spread throughout the world, with over 90,000 classified stock in thirty states of the U.S.A. and in thirty-three other countries. Its breeding programme is still largely influenced by the King Ranch in Texas, and breeding bulls from this centre fetch prices which average between 6,000 and 7,000 US dollars. The breed society has adopted an open herd-book policy, and four top-crosses with registered Santa Gertrudis bulls enable animals from other breeds to qualify for eventual registration as pedigree Santa Gertrudis cattle. The breed society has regarded liveweight gain and food conversion efficiency as essential selection requirements, but has also endeavoured to stamp the breed with easily recognizable characteristics, such as a deep red hair colour, pigmented skin, loose hide and a small hump on the male.

The Jamaica Hope and the Santa Gertrudis are good examples of cross-breeds which have been successful in achieving a high degree of adaptation to a tropical environment whilst incorporating the high productive capacity normally associated with European-type cattle. Each breed is still evolving and, with all due respect to their protagonists, neither breed has yet achieved the pure-breeding 'stability' which is generally associated with well established temperate breeds such as the Shorthorn, Hereford, Friesian or Jersey. Other examples of successful starts to the formation of new tropical cross-breeds are provided by the Bonsmara (Afrikander × Shorthorn); the Beefmaster (Shorthorn × Hereford × American Brahman); the Braford (Brahman × Hereford); the Jamaica Red (Red Poll × zebu); the Nelthropp or Senepoll (Senegal zebu × Red Poll); the Achiote (Criollo × Shorthorn); the Jamaica Black (Aberdeen Angus × zebu); the Charbray (Charolais × American Brahman see Plate 41). The essential feature of the stabilization programmes of these new cross-breeds is that the animals are mated *inter se* in a closed, or relatively closed, population once sufficient genetic variation has been introduced from the original parental foundation stocks. These stocks may be two in number, as with the Bonsmara, or may be the result of triple or quadruple crosses, as with both the Jamaica Hope and the Santa Gertrudis. Once sufficient numbers of animals of approximately the desired genotype have been obtained, all future breeding work consists of genetic selection within the inter-breeding population, with only rare and exceptional 'reference back' for outside blood from other genetic sources. This fixation of the breed requires high levels of inbreeding to reduce the excessive heterozygosity and to allow the easy recognition of undesirable recessive genes, followed by heavy culling rates to ensure that as much of the population as possible consists of superior breeding-stock.

In the majority of the cross-breeding programmes the *B. taurus* blood is introduced via the male parent and the tropical blood via the female. In certain tropical countries, it is difficult or impracticable to import purebred European bulls because disease or climatic stress would severely curtail the effective breeding life of any 'exotic' stock. The use of artificial insemination enables cross-breeding programmes to take place without importing animals of temperate breeds into the tropics, and also substantially reduces costs. The usual technique of A.I., in which semen is obtained from stud bulls once or twice a week and stored for limited periods after suitable dilution with media such as egg-yolk, is inapplicable in these cases. A 'bank' of semen from temperate bulls is required, and the semen stocks held by the bank require to be replenished at long intervals (e.g. annually). This requirement can be met by the newly evolved technique of freezing semen and maintaining it for long periods, extending to several years if necessary, in a deep-frozen state. The two most commonly employed freezing materials are liquid nitrogen, which will store semen at $-193°C$, ($-379.4°F$) and solid carbon dioxide, which will maintain the semen temperature at $-79°C$ ($-150.2°F$). Semen can be air-freighted in special containers, packed in liquid nitrogen or solid carbon dioxide inside a large thermos-flask, and can reach the most remote of tropical countries in less than two days, instead of the month or so it would take to ship out a live bull. The cost of sending 100 ampoules of semen by this means from, say, the United Kingdom to Fiji would be approximately £20, compared to about £200 sea-freight on a live animal. Semen exports have now been successfully made to most tropical countries to enable them to embark upon cross-breeding programmes, examples being Friesian semen to Northern Nigeria for use on Fulani cows, and to Trinidad for use on Nellore zebus and Holstein × zebus; Jersey semen to the Buganda province of Uganda for use on Uganda shorthorn zebus; Devon, Sussex and beef-shorthorn semen to Kenya for use on a variety of indigenous breeds and cross-breeds; Devon and Sussex semen to Paraguay for use on Criollo cows; Galloway semen to India for use on Asian zebu cows.

A further advantage in introducing exotic blood by means of air-freighted semen instead of by natural service with an imported bull is that the semen can be obtained from older, progeny-tested sires. The prepotency of the sire in respect to productive characters is therefore known from the start. However, it must be realized that the progeny test of the bull in his home country will have been carried out with different cows and in an environment different from that in which his tropical offspring will perform. There may well be a 'bull × place' interaction (and possibly a species × species interaction also) so that although the bull showed good promise in his homeland his tropical progeny may be disappointing. It is therefore prudent to use the semen of imported bulls sparingly in the first instance, and to 'bank' the greater part of the semen received until such time as performance records from his tropical progeny are available. This latter test may be described as an 'adaptability test' rather than as a 'progeny test', and it can be incorporated into a breeding programme with very little extra expense or effort, other than the loss of two to three years at the start.

PROGENY TESTING AND PFRFORMANCE TESTING

The maximum rate of genetic improvement can only be achieved when breeding-stock are selected on their ability to beget offspring which are above the average of their contemporaries in performance. This ability cannot be judged phenotypically by eye. Objective tests must be made in order to evaluate an animal's breeding potential.

As male animals are capable of producing many times more offspring than females, more attention is devoted to the genetic testing of sires than to that of dams, but it should be remembered that both sire and dam are of equal importance in determining the genotype of their offspring, since each parent contributes half the genes present in the fertilized ovum.

The classic method of assessing the genetic value of a male animal is by means of a progeny test. These tests take many forms, but the basic feature is that each male is mated to a random selection of females, and the resultant offspring are reared under comparable conditions so as to enable an objective estimate to be made of the relative performance of the offspring sired by the various males on test. If males are being progeny tested for milk, then their daughters must be bred before any data can be collected on their milking performance. The male to which they are bred is completely irrelevant to their sire's progeny test results. Progeny testing for milk yield, twinning percentage of sheep and goats, litter number in pigs and other factors associated with the reproductive performance of the offspring of the male on test, is essentially a long-term process, and there is a real danger that the male may be well past his prime by the time his progeny test results are known. This point is very important in the tropics, where the turnover of livestock generations is longer due to slower growth rates, but where longevity is less and effective breeding lives are shorter. Another difficulty operating against the success of progeny testing in many tropical countries is that there are very few herds large enough to provide sufficient females for mating to the various bulls on test. An absolute minimum of ten daughters is required to progeny test a bull for milk. It will probably be necessary to mate the bull to at least twenty-five females in order to provide a minimum of ten daughters. If there are ten bulls on test at any one time, then 250 females and 100 daughters (a total of 350 head of female stock) will be required in order to accommodate an objective progeny test. If this programme is repeated each year, with consequent overlap of the rearing periods for the test daughters, than the size of farm required will be very large indeed, and not far short of 1000 head of stock will need to be maintained.

The temperate countries are now making increased use of A.I. for extensive progeny testing purposes. In this way, several thousand recorded cows can be incorporated into the test results of bulls which show promise at a single testing centre (Robertson, 1954). Unfortunately this method is only applicable when a large number of the farmers using the A.I. service keep accurate livestock records. This situation is uncommon in the tropics where the vast majority of herds are unrecorded. There are, however, limited opportunities for utilizing the resources of recorded herds attached to such institutions as universities, farm schools, prisons and missions in conjunction with an A.I. programme, so that a compromise version of Robertson's suggested scheme could be used.

There are three different approaches to the progeny testing of bulls for milk yield. First, the bulls can be evaluated by comparing the average performance of their daughters. This method is invalid when the daughters are dispersed, in different numbers, over a range of different farms and stations. Secondly, the bulls can be evaluated by comparing their daughters' performance with that of their dams' which are milking alongside them over roughly the same period of time. Such a test is known as a 'daughter-dam comparison' but it is limited in its application as it is essential that both daughter and dam, should be in milk together. It also necessitates the use of correction factors to allow for the expected differences in yield between the daughters' heifer lactations and their dams' adult lactations. Lastly, the bulls can be evaluated by simply comparing their daughters' yields with those of their contemporaries milking alongside them for the same lactation period. This test is known as the 'contemporary comparison' test, and it has found much favour in the United Kingdom. It has been fully described by Robertson *et al.* (1956) and an example of the calculation of a progeny test result using this method is provided by Maule (1962).

The progeny testing of livestock other than cattle follows the same basic principles, and the minor variations to suit the peculiar needs of the different species will not be discussed here. However, an interesting new development with regard to the progeny testing of pigs is worthy of mention and could be directly incorporated into tropical breeding programmes with swine. The development consists of inseminating sows with mixed semen, part from a control boar of known breeding potential bearing a distinctive feature such as a colour-mark, and part from a boar on test. The offspring of the two boars will have identical uterine environment during pregnancy, and the differences in their postnatal growth and development can be attributed solely to the genetic differences of their respective sires.

Progeny testing must be regarded as the only reliable method for assessing the genetic value of an animal. However, in the case of those characters which have relatively high heritability coefficients, of the order of 0·5 or greater, it is possible to make a fair assessment of the breeding value of an animal based on its own phenotypic performance. Thus the genetic value of a beef bull from the standpoint of liveweight gain can be roughly predicted, in very general terms, from a knowledge of the bull's own growth-rate during its first few years of life. This indirect approach at forecasting the breeding value of animals is known as 'performance testing', and it can be a most helpful technique in the case of meat animals in which liveweight gain, food conversion efficiency and general meat-type conformation can be objectively measured. It is quite possible, by virtue of the limitations imposed by a performance test, to rate an animal highly when in fact it is infertile or carrying undesirable genes or incapable of transmitting its own excellence to its offspring. Performance testing is no substitute for progeny testing, but where progeny tests cannot be operated performance tests may have decided merit. In the absence of either progeny test results or performance test data the breeder may be forced to make a personal, subjective assessment of an animal's potential value, relying solely on the animal's physical appearance and his own 'eye for stock'. With tropical livestock, where adaptation is of vital importance and where breed points are of comparatively little significance, a month's performance records may well be worth more than the opinion of the best livestock judge.

For many years research workers have sought in vain for a clearcut correlation between some easily identifiable morphological character and economic production factors, such as milk yield or meat quality. Many links between visible factors and hidden economic factors have been suggested. For instance, it was thought that the size of the 'milk vein' indicated an animal's potentiality for high milk yield. However, most of the suggested linkages proved valueless, and it appeared that the productive factors of greatest economic importance were multifactorially inherited and that the chances of correlating some or all of the genes responsible for production factors with visible characters were remote. Recent work with the blood groupings of livestock has thrown new light on this question, and there is now some promise that the blood type of an animal may be correlated to some item of its productive capacity. Thus Rendel (1960) has shown that a definite relationship between the B locus and the fat percentage of milk is indicated, such that cows with the genotype known as $B^{BO}_1Y^D_2$ produce milk with 0·16 units of extra butterfat compared to cows which do not carry this particular allele. Other workers have reported similar relationships between blood group and total milk yield. Further studies have indicated that useful correlations between blood groups and productive indices also exist in swine (Bouquet, 1962; Brucks, 1964). If this work is confirmed with tropical breeds, it may be possible in the future to predict the breeding value of a male or female animal even for characters, such as milk yield or litter size, which are sex-limited in their expression, by merely typing their blood, or protein components of the blood, such as the haptoglobins (Bangham and Blumberg, 1958). Further research work in this field should be closely followed by all those with a real interest in the improvement of livestock.

REFERENCES

AAMDAL, J. (1964) 'Artificial insemination in the pig.' *Ve Cong. int. Reprod. anim. Insémin. artif.* **4**, 147–77.

ABRAMS, J.T., ed. (1950) *Linton's Animal Nutrition and Veterinary Dietetics*. 3rd edn. Edinburgh, W. Green.

ACOCKS, J.P.H. (1953) *Veld types of South Africa*. Bot. Survey of South Africa, No. 28.

ACQUAYE, D.K. (1963) 'Some significance of soil organic phosphorus mineralization in the phosphorus nutrition of cocoa in Ghana', *Plant & Soil*, **19**, 65–80.

ADRIANCE, C.W. and BRISON, F.R. (1953) *Propagation of Horticultural Plants*. 2nd edn. New York, McGraw-Hill.

A.R.C. (1963) *The nutrient requirements of farm livestock, No. 1. Poultry*. London, H.M.S.O.

ALBERDA, T. (1953) 'Growth and development of lowland rice and its relation to oxygen supply.' *Plant & Soil*, **1**, 1–28.

ALIM, K.A. (1960) 'Reproductive rates and milk yield of Kenana cattle in the Sudan.' *J. agric. Sci.* **55**, 183–8.

—— (1962) 'Environmental and genetic factors affecting milk production of Butana cattle in the Sudan.' *J. Dairy Sci.* **45**, 242–7.

ALLEN, E.F. (1956) 'The effect of crop rotation on the growth and yield of padi.' *Malay. agric. J.* **39**, 133–9.

ALVIM, P. DE T. (1958) 'Recent advances in our knowledge of coffee trees. – I. Physiology.' *Coffee and Tea Ind.* **81**, 17–25.

ANDERSON, J. (1943) 'The periodicity and duration of oestrus in Zebu and Grade cattle.' *J. agric. Sci.* **34**, 57–68.

ANDREWS, C.S. (1962) 'Influence of nutrition on nitrogen fixation and the growth of legumes', in *A review of nitrogen in the tropics with particular reference to pastures*. Com. Bur. Pastures and Field Crops, Bull. **46**, 136–46.

ANDREWS, F.W. (1948) 'The vegetation of the Sudan', in *Agriculture in the Sudan*, ed. Tothill. Oxford University Press.

ANKER-LADEFOGED (1955) 'The role of grassland in Ceylon's agriculture.' *Trop. Agriculturist*, **111**, 257–66.

ANTHONY, K.R.M. and WILLIMOT, S.G. (1957) 'Cotton interplanting experiments in the South West Sudan.' *Emp. J. exp. Agric.* **25**, 29–36.

ARMSBY, H.P. (1887) *Manual of cattle feeding*. New York, J. Wiley.

—— (1917) *The nutrition of farm animals*. New York, Macmillan.

ARNOTT, G.W. (1957) 'Soil Survey reports, No. 6. The Kelantan Deficiency Area.' *Malay agric. J.* **40**, 60–91.

ASHBY, D.G. and PFEIFFER, R.K. (1956) 'Weeds: a limiting factor in tropical agriculture.' *World Crops.* **8**, 227–9.

ASPREY, G. F. (1959) 'Vegetation in the Caribbean area.' *Caribbean Quart.* **5**, 245.

—— and ROBBINS, K. G. (1943) *The vegetation of Jamaica.* Ecol. Monogr. No. 23.

AUBERT, G. and DUCHAUFOUR, P. (1956) 'Projet de classification des sols.' *Trans. VIth Internat. Cong. Soil Sci.* Vol. E, 597–604.

AUBREVILLE, A. (1938) 'La Forêt coloniale: les forêts de l'Afrique Occidentale Francaise.' *Ann. Acad. Sci. Colon.* Paris, **9**, 1–245.

—— (1949) *Climats, forêts et desertification de l'Afrique tropicale,* Paris.

BAKER, H. K. (1961) 'The influence of grazing management on grassland production.' *N.A.A.S. Quart. Rev.* **53**, 25–9.

BALDWIN, S. P. and KENDLEIGH, S. C. (1932) *Sci. Publ. Cleveland Mus. Nat. Hist.* **3**, 1.

BANGHAM, A, D. and BLUMBERG, B. S. (1958) 'Distribution of electrophoretically different haemoglobin among some cattle breeds of Europe and Africa.' *Nature, Lond.* **181**, 1551–2.

BARBOUR, K. M. (1961) *The Republic of the Sudan: a Regional Geography.* Univ. of London Press.

BARBOUR, W. R. (1942) 'Forest types of tropical America.' *Caribbean For.* **3**, 137–50.

BARRAU, J. (1959) 'The bush fallowing system of cultivation in the continental islands of Melanesia.' *Proc. 9th Pacifie Sci. Cong.* **7**, 53–5.

BARRETT, S. F. and BAILEY, K. P. (1955) 'The duration of immunity in Zebu cattle recovering from a single infection as calves.' *E. Africa High Commission. Ann Rep. E. Afr. Vet. Res. Organ. 1954–1955,* 65–74.

BARUA, D. N. and DUTTA, K. N. (1961) 'Root growth of cultivated tea in the presence of shade trees and nitrogenous manure.' *Emp. J. exp. Agric.* **29**, 287–93.

BEARD, J. S. (1944) 'Climax vegetation in tropical America.' *Ecology,* **25**, 127.

—— (1946) 'The natural vegetation of Trinidad.' *Oxford For. Mem.* **20**.

—— (1949) *The natural vegetation of the Windward and Leeward Islands.* Oxford For. Mem. **12**.

—— (1955) 'The classification of tropical American vegetation types.' *Ecology,* **36**, 89–100.

BECKINSALE, R. P. H. (1957) 'The nature of tropical rainfall.' *Trop. Agric. Trin.* **34**, 76–87.

BENNEMA, J. (1963) 'The red and yellow soils of the tropical and subtropical uplands.' *Soil Sci.* **95**, 250–7.

BENNETT, G. H. and O'HAGAN, C. (1964) 'Factors influencing the success of artificial insemination in pigs.' *Ve. Cong. int. Reprod. anim. Insémin. artif.* **4**, 481–7.

BENNETT, H. H. (1947) *Elements of soil conservation.* New York, McGraw-Hill. 2nd edn. 1955.

—— (1959) *Soil conservation.* New York, McGraw-Hill.

BERLIER, Y., DABIN, B. and LENEUF, N. (1956) 'Physical, chemical and microbiological comparison of forest and savanna soils on the tertiary sands of the lower Ivory Coast.' *Trans. 6th Int. Cong. Soil Sci.* Vol. E. 499–502.

BHARATH, S. (1958) 'Mango propagation—a modified system of approach grafting.' *Trop. Agriculture, Trin.* **35**, 190–4.

BIRCH, H. F. (1958) 'The effect of soil drying on humus decomposition and nitrogen availability.' *Plant & Soil.* **10**, 9–31.

—— and FRIEND, M. T. (1956a) 'The organic matter and nitrogen status of East African soils.' *J. Soil Sci.* **7**, 156–67.

—————— (1956b) 'Humus decomposition in East African soils.' *Nature, Lond.* **178**, 500–1.

BIRCH, W. R. (1959) 'High altitude ley agronomy in Kenya, II. The effects of phosphate and nitrogen.' *E. Afr. agric. for. J.* **25**, 113–20.

BLAXTER, K. L. (1962) *The Energy Metabolism of ruminants.* London, Hutchinson.

BLIGH, J. (1955) 'Comparison of rectal and deep body temperature in the calf.' *Nature, Lond.* **176**, 402–3.

BLUMENSTOCK, D. I. (1958) 'Distribution and characteristics of tropical climates.' *Proc. 9th Pacific Sci. Cong.* **20**, 3–21.

BOA, W. (1958) 'Development of the N.I.A.E. ditch cleaner.' *J. Agric. Eng. Res.* **3**, 17.

BOGDAN, A. V. (1955) 'Bush clearing and grazing trial at Kisikon, Kenya.' *E. Afr. agric. for. J.* **19**, 253.

—— (1956) 'Indigenous clovers of Kenya.' *E. Afr. agric. for. J.* **22**, 40.

—— (1959) 'The selection of tropical ley grasses in Kenya.' *E. Afr. agric. for. J.* **24**, 206–17.

—— and STORRAR, A. (1954) 'The control of *Aristida* and other annuals in Kenya Rift Valley pastures.' *Emp. J. exp. Agric.* **22**, 211.

BONSMA, J. C. (1940) 'The influence of climatological factors on cattle.' *Fmg. S. Afr.* **15**, 373–86.

—— and JOUBERT, D. M. (1952) 'The Africander Breed.' *Fmg. S. Afr.* **27**, 99–104.

—— and PRETONIUS, A. J. (1943) 'Influence of colour and coat cover on adaptability of cattle.' *Fmg. S. Afr.* **18**, 101.

—— SCHOLTZ, G. D. J. and BODENHORST, F. J. G. (1940) 'The influence of climate on cattle.' *Fmg. S. Afr.* **15**, 7–12.

BOTELHO DA COSTA, J. V. and AZEVEDO, A. L. (1959) *Carta geral dos solos de Angola. I. Distrito da Huila* (with French and English summaries). Lisbon.

—— (1960). 'Generalised soil map of Angola.' *Trans. 7th Int. Cong. Soil Sci.* **4**, 56–62.

BOUQUET, Y-H. (1962) 'Blood groups in pigs.' *Vlaams diergeneesk. Tijdschi.* **31**, 253–78.

BRAAK, C. (1929) *The climate of the Netherlands Indies.* Batavia.

BRADFIELD, R. and MILLER, R. D. (1954) 'Soil structure.' *Trans. 5th Int. Cong. Soil Sci.* **1**, 131–45.

BRAMAO, D.L. (1962) 'Considerations on the potentialities of the soils of the world.' *Ann. Accad. Italiana di Scienzi Forestali*, **11**, 333–47.
—— and DUDAL, R. (1958) 'Tropical soils.' *Proc. 9th Pacific Sci. Cong.*, **20**, 46–50.
BRAMMER, H. (1962) 'Soils', in *Agriculture and Land Use in Ghana*, ed. J.B. Wills, Ghana Ministry of Food and Agriculture, pp. 88–126.
BRODY, S. (1945) *Bioenergetics and growth*. New York, Reinhold.
BROOK, A.H. and SHORT, B.F. (1960a) 'Regulation of body temperature of sheep in a hot environment.' *Aust. J. agric. Res.* **11**, 402–7.
—— (1960b) 'Sweating in sheep.' *Aust. J. agric. Res.* **11**, 557–69.
BROWN, G.M. (1960) 'Biosynthesis of water-soluble vitamins and derived co-enzymes.' *Physiol. Rev.* **40**, 331.
BRUCKS, R. (1964) 'Blood groups of the pig with special reference to the L system.' *Z. Tierz. Zücht. Biol.* **80**, 66–80.
BURTT, B.D. (1942) 'Some East African vegetation communities.' *J. Ecol.* **30**, 65–146.
BURTT-DAVY, J. (1938) *The classification of tropical woody vegetation types*. Imp. For. Inst., Oxford, paper No. 13.
BUTTERWORTH, M.H. (1961) 'Studies on pangola grass. II. The digestibility of pangola grass at various stages of growth.' *Trop. Agriculture, Trin.* **38**, 189–93.
—— (1962) 'The digestibility of sugar-cane tops, rice aftermath, and bamboo grass.' *Emp. J. exp. Agric.* **30**, 77–81.
—— (1963) 'Digestibility trials of forages in Trinidad and their use in the prediction of nutritive value.' *J. agric. Sci.* **60**, 341–6.
—— GROOM, C.G. and WILSON, P.N. (1961) 'The intake of Pangola grass under wet- and dry-season conditions in Trinidad.' *J. agric. Sci.* **56**, 407.
BUTTERY, B.R. (1961) 'Investigations into the relationship between stock and scion in budded trees of *Hevea brasiliensis*.' *J. Rubb. Res. Inst. Malaya*, **17**, 46–76.
CABORN, J.A. (1957) 'The value of shelter belts to farming.' *Agric. Rev., Lond.* **2**, 14.
CAIN, S.A. and CASTRO, G.M. DE O. (1959) *Manual of vegetation analysis*. New York, Harper.
CANNON, W.A. (1925) 'Physiological features of roots with especial reference to the relation of roots to the aeration of the soil.' *Carnegie Inst. Washington*. Pub. No. **368**, 119–22.
CARO-COSTAS, R. and VINCENTE-CHANDLER, J. (1956) 'Comparative productivity of Merker grass and of a Kudzu-Merker mixture as affected by season and cutting height.' *J. Agric. Univ. Puerto Rico*, **40**, 144–51.
CARTER, D.B. (1954) 'Climates of Africa and India according to Tnornthwaite's 1948 classification.' *Johns Hopkins Univ. Pub. in Climatology*, **7**, 453–74.
CARTER, G.F. and PENDLETON, R.L. (1956) 'The humid soil: process and time.' *Geogr. Rev.* **46**, 488–507.

CASTILLO, L.S., PALAD, O.A., NAZERENO, L.E., CLAMOHOY, L.L. and SARAO, F.B. (1960) 'The effect of season of calving and advancing lactation on milk production of dairy Carabaos and cows.' *Philipp. Agric.* **43**, 604–12.

CHALMERS, A.W. (1954) 'Notes on the livestock of the Sudan.' MS.

CHAMNEY, A. (1928) 'Climatology of the Gold Coast.' *G. Coast Dep. Agric. Bull.* **15**.

CHAMPION, H.G. (1936) 'A preliminary survey of forest types of India and Burma.' *Indian For. Record.* N.S. **1**, 1–286.

CHEBNIL, W.S. (1957) 'Erosion of soil by wind', in *Soil, the yearbook of agriculture for 1957*, 308. U.S. Dept. Agric. Washington, D.C.

CHEMISTRY BRANCH, S. RHODESIA. (1947) 'Analyses of Rhodesian Foodstuffs.' *Rhod. Agric. J.* **44**, 463.

CLARKE, R.T. (1962) 'The effect of some resting treatments on a tropical soil.' *Emp. J. exp. Agric.* **30**, 57–62.

CLEGHORN, W.B. *et al.* (1958) 'Results of preliminary arboricide and herbicide trials on a number of species limiting livestock carrying capacity of veld in S. Rhodesia.' *Proc. Afr. Weed Control Conf.*, July 1958.

COBBLE, J.W. and HERMAN, H.A. (1951) 'Influence of environmental temperatures on the composition of milk of the dairy cow.' *Res. Bull. Mo. Agric. Exp. Sta.* 485.

COCONUT RESEARCH INSTITUTE CEYLON (1953) *Annual Report*.

COETZEE, PAGE, A.I. and MEREDITH, D. (1946) 'Root studies in high veld pasture communities.' *S. Afr. Sci. J.* **42**, 105.

COLWELL, J.O. (1958) 'Observations on the pedology and fertility of some Kraznosems in northern New South Wales.' *J. Soil Sci.* **9**, 46–57.

COMMONWEALTH AGRIC. BUREAUX (1947) *The use and misuse of shrubs and trees as fodder*. Joint Pub. No. 10.

CONKLIN, H.C. (1957) *Hanunoo agriculture*. F.A.O. Forestry Development Paper No. 12.

COPE, F.W. and BARTLEY, B.G.D. (1957) 'Some early observations on seedling progenies.' *Rep. on cacao res. 1955–6*, pp. 7–8. I.C.T.A. T'dad.

CORBETT, A.S. (1934) 'Studies in tropical soil microbiology. II. The bacterial numbers in the soil of the Malay Peninsula.' *Soil Sci.* **38**, 407–16.

CORMACK, J.M. (1948) 'The construction of small earth dams.' *Rhod. agric. J.* **46**, 355–62.

CORMACK, R.M. (1951) 'The mechanical protection of arable land.' *Rhod. agric. J.* **48**, 135.

—— (1953) 'The vlei areas of S. Rhodesia and their uses.' *Rhod. agric. J.* **50**, 465–83.

COULTER, J.K. (1950) 'Peat formation in Malaya.' *Malay agric. J.* **33**, 63–81.

—— (1957) 'Development of the soils of Malaya.' *Malay agric. J.* **40**, 161–75.

COWGILL, W.H. (1958) 'The sun hedge system of coffee growing.' *Coffee and Tea Ind.* **81**, 87–90.

CRAMER, P. J. S. (1957) *A review of literature of coffee research in Indonesia.* Inter-Amer. Inst. of Agric. Sci. Turrialba Misc. Pub. No. 15.

CRAMPTON, E. W., LLOYD, L. E. and MACKAY, V. G. (1957) 'The calorie value of TDN.' *J. anim. Sci.* **16**, 541–5.

CRANE, J. C., REINER, M. and SCOTT BLAIR, G. W. (1960) 'The rheology of cervical secretions of cows.' *Bull. Res. Coun. Israel*, **8**, 87–100.

CROWTHER, F. (1943) 'Influence of weeds on cotton in the Sudan Gezira.' *Emp. J. exp. Agric.* **11**, 1–14.

CUNNINGHAM, P. K. and LAMB, J. (1958) 'Cocoa shade in a manurial experiment in Ghana.' *Nature, Lond.* **182**, 119.

DABIN, B. (1951) 'Contribution à l'étude des sols du Delta Central Nigerien.' *L'Agronomie Tropicale*, **6**, 606–37.

—— (1954) 'Les problèmes de l'utilisation des sols à l'office du Niger.' *Proc. 2nd Inter-Afr. Soils Conf.* Vol. II, 1165–76.

DAGG, M. (1962) 'Grazing control in semi-arid ranch land. Physical properties of the surface soil; infiltration and availability of soil moisture.' *E. Afr. agric. for J.* **27**, 68.

DAMES, T. W. G. (1959) *Soil survey of the Pangani Valley.* F.A.O. Rome (mimeographed).

DASTUR, N. A. (1956) *Dairy Sci. Abstr.* **18**, 967.

DAVIDSON, H. R. and COEY, W. E. (1966) 3rd Ed. *The Production and Marketing of Pigs.* London, Longmans.

DE, P. K. (1939) 'The role of blue-green algae in nitrogen fixation in rice fields.' *Proc. roy. Soc. B.* **127**, 121–39.

—— and MANDAL, L. N. (1956) 'Fixation of nitrogen by algae in rice soils.' *Soil Sci.* **81**, 453–9.

—— and SULAIMAN, M. (1950) 'Influence of algal growth in the rice fields on the yield of crop.' *Ind. J. agric. Sci.* **20**, 327.

DENNISON, E. B. (1959) 'The maintenance of fertility in the Southern Guinea Savanna Zone of Northern Nigeria.' *Trop. Agriculture, Trin.* **36**, 171–8.

—— (1961) 'The value of farmyard manure in maintaining fertility in Northern Nigeria.' *Emp. J. exp. Agric.* **29**, 330.

DEPARTMENT OF LANDS AND SURVEYS, TANGANYIKA (1956) *Atlas of Tanganyika*, 3rd edn.

DESPANDE, T. L., GREENLAND, D. J. and QUIRK, J. P. (1964) 'Role of iron oxide in the bonding of soil particles.' *Nature, Lond.* **201**, 107–8.

DIERENDOCK, F. J. E. VAN (1959) *The manuring of coffee, cocoa, tea and tobacco.* Centre de l'Etude de l'Azote, Geneva.

DJOKOTO, R. K. and STEPHENS, D. (1961) 'Thirty long-term fertiliser trials under continuous cropping in Ghana.' *Emp. J. exp. Agric.* **29**, 181–95; 245–57.

DOBBY, E. H. C. (1942) 'Settlement patterns in Malaya.' *Geogr. Rev.* **32**, 24–32.

DOMMERGUES, Y. (1952) 'Influence du défrinchement de foret suivi d'incendie sur l'activite biologique du sol.' *Mem. Inst. Sci. Mad.* D.**4**, 273–96.

—— (1954) 'Action de feu sur la microflora des sols de prairie.' *Mem. Inst. Sci. Mad.* D. **6**, 149–58.

DOUGALL, H.W. (1954) 'Fertiliser experiments on grassland in the Kenya Highlands.' *E. Afr. agric. for. J.* **19**, 171; 212.
—— (1960) 'Average nutritive value of Kenya feeding stuffs.' *E. Afr. agric. for. J.* **26**, 119.
—— and BOGDAN, A.V. (1958a) 'The chemical composition of the grasses of Kenya.' *E. Afr. agric. for. J.* **23**, 285.
—— (1958b) 'Browse plants of Kenya.' *E. Afr. agric. for. J.* **23**, 236.
DOUGHTY, L.R. (1953) 'The value of fertilisers in African agriculture: field experiments in E. Africa, 1947–51.' *E. Afr. agric. for. J.* **19**, 30–1.
DOWLING, D.F. (1956) 'An experimental study of heat tolerance of cattle.' *Aust. J. agric. Res.* **7**, 469–81.
—— (1958) 'The significance of sweating in heat tolerance of cattle.' *Aust. J. agric. Res.* **9**, 579–86.
—— (1959) 'Medullation in heat tolerance of cattle.' *Aust. J. agric. Res.* **10**, 736–43.
DUCKER, H.C. and HOYLE, S.T. (1947) 'Some studies in cultivation practices, food crops and the maintenance of fertility in Nyasaland.' *E. Afr. agric. for. J.* **13**, 107–13.
DUDAL, R. and SOEPRAPTOHARDJO, M. (1960) 'Some considerations on the genetic relationship between latosols and andosols in Java (Indonesia).' *Trans. VIIth Int. Cong. Soil Sci.* **4**, 229–37.
DUNIEC, H., KIELANOWSKI, J. and OSINSKA, ZOFIA (1961 'Heritability of chemical fat content in the loin muscle of baconers.' *Anim. Prod.* **3**, 195–8.
DUNNE, L.W. (1959) *Diseases of Swine.* Iowa State Univ. Press, Iowa, U.S.A.
DUTHIE, D.W. (1957) 'The first E. African Herbicide Conference: summaries of papers.' *E. Afr. agric. for. J.* **23** 6–21.
DUTTA, S.A., BASU, S. and SHARMA, K.N. (1958) *Rep. Ind. Tea Ass. Sci. Dep. (Tocklai) for 1957*, 71–114.
DUTTA-ROY, D.K. and KORDOFANI, A. (1961) 'Study of long term rotation effects in the Sudan Gezira.' *J. agric. Sci.* **57**, 387–92.
EAST AFRICA ROYAL COMMISSION (1955) Report 1953–5. London, H.M. Stationery Office, Cmd. 9475.
EATON, O.N. and SIMMONS, U.L. (1953) *Inducing extra-seasonal breeding in goats and sheep by controlled lighting.* U.S.D.A. Circ. No. 933.
ECONOMIC ADVISORY COUNCIL (1939) *Nutrition in the Colonial Empire.* London, H.M. Stationery Office, Cmd. 6050.
EDELMAN, C.H. and VAN STAVEREN, J.M. (1958) 'Marsh soils in the United States and in the Netherlands.' *J. Soil and Water Cons.* **13**, 5.
EDEN, T. (1940) 'Studies in the yield of tea. IV. Effect of cultivation and weeds on crop growth.' *Emp. J. exp. Agric.* **8**, 269–79.
—— (1958) *Tea.* London, Longmans (2nd edn. 1965).
EDWARDS, D.C. (1942) 'Grass burning.' *Emp. J. exp. Agric.* **10**, 219–31.
—— (1956) 'The ecological regions of Kenya.' *Emp. J. exp. Agric.* **24** 89–108.
—— and BOGDAN, A.V. (1951) *Important grassland plants of Kenya,* London, Pitman.

EDWARDS, R.L., TOVE, S.B., BLUMER, T.N. and BARRICK, E.R. (1961) 'Effects of added dietary fat on fatty acid composition and carcass characteristics of fattening steers.' *J. anim. Sci.* **20**, 712–20.

ELLIS, B.S. (1953) 'Soil and farming systems in Southern Rhodesia with special reference to grass leys.' *Rhod. agric. J.* **50**, 5–11.

ELLISON, W.D. (1952) 'Raindrop energy and soil erosion.' *Emp. J. exp. Agric.* **20**, 81.

EMPIRE COTTON GROWING CORP. (1951–6) Progress reports from Experimental Stations, Namulonge, Uganda: Lake Province, Tanganyika.

ENLOW, C.R. (1961) *Some observations on agriculture in Kenya.* Nairobi, Kenya, Dept. of Agric.

EPSTEIN, H. (1953) 'The dwarf goats of Africa.' *E. Afr. agric. for. J.* **18**, 123–32.

—— (1957) 'The Sanga cattle of E. Africa.' *E. Afr. agric. for J.* **22**, 149–64.

ESPFRANDIEU, G. (1952) *Congrés Panafricain de Préhistoire, Alger. Actes de la 11ᵉ Session.*

EVANS, A.C. (1955) 'A study of crop production in relation to rainfall reliability.' *E. Afr. agric. for. J.* **20**, 263.

—— (1960) 'Studies of intercropping: I. Maize or sorghum with groundnuts.' *E. Afr. agric. for. J.* **26**, 1–10.

—— and SREEDHARAN, A. (1962) 'Studies of intercropping: II. Castor bean with groundnuts or soya bean.' *E. Afr. agric. for. J.* **28**, 7–80

EVANS, E. and LEMON, E. (1957) 'Conserving soil moisture', in *Soil, the year book of agriculture for 1957.* Washington, D.C., U.S. Dept. Agric.

EVANS, H. (1953) 'Recent investigations on the propagation of cacao.' *Rep. on Cacao Res. 1945–51,* pp. 29–37. I.C.T.A. T'dad.

—— and MURRAY, D.B. (1953) 'A shade and fertiliser experiment on young cacao.' *Rep. on Cacao Res. 1945–51,* pp. 67–76. I.C.T.A. T'dad.

FALCONER, D.S. (1961) *Introduction to Quantitative Genetics.* Edinburgh, Oliver & Boyd.

FANSHAWE, D.B. (1952) *The vegetation of British Guiana.* Imp. For. Inst. Oxford, paper No. 29.

FAO (1948a) 'Nutrition problems of rice-eating countries in Asia.' *Rep. Nutrition Cttee., Baguio, Philippines.*

—— (1948b) *Rice and rice diets.* (Nutritional Studies No. 1).

—— (1952) *Kwashiorkor in Africa.* (Nutritional Studies No. 8).

—— (1953) *Maize and maize diets.* (Nutritional Studies No. 9).

—— (1955) Meeting Report No. 1955/19. Rome, Agric. Div. F.A.O.

FAO–UNICEF (1958) Rep. FAO–UNICEF Regional School Feeding Seminar for Asia and the Far East.

FAO–WHO (1950–1953) Reps. Joint FAO–WHO Expert Cttee. on Nutrition.

FARBROTHER, H.G. and MANNING, H.L. (1952) Emp. Cott. Growing Corp., Progress Reps. from Expt. Sts.; Namulonge, 1951–2, pp. 3–6.

—— *Ibid.* 1955–6, p. 16.

References

FARMER, B.H. (1957) *Pioneer Peasant Cultivation in Ceylon*. Oxford U.P.

FAULKNER, D.E. (1951) 'The improvement of cattle in Kenya with special reference to the indigenous types.' F.R.C.V.S. Thesis. London.

—— and BROWN, J.D. (1953) *The improvement of cattle in British Colonial territories in Africa*. Publ. Colon. Adv. Coun. Agric. Anim. Hlth. For. No. 3. London, H.M. Stationery Office.

—— and EPSTEIN, H. (1957) *The indigenous cattle of the British Dependent territories in Africa*. Publ. Colon. Adv. Coun. Agric. Anim. Hlth. For. No. 5. London, H.M. Stationery Office.

FERNANDO, G.W.E. (1961) 'Preliminary studies on the associated growth of grasses and legumes.' *Trop. Agriculturist*, **97**, 167–79.

FERNIE, L.M. (1958) 'Supply of better planting material – 3. Asexual propagation of coffee.' *Coffee and Tea Ind.* **81**, 17.

FERWERDA, F.P. (1958) 'The supply of better planting material (2) Canephoras.' *Coffee and Tea Ind.* **81**, 58–63.

FIELDEN, G. ST. C. and GARNER, R.J. (1936) 'Vegetative propagation of tropical and subtropical fruits.' *Imp. Bur. of Fruit Production* Tech. Comm. No. 7.

—— (1940) 'Vegetative propagation of tropical and subtropical plantation crops.' *Imp. Bur. of Fruit Production* Tech. Comm. No. 13.

FINCK, A. and OCHTMANN, L.H.J. (1961) 'Problems of soil evaluation in the Sudan.' *J. Soil Sci.* **12**, 87–95.

FINDLAY, J.D. (1950) 'The effects of temperature, humidity, air movement and solar radiation on the behaviour and physiology of cattle and other farm animals.' *Bull. Hannah Dairy Inst.* **9**.

—— (1954) *Meteorological Monographs*, **2**, 9.

FISHER, R.A, and YATES, F. (1953) *Statistical Tables*. London, Oliver and Boyd.

FOCAN, P. et al. (1954) 'L'influence de l'incineration sur l'incidence des malades radiculaires.' *Bull. agric. Congo belge.* **41**, 921–4.

FOGGIE, A. (1947) 'On the definition of forest types in the closed forest zone of the Gold Coast.' *Farm and For.* **8**, 50–5.

FORSTER, R.H., WILSON, P.N. and BUTTERWORTH, M.H. (1961) 'Pasture grass investigations in Trinidad with special reference to pangola grass.' *Proc. 8th Int. Grassland Cong., 1960.* 390–2.

FOSTER, L.J. and WOOD, R.A. (1963) 'Observations on the effect of shade and irrigation on soil moisture utilisation under coffee in Nyasaland.' *Emp. J. exp. Agric.* **31**, 108–14.

FOSTER, W.H. and MUNDY, E.J. (1961) 'Forage species in Northern Nigeria.' *Trop. Agriculture, Trin.* **38**, 311–8.

FRANCIS, C.S.L. (1957) 'Jungle clearance and reclamation.' *World Crops*, **9**, 187–90.

FRENCH, M.H. (1943a) 'The composition and nutritive value of Tanganyika feeding stuffs.' *E. Afr. agric. for. J.* **8**, 126.

—— (1943b) 'Feeding values of stover from maize, millet and bullrush millet.' *E. Afr. agric. for. J.* **9**, 88.

FRENCH, M.H. (1950) 'The nutritive value of East African grass and fodder plants.' *E. Afr. agric. for. J.* **15**, 214.

—— (1956a) 'The effect of infrequent water intake on the consumption and digestibility of hay by Zebu cattle.' *Emp. J. exp. Agric.* **24**, 128–36.

—— (1956b) 'The importance of water in the management of cattle.' *E. Afr. agric. for. J.* **21**, 171–81.

—— (1956c) 'The nutritive value of East African hay.' *Emp. J. exp. Agric.* **24**, 53.

—— (1957) 'Nutritional value of tropical grasses and fodders.' *Herbage Abstr.*, **27**, 1.

—— GLOVER, J. and DUTHIE, D.W. (1960) 'The apparent digestibility of crude protein in the ruminant. 1. A synthesis of the results of digestibility trials with herbage and mixed feeds.' *J. agric. Sci.* **48**, 373–83.

FRIPIAT, J.J. and GASTUCHE, M.C. (1952) *Physico-chemical study of clay surfaces: combinations of kaolinite with trivalent iron oxides.* Pub. I.N.E.A.C., Series Scientifique, No. 54.

GARBELL, M.A. (1947) *Tropical and equatorial meteorology.* London, Pitman.

GARFITT, J.E. (1941) 'Malayan forest types.' *Malay. For.* **10**, 136–40.

GARMANY, H.F.M. (1954) 'Plough deep and early for higher yields.' *Rhod. Tobacco.* **6**, 12–4.

GARNER, R.J. (1958) *The Grafter's Handbook*, 2nd edn. London, Faber & Faber.

GETHIN JONES, G.H. (1942) 'The effect of a leguminous cover crop in building up soil fertility.' *E. Afr. agric. for. J.* **8**, 48.

GHOSE, R.L.M., GHATZE, M.B. and SUBRAMANYAM, P.V. (1956) *Rice in India.* Indian Council for Agric. Res.

GILLMAN, C. (1949) 'A vegetation type map of Tanganyika.' *Geogr. Rev.* **39**, 7–37.

GLEASON, H.A. and COOK, W.G. (1926) *Plant ecology of Puerto Rico: Scientific Study of Puerto Rico.* vol. 7. New York, Acad. Sci.

GLOVER, J. (1957) 'The relationship between total seasonal rainfall and the yield of maize in the Kenya Highlands.' *J. agric. Sci.* **49**, 285.

—— and FRENCH, M.H. (1957) 'The apparent digestibility of crude protein in the ruminant. IV. The effect of crude fibre.' *J. agric. Sci.* **49**, 78–80.

—— and ROBINSON, H.C. (1953) 'A simple method for assessing the reliability of rainfall.' *J. agric. Sci.* **43**, 275.

—— ROBINSON, H.C. and HENDERSON, J.P. (1954) 'Provisional maps of reliability of annual rainfall in East Africa.' *Quart. J.R. met. Soc.* **80**, 602.

GORRIE, R.M. (1946) *Soil and water conservation in the Punjab.*

GOUROU, P. (1953) *The tropical world*, trans C.D. Laborde. London, Longmans.

GRASSLAND RESEARCH STATION, KITALE (1959) 'Notes on Kenya agriculture. VI. Grass leys and grassland plants.' *E. Afr. agric. for. J.* **20**, 54–6.

GREENE, H. (1928) 'Soil permeability in the eastern Gezira.' *J. agric. Sci.* **18**, 531–43.

—— and SNOW, O.W. (1939) 'Soil improvement in the Sudan Gezira.' *J. agric. Sci.* **29**, 1–34.

GREENLAND, D. J. and NYE, P. H. (1959) 'Increases in the carbon and nitrogen contents of tropical soils under natural fallows.' *J. Soil Sci.* **9**, 284–99.

GREENWOOD, M. (1951) 'Fertiliser trials with groundnuts in Northern Nigeria.' *Emp. J. exp. agric.* **19**, 225–41.

GRIFFITHS, A. L. (1947) 'Effects of artificial soil regeneration by burning.' *Indian For.* **73**, 526–8.

GRIFFITHS, J. F. (1959) 'Climatic zones of East Africa.' *E. Afr. agric. for. J.* **23**, 179.

GRIMES, R. C. (1963) 'Intercropping and alternate row cropping of cotton and maize.' *E. Afr. agric. for. J.* **28**, 161–3.

—— and CLARKE, R. T. (1962) 'Continuous arable cropping with the use of manure and fertilisers.' *E. Afr. agric. for. J.* **28**, 74–80.

GRIST, D. H. (1965) *Rice*, 4th edn. London, Longmans.

GULLIVER, P. H. (1955) *The family herds*. London, Routledge & Kegan Paul.

HADDON, A. V. and TONG, Y. L. (1959) 'Oil palm selection and breeding: a progress report.' *Malay. agric. J.* **42**, 124–56.

HAFEZ, E. S. E. (1952) 'The buffalo; a review.' *Indian J. Vet. Sci.* **22**, 257–63.

HALE, R. W. (1961) 'Heritabilities and genetic correlations of egg production and other characters in a White Wyandotte flock.' *Anim. Prod.* **3**, 73–88

HALL, T. D., MEREDITH, D. and ALTONA, P. E. (1950) 'Production from grassland in South Africa.' *Emp. J. exp. Agric.* **18**, 8.

HAMILTON, R. A. and ARCHBOLD, J. W. (1945) 'Meteorology of Nigeria and adjacent territory.' *Quart. J.R. met. Soc.* **71**, 231–62.

HANCOCK, J. (1954) 'Studies of grazing behaviour in relation to grassland management.' *J. agric. Sci.* **44**, 420–9.

—— and PAYNE, W. J. A. (1955) 'The direct effect of tropical climate on the performance of European-type cattle. I. Growth.' *Emp. J. exp. Agric.* **23**, 55–74.

HANCOCK, J. L. (1962) 'Fertilization in farm animals.' *A.B.A.* **30**, 285–310.

HANSEL, W. (1961) 'Oestrus cycle and ovulation control in cattle.' *J. Dairy Sci.* **44**, 2307–14.

HARDY, F. (1951) 'Soil productivity in the British Caribbean region.' *Trop. Agriculture, Trin.* **28**, 3–23.

—— AKHURST, C. G. and GRIFFITH, G. (1931) *The cacao soils of Tobago. Studies in West Indian Soils No. 3.* I.C.T.A., T'dad.

—— and BEARD, J. S. (1949) *Caribbean Soil Conf. Proc.* Puerto Rico. Mimeographed: Caribbean Commission, Port of Spain, Trinidad.

—— ROBINSON, C. K. and RODRIGUES, G. (1934) *The agricultural soils of St Vincent. Studies in West Indian Soils No. 8.* I.C.T.A., T'dad.

—— RODRIGUES, G. and NANTON, W. R. (1949) *Agricultural soils of Montserrat. Studies in West Indian Soils No. 12.* I.C.T.A., T'dad.

HARKER, K. W. (1954) 'The establishment of a *Chloris gayana* ley under a sorghum silage crop.' *E. Afr. agric. for. J.* **20**, 54–6.

—— TAYLOR, J. I. and ROLLINSON, D. H. L. (1954) 'Studies on the habits of Zebu cattle, 1.' *J. agric. Sci.* **44**, 193–8.

HARRIS, L.E. (1962) *Glossary of energy terms.* Nat. Acad. Sci. N.R.C. Publ. 1040.

HARRISON, M.N. and JACKSON, J.K. (1955) *Vegetation map of the Anglo-Egyptian Sudan.* C.S.A. Report on phytogeology.

HART, M.G.R., CARPENTER, A.J and JEFFERY, J.W.O. (1963) 'Problems in reclaiming saline mangrove soils in Sierra Leone.' *L'agron. Tropicale* (Serie 1) **18**, 800–2.

HARTLEY, C.W.S. (1947) 'Experiments on the growing of off-season crops on padi land in Province Wellesley.' *Malay. agric. J.* **30**, 114–20.

—— (1949) 'The felling and disposal of old oil palms prior to replanting.' *Malay agric J.* **32**, 223–30.

—— (1957) 'Oil palm breeding and selection in Nigeria.' *J.W. Afr. Inst. for Oil Palm Res.* **2**, 108.

HARTLEY, K.T. (1937) 'The effect of farmyard manure in Northern Nigeria.' *Emp. J. exp. Agric.* **5**, 254–63.

HAWKINS, J.C. (1960) *The mechanisation of agriculture in the West Indies.* Nat. Inst. Agric. Eng. G. Britain.

HAWS, R.C. (1959) *The conservation of natural resources.* London, Faber & Faber.

HAYLETT, D.G. (1943) 'Crop residues and soil fertility.' *Fmg. in S. Afr.* **18**, 627–36.

HEMMING, C.F. and TRAPNELL, C.G. (1957) 'A reconnaissance classification of the soils of the Turkhana Desert.' *J. Soil Sci.* **8**, 167–83.

HENDERSON, R. (1955) 'The cultivation of fodder grasses in Malaya.' *Malay. Agric. J.* **38**, 71; 141; 250.

HENKEL, J.S. (1931) 'A vegetation map of Southern Rhodesia.' *Proc. Rhod. Sci. Ass.* **30**, 1.

HICKLING, C.F. (1961) *Tropical inland fisheries.* London, Longmans.

HODNETT, G.E. and SMITH, C.A. (1962) 'Compensatory growth of cattle on the natural grasslands of Northern Rhodesia.' *Nature, Lond.* **195**, 919.

HOFMEYER, J.H. (1942) 'Soil cultivation and increased production.' *Fmg. in S. Afr.* **17**, 721–6.

HOGG, W.H. and IBBET, W.C. (1955) 'Wind shelter for horticultural crops.' *Agriculture, Lond.* **62**, 587.

HOLMES, C.M. (1951) *The grass, fern and savanna lands of Ceylon.* Imp. For. Inst. Oxford, Paper No. 28.

—— (1958) 'The broad pattern of climate and vegetation distribution in Ceylon', in *Study of tropical vegetation,* UNESCO, p. 99.

D'HOORE, J.L., FRIPIAT, J.J. and GASTUCHE, M.C. (1954) 'Tropical clays and their iron oxide covering.' *Proc. 2nd Inter-Afr. Soils Conf.* **1**, 157–60.

—— (1960) 'Soils map of Africa on the scale of 1/5,000,000.' *Afr. Soils,* **5**, 55–64.

HOPKINS, A.L. and AUMER, W. (1943) Quoted in *Soils and Fert.* (1949) **12**, 155.

HORAK, I.G. (1960) 'Artificial insemination under ranching conditions.' *J. S. Afr. vet. med. Assn.* **31**, 99–106.

HORNBY, H. E. and VAN RENSBURG, H. J. (1948) 'The place of goats in Tanganyika Farming systems. 1. In deciduous bushland formation.' *E. Afr. agric. for. J.* **14**, 94–8.

HOUGHTON, T. R. (1960) 'The water buffalo in Trinidad.' *J. agric. Soc. Trin. & Tob.* **60**, 339–56.

—— BUTTERWORTH, M. H., KING, D. and GOODYEAR, R. (1964) 'The effects of different levels of food intake on fattening pigs in the humid tropics.' *J. agric. Sci.* **63**, 43–50.

HOWE, G. M. (1953) 'The climates of the Rhodesias and Nyasaland according to Thornthwaite's classification.' *Geogr. Rev.* **43**, 525.

HUDSON, N. W. (1957) 'Erosion control research.' *Rhod. agric. J.* **54**, 297.

HUMPHREY, D. (1947) *The Liguru and the land.* Nairobi, Govt. Printer.

HUROV, H. R. (1961) 'Green strip budding of two- to eight-month-old rubber seedlings.' *Proc. Nat. Rub. Res. Conf. Kuala Lumpur, 1960*, 419–28.

HUTCHINSON, J. B., FARBROTHER, H. G. and MANNING, H. L. (1958) 'Crop water requirements of cotton.' *J. agric. Sci.* **51**, 177.

HUTCHINSON, J. C. D. (1954) 'Heat Regulation in birds', in *Progress in the Physiology of Farm Animals*. London, Butterworths. Ch. 7.

I.C.A.R. (1950) *Indian Coun. Agric. Res. Bull.* **8**, 22. Delhi; Manager of Publications.

IGNATIEFF, V. and LEMOS, P. (1963) 'Some management aspects of more important tropical soils.' *Soil Sci.* **95**, 243–9.

—— and PAGE, H. J. (1958) *Efficient use of fertilizers*. Rome, F.A.O.

INTERNATIONAL BANK FOR RECONSTRUCTION AND DEVELOPMENT (1961) *The economic development of Tanganyika*. Baltimore, Johns Hopkins Press.

IRVING, H. (1954) *Fertilizer studies in eastern Nigeria 1947–51*. Eastern Region Nigeria Tech. Bull. No. 1, Enugu.

ISLAM, M. A. and ELAHI, M. A. (1954) 'Reversion of ferric iron to ferrous iron under waterlogged conditions and its relation to available phosphorus.' *J. agric. Sci.* **45**, 1–9.

IVENS, G. W. (1958–60) 'The effect of arboricides on East African trees and shrubs.' *Trop. Agriculture, Trin.* **35**, 257–71; **36**, 52–64, 219–29; **37**, 143–52.

JACKS, G. V. (1934) *Soil, vegetation and climate.* Comm. Bur. Soil Sci., Tech. Comm. No. 29.

—— BRIND, W. D., and SMITH, G. A. (1955) *Mulching*. Comm. Bur. Soil Sci., Tech. Comm. No. 49.

JACKSON, G. (1955) Quoted by Ivens, G.W. (1959).

JAMAICA, MIN. OF AGRIC. AND LANDS (1960, 1961). Livestock Division Annual Report. *Ann. Rep. Min. Agric. and Lands, Jamaica*, 1958, 1959. Kingston, Govt. Printer.

JAMESON, J. D. and KERKHAM, R. K. (1960) 'The maintenance of soil fertility in Uganda. I. Soil fertility experiments at Serere.' *Emp. J. exp. Agric.* **29**, 179–92.

JEFFREYS,M.D.W. (1953) '*Bos brachyceros* or dwarf cattle.' *Vet. Rec.* **65**, 393-6.
JESSUP,R.W. (1961) 'A Tertiary-Quaternary pedological chronology for the S.E. portion of the Australian arid zone.' *J. Soil Sci.* **12**, 199-213.
JEWITT,T.N. and MIDDLETON,K.R. (1955) 'Changes in Sudan Gezira soil under irrigation.' *J. agric. Sci.* **45**, 277-9.
JONES,P.A. and WALLIS,J.A.N. (1963) 'A tillage study in Kenya coffee. III. The long term effects of tillage practices upon yield and growth of coffee.' *Emp. J. exp. Agric.* **31**, 243-54.
JONES,T.A., LOXTON,R.F., RUTHERFORD,G.K. and SPECTOR,J. (1958) *Soil and Land-Use Surveys No. 2. British Guiana. The Rupununi Savannas.* I.C.T.A., T'dad.
—— STARK,J., RUTHERFORD,G.K. and SPECTOR,J. (1959) *Soil and Land-Use Surveys No. 6. British Guiana. The Rupununi Savannas* (contd.). I.C.T.A., T'dad.
JORDAN,H.D. (1963) 'Development of mangrove swamp areas in Sierra Leone.' *L'agron Tropicale* (Serie 1) **18**, 798-9. ·
JORDAN,S.M. (1957) 'Reclamation and pasture management in the semi-arid areas of Kitui District, Kenya.' *E. Afr. agric. for. J.* **23**, 84.
JOSHI,N.R. and PHILLIPS,R.W. (1953) *Zebu cattle of India and Pakistan.* F.A.O. agric. Stud. (Rome) No. 19.
—— MCLAUGHLIN,E.A. and PHILLIPS,R.W. (1957) *Types and breeds of African cattle.* F.A.O. agric. Stud. (Rome) No. 37.
JUKO,C.D. and BREDON,R.M. (1961) 'The chemical composition of leaves and whole plant as an indicator of the range of available nutrients for selective grazing by cattle.' *Trop. Agriculture, Trin.* **38**, 179-87.
JULL,M.A. (1951). *Successful Poultry Management.* 2nd edn. New York, McGraw-Hill.
JURION,F. and HENRY,J. (1951) 'Cropping systems in the equatorial forest region of the Belgian Congo.' *U.N. Sci. Conf. on Conservation and Utilisation Resources*, **6**, 255-8.
KANAPATHY,K. (1957) 'Nitrogen and phosphorus uptake of padi compared with that of dry land crops.' *Malay. agric. J.* **40**, 110-20.
—— (1962) Personal communication.
KANITKAR,N.V. (1944) *Dry farming in India.* Ind. Council of Agric. Res., Sci. Monograph No. 15.
KANNO,I. (1956) 'A pedological investigation of Japanese volcanic ash soils.' *Trans. VIth Int. Cong. Soil Sci.*, vol. E. 105-9.
KARTHA,K.P.R. (1959) Chapter 2 in J.C.Williamson and W.J.A. Payne, *An introduction to animal husbandry in the tropics.* London, Longmans.
KAURA,R.L. (1943) *Indian Fmg.* **4**, 549-56.
KEAY,R.W.J. *et al.* (1949) *An outline of Nigerian vegetation.* Lagos, Govt. Printer.
—— (1951) 'Notes on the ecological status of savanna vegetation in Nigeria', in *Management and conservation of vegetation in Africa*, Comm. Bur. Pastures, Bull. No. 41.

References

KEAY, R. W. J. et al. (1959) *Explanatory notes: Vegetation map of Africa south of the Tropic of Cancer.* Oxford University Press.

KEEN, B. A. (1949) 'The effects of mechanical cultivation on soil degradation.' *Bull agric. Congo belge.* **40**, 2003–9.

KEEPING, G. S. (1951) 'A review of fodder grass investigations in Malaya.' *Malay agric. J.* **34**, 65–75.

KELLNER, O. (1905) *Die Ernahrung der landwirtschaftlichen Nutztiere.* Berlin, Paul Parey.

KELLOGG, C. E. (1949) 'Preliminary suggestions for the classification and nomenclature for great soil groups in tropical and equatorial regions.' *Comm. Bur. Soils Tech. Comm.* **46**, 76-85.

—— and DAVOL, F. D. (1949) *An exploratory study of soil groups in the Belgian Congo.* Pub. I.N.E.A.C., Series Scientifique, No. 46.

KENDREW, W. G. (1953) *The climates of the continents.* 4th edn. Oxford University Press.

KENNAN, T. C. D. (1950) 'Preliminary report on a comparison of several heavy yielding perennial grasses for the production of silage.' *Rhod. agric. J.* **47**, 531.

—— STAPLES, R. R. and WEST, O. (1955) 'Veld management in Southern Rhodesia.' *Rhod. agric. J.* **52**, 4.

KENYA, DEPT. AGRIC. (1953-5) *Ann. Reps.* 1953–5. Nairobi, Govt. Printer.

—— (1955, 1956, 1957, 1959, 1960) *Ann. Reps.* vol. 2. Nairobi, Govt. Printer.

KENYA, DEPT. VET. (1961) 'Production of Sahiwal crosses.' *Rep. Vet. Dept. Kenya, 1960.* Nairobi, Govt. Printer.

KERFOOT, O. (1962) 'Tea root systems.' *World Crops*, **14**, 140–4.

KING, F. H. (1927) *Farmers of forty centuries.* London, Cape.

KOSTERMANS, A. J. G. H. (1958) 'Notes on lowland vegetation in equatorial Borneo', in *Study of tropical vegetation.* UNESCO, p. 154.

KOTHAVALA, Z. R. (1935) *Agriculture and livestock in India*, **5**, 47–51.

KRUG, C. A. (1958) 'Supply of better planting material – 1. Arabicas.' *Coffee and Tea Ind.* **81**, 52–7.

KUBIENA, W. L. (1953) *The soils of Europe.* London, Allen & Unwin.

KURTZ, L. T. (1953) 'Inorganic phosphorus in acid and neutral soils', in *Soil and Fertilizer Phosphorus*, ed. W. H. Pierre and A. G. Norman. New York, Academic Press, pp. 59–88.

LANGDALE-BROWN, J. (1959a) *The vegetation of the Eastern Province of Uganda.* Dept. Agric. Uganda, Mem. Res. Division, Series 2, No. 1.

—— 1959b) *The vegetation of Buganda. Ibid.* Series 2, No. 2.

LAPAGE, G. (1956) *Veterinary parasitology.* Springfield, Chas. C. Thomas.

LAWES, D. A. (1961) 'Rainfall conservation and the yield of cotton in Northern Nigeria.' *Emp. J. exp. Agric.* **29**, 307.

LAYCOCK, D. H. and WOOD, R. A. (1963) 'Some observations on soil moisture use under tea in Nyasaland.' *Trop. Agriculture, Trin.* **40**, 35–48.

LEACH, R. (1936) *Tea seed management.* Dep. Agric. Nyasaland Bull. 14.

LEDGER, H.P. (1950) 'The effect of lime on a Chloris gayana ley.' *E. Afr. agric. for. J.* **16**, 43.

LEESE, A.S. (1927) *A treatise on the one-humped camel in health and in disease.* Stanford, Haynes & Son.

LEITCH, I. and THOMSON, J. (1944) 'The water economy of farm animals.' *Nutr. Absts.* **14**, 197–223.

LIVENS, P.H. (1949) 'Characteristics of some soils of the Belgian Congo.' *Comm. Bur. Soils Tech. Comm.* **46**, 29–35.

LIYANAGE, D.V. (1953) 'An isolated seed garden for coconuts.' *Ceylon Coconut Quart.* **4**, 59.

LOCK, G.W. (1957) *Ann. Rep. Sisal Res. Stn. 1956*, Tanganyika Sisal Growers' Association.

LUSH, J.L. (1945) *Animal breeding plans.* Iowa, Collegiate Press.

—— (1949) 'Heritability of quantitative characters in farm animals.' *Proc. 8th Int. Cong. Genet. (Hereditas, Supply Vol.)* 356–75.

LYDEKKER, R. (1912) *The ox and its kindred.* London, Methuen.

MACGREGOR, R. (1941) 'The domestic buffalo.' *Vet. Rec.* **53**, 443–50.

MACKINTOSH, W.L.S. (1938) *Some notes on the Abahima and the cattle industry of Ankole.* Entebbe, Govt. Printer.

MCWILLIAM, A.P. and DUCKWORTH, J. (1949) 'The preparation of elephant grass silage and its feeding value for tropical dairy cattle.' *Trop. Agriculture, Trin.* **26**, 16–23.

MAHADEVAN, P. (1966) 2nd Ed. 'Population and Production characteristics of Red Sindhi cattle.' *J. Dairy Sci.* **38**, 1231–41.

—— (1958) *Dairy cattle breeding in the tropics.* Comm. Agric. Bur. Tech. Com. 11. Edinburgh, Oliver & Boyd.

MAHER, C. (1945) 'The goat: friend or foe?' *E. Afr. agric. for. J.* **11**, 115–21.

MAINSTONE, B.J. (1963) 'Residual effects of type of ground cover and duration of nitrogenous fertiliser treatments on the growth and yield of Hevea.' *Planters' Bull. Rubb. Res. Inst., Malaya.* **68**, 130–8.

MAMMERICKS, M. (1960) 'The buffalo: a monograph on the genus Bubalus.' *Bull. agric. Congo belge.* **51**, 171–211.

MANNING, H.L. (1950) 'Confidence limits of expected rainfall.' *J. agric. Sci.* **40**, 169.

—— (1956a) 'Calculation of confidence limits of monthly rainfall.' *Ibid.* **47**, 154.

—— (1956b) 'The statistical assessment of rainfall probability and its application to Uganda agriculture.' *Proc. roy. Soc.* **B. 144**, 460.

—— (1958) 'The relationship between soil moisture and yield variance in cotton.' *Rep. Conf. Directors &c. of Overseas Depts. Agric. Col. Office Misc. No. 531*, p. 47. London, H.M. Stationery Office.

MARES, R.G. (1954) 'Animal husbandry, animal industry and animal disease in the Somaliland Protectorate.' *Brit. vet. J.* **110**, 411–23.

MARTIN, W.S. and BIGGS, G. (1937) 'Experiments on the maintenance of soil fertility in Uganda.' *E. Afr. agric. for. J.* **2**, 371–5.

MASEFIELD, G.B. (1952) 'The nodulation of annual legumes in England and Nigeria.' *Emp. J. exp. Agric.* **20**, 175.

MASON, I.L. (1951) *The classification of West African livestock.* Comm. Agric. Bur. Tech. Comm. No. 7.
MASON, I.L. and MAULE, J.P. (1960) *The indigenous livestock of Eastern and Southern Africa.* Comm. Agric. Bur. Tech. Com. No. 14.
MAULE, J.P. (1953–54) Brit. agric. Bull. **6**, 244–52.
—— (1962) 'Objectives in cattle breeding in the tropics.' *Paper IV. 1. Rep. of a Tech. Conf. of Directors of Agric. Sept. 1961. Dept. of Tech. Co-operation Misc. No. 2.* London, H.M. Stationery Office.
MAY, L.H. (1961) 'The utilisation of carbohydrate reserves in pasture plants after defoliation.' *Herbage Abstr.*, **30**, 239–45.
MEIKLEJOHN, J. (1955) 'The effect of bush burning on the microflora of a Kenya upland soil.' *J. Soil Sci.* **6**, 111–8.
MEREDITH, D. (1947) *The grasses and pastures of South Africa.* Johannesburg, Central News Agency.
MERWE, C.R. VAN DER (1941) *Soil groups and sub-groups of South Africa.* Science Bull. 231. Pretoria, Dept. Agric.
MILLER, R. (1952) 'The climate of Nigeria.' *Geography*, **37**, 204.
MILLS, H.D. (1953) '*Bos brachyceros* in Africa.' *Vet. Rec.* **65**, 587–8.
MILNE, A.H. (1955) 'The humps of East African Cattle.' *Emp. J. exp. Agric.* **23**, 234–9.
MILNE, G. (1947) 'A soil reconnaissance journey through parts of Tanganyika Territory, December 1935 to February 1936. *J. Ecol.* **35**, 192–265.
—— BECKLEY, V.A., GETHIN JONES, G.H., MARTIN, W.S., GRIFFITH, G. and RAYMOND, L.W. (1936) *A provisional soil map of East Africa.* Amani Memoirs.
MITSUI, S. (1960) *Inorganic nutrition, fertilisation and soil amelioration for lowland rice.* Tokyo, Yokendo.
MOHR, E.C.J. and VAN BAREN, F.A. (1954) *Tropical soils: a critical study of soil genesis as related to climate, rock, and vegetation.* Amsterdam, Royal Tropical Institute.
MONNIG, H.O. (1949) *Monnig's Veterinary Helminthology and Entomology.* 4th edn. ed by Lapage. Baltimore, Maryland.
MOORE, A.W. (1962) 'The influence of a legume on soil fertility under a grazed tropical pasture.' *Emp. J. exp. Agric.* **30**, 239–48.
MOORMAN, F.R. (1963) 'Acid sulphate soils (cat clays) of the tropics.' *Soil Sci.* **95**, 271–5.
—— and PANABOKKE, C.R. (1961) 'Soils of Ceylon: a new approach to the identification and classification of the most important soil groups of Ceylon.' *Trop. Agriculturist*, **117**, 5–65.
MORGAN, B.B. and HAWKINS, P.A. (1949) *Veterinary helminthology.* Minneapolis, Burgess Publ. Co.
MORRIS, L.E. (1934) 'A note on transplanting budded stumps of Hevea.' *J. Rubb. Res. Inst., Malaya*, **5**, 145–51.
MORRISON, A.B. (1964) 'Calorific intake and nitrogen utilization.' *Fed. Proc.* **23**, 1083–6.
MORRISON, C.G.T. (1949) 'The catena concept and the classification of tropical soils.' *Comm. Bur. Soil Sci. Tech. Comm.* **46**, 124–8.

MORRISON, F.B. (1956) *Feeds and Feeding.* 22nd edn. Ithaca, New York, Morrison Pub. Co. Inc.

MOSER, F. (1942) Quoted in *Soils and Fert.* **12**, 155.

MUIR, A., ANDERSON, B., and STEPHEN, I. (1957) 'Characteristics of some Tanganyika Soils.' *J. Soil Sci.* **8**, 1–18.

MUNRO, J.M. (1959) *Prog. repts. from Exp. Stns. Nyasaland 1958–9.* Emp. Cot. Growing Corp.

MURRAY, D.B. (1953–5) 'A shade and fertiliser experiment with cacao: II–IV.' *Rep. on Cacao Res.* 1952, 11–21; 1953, 30–34; 1954, 32–36; I.C.T.A. T'dad.

—— and COPE, F.W. (1953, 1955, 1959) 'A stock scion experiment with cacao. – I–III. *Rep. on Cacao Res. 1952, 34–40; 1954,* 37–42; *1957–8,* 29–35. I.C.T.A. T'dad. 1953.

MURRAY, W. (1954) 'Mechanised replanting.' *Plant. Bull. R.R.I. (N.S.)* **13**, 68–73.

NAGLESCHMIDT, G., DESAI, A.D. and MUIR, A. (1940) 'The minerals in the clay fraction of a Black Cotton Soil and a Red Earth from Hyderabad (Deccan) State, India.' *J. agric. Sci.* **30**, 639.

N.R.C. (1960) *Nutrient requirements of Poultry.* Pub. 827 of the National Academy of Science, Washington.

NAY, T. (1959) 'Sweat glands in cattle: histology, morphology and evolutionary trends.' *Aust. J. agric. Res.* **10**, 121–8.

NESTEL, B.L. (1964) 'Animal production studies in Jamaica. IV. The costs of developing and maintaining Pangola grass pastures.' *J. agric. Sci.* **62**, 178–86.

—— and CREEK, M.J. (1964a) 'Animal production studies in Jamaica. I. Introduction.' *J. agric. Sci.* **62**, 151–5.

—— —— (1964b) 'Animal production studies in Jamaica. V. Liveweight production from Pangola pastures used in rearing and fattening beef cattle.' *J. agric. Sci.* **62**, 187–97.

NORRIS, D.O. (1956) 'Legumes and the Rhizobium symbiosis.' *Emp. J. exp. Agric.* **24**, 247.

NORTHERN NIGERIA (1955–6) *Dep. Agric. Ann. Rep.* 1955–6.

NUTMAN, F.J. (1933) 'The root system of *Coffea arabica.*' *Emp. J. exp. Agric.* **1**, 271–84, 285–96; **2**, 293–302.

NYASALAND (1957–8) Dep. Agric. Ann. Rep. 1957–8, vol. 2.

NYE, P.H. (1954) 'Some soil-forming processes in the humid tropics. I. A field study of a catena in the West African forest.' *J. Soil Sci.* **5**, 7–21.

—— (1955) 'Some soil-forming processes in the humid tropics. II. The development of the upper-slope member of the catena. III. Laboratory studies on the development of a typical catena over granitic gneiss. IV. The action of the soil fauna.' *J. Soil Sci.* **6**, 51–62, 63–72, 73–83.

—— (1958) 'The relative importance of fallows and soils in storing plant nutrients in Ghana.' *J. W. Afr. Sci. Ass.* **4**, 31–41.

—— and BERTHEUX, M.N. (1957) 'The distribution of phosphorus in forest and savanna soils of the Gold Coast.' *J. agric. Sci.* **49**, 145–59.

NYE, P. H. and FORSTER, W. N. M. (1958) 'A study in the mechanism of soil phosphate uptake in relation to plant species.' *Plant and Soil*, **9**, 338–52.
—— —— (1960) 'The relative uptake of phosphorus by crops and natural fallows from different parts of their root zone.' *J. agric. Sci.* **56**, 299–306.
NYE, P. H. and GREENLAND, D. J. (1960) *The soil under shifting cultivation.* Comm. Bur. Soils, Tech. Comm. No. 51.
—— and STEPHENS, D. (1962) 'Soil fertility', in *Agriculture and land use in Ghana.* Oxford University Press.
OAKES, H. and THORP, J. (1950) 'Dark clay soils of warm regions variously called Rendzina, Black Cotton Soils, Regur and Tirs.' *Soil Sci. Soc. Amer. Proc.* **15**, 347–54.
OATES, A. J., BARNES, R. M. and SKOB, O. (1959) 'Pangola grass in the U.S. Virgin Islands.' *Trop. Agriculture, Trin.* **36**, 130–7.
OATES, A. V. (1956) 'Effect on vegetation recovery of ripping denuded veld.' *Rhod. agric. J.* **53**, 551.
OLVER, A. (1938) *A brief survey of some of the important breeds of cattle in India.* Misc. Bull. Imp. Coun. Agric. Res. India 16, Delhi, I.C.A.R.
ORCHARD, P. and GREENSTEIN, A. L. (1949) 'Green manure as a source of phosphate for crops.' *Union S. Afr., Dept. Agric. Sci. Bull. No. 290.*
OSBORNE, H. G. (1961) 'The use of pregnancy diagnoses in the management of extensive beef herds.' *Vet. Rec.* **73**, 1121–4.
OVERSEAS FOOD CORPORATION (1956) *Overseas Food Corp. Rep. and Accounts.* London, H.M. Stationery Office.
OVINGTON, J. D. (1963) *The better use of the world's fauna for food.* Symp. Inst. of Biology No. 11., London.
OYENUGA, V. A. (1959) 'Effect of frequency of cutting on the yields and composition of some fodder grasses in Nigeria.' *J. agric. Sci.* **53**, 25–33.
PARSONS, D. J. (1958) 'Problems of increasing livestock production in the intensive farming areas of Uganda.' *E. Afr. agric. for. J.* **23**, 167–71.
PATERSON, D. D. (1936) 'The cropping qualities of certain tropical fodder grasses.' *Emp. J. exp. Agric.* **4**, 6.
—— (1938) 'Further experiments with cultivated tropical fodder crops.' *Emp. J. exp. Agric.* **6**, 323–40.
—— (1939) 'The cultivation of perennial fodder grasses in Trinidad.' *Trop. Agriculture, Trin.* **16**, 55–7.
PATON, T. R. (1961) 'Soil genesis and classification in Central Africa.' *Soils and Ferts.* **24**, 249–57.
PAYNE, W. J. A. (1964) 'The origin of domestic cattle in Africa.' *Emp. J. exp. Agric.* **32**, 97–173.
—— and HANCOCK, J. (1957) 'The direct effect of tropical climate on the performance of European-type cattle. II. Production.' *Emp. J. exp. Agric.* **25**, 321.
—— LAING, W. I. and RAIVOKA, E. N. (1951) 'Grazing behaviour of dairy cattle in the tropics.' *Nature, Lond.* **167**, 610–11.
—— (1952) 'Breeding Studies.' *Agric. J. Fiji*, **23**, 9–13.
PEARSALL, W. H. (1950) 'The investigation of wet soils and its agricultural implications.' *Emp. J. exp. Agric.* **18**, 289–98.

PEAT, J.E. and BROWN, K.J. (1962) 'The yield response of rain grown cotton at Ukiriguru in the Lake Province of Tanganyika. I: The use of organic manure, inorganic fertilizers and cotton seed ash.' *Emp. J. exp. Agric.* **30**, 215–31. 'II: Land resting and other rotational treatments contrasted with the use of organic manures and inorganic fertilizers.' *Ibid.* **30**, 305–14.

PENDLETON, R.L. (1943) 'Land use in north-eastern Thailand.' *Geogr. Rev.* **33**, 14–41.

PENMAN, H.L. (1948) 'Natural evaporation from open water, bare soil and grass.' *Proc. roy. Soc.* A. **193**, 120.

—— (1949) 'The dependence of transpiration on the weather and soil conditions.' *J. Soil Sci.* **1**, 74.

—— (1950) 'Evaporation over the British Isles.' *Quart. J. R. met. Soc.* **75**, 293.

—— (1956) 'Evaporation: an introductory survey.' *Neth. J. agr. Sci.* **4**, 9.

PEREIRA, H.C. (1953) *Ann. Rep. E. Afr. Agric. & For. Res. Organisation.* Nairobi, 1953.

—— (1954a) 'Soil water storage under catchment area vegetation.' *Proc. 2nd Int.-Afr. Soils Conf.* 158–61.

—— (1954b) *Proc. 2nd Int.-Afr. Soils Conf.* 427–8.

—— (1955a) 'The assessment of structure in tropical soils.' *J. agric. Sci.* **45**, 401–10.

—— (1955b) *Ann. Rep. E. Afr. Agric. & For. Res. Organization*, Nairobi, 1955.

—— (1957) 'Field measurements of water use for irrigation control in Kenya coffee.' *J. agric. Sci.* **49**, 459–66.

—— and BECKLEY, V.A. (1953) 'Grass establishment on eroded soil in a semi-arid African area.' *Emp. J. exp. Agric.* **21**, 1–19.

—— CHENERY, E. and MILLS, W.R. (1954) 'The transient effects of grasses on the structure of tropical soils.' *Emp. J. exp. Agric.* **22**, 148–60.

—— DAGG, M. and HOSEGOOD, P.H. (1964) 'A tillage study in Kenya coffee. IV. The physical effects of contrasting tillage treatments over thirty consecutive cultivation seasons.' *Emp. J. exp. Agric.* **32**, 31–4.

—— HOSEGOOD, P.H. and THOMAS, D.B. (1961) 'The productivity of semi-arid thorn scrub country under intensive management.' *Emp. J. exp. Agric.* **29**, 269–86.

—— and JONES, P.A. (1954a) 'Field responses by Kenya coffee to fertilizers, manures and mulches.' *Emp. J. exp. Agric.* **22**, 23.

—— and JONES, P.A. (1954b) 'Tillage studies in Kenya coffee.' *Emp. J. exp. Agric.* **22**, 231–40, 323–31.

—— WOOD, R.A., BRZOSTOWSKI, H.W. and HOSEGOOD, P.H. (1958) 'Water conservation by fallowing in semi-arid East Africa.' *Emp. J. exp. Agric.* **26**, 213–28.

—— et al. (1962) 'Hydrological effects of change in land use in some East African catchment areas.' *E. Afr. agric. for. J.* **27**, 1–131.

PHILLIPS, G.D. (1961) 'Dry matter and water intakes of tropical cattle.' *Res. Vet. Sci.* **2**, 202–9.

References

PHILLIPS, J. (1930) 'Some important vegetation communities in the Central Province of Tanganyika.' *J. Ecol.* **18**, 193.
—— (1959) *Agriculture and ecology in Africa.* London, Faber & Faber.
PHILLIPS, T. A. (1956) *An agricultural note book.* London, Longmans.
PICHI-SERMOLLI, R. E. G. (1955) 'Tropical East Africa (Ethiopia, Somaliland, Kenya, Tanganyika).' *UNESCO Arid Zone Research – VI: Plant Ecology*, 302–60.
PILLAI, M. S. (1958) *Cultural trials and practices of rice in India.* Indian Council for agric. Res. Monograph No. 27
POPENOE, H. (1959) 'The influence of the shifting cultivation cycle on soil properties in Central America.' *Proc. 9th Pacific Sci. Cong.* **7**, 148–60.
PRESCOTT, J. A. and PENDLETON, R. L. (1952) *Laterite and lateritic soils.* Comm. Bur. Soils Tech. Comm. 47.
PRIANISHNIKOFF, A. (1940) Quoted in *Soils and Ferts.* **12**, 155.
QUARTERMAN, J. (1961) 'The digestibility of crude fibre in the tropics.' *Emp. J. exp. Agric.* **22**, 102–9.
RAALTE, M. H. VAN (1940) 'On the oxygen supply of rice roots.' *Ann. Jard. Bot. Buitenzorg*, **50**, 99–113.
—— (1944) 'On the oxidation of the environment by the roots of rice.' *Ann. Jard. Bot. Buitenzorg.* Hors. Serie, 15–84.
RADWANSKI, S. A. (1959) 'Soil survey and the problems of soil classification in the Kingdom of Buganda.' *Proc. 3rd Int.-Afr. Soils Conf.* vol. I, 283–9.
—— and OLLIER, C. D. (1959) 'A study of an East African catena.' *J. Soil Sci.* **10**, 149–68.
RANDELL, J. R. (1964) 'The Manaquil Extension to the Sudan Gezira Scheme.' *20th Int. Geographical Cong. Papers.*
RATTRAY, A. (1956) *Maize investigations at the Agricultural Experiment Station, Salisbury.* Proc. 2nd Ann. Conf. Professional Officers, Min. Agric. Rhodesia, Nyasaland.
—— and ELLIS, B. S. (1953) 'Maize and green manuring in S. Rhodesia.' *Rhod. agric. J.* **49**, 188–99.
RATTRAY, J. M. (1957) 'The grasses and grass associations of Southern Rhodesia.' *Rhod. agric. J.* **54**, 197.
REE, W. O. (1954) *Handbook of channel design for soil and water conservation.* U.S. Dept. Agric., Soil Conservation Service.
REES, A. R. (1959a) 'The germination of oil palm seed: the cooling effect.' *J.W. Afr. Inst. for Oil Palm Res.* **3**, 76–82.
—— (1959b) 'Germination of oil palm seed: large scale germination.' *J.W. Afr. Inst. for Oil Palm Res.* **3**, 83–94.
RENDEL, J. (1960) 'A study on relationships between blood groups and production characters in cattle.' *Rep. VI Int. Bloodgr. Cong. Munich (1959)* 8–23.
RENSBURG, H. J. VAN (1956) 'Comparative values of fodder plants in Tanganyika.' *E. Afr. agric. for. J.* **22**, 14–19.
RENSBURG, J. A. VAN and TIDMARSH, C. E. M. (1959). 'Land use in the Karroo region.' *Proc. 3rd Int.-Afr. Soils Conf.* vol. 2, 909–14.

RENSBURG, P.J.J. VAN (1938) 'Boer goats.' *Fmg. S. Afr.* **13**, 133–4.
RHOAD, A.O. (1940a) 'A method of assaying genetic differences in the adaptability of cattle to tropical and subtropical climates.' *Emp. J. exp. Agric.* **8**, 190–8.
—— (1940b) 'Absorption and reflection of solar radiation in relation to coat colour in cattle.' *Proc. Amer. Soc. Anim. Prod. 33rd Ann. Meet:* 291–301.
—— (1944) 'The Iberia Heat Tolerance Test for cattle.' *Trop. Agriculture, Trin.* **21**, 162–4.
RICE, V.A., ANDREWS, F.W., WARRICK, E.J. and LEGATES, J.E. (1957) *Breeding and improvement of farm animals.* New York, McGraw-Hill.
RICHARDS, P.W. (1952) *The tropical rain forest.* Cambridge University Press.
RICHARDSON, H.L. (1946) 'Soil fertility maintenance under different systems of agriculture.' *Emp. J. exp. Agric.* **14**, 1–17.
RIEHL, H. (1954) *Tropical meteorology.* London, McGraw-Hill.
RIPPERTON, J.C. and TAKAHASHI, M. (1948) 'Nitrogen fertilization of pasture.' *Univ. Hawaii Agric. Exp. Sta. Rep. 1946–8*, 31.
RIQUIER, J. (1953) 'Etude d'un sol de tary et d'un sol de forêt à Perinet.' *Mem. Inst. Sci. Madagascar*, Series D5, 75–92.
ROBERTS, R.C. (1942) *Soil Survey of Puerto Rico.* U.S.D.A.
ROBERTSON, A. (1950) 'A preliminary report on the herd of Fulani cattle at Shika, Nigeria.' *Conf. on the improvement of livestock under trop. conditions. Dec. 1950. Edinburgh (MS).*
—— (1954) 'Artificial insemination and livestock improvement.' *Advances in Genet.* **6**, 451–72.
—— STEWART, A. and ASHTON, E.D. (1956) 'The progeny assessment of dairy sires for milk: the use of contemporary comparisons.' *Proc. Brit. Soc. Anim. Prod. 1956*, 43–50.
ROBINSON, G.W. (1951) *Soils, their origin, constitution, and classification.* 3rd edn. London, Allen & Unwin.
ROBINSON, J.B.D. (1955) 'A cultivation system for ground water (vlei) soils.' *E. Afr. agric. for. J.* **21**, 69.
—— (1959a) 'Camber bed cultivation of ground water (vlei) soils. I. Experimental crop yields.' *E. Afr. agric. for. J.* **24**, 184–91.
—— (1959b) 'Camber bed cultivation of ground water (vlei) soils. II. Modifications of the system.' *E. Afr. agric. for. J.* **24**, 192–5.
ROBINSON, K.W. and LEE, D.H.K. (1947) 'The effect of the nutritional plane upon the reactions of animals to heat.' *J. anim. Sci.* **6**, 182–94.
ROBSON, J. and WILDE, J.K.H. (1954) 'Prophylaxis against trypanosomiasis in Zebu cattle using Antrycide and Dimidium bromide.' *Brit. Vet. J.* 459–69.
ROCHE, B. and JOLIET, B. (1953) 'Observations sur l'essais antierosifs.' *Rech. agron. Madagascar*, **2**, 11–6.
ROGERSON, A. (1960) 'Effect of environmental temperature on the energy metabolism of cattle.' *J. agric. Sci.* **55**, 359–65.
ROLLINSON, D.H.L. (1955) 'Oestrus in Zebu cattle in Uganda.' *Nature, Lond.* **176**, 352.

References

ROLLINSON, D. H. L. (1956) 'The use of electro-ejaculation in the development of artificial insemination in African cattle.' *Proc. III Int. Cong. on Anim. Reprod. June*, 1956, Cambridge.

ROMNEY, D. H. (1961) 'Productivity of pasture in British Honduras.' *Trop. Agriculture, Trin.* **38**, 39–47, 161–71.

ROSAYRO, C. (1950) 'Ecological conceptions and vegetation types with reference to Ceylon.' *Trop. Agriculturist*, **56**, 108–21.

ROSE, C. J. (1952) 'The economics of fertilizing natural veld.' *Emp. J. exp. Agric.* **20**, 35.

ROSEVEARE, J. M. (1948) *The grasslands of Latin America*. Comm. Bur. Pastures, Bull. No. 36.

ROSS, J. G. (1958) 'A classification of Zebu cattle types in Teso district, E. P. Uganda.' *Emp. J. exp. Agric.* **26**, 298–308.

ROSSETTI, G. and CONGIU, S. (1955) *Richerche zootechnia-veterinarie su gli animali domestice della Somalia*. Inspetorato Veterinario, Amministrazione Fiduciania Italiana Della Somalia, Mogodishu.

ROWLAND, W. S. (1955) Personal communication.

RUBBER RESEARCH INSTITUTE OF MALAYA (1962) *Annual Report*, 1961.

RUSSELL, E. W. (1956) 'Effects of very deep ploughing and of subsoiling on crop yields.' *J. agric. Sci.* **48**, 129–44.

—— (1957) 'The first African Herbicide Conference: A general introduction and review of the proceedings.' *E. Afr. agric. for. J.* **23**, 1–3.

—— (1958) 'The fertility of tropical soils.' *Report of a conference of Directors and Senior Officers of Overseas Departments of Agriculture*. Colonial Office, London, Misc. No. 531. 33–40. H.M.Stationery Office.

—— (1961) *Soil conditions and plant growth*. 9th edn. London, Longmans

—— (1962) 'Hydrological effects of change in land use in some East African catchment areas' *E. Afr. agric. for. J.* **27**, 1.

—— and KEEN, B. A. (1938) 'Studies in soil cultivation: VII. The effect of cultivation on crop yield.' *J. agric. Sci.* **28**, 212–33.

—— —— (1941) 'Studies in soil cultivation: IX. The results of a six year cultivation experiment.' *Ibid.* **31**, 326–47.

—— and MEHTA, N. P. (1938) 'Studies in soil cultivation: VIII. The influence of the seed bed on crop growth.' *J. agric. Sci.* **28**, 272–98.

SAHASRABUDDE, D. C. (1936) *3rd Int. Cong. Soil Sci.* **3**, 111–3.

ST CROIZ, F. W. DE (1945) *The Fulani of Northern Nigeria*. Lagos, Govt. Printer.

SANDERS, F. R. (1951) Ann. Rep. Coffee Res. Stn. Lyamungu, Tanganyika.

SCHLIPPE, P. DE (1956) *Shifting cultivation in Africa: The Zande system of agriculture*. London, Routledge.

SCOTT, J. D. (1947) *Veld management in S. Africa*. Dept. Agric. Union S. Afr. Bull. No. 278.

SCOTT, V. A. (1956) 'Relative infiltration rates of burned and unburned soils.' *Trans. Amer. Geophys. Un.* **37**, 67–9.

SEARS, P. D. (1953) 'Pasture growth and soil fertility.' *N.Z. J. Sci. Technol.* **35** Seca. Suppl. 1, 1–29, 42–77, 190–220.

SEATH, D.M. and MILLER, G.D. (1947) 'The effect of hay feeding in summer on milk production and grazing performance of dairy cows.' *J. Dairy Sci.* **30**, 921–32.
SEMPLE, A.T. (1952) *Improving the world's grasslands*. London, Leonard Hill.
SETHI, D.R. (1951) 'Land reclamation.' *U.N. Sci. Conf. on Cons. and Utilization of Resources*, **6**, 566.
SETHI, R.L. (1930) 'Root development in rice under different conditions of growth.' *Ind. Dept. agric. Mem. Bot. Series.* **18**, 57–80.
SHANTZ, H.L. and MARBUT, C.F. (1923) *The vegetation and soils of Africa*. Amer. Geog. Soc.
SHAW, T. and COLVILLE, G. (1950) *Report of Nigeria Livestock Mission*. London, H.M. Stationery Office.
SIKKA, L.C. (1933) 'Statistical studies of records of Indian dairy cattle. II. Reliability of different lactation yields as measures of a cow's milking capacity.' *Ind. J. vet. Sci.* **3**, 240–53.
SIMMONDS, N.W. (1959) *Bananas*. London, Longmans.
SIMONSON, R.W. (1954a) 'Morphology and classification of the regur soils of India.' *J. Soil Sci.* **5**, 275–88.
—— (1954b) 'The regur soils of India and their utilization.' *Soil Sci. Soc. Amer. Proc.* **18**, 199–203.
SINGH, S. (1954) 'A study of the Black Cotton soils with special reference to their colouration.' *J. Soil Sci.* **5**, 280–99.
SIVARAJASINGHAM, S., ALEXANDER, L.T., CADY, J.G. and CLINE, M.G. (1962) 'Laterite.' *Adv. in Agronomy*, **14**, 1–60.
SMITH, C., KING, J.W.B. and GILBERT, N. (1962) 'Genetic parameters of British Large White bacon pigs.' *Anim. Prod.* **4**, 128–43.
SMITH, C.A. (1961) 'The utilization of Hyparrhenia veld for the nutrition of cattle in the dry season. II. Veld hay compared with in situ grazing of mature forage and the effects of feeding supplementary nitrogen.' *J. agric. Sci.* **57**, 34.
SMITH, G.W. (1954) 'Some physical aspects of the cacao shade experiment.' *Rep. on Cacao Res. 1953*, 38–44, I.C.T.A., T'dad.
SMITH, R.M., SAMUELS, G. and CERNUDA, C.F. (1951) 'Organic matter and nitrogen build-ups in some Puerto Rican soil profiles.' *Soil Sci.* **72**, 409–27.
SPARNAAIJ, L.D. and GUNN, J.S. (1959) 'The development of transplanting techniques for the oil palm in West Africa.' *J.W. Afr. Inst. for Oil Palm Res.* **2**, 281–309.
SQUIRE, H.A. (1964) 'The costs and returns of five selected dairy farms in the West Indies.' *Surinamse Landbouw*.
STALLINGS, J.H. (1957) *Soil conservation*. New Jersey, Prentice Hall.
STAMP, L.D. (1925) *The vegetation of Burma from an ecological standpoint* Calcutta.
STAPLES, R.R. (1945) 'Bush control and deferred grazing.' *E. Afr. agric. for. J.* **9**, 217; **10**, 43.
—— HORNBY, H.E. and HORNBY, R.M. (1942) 'Comparative effects of goats and cattle on a mixed grass-bush pasture.' *E. Afr. agric. for. J.* **8**, 62–80.

STEENIS, C. G. G. J. VAN (1958) 'Tropical lowland vegetation.' *Proc. 9th Pacific Sci. Cong.* vol. 20, 25.
—— (1958a) 'Rhizophoraceae,' in *Flora Malaysiana*, Series 1, vol. 5, p. 431.
—— (1958b) *Vegetation map of Malaysia*. UNESCO.
STÈHLÉ, H. (1935) *Essai d'ecologie et de geographie botanique: Flora de la Guadaloupe*, Guadaloupe.
—— (1939) Esquiets des associations végétales de la Martinique, *Bull. Agric. Martinique.*
—— (1941) 'Conditions eco-sociologiques et evolution des forêts des Antilles Francaises.' *Caribbean For.* 2, 154–259.
STEPHENS, C. G. (1953) *A manual of Australian soils*. C.S.I.R.O., Melbourne.
—— (1961) 'Laterite at the type locality, Angadipuram, Kerala, India.' *J. Soil Sci.* 12, 214–7.
—— and DONALD, C. M. (1958) 'Australian soils and their responses to fertilizers.' *Adv. in Agronomy*, 10, 168–256.
STEVENS, N. S. (1942) 'Forest associations of British Honduras.' *Caribbean For.* 3, 164–71.
STEWART, O. C. (1956) 'Fire as the first great force employed by man', in Thomas, W. L. *Man's role in changing the face of the earth.*
STONAKER, H. H. (1953) 'Estimates of genetic changes in an Indian herd of Red Sindhi dairy cattle.' *J. Dairy Sci.* 36, 688–97.
STRAHLER, A. N. (1960) *Physical Geography*, New York, Wiley.
STRANGE, R. (1955) 'Forage legumes for medium altitudes in Kenya.' *E. Afr. agric. for. J.* 20, 221.
—— (1961) 'Effects of legumes and fertilizers on yields of temporary leys.' *E. Afr. agric. for. J.* 26, 231.
STURGIS, M. B. (1936) *Changes in the oxidation – reduction equilibrium in soils as related to the physical properties of the soils and the growth of rice*. Louisiana Agric. Exp. Stn. Bull. 271.
SUGAR MANUFACTURERS' ASSOCIATION OF JAMAICA LTD. and ALCON JAMAICA LTD. (1963) *The production costs and returns on eleven beef farms.* Kingston, Sugar Manufacturers' Association.
SULAIMAN, M. (1944) 'Effect of algal growth on the activity of Azotobacter in rice soils.' *Ind. J. agric. Sci.* 14, 277–83.
SYLVAIN, P. (1952) 'Effect of shade upon growth and differentiation of coffee seedlings.' *Rep. Inter-Amer. Inst. agric. Sci. Turrialba*, 1952.
SYS, C. (1960) 'Principles of soil classification in the Belgian Congo.' *Trans. VIIth Int. Cong. Soil Sci.* IV, 112–8.
TANADA, T. (1946) 'Utilisation of nitrates by the coffee plant under different sunlight intensities.' *J. agric. Res.* 72, 245.
TAYLOR, D. L. (1942) 'Influence of oxygen tension on respiration, fermentation and growth in wheat and rice.' *Amer. J. Bot.* 29, 721–38.
TEMPANY, H. A. (1949) *The practice of soil conservation in the British Colonial Empire*. Comm. Bur. Soil Sci. Tech. Comm. No. 45.
—— and GRIST, D. H. (1958) *An introduction to tropical agriculture*. London, Longmans.

TEMPLIN, E.H., MOWERY, I.C. and KUNZE, G.W. (1956) 'Houston Black Clay, the type grumusol. I. Field morphology and geography.' *Soil Sci. Soc. Amer. Proc.* **20**, 88–90.

TERRA, C.J.A. (1958) 'Farm systems in south-east Asia.' *Netherlands J. agric. Sci.* **6**, 157.

THOMAS, A.S. (1944) 'Observations on the root systems of Robusta coffee and other tropical crops in Uganda.' *Emp. J. exp. Agric.* **12**, 191–205.

THORNTHWAITE, C.W. (1948) 'An approach towards a rational classification of climate.' *Geogr. Rev.* **38**, 55.

THORP, J. (1949) *A review of tropical and sub-tropical soils.* Comm. Bur. Soils Tech. Comm. **46**, 61.

—— BELLIS, E., WOODRUFF, G.A. and MILLER, F.T. (1960) *Soil survey of the Songhor area.* Nairobi, Dept. of Agric., Kenya.

—— BELLIS, E., WOODRUFF, G.A., and MILLER, F.T. (1961) *Soil Survey of the East Konyango area.* Nairobi, Dept. of Agric., Kenya.

—— and SMITH, G.D. (1949) 'Higher categories of soil classification: Order, Sub-Order, and great Soil Groups.' *Soil Sci.* **67**, 117–26.

THORPE, B. (1953) 'The hump of a Zebu.' *Aust. vet. J.* **29**, 79–81.

TINLEY, G.H. (1961) 'Vegetative propagation of clones of *Hevea brasiliensis* by cuttings.' *Proc. Nat. Rub. Res. Conf., Kuala Lumpur, 1960*, 409–16.

—— (1962) 'Propagation of *Hevea* by budding young seedlings.' *Planters' Bull. Rub. Res. Inst. Malaya*, **62**, 136–47.

—— and GARNER, R.J. (1960) 'Developments in the propagation of clones of *H. brasiliensis* by cuttings.' *Nature, Lond.* **186**, 407–8.

TOMAR, N.S. and MITTAL, K.K. (1960) 'Significance of the calving season in Hariana cows.' *Ind. vet. J.* **37**, 367–70.

TOTHILL, J.D. (ed.) (1940) *Agriculture in Uganda.* Oxford University Press.

—— (1948) *Agriculture in the Sudan.* Oxford University Press.

TRAPNELL, C.G. (1943) *The soils, vegetation and agriculture of North Eastern Rhodesia.* Lusaka, Govt. Printer.

—— and CLOTHIER, J.N. (1937) *The soils, vegetation and agriculture of North Western Rhodesia.* Lusaka, Govt. Printer.

—— and GRIFFITHS, J.F. (1960) 'The rainfall-altitude relation and its ecological significance in Kenya.' *E. Afr. agric. for. J.* **25**, 207–13.

TUBBS, F.R. (1947) 'Tea Selection. III. The vegetative propagation of selected bushes.' *Tea Quart.* **18**, 91.

TURNER, H.G. and SCHLEGER, A.V. (1960) 'The significance of coat type in cattle.' *Austr. J. agric. Res.* **11**, 645–63.

U.S. DEPT. AGRIC. (1951) *Soil survey manual.* Washington.

—— (1960) *Soil classification: a comprehensive system: 7th approximation.* Washington.

—— SOIL CONS. SERVICE (1954) *A manual on conservation of soil and water.* Agriculture Handbook, No. 61.

—— —— (1958) *Engineering handbook for soil conservation in the cornbelt.* Agriculture Handbook, No. 135.

UPPAL, B.N., PATEL, M.K. and DAJI, J.A. (1939) 'Nitrogen fixation in rice soils.' *Ind. J. agric. Sci.* **9**, 689–702.

VERNON, K. C. (1958) *Soil and land use surveys, No. 1. Jamaica, The Parish of St. Catherine.* I.C.T.A., T'dad.
VINE, H. (1953) 'Experiments on the maintenance of soil fertility at Ibadan, Nigeria.' *Emp. J. exp. agric.* **21**, 65–85.
—— (1954) 'Latosols of Nigeria and some related soils.' *Proc. 2nd Int.-Afr. Soils Conf.* vol. I, 295–308.
—— (1956). 'Studies of soil profiles at the W.A.I.F.O.R. Main Station and at some other sites of oil palm experiments.' *J.W. Afr. Inst. Oil Palm Res.* **1**, 8–59.
VLAMIS, J. and DAVIS, A. R. (1943) 'Germination, growth and respiration of rice and barley seedlings at low oxygen pressures.' *Plant Physiol.* **18**, 685–92.
—— —— (1944) 'Effect of oxygen tension on certain physiological responses of rice, barley and tomato.' *Ibid.* **19**, 33–51.
—— SCHULTZ, A. M. and BISWELL, H. H. (1955) 'Burning and soil fertility.' *Calif. Agric.* **9**, 3–7.
WAGNON, K. A. and ROLLINS, W. C. (1959) 'Heritability estimates of post-weaning growth of range-fed beef cows.' *J. anim. Sci.* **18**, 918–24.
WAITES, G. M. H. (1961) 'Polypnoea evoked by heating the scrotum of the ram.' *Nature, Lond.* **190**, 172.
WALDOCK, E. A., CAPSTICK, E. S. and BROWNING, A. J. (1951) *Soil conservation and land use in Sierra Leone.* Freetown, Govt. Printer.
WALKER, C. A. (1960) 'The population, morphology and evolutionary trends of apocrine sweat glands of African indigenous cattle.' *J. agric. Sci.* **55**, 119–26.
WALKER, T. W., ORCHISTON, H. D. and ADAMS, A. F. (1954) 'The nitrogen economy of grass-legume associations.' *J. Brit. Grassland Soc.* **9**, 249.
WALTON, P. D. (1962a) 'Cotton agronomy trials in the Northern and Eastern Provinces of Uganda, 1956–61.' *Emp. Cott. Gr. Rev.* **39**, 114–24.
—— (1962b) 'The effect of ridging on the cotton crop in the Eastern Province of Uganda.' *Emp. J. exp. Agric.* **30**, 63–76.
—— (1962c) 'Estimates of the water use by cotton crops at Serere, Uganda.' *Emp. Cott. Gr. Rev.* **39**, 241–51.
WAMBEKE, A. R. VAN (1962) 'Criteria for classifying tropical soils by age.' *J. Soil Sci.* **16**, 124–32.
WARMKE, H. E. *et al.* (1952) 'Evaluation of some tropical grass-legume associations.' *Trop. Agriculture, Trin.* **29**, 115–24.
WATANABE, A., NISHIGAKI, S. and KONISHI, C. (1951) 'Effects of nitrogen-fixing blue-green algae on the growth of rice plants.' *Nature, Lond.* **168**, 748–9.
WATSON, G. A. (1955) 'Cover plants in rubber cultivation.' *J. Rub. Res. Inst. Malaya,* **15**, 2–18.
—— (1957) 'Nitrogen fixation by *Centrosema pubescens*.' *J. Rub. Res. Inst. Malaya,* **15**, 168–74.
—— WONG, P. W. and NARAYAN, R. (1963), 'Effect of cover plants on growth of *Hevea*. IV. Leguminous cover crops compared with grasses, *Mikania scandens* and mixed indigenous covers.' *J. Rub. Res. Inst. Malaya,*

WATSON, S. J. (1951) *Grassland and grassland products.* London, Edward Arnold.
WATTS, I. E. M. (1955) *Equatorial weather.* London, University of London Press.
WAYLE, S. R., MEHTA, M. and JOHNSON, B. C. (1958) 'Vitamin B_{12} and protein biosynthesis.' *J. Biol. Chem.* **233**, 619.
WEBSTER, C. C. (1938) 'Experiments on the maintenance of soil fertility by green manuring.' *Proc. 3rd W. Afr. Agric. Conf.* 299–321.
—— (1942) 'Transplanting budded plants of *Aleurites Montana*.' *E. Afr. agric. for. J.* **8**, 39–41.
—— (1948) 'The effect of seed treatments, nursery technique and storage methods on the germination of tung seed.' *E. Afr. agric. for. J.* **14**, 38.
—— (1950) 'The improvement of yield in the tung oil tree (*A. Montana*).' *Trop. Agriculture, Trin.* **21**, 179–220.
WEBSTER, R. (1960) 'Soil genesis and classification in Central Africa.' *Soils and Ferts.* **23**, 77–9.
WEINMANN, H. (1946) 'Some fundamental principles of modern pasture management.' *Rhod. agric. J.* **43**, 418.
—— (1948a) 'Seasonal growth and changes in composition of herbage.' *Rhod. agric. J.* **45**, 119.
—— (1948b) 'Underground development and reserves of grasses.' *J. Brit. Grassland Soc.* **3**, 115–40.
WEST, O. (1947) 'Thorn Bush encroachment in relation to the management of veld grazing.' *Rhod. agric. J.* **44**, 488.
—— (1948) 'Programme and progress report, Central Pasture Station, Matabeleland.' *Rhod. agric. J.* **45**, 366.
—— (1950) 'Indigenous tree crops for Southern Rhodesia.' *Rhod. agric. J.* **47**, 214–7.
—— (1956) 'Pasture improvement in the higher rainfall regions of S. Rhodesia.' *Rhod. agric. J.* **53**, 439.
—— (1958) 'Bush encroachment, veld burning and grazing management.' *Rhod. agric. J.* **55**, 407.
WHITLOCK, J. D. (1960) *Diagnosis of veterinary parasites.* London, Kempton.
WHYTE, R. O. and TRUMBLE, H. C. (1953) *Legumes in agriculture.* Rome, F.A.O. Agricultural Studies No 21.
—— et al. (1959) *Grasses in agriculture.* Rome, F.A.O. Agriculture Studies No. 42.
WIGHT, W. (1942) *Selection and propagation of the tea plant.* Ind. Tea Assn. Memo. No. 15.
—— (1959) 'The shade tea tradition in tea gardens of North India.' *Rep Ind. Tea Assn. Sci. Dept. (Tocklai) for 1958*, 75–122.
WILLIAMS, E. and BUNGE, V. A. (1952) 'Notes on the development of the Serere herd of Nkedi shorthorned Zebu cattle.' *Emp. J. exp. Agric.* **20**, 142–160.
WILLIAMS, F. R. (1953) 'Pasture research sub-station Melsetter: Report on experimental work.' *Rhod. agric. J.* **50**, 391.

WILLIAMSON, G. and PAYNE, W. J. A. (1965) *An introduction to animal husbandry in the tropics*. London, Longmans.
WILLIS, W. H. and GREEN, V. E. (1948) 'Movement of nitrogen in flooded soils planted to rice.' *Proc. Soil Sci. Soc. Amer.* **13**, 229–37.
WILSON, P. N. (1957a) 'Studies on the browsing and reproductive behaviour of the East African dwarf goat.' *E. Afr. agric. for. J.* **23**, 138–47.
—— (1957b) 'Further notes on the development of the Serere herd of East African Shorthorned Zebu cattle.' *Emp. J. exp. Agric.* **25**, 263.
—— (1958a) 'An agricultural survey of Moruita Erony, Teso, Uganda.' *Uganda J.* **22**, 22.
—— (1958b) 'Effect of plane of nutrition on the carcass development and composition of the East African Dwarf goat, I.' *J. agric. Sci.* **50**, 198–210.
—— (1958c) 'Effect of plane of nutrition on the carcass development of the East African Dwarf goat, II." *J. agric. Sci.* **51**, 4–21.
—— (1960) 'Effect of plane of nutrition on the growth and development of the East African Dwarf goat, III.' *J. agric. Sci.* **54**, 104–30.
—— (1961a) 'The grazing behaviour and free water intake of East African shorthorned Zebu cattle at Serere, Uganda.' *J. agric. Sci.* **56**, 351–64.
—— (1961b) 'Observations on the grazing behaviour of crossbred Zebu Holstein cattle managed on Pangola pastures in Trinidad.' *Turialba*, **11** 55–71.
—— (1961c) 'Palatability of water buffalo meat.' *J. agric. Soc. Trin. and Tob.* **61**, 461–88.
—— (1963) 'The need for more research on the economics of grassland and livestock production in the Caribbean.' *Caribbean Agric.* **1**, 95.
—— BARRATT, M. A. and BUTTERWORTH, M. H. (1962) 'The water intake of milking cows grazing Pangola pastures under wet and dry season conditions in Trinidad.' *J. agric. Sci.* **58**, 257–64.
—— and HOUGHTON, T. R. (1962) 'Notes on the development of the herd of Holstein-Zebu cattle at the I.C.T.A., Trinidad.' *Emp. J. exp. Agric.* **30**, 160.
—— and OSBOURN, D. F. (1960) 'Compensatory growth after undernutrition in mammals and birds.' *Biol. Rev.* **35**, 324.
—— and WATSON, J. M. (1956) 'Two surveys of Kasilang Erony, Teso, 1937 and 1953.' *Uganda J.* **20**, 182.
WOOLDRIDGE, W. R. (1954) *Farm animals in health and disease*. London, Crosby Lockwood.
WORRALL, G. (1961) 'A brief account of the soils of the Sudan.' *Afr. Soils*, **6**, 53–65.
WORSTELL, D. M. and BRODY, S. (1953) *Comparative physiological reactions of European and Indian cattle to changing temperature*. Res. Bull. Mo. Agric. Exp. Sta. No. 515.
WRIGHT, C. H. (1938) *Agricultural analysis. Methods excluding those for soils*. London, Thomas Murby.
WYCHERLEY, P. R. (1963) 'The range of cover plants.' *Planters' Bull. Rub. Res. Inst. Malaya*, **68**, 117–22.

Index

Note: References in bold type refer to the major reference connected with the subject. Usually these major references represent a sub-headed section of a Chapter.

Abahima, **370**, 420
Abattoir, 279, 297, 390
Aberdeen Angus (cattle breed), 357, 359
Absorbtivity, mean effective, 358
Acaricide, 422
Acclimatization, of livestock, **Chap. 15**
Achiote (cattle breed), 443
Adaptation, thermal, of livestock, **Chap. 15**
African dwarf (goat type), 342
Africander (cattle breed), 328, 357, 359
Air masses, 1, 2, 5
Algae, blue-green, 209
Alpacas, 253, 334
Alpine (goat breed), 340
Aluminium, in plant nutrition, 153
Amino acids, in animal nutrition, 401, 415–17
Amino triazole (herbicide), 150
Andosoils, 32, 39
Andropogon gayanus, 316
Anglo-Nubian (goat breed), 343, 436
Angora (goat breed), 341
Animal Protein Factor (APF), 409
Ankole (cattle breed), 322, 325, 328, 370
Antibiotic, 419
Antibody, 421
Arboricide, 252, 264
Armsby, net energy value, see Energy, net
Artificial insemination (A.I.), 383, 424, 426–7, 429, 430, 436, 443–4, 446
Auchenia, 334
Azotobacter, 209

Bacteria: in animal disease, 421; nitrifying, in soil, 132, nitrogen fixing, in soil, 310, 319; symbiotic, 311
Bail, portable dairy, 376
Banana (*Musa spp.*), 2, 7, 93, 156, 164, 213, 219, 221, 230, 235, 237, 239
Banteng, 323

Beans (*Phaseolus* spp.), 8, 93, 163, 167, 169, 177, 182, 187, 230
Beefmaster (cattle breed), 443
Behaviour, animal; cattle; grazing, 361
Benedir (goat breed), 342
Bermuda grass, see *Cynodon* spp.
Berseem, see *Trifolium* spp.
Biological value, of protein, 415
Bison, 320, 324
Blight, South American leaf (*Dothidella ulei*), 246–7
Blood-groups, of livestock, 447
Boer (goat breed), 343
Bokhara sweet clover (*Melilotus* spp.), 295, 310
Bonsmara (cattle breed), 443
Boron, in plant nutrition, 237
Bos bison, see Bison; *B. bonasus*, see Bison; *B. brachyceros* (humpless shorthorn cattle), 323, 328; *B. bubalis*, see Buffalo; *B. caffer*, 324; *B. depressicornis*, 324; *B. frontalis*, see Gayal; *B. gaurus*, see Gaur; *B. grunniens*, see Yak; *B. indicus* (Zebu), **Chap. 14**, 320, 323–4, 326, 348, 353, 363, 376, 383–4, 389, 424, 426, 429, 432, 434–5, 438–9, 441–2; *B. mindorensis*, 324; *B. nomadicus*, 323; *B. opisthonomus*, 323, 327; *B. primigenius*, 323; *B. sondaicus*, see Banteng; *B. taurus*, **Chap. 14**, 320, 323–4, 326, 329, 348, 353, 363, 376, 383, 388–9, 426, 429, 432, 434, 438–9, 441–3
Bothriochloa insculpta, 308–9
Brachiaria brizantha, 308; *B. mutica*, 287–8, 290–1, 293–4, 296, 312–13
Braford (cattle breed), 443
Brahman (cattle breed), 359, 435, 442
Breed: society, 441–3; stabilization, 438–40
Breeding: animal, 424–48; animal, seasonal, 427; fractional, 439; plant, 218–20; systems, animal, 436–41

Bride price, 88, 273, 369
Brown earths, 40–1
Browse; browsing, 261, 407
Budding; budgrafting, 218, 222, **224–5**
Buffalo, water: 203, 212, 320, 324, **330–4**, 406, 427; riverine, **331–2**; swamp, **330–1**
Buffel grass, see *Cenchrus* spp.
Burning, see Fire
Bush control: by fire, **262–4**; by other means, **264–5**
Bush crop culture, **Chap. 10**
Butterfat, in milk, 363, 404, 413–14, 432–4, 442, 448

Cacao, see Cocoa
Calcium: in animal nutrition, **406–7, 412–17**; in soils and plant nutrition, 31, 92, 119, 153–3, 158–9, 208–9, 284, 293–4, 311–12
Calf: mortality, 274, 276; nutrition, **411–12**; rearing, 387–9
Calorie, **399–461**
Calving: index, interval, 333, 364, 383–4, 434–5, 439; percentage, 276, 427
Cambered bed, **141–2**
Camel: Bactrian, 335; Dromedary, see Dromedary
Capital, **96–7**, 381, 384, 391–2
Capping, 18
Capra, see Goat
Carabao, 324
Carbohydrate, 289–93, 396, **398–401**, 405, 415–16
Carrying capacity, of pasture, 9, 252–3, 255–7, 260, 266, 272–5, **277–8**, 281, 283, 301, 378–80, 392
Cashew nut (*Anacardium occidentale*), 8, 227
Cassava (*Manihot utilissima*), 2, 7, 8, 164–8, 178, 181, 190, 196, 203, 239, 316, 398
Castration, 275, 284, 379
Catchment areas, 113
Cattalo, 320
Cattle: general, **Chaps. 12, 14, 16**; 16, 82, 253, 260, 267, 302–21, 411–14, 431–48; beef, **375–81, 413–14, 434–8**; Bibovine sub-group, **323–4**; Bisontine sub-group, **324**; breeding, 329, **424–48**; Bubaline sub-group, 324; evolution, **323–30**; dairy, nutrition of, 412–15; growing, nutrition of, 412–13; housing, 386–7; management, **Chap. 16**; migrations, 327–9; population, 321, skin of, 326; Taurine sub-group, 323–7
Cenchrus ciliaris, 297, 308–9
Centrosema pubescens, 312–13, 318
Charbray (cattle breed), 443
Chernozem, 37, 43–4
Chitemene system, 162, 166–7
Chloris gayana, 306, 308–9, 312, 315–16, 318
Citrus, 2, 10, 220, 222, 224, 229, 235, 237
Clay, 31–4
Clearing: land, 75, **129–31**; chain, 129; mechanical, **129–30**
Climate: general, **Chap. 1**; alternate wet/dry, **5, 9**; dry, of tropical and sub-tropical West coasts, **11**; dry tropical, **5**; indirect effects on livestock, **368**; monsoon, **5–9**, 12, 24, 138, 193, 314; wet equatorial, **1, 5**; wet windward coast, **9, 11**
Clitoria ternata, 310
Clones, 218–20, 225
Coat: characteristics, of cattle, 358–60; score, 359
Cobalt, in animal nutrition, **408**
Cocoa (*Theobroma cacao*), 2, 7, 10, 26, 126, 218, 222–3, 225, 227–8, 234–7, 239–43
Coconut (*Cocos nucifera*), 2, 10, 128, 213, 219, 221, 226–7, 237
Coconut cake (copra meal), 271, 386
Coffee (*Coffea* spp.), 2, 7, 8–10, 26, 147, 213, 220, 222–3, 225, 229, 236, 237–41, 243, 245, 247
Colostrum, 338, 411
Comfort zone, of livestock, 327, 348, 362
Compost, **192–4**, 207, 210, 238, 420
Continuous cropping, **212–13**
Copper, in animal nutrition, **408–9**, 417
Cotton (*Gossypium* spp.), 7, 9–10, 15–17, 23, 25, 139–40, 143, 146, 148, 170, 176, 181–3, 190, 196–8, 317, 375
Cotton seed cake, 271
Cover crop; plant, 56, 102, **117–18, 230–3**

Index

Cowpeas (*Vigna sinensis*), 167, 184, 186, 211
Credit, agricultural, 96, 214
Criollo (cattle breed), 376, 444
Critical periods of growth, of grasses, 258
Critical temperature, 348
Crop: rotation, see Rotation, crop; improvement, **218–20**
Crossbred cattle; crossbreeding, 376, 394, 429, 437–9, 441–4
Crude fibre; protein, see Fibre; Protein
Culling, of livestock, 275, 284, 379, 388, 428, 443
Cutchi (goat breed), 341
Cutting height, of grasses and fodders, 291
Cutting interval, of grasses and fodders, 289–90
Cynodon dactylon, 308–9; *C. plectostachyum*, 300, 304, 306, 312, 315–16

Dalapon (herbicide), 149
Dams, 205, 283, 380
Dasheen (*Colocasia* spp.), 2, 10, 164, 239
Day length, 26, **364–5**
DDT (insecticide), 282
Delhi (water buffalo breed), **331**
Dentition, of camel, 336
Desert, 5, 11
Digestibility coefficient: general, 284, **402–3**; of grasses, 259
Digitaria decumbens, Stent. 10, 306–8, 385
Dipping; dips, 275, 376, 379, 421–2
Diseases: animal, 284, **418–23**, 428, 436–7; plant, **246–7**
Ditch: diversion, **113**, 141; hillside, 109–10
Dolichos lablab, 295
Drain, storm, 112–15
Drainage, 135–6, 204, 206, 215, 420
Dressing percentage, 384
Dromedary, 320, **334–9**, 348, 373–4
Drought resistance, of crops, 8, 313, 378
Dry-season feeding; grazing, see Feed; Grazing, dry season

East Coast Fever (*Theilariasis*), 375, 421–2
Eddoe (*Colocasia* spp.), 2, 164, 239
Electric fence, 303, 382
Electro-ejaculation (in A.I.), 426
Elephant grass, see *Pennisetum purpureum*
Energy: gross, **399**, 403; metabolizable, **399**; net, **399**
Ensilage, see Silage
Epiphytes, 73–4, 76, 82
Erosion, see Soil erosion
Evaporative cooling, 351, 365, 367
Evaporimeter, 21
Evapo-transpiration, 11, **20–2**, 234

Fallow, 57, 90, 125, 138, 152, **155–8**, 162, 164–6, 170–6, **178–82**, 206, 314, 316–17
Farmyard manure, see Manure, farmyard
Fat, 398, 404–6
Feed, dry-season, **269–71, 384–6**
Feedlot, 442
Feeding standard, 410–18
Felling, tree, 129, 158, 162, 166
Fencing, 279–80, 286, **382–3**, 431
Fertilizer: general, **195–201, 304–7**, 380, 408; compound, 199; inorganic, 190, **268**, 292, 305; nitrogenous, 190, 192, 196, 236–7, 241, **268**, 292, 296, 306, 310, 314, 318, 392; phosphatic, 190, 192, 196, 198–9, 236–7, **268**, 292, 305–6, 316–17; placement, 299; potassic, 190, 192, 196, 199, 236–7, 305; sulphate of ammonia, 191, 194–5, 197, 268, 292, 306, 392; super phosphate, 194, 198, 268, 304–6
Fibre, crude, 258–9, 269–70, 289–90, 293–5, 385, 398, 403
Field: capacity, 39; size, **382**
Fire: general, 71, 75–7, 82, 101, 127, 133–4, 154–5, 157, **160–1**, 166, 175, 233, 248–50, 252–5, **262**, 265–6, 274, 279; effects of, **131–4, 162–3**, 185, 233, 251
Fish culture, 202, 214
Fluorine; fluorosis, 407
Fodder: crops, **Chap. 13**, 398; grasses, **288, 294**; legumes, 294–6
Foggage, 378, 385, 403

Forest, see Vegetation, forest zone
Formation series, 70
Fragmentation, of holdings, 95, 297
Friesian (cattle breed), 360, 376, 436, 441–4
Fulani (cattle breed), 328, 373–4, 434–5, 444

Galla (goat breed), 342
Gamete, 431
Gammaxane (insecticide), 282
Gaur, 323
Gayal, 323
Gene, 431, 443, 447
Genetic improvement: of crops, 219–20, of livestock, **284–5**, **431–48**
Germination capacity, 282, 300
Gestation period, 345, 358, **425**
Ghee, 377
Glycine javanica, 231–2, 296, 310; *G. max*, see Soya bean
Goat: general, 10, 82, 203, 260–1, 267, 280, 320, **339–46**, 414, 436, 445; African type, **342–4**, Asiatic type, **341–4**; browsing, 267, **280–1**, 345–6; European type, **339**, 344–5; Oriental type, **340**; population, distribution, **339**
Grade; grading-up, of cattle, 428, 430
Grafting, 218, 222, **224–5**
Gramoxone (herbicide), 150
Grass: composition, 259–60, 289–90, 293; legume mixtures, **294–6**, 313; rest, 257, 262, 265, 274, 302, 380; strip, 115; pastures, species and strains, **307–9**
Grassland: management, **255–71**, **279–83**; natural, Chap. 11, 7–10; natural, types of, **249–55**; natural, use by pastoralists, Chap. 12
Grazing: behaviour, 361, 387; communal, 280; deferred, 303; dry-season, 302; management, 255, **265**, **279–83**, 301, 315; mixed, **267**, 280; rotational, 257, 265, 279, 283, 302–3, 422; seasonal, **254–9**, 373; selective, 302; strip, 302; zero, 383
Green manure, see Manure, green
Groundnuts (*Arachis hypogaea*), 8–10, 93, 125, 140, 142, 146, 167–8, 182–3, 190, 196, 200, 203, 211

Growth, compensatory, 378
Grumusols, **31–2**, **36**, 44
Guatemala grass, see *Tripsacum laxum*
Guinea grass, see *Panicum maximum*
Gully, 108, 115; see also Soil erosion, gully

Hair: colour, of cattle, 355, 360–1, 443; fibre, of cattle, 327, 359; follicle, of cattle, 354
Hamitic longhorn (cattle type), 325, 327
Hard pan, see Soil, pan, hard
Hariq system, 133
Harrow, harrowing, 136–8, 141, 206, 229, 282, 300
Hay: general, 269, 284, 287, 295, 302, 305, 378, 385, 398, 402; composition of, 270
Heat, in mammals, see Oestrus
Heat: load, stress, 304, 347–52, 357, 360, 366; regulation, in dromedary, 337–8; tolerance, of livestock, 351, 353, **354–8**, 442; tolerance index, 356–8
Helminth, 422–3
Herbicide, see Weed control, chemical
Hereford (cattle breed), 357, 361
Heritability, of livestock characters, 432–5, 447
Heterosis, 429
Hoe, hoeing, 126, 136–7, 139, 141, 148, 164, 166, 169, 172, 177, 229, 236, 282, 382
Holstein (cattle-breed), 361–2, 366; see also Friesian
Homeothermy, 348–9, 354, 356
Horse; horsekind, 320
Humidity, relative, **24**, 437
Hump: in camel, 335, 337; in water buffalo, 330; in crossbred and Zebu cattle, **325–6**, 370, 443
Humus, **55–8**, 172–5, 185, 189; see also Manure, organic
Hybrid vigour, 429
Hyparrhenia spp. 250, 306

Indigofera spp., 310
Inheritance: of farm holdings, **94**, 156, 371; of genetic characters; see Heritability

Inter-cropping; inter-planting, 164, **228–30**
Iodine, in animal nutrition, **408**, 417
Iron: in animal nutrition, **406–8**, 417; in plant nutrition and soil, 153, 208
Irrigation, in rice culture, 205–6, 215; of grass and fodder crops, 254, 270, 287, 292–4, 302

Jaffrabadi (water buffalo breed), **331**
Jamaica Black (cattle breed), 443
Jamaica Hope (cattle breed), **441–3**
Jamaica Red (cattle breed), 443
Jumna Pari (goat breed), **341**, 343

Kashmir (goat breed), **341**
Kellner, 400
Kenana (cattle breed), 435
Killing out percentage, 342
Kjeldhal method, for nitrogen determination, 401
Kraal, cattle, 316, 322, 370, 374, 379, 386
Kudzu, tropical, see *Pueraria* spp.

Land: clearing, **Chap. 6**, 154, **162**; preparation, 136; tenure, **94**, 155, 157, 297; usage, use, 93–4, **104–7**, 121, 285
Laterite, **32**, 48, 79
Latosol, **30–3**, 41, **46–8**, **55**, 76; humus content of, 55–8; pale yellow, 47; red and red-yellow, 48–53
Layering, of plant stems, 222, **224**
Leaching, 5, 58–60, 131–3, 158–62, 170, 173–5, 189, 192, 213, 229, 231, 235, 253, 305
Leaf: stem ratio, 290
Legume, **Chap. 13**, 392
Lehmann system, 415
Ley, 8, 10, 296–9, 303–5, 309, **314–19**, 382
Lianas, 73–6
Light, general, 25–7; intensity, 223, 240, 299
Lime; liming, 59–60, 151, 174, 199, 305
Livestock, see Cattle, Goats, Sheep, etc.
Llama, 253, **334**
Llanos, 42, 254
Lucerne (*Medicago sativa*), 287, 295–6, 309, 311

MCPA (herbicide), 149
Magnesium: in animal nutrition, 406, 417; in plant nutrition and soil, 152, 158–9, 174, 208–9, 236–7
Maintenance ration; requirements, 261, 290, 293, 395, 412–14
Maize (*Zea mais*), 5, 7–10, 13–17, 25–6, 92, 103, 125, 140, 144–6, 149, 164–70, 176, 181, 183, 185–7, 196–8, 203, 213, 230, 270, 299–300, 391, 405
Manganese: in animal nutrition, **408**, 417; in plant nutrition and soil, 208, 238
Mango (*Mangifera indica*), 2, 213, 224
Mangrove, 62–3, 83
Manure; manuring: general, 236, 291; farmyard, **188–92**, 237–8, 315–16, 420; green, 56, 66, 103, 184–8, 210; kraal, 188; organic, 194–5, 207, 238; see also Compost
Marcotting, 222, 224
Marketing, of stock and animal products, 278, 378, 390
Mating, seasonal, 274
Mbuga, 63, 86, 135
Meat: general, 273, 275, 278, 297, 377, 390–1, 404, 432, 447; of dromedary, 338; of water buffalo, 332
Medicago sativa, see Lucerne
Medullation, of cattle hair, 327, 359
Melilotus alba, 295, 310
Melinis minutiflora, 308–9, 312
Merker grass, see *Pennisetum purpureum*
Metabolism; metabolic rate, 348–9, 354–6, 395
Migration of stock, 280
Milk: bovine, 273, 275, 278, 371, 376–7, 387, 390, 392, 395, 430; composition, of bovine, 344, 363, 404; of dromedary, 338; of goat, 340–1, 343–4, 414; of water buffalo, 333, 344; production, effect of climate on, **362**; yield, 327, 344, 363, 383–4, 394, 413–14, 426, 431–5, 438–9, 442, 445, 447–8
Milking procedure; routine, **387–9**
Millet: general, 8–9, 25, 92, 165, 168, 178, 181, 196, 198, 374; bulrush (*Pennisetum typhoideum*), 143–4, 270; finger (*Eleusine coracana*), 163, 167–8, 176, 190, 317; pearl (*Pennisetum typhoideum*), 186

Mineral, bricks; licks; supplements, 275, 284, 396, **406**, 416
Mixed cropping, 163-4, **183-4**, 207
Mixed farming, 286-7, **381**
Mixed stocking, 267, 346
Mixed tree culture, 213
Mohair, 341
Molasses grass, see *Melinis minutiflora*
Monsoon, see Climate, monsoon
Mucuna utilis, 287
Mulch; mulching, 8, 18-19, 56-8, **118**, 124-5, 146, 154, 156, 217, 229, **234-6**, 239, 242, 282
Multiplication, animal, **428-31**
Murrah (water buffalo breed), **331**

Nagpuri (water buffalo breed), 332
Napier grass, see *Pennisetum purpureum*
Natural regeneration, 117, 155, 178, **233-4**, 315-16
Nelthropp (cattle breed), 443
Nematodes, 422
Net assimilation rate, 241
Nigerian (goat breed), **342**
Night soil, 194, 210
Nili and Ravi (water buffalo breed), **331**
Nitrogen: content in milk, 363; in soils and plant nutrition, 59, 157, 186, 208, 231, 237, 268, 288, 292, 304-5; fixation, in soil, 310, 319; free extract (NFE), 290, **399**
Non-protein nitrogen (NPN), **401-2**
Nomad; nomadism, 9, 272-3, **372-3**, 396
Nubian (goat breed), **340**
Nutmeg (*Myristica fragrans*), 10, 221, 224
Nutrient ratio; nutritive ratio, 293, **403**, 415
Nutrition: animal, **359-418**; human, **90-3**

Oestrometer, 424
Oestrus; oestrous cycle, 345, 364, **424-5**
Oil, see Fat
Oil palm (*Elaeis guineenis*), 2, 7, 10, 128, 219-20, 226, 237-8
Old English (goat breed), **340**
Onobrychis viciifolia, see Sainfoin
Outcrossing, 394

Overgrazing, 123, 255, 257, 272, 274-8, 281-3, 345
Overstocking, 88, 272, 275-6, 278, 371, 380; see also Stocking rate
Ovulation, **424-5**, 427
Oxen, nutrient requirements when working, **413-14**
Oxygen, 208-9
Oxysol, 79, 132, 196

Padi, see Rice
Pan, see Soil pan
Panicum maximum, 287-9, **290-2**, **294**, 304, 306
Pangola grass, see *Digitaria decumbens*, Stent.
Para grass, see *Brachiaria mutica*
Paranas, 253
Paspalum dilatatum, 306, 310
Pastoralism; pastoralists, **Chap. 12**, 98
Pastures: general, **Chaps. 11, 13**; 5, 10, 247; cultivated, **296-314**; establishment, **298-300**; legumes, **309-14**; management, **279-83**, **301-4**
Patanos, 249, 253
Pedalfer, **31-6**
Pedocal, **31-2**, **43-6**
Pennisetum purpureum, 287-96, 303, 307-8, 431
Performance test, 442, **444-7**
Pests, of crops **246-7**
Phosphorus: in animal nutrition, 406-7, 412-13, 415, 417; phosphate in soils and plant nutrition, 55, 60-1, 92, 185-6, 231, 235, 258, 284, 293-4, 304, 312
Photoperiodism, of plants, 26
Photosensitivity: of animals, 423; of plants, 26
Photosynthesis, 243
Pig: general, 10, 203, 212, 214, 320, 395, 397-8, 405, 407-10, 433, 445-8; heritability, **432**; nutritional requirements, **415**
Pigeon pea (*Cajanus cajan*), 165-8, 178, 181-2, 310
Pigmentation, see Skin colour
Piling, of felled trees, **130**
Pineapples (*Ananas comosus*), 25, 197, 213, 217, 230, 235
Plant breeding, see Crop improvement
Plant density, see Spacing

Planting: depth, 300; time, 139
Plough; ploughing, 115, 136–8, 140–1, 149, 177, 187, 206, 229, 282–3, 298, 300, 317, 380; depth, **139**
Podsols, 31, 35, 43
Poll glands, of dromedary, 336
Pollination, 218, 244
Population density: human, **89–90**, 214, 297; plant, see Spacing
Potash, potassium: in plant nutrition and soil, 185–6, 235, 258; see also Fertilizers, potassic; in animal nutrition, 406, 417
Poultry: general, 10, 203, 320, **366–8**, 395–8, 408–10, 418–20, 429, 433; nutrition, **416–18**
Pregnancy diagnosis, 426
Pressure belts; cells, 1, 5
Production requirement, of livestock, **395**, 412–14
Progeny test, 436, **444–7**
Propagation: seed, **220–1**; vegetative, **221–5**
Protein: in animal nutrition, 258–60, 269–70, 289–90, 293–6, 306, 310–14, 385–6, 398, **401–3**, 412–15; equivalent, **402**; in human nutrition, 92
Pruning, of bush and tree crops, 217
Pueraria spp., 296, 310, 312–13

Radiation, solar, 354–5, 358, 361, 363, 410, 423
Rainfall: general, **11–22**, **100–1**, 287; annual, **12–17**, confidence limits; fiducial limits, **12–17**, 23; convectional, 9; effective, 11, 18; frequency distribution, 3–4, 11–17, 23, 100, 143, 427; intensity, **17–20**, 100–1, 140; orographic, 9; percolation rate, 18, 316; probability, **12–17**; reliability, 8, 12–17, 22
Rain forest, see Vegetation, rain forest
Rains, grass, 262
Ranch; ranching, 13, **255–71**, 276, 372, **375–81**, 408, 426, 442
Red oat grass, see *Themeda trianda*
Regosols, 31–6, 50, 61–2
Religion; religious beliefs and taboos, 88, 322, 372
Reseeding, 282, 380
Respiration rate, **338, 350**, 358, 367

Resting period, see Fallow; Grass rest
Rhodes grass, see *Chloris gayana*
Rice (*Oryza sativa*), **Chap. 9**, 2, 7, 9–10, 26, 137, 164–5, 196, 287
Ridge: general, 120, 127, **142**, 145–6, 164, 166; and furrow cultivation, **141**, contour, 116, 120; tie, 116, 124, **142–4**
Rinderpest, 421
Ring-barking, of trees, **128–9**, 224
Root cutting; ripping, **130**
Rootstock, 218, 220, 225
Rotation, crop, 156, 167–9, 178, **182–4**, 190, 197, 202, 204, 286, 314, 318
Rotational grazing, see Grazing, rotational
Rubber (*Hevea brasiliensis*), 2, 10, 25, 127–8, 202, 217, 220, 222–8, 231–3, 236–8, 246–7
Rumen, 398, 401
Run-off: general, 11, 18–19, 98–9, 101–5, 108, 113–16, 120–4, 131, 134–7, 158, 171, 228, 283; coefficient, 114, 256
Rust, leaf, of coffee (*Hemelia vestatrix*), 246–7

Saanen (goat breed), **340**
Sahiwal (cattle breed), 329, 430–1, 436, 441
Sainfoin (*Onobrychis viciifolia*), 309
Salt brick; lick, 157
Sanga (cattle breed), 325–6, 328–9, 370
Santa Gertrudis (cattle breed), 357, 376, 436, 442–3
Savanna, see Vegetation, savanna
Scion, 218, 221, **224–6**
Screwworm (*Callitroga hominivorax*), **422**
Seasonality: of calving, 392; of growth, **378**
Seedbed, 136, 138, 140, 168
Seed: propagation, 220; rate, 233, 300; treatment, 233
Selenium, in animal nutrition, **408**
Semen, of cattle, 329, 419, 426, 444, 446
Senepoll (cattle breed), 443
Sesame (*Sesamum orientale*), 9, 167, 180, 189–90
Setaria spp., 287–9, 292, **294**, 308–9, 312

Sexual maturity, 425; reproduction, **424–7**
Shade, 8, 25, 57, 168, 217, **239–44**
Sheep, 82, 203, 253, 260, 267, 320, 352, 397, **414**, 427, 433, 445
Shelter belt, 120, **244–5**
Shifting cultivation, **Chaps. 7, 8**; 90, 98, 133, 202, 213
Shorthorn (cattle breed), 359
Sierozem, 43
Signal grass, see *Brachiaria* spp.
Silage, 270, 287, 292, 294, 296, 299, 300, 302, 304–5, 378, 398, 409
Sindhi (cattle breed), 329, 434
Sisal (*Agave sisalana*), 8–9, 25, 140, 197
Skin: of cattle, 274; colour, of cattle, **360**; thickness, of cattle, 326
Sleeping sickness, see *Trypanosomiasis*
Slope of ground, **101**, 104, 107, 111
Sodium: in animal nutrition, 406, 417; arsenite (herbicide), 128, 147, 264; chlorate (herbicide), 147, 264
Soil: general, **Chap. 2**; acidity, 29, **58–60**, 132–3, 151, 160, 173, 197, 200, 305, 311; aggregates, 18, 115, 137; allophane, 31, **39–41**; alluvial, 62, 66; association, 29; azonal, 31, **61–3**; black cotton, **31**; calcimorphic, 31, **41**, 66; catena, **29**, 51, 63; classification, **30**; conservation, **Chap. 5**, 217; crumb structure, 140, 152; cultivation after planting, **145–7**; cultivation depth, **139–40**; erosion, **Chap. 5**, 19, 37–40, 134, 148, 158, 160–1, 166, **171**, 175–7, 182, 201, 228, 255, 262, 274–5, 281, 283; erosion, by goats, 345; erosion, gully, 99, **114–16**, 135, 171, 281; erosion, rill or shoestring, 99; erosion, sheet, 99, 114; erosion, wind, 98, **118–19**; exchange capacity, 33, 153; exposure, **134**, 139, 154, 174; ferralitic, **31–6**, 47–8; fersiallitic, 31, 35–6, **53–5**, **58–60**; ferruginous siallitic, 31, 35–6, **53–5**; fertility, **Chaps. 7, 8, 9**; 5, **151–4**, **241–3**, 314; fertility, decline in, **169–70**; gley, 31, **41–3**; halomorphic, 31; horizons, 32, 35, 159; hydromorphic, 30–1, 35, **41–4**, 66–7; intrazonal, 31, 41, **66–7**; management, 30, **228–9**; mapping, 28–9; moisture, water, 11, 20–3, 120, 124, 197, 230, 235, 242; nitrate, nitrogen, 56–9, 131–3, 152–3, 158–61, 166, 168, **173–4**, 185, 190, 317; see also Fertilizer, nitrogenous; nutrient status, **172**; organic matter, 131, 134, 151–4, 157–9, 170–1, 173, 177, 185–6, 189, 191, 200, 228–9, 232, 241; oxygen, 151; pH, see Soil acidity; pan, hard, 140, 227, 242; particles, 119, 152; phosphate, 131, 133, 152–3, 161–2, 170, 173, see also Fertilizers, phosphatic; see also Phosphorus; physical condition, 163, **170–1**; potash, 152, 174, see also Fertilizer, potassic; see also Potassium; saline, 64–5; series, 28, 37, 41, 47, 49, 53; structure, 34, 57–8, 152, 171, 230, 313, 316; survey, 28–30, temperature, 132, 134; terra rosa, 47–8, 54; type, 104, 107
Solar radiation, see Radiation, solar
Solids-not-fat, in milk, 363
Somal (goat breed), **342**
Sorghum (*Sorghum vulgare*), 8–9, 13, 15, 25–6, 125, 143–4, 149, 163, 165, 167–8, 176, 180–2, 189–90, 196, 198, 270, 300, 374
Sorghum sudanense, 304
Sowing date, see Planting time
Soya bean (*Glycine max*), 26, 93, 182, 184, 187, 230, 287, 295, 386, 402
Spacing, of crop plants, 217, 221, **227**, 242–3, 288
Spermatogenesis, 364, 431
Speying, **379–80**
Spillway, 114–15
Star grass, see *Cynodon* spp.
Starch equivalent (Kellner), 260, 270, 293, **400–1**
Stizolobium spp. see Velvet bean
Stocking density; rate, 9, 252, 255, 260, 265, 277–8, 301–2, 376
Stomoxys fly, 273, 371
Stratification, of cattle industry, 377
Strip cropping, 106, **116**, 120
Stumping, of trees and bushes, 126, 129, 134
Stylosanthes spp., 310
Subsistence farming, 97
Subsoiling, 139–41, 288
Sudan grass, see *Sorghum sudanense*

Sugar-cane (*Saccharum officinarum*), 10, 141–2, 147, 197, 391
Sulphate of ammonia, see Fertilizer
Sulphur: in animal nutrition, 406; in plant nutrition and soil, 200, 237
Sunnhemp (*Crotalaria juncea*), 185–7
Sunlight; sunshine, 22, 26, 420
Superphosphate, see Fertilizer
Surti (water buffalo breed), **331**
Sweat; sweating, 349, 352, 406; glands, 349, **352–4**, 367
Sweet pitted grass, see *Bothriochloa* spp.
Sweet potato (*Ipomoea batatas*), 8–10, 165, 167, 190, 197, 199, 203, 211, 398
Swine, see Pigs
Sybokke (goat breed), **341**
Symbiosis, 311

Trichloracetic acid; TCA (herbicide), 149
Tamarao, 324
Tannia (*Xanthosoma sagitifolium*), 2, 10, 164–5, 239
Tea (*Camellia sinensis*), 7, 9, 25–6, 220–3, 228–9, 237–43
Teak, 382
Temperament, of cattle, **379**
Temperature: body, 326, 237, 347–50, 356, 358–60, 363, 366; mean ambient, **24–5**, 223, 437; monthly distribution, 3, 4, 7; range, 10
Tensometer, 23
Terrace: bench, **110–12**, 117, 124; broad based ridge, **110–11**, 117, 120, 204; contour, 281; mangum, **110**; narrow based ridge, **109–10**, 117
Theilariasis, see East Coast Fever
Themeda trianda, 370
Theobroma, see Cocoa
Thermal stress, see Heat load
Thiamine, see Vitamin B_1
Ticks, cattle, 275, 339, 421–2, 431, 436
Tie-ridging, see Ridge, tie
Tillage; tilling, **Chap. 6, 115**
Tiller; tillering, 301
Tilth, 136, **140**, 282, 298
Tobacco (*Nicotiana tabacum*), 10, 26, 140, 148, 170, 183, 199, 212, 230
Toggenburg (goat breed), **339**
Tolerance: heat, see Heat tolerance; crop, to herbicides, 148

Total Digestible Nutrients (TDN), 385, **403–4, 412–16**
Tractor, 138, 216, 382
Trace elements, in animal nutrition, 406–8; in plant nutrition, 152, 237–8
Trade winds, see Wind
Transplanting, 217, **225–7**
Transpiration: of animals, 349; rate, of plants, 20, 22, 122
Tree culture, **Chap. 10**
Trematoda; trematodes, 422
Trifolium spp., 294, 309–11
Tripsacum laxum, 287, 290–6
Trypanosomiasis (sleeping sickness), 418, 421
Tsetse fly (*Glossina* spp.), 8, 91, 96, 191, 250, 255, 272–3, 277, 281, 285, 418, 421
Tung oil (*Aleurites* spp.), 9, 25, 218, 220–1, 226–7, 230
2,4-Dichlorophenoxyacetic acid; 2,4-D 148–9, 264, 301
2,4,5-Trichlorophenoxyacetic acid; 2,4, 5-T, 128–9, 264, 301

Undersowing, 299

Vaccination; vaccine, 375, 421, 431
Vegetation: general, **Chap. 3, 101**; broadleaved woodlands, **77**, 272; deciduous seasonal forest, 6, **76**, 155, 160, 170–3, 178; dry evergreen, 70, **80**; evergreen seasonal, **73**, 160, 170–1, 173; forest zone, **158**, 163, 171–4; high altitude grassland, 81; littoral woodland, 80; llanos, 85; lowland evergreen rainforest, 70, **72**, 155; mangrove swamp and forest, 83; Miombo, 45, 78, 162, 250; montane forest, 70, **80–2**, 155, 161, 172, 197, **252**, 255, 286; montane, high altitude woodlands, thickets and grasslands, **82**; montane, lower and higher evergreen forest, **81–2**, 272; open broadleaved woodland, 6; peat forest, 84; savanna, 42, 45, 71, **85**, 126, 131–2, 136, 138, 154–5, 157, **160–1**, 166, 168–75, 178, 197, 250, 254; semi-evergreen seasonal forest, 6, **74**; steppe, 71; sub-desert scrub and grass, **80**, 272; swamp, 70, **83**, **85**; swamp, herbaceous, **85**; swamp,

seasonal, **85**; swamp forest, **83**; swamp woodland and thicket, **84**; thorn woodland and thicket, 6, **79**, **251**, 272; tropical Alpine meadow, 81; woodland zone, 160, 166

Veld: sour, 250, 263; sweet, 251

Velvet bean (*Stizolobium deeringianum*), 287, 295, 315

Vigna spp., see cowpea

Vitamin: general, 396, **408–10**, **416–17**; A, 92, **408–9**, 417; B_1(Thiamine), 92, **409**, 417; C, **410**; D, **410**, 417; E, **410**, 417; K, **410**, 417

Vlei, 86, 135–6, 141, 254

Water: general, 396; catchment, **120–2**; conservation, **Chap. 5**, **120**, **123**, 135; consumption by crops, 122; consumption by stock, 380, 397; consumption by stock, effect of climate on, **365**; percolation, 118; requirement, of crops, 17, **22**, 392; requirement, of cattle, **379**, 397–8; requirement of dromedary, 337; supplies, 105, 272, 274, 280, **283**, 355, 371, 373, 376

Weed: competition, 138, 168; control, **Chap. 6**, 5, **140**, **145**, 165–6, 168, 170, 174–5, 216–17, 288; control, chemical, 125, **147**, 216, 236, 301; control, selective, 134

Weeding, clean, 146, 229, 236

Winch, **128**

Wind, 1; break, 120, 217, 244; speed, 244; stripping, 120; trade, 6, 9

Woodland, see Vegetation, woodland

Yak, 324

Yam (*Dioscorea* spp.), 2, 7, 9–10, 164–6, 168, 180, 189, 196–7, 199, 398

Yield, of crops, **169**

Zinc: in animal nutrition, **408**, 417; in plant nutrition and soil, 237